The Metallurgy of Anodizing Aluminum

Jude Mary Runge

The Metallurgy of Anodizing Aluminum

Connecting Science to Practice

Springer

Jude Mary Runge
Chicago, IL, USA

ISBN 978-3-319-72175-0 ISBN 978-3-319-72177-4 (eBook)
https://doi.org/10.1007/978-3-319-72177-4

Library of Congress Control Number: 2017960892

Illustrations by Joy M. Kaufman, Illustrator & Graphic Designer: Joyjoycreations, Geneva, IL, USA

Printed on acid-free paper

This Springer imprint is published by Springer Nature
The registered company is Springer International Publishing AG
The registered company address is: Gewerbestrasse 11, 6330 Cham, Switzerland

With deepest love and gratitude, I dedicate this book to my parents.

First to my mother, Dorothy Koop Runge, whose constant encouragement gave me the grit to find my passion, to practice it, and to never let go.

And to my dad, Fred Runge, who showed me that nonfiction books are interesting, stimulating, and necessary for answering the questions of how and why. He gave me wonder and cultivated my imagination, which taught me to ask the next important question: What if?

Preface

For this book, *The Metallurgy of Anodizing Aluminum*, three goals were set: first, to make a clear connection between corrosion science and the nucleation and growth of the anodic oxide; second, to show that the composition and microstructure of the base metal, as the source of the oxide, are inextricably linked to the process of oxide growth and development; and third, more than anything else, to show that the quality of the aluminum substrate defines the quality of the anodic oxide finish. This is not a book that teaches one how to anodize, but a book that explains what happens when one anodizes.

However simple the anodizing process may seem, growing the anodic aluminum oxide is a complex and multidisciplinary process that combines basic chemistry and electrochemistry with the physical metallurgy of the substrate. Therefore, the process of anodizing comprises engineering issues and procedures, not only in the anodizing plant but also prior to anodizing, during substrate manufacturing. The main ramification of specifying the anodic oxide as a design feature for an engineering application is that the component surface must be considered from alloy formulation through all the various casting, deformation, and mechanical finishing processes required to manufacture it. The interactive factors in manufacturing the aluminum anode complicate the anodizing process, making some alloys more difficult to anodize than others.

To date, studies of the anodic oxide have only been undertaken on finished oxides, under standard temperature and pressure, and as a result, most consider anodic oxide growth an equilibrium process. The actual dynamic nature of the growth and development conditions for anodic oxidation is, by definition, a non-equilibrium process. Due to this misunderstanding, many anodizing phenomena have not been accurately interpreted, and the anodizing industry has relied on a substantial body of empirical knowledge.

In this book, much of the empirical knowledge that has sustained the operation of many anodizing lines is explained by viewing anodic oxidation in terms of corrosion science, which elevates anodizing to an engineering process. Metallurgical concepts bring a new level of understanding to the process, especially in terms of the functionality and appearance of the anodic aluminum oxide (AAO). By taking a metallurgical approach to the formation and growth of the anodic oxide, the reader is invited first, to develop a virtual image of the various possible types of aluminum and aluminum alloy substrate surfaces that are used in industry, and second, to link the surface microstructure from which the oxide is grown, by way of the reactions that occur at the interface, to the and amazing oxide structure. In so doing, an understanding of the relationships between the substrate microstructure and the structure of the AAO can be developed.

In order to develop a complete picture of the various phenomena that occur during the dynamic process that leads to the beautiful and uniquely ordered AAO, I felt compelled to research the history of scientific progress that enables the understanding on the mechanisms that occur during anodic oxidation. A fascinating research of supporting literature that spans over 200 years, covering significant events over the last 5000 years, provided the foundation for Chaps. 1 and 2, which present the history of the science and technology of aluminum and anodizing. The history of aluminum and anodizing is presented from their alchemic origins to show the strides in extractive metallurgy, smelting, alloy development, and even electrochemical coating processes before electricity was discovered. As the practice of alchemy bifurcated between magic and science, the part that evolved into science, gave birth to the Scientific Revolution at the end of the Renaissance. From this point in time, the concurrent technological developments in chemistry, electrical power, and metallurgy, together with the related sciences of electrochemistry, thermodynamics, atomic theory, crystallography, and diffusion continued into the Industrial Revolution, which culminated with the Aluminum Age that brought us the anodizing industry.

In doing the research to learn and understand the history of aluminum and anodizing, I found that aluminum, as with most revolutionary discoveries, has a past checkered with much drama. From its very beginnings, aluminum has caused national rivalries, ambition, and strife; indeed, if we accept the suspicion that the strange metal Pliny the Elder reported as "metal from clay" is aluminum, two men lost their heads while researching and developing aluminum, one for greed in ancient Rome (which is most likely a legend) and the other during the most famous political revolution in France, not for his revolutionary scientific work for decomposing elements from their mineral states but for collecting taxes on tobacco (and that really happened).

By making this journey through time, I entered another world where science was truly hands-on and observation-based. By researching copies of the original documents, from the first publication of the idea that aluminum could exist as an element by Lavoisier (1784), who knew that the limits of technology failed to provide enough energy to reduce it, to the trials and failures to successfully reduce it by Davy, Berzelius and Ørsted, to the time when aluminum was first successfully reduced and characterized by Wöhler in 1827, a scientific genealogy for the discovery of aluminum and ultimately anodizing developed. During this time, the failure or success of scientific experiments was communicated and shared like a relayer's baton, enabling continued research and development. Original letters between the scientists and reports from periodicals, books, and scientific meetings from esteemed scientific organizations such as the Royal Society of London and the Electrochemical Society are the basis for much of the history presented.

Once aluminum was reduced, the baton was passed to researchers who had the vision to realize that characterization of the properties of the element was key to unlocking its potential as an industrial material. The determination by Wöhler and other colleagues that aluminum has a low density and high strength to weight ratio indicated the need to develop an industrial process to extract aluminum. Government interest in new materials, for the purpose of defense, funded the research over 20 years that followed the successful extraction of pure aluminum in 1827. Aluminum extraction was brought, with some difficulty, to production as an industrial process by Deville in 1855 and later refined because of improvements in electrical power, by the Hall-Héroult process in 1886, to the extraction process used today.

In presenting the history of anodizing, another timeline develops that bridges electricity and physics with chemistry and forms the new science of electrochemistry. Along this timeline, impressive names such as Volta, Faraday, Helmholtz, Ostwald, Nernst, and Tafel are encountered, just to name a few. Prior to the discovery of electricity, in order to identify the unique behavior of the newly discovered elements, scientists characterized them by way of thermal treatment and exposure to various acid and alkaline solutions. It was during this time that Schönbein identified the characteristic we have come to know as passivity, which, with the discovery of electricity, became the basis for anodizing. It was soon determined that anodic polarization could produce a change in the corrosion behavior of some metals and that cathodic polarization could reverse it. Regarding the relationship of the discovery and development of aluminum to that of anodizing, Wöhler, in 1827, determined that aluminum is conductive, and in 1857 Heinrich Buff characterized aluminum as a passive metal, showing that impurities in the base material changed the appearance of the oxide that formed, publishing what appears to be the first paper on an anodic aluminum oxide.

If one considers aluminum to be a rather new metallic element, then one must consider that anodizing is an even newer process. In order for anodizing to become an industrial process, the need for functional anodic oxides in engineering designs had first to be established. Therefore, in deciding on how to pursue the research for presenting the history of anodizing, it became clear that each step of the path to anodizing, from the discovery of aluminum to the discovery of its ability to develop a functional oxide by way of anodic polarization, had to be explored. This, as a matter of course, led to the development of an industrial scale electrochemical anodic oxidation operation, or anodizing process. Consequently, the sections in Chap. 2 that regard the history of anodizing are devoted to the various aspects of science and engineering that are part of the explanations for the reactions that occur during the anodic oxidation of aluminum, as well as the engineering developments that determined the applications for the AAO.

As the timeline for sciences of electrochemistry, electrical engineering, and metallurgical analysis developed concurrently with the technologies extracting aluminum, alloy development, industrial application, and anodizing, I found myself admiring the people who made and supported these discoveries and therefore made tremendous strides in basic sciences, but most importantly in the young science and industry of aluminum metallurgy and even younger science and industry of anodizing. I hope that by presenting some of the important developments in metallurgy, chemistry, and electricity as they connect with anodizing aluminum, I am honoring the memories of the scientists and engineers by sufficiently acknowledging their accomplishments and by showing that what was learned about aluminum and anodizing set the foundation of the industry that we have today.

It was important to me to understand how the anodizing industry developed from the struggle to determine an accepted theory for the passive behavior of metals, to characterization experiments with aluminum anodes, to the application in capacitors and rectifiers, which use an entirely different type of oxide, grown with entirely different types of electrolytes than are commonly used today for different applications. This required deep dives into old journals and proceedings of the Royal Society and the Transactions of the American Electrochemical Society and reports of the Electro-physical Laboratories of the Electro-technical Institute at the Technical University of Munich, among other periodicals, proceedings, and books. It became just as important to review the references and citations at the end of pertinent articles as it was to read them, to find new original sources for ideas that translated in more and better detail, the reasons we have accepted for theories of anodic oxidation. Regardless of the conclusions drawn, all of the research consulted was tremendously insightful and germane to the process of anodizing.

In every publication that I have reviewed, from Setoh and Miyata (1932), to Wernick, Pinner, and Sheasby (as early as 1953), to Diggle (1968), to Lee (2014), references to the work of Adolf GüntherSchulze are made, and very difficult to locate. In constructing the history of anodizing, I was determined to find his work, in order to understand its importance. What I found is that over the period of time from about 1903 to 1910, Adolf GüntherSchulze worked with a simple electrochemical cell, characterizing a variety of metal anodes, in a variety of electrolytes. In 1906, he published his treatise, "Über das Verhalten von Aluminiumanoden" (On the Behavior of Aluminum Anodes) in Annalen der Physik. In 1907, he published studies on Tantalum and in 1909 on Niobium. In his last paper directly written on the subject in 1909, the description of both a passive film and an anodic oxide was fully developed. This description and explanation of anodic oxide formation, with only minor modifications, is the same description and explanation considered by most familiar with anodizing processes today.

By coincidence, in researching GüntherSchulze, I found reference to one of my own ancestors, a German physicist and mathematician by the name of Iris Runge. Dr. Runge initially could only attend lectures at the university because women weren't allowed to formally study until 1908. She began her career as a teacher and nevertheless continued her research and studies, ultimately receiving a university degree and doctorate in 1921. Her dissertation was entitled "Über Diffusion in Feste Zustände," "On Diffusion in Solids." She is the only female researcher that I found with some connection to anodic oxidation, who, as a colleague of Nernst and GüntherSchulze, contributed to the sciences of electrochemistry and diffusion.

Social and political issues certainly appear to have impacted how we generally view the theory of anodic oxidation. Brilliant treatises that focus on anodic alumina growth and structure were written in Germany during the 1930s that exceed the foundation provided by GüntherSchulze. Under Schumann, Baumann and Rummel wrote significant papers that identify the columnar structure of the oxide and suggest its function as a semiconductor. These papers even discuss the nature of plasma electrolytic oxidation and the effects of pulse anodizing and anodizing with AC current. In spite of the depth of this research, these papers are given only a cursory acknowledgement in American and British anodizing publications that follow the Second World War. Once I realized the wealth of interesting and important data and information that were contained in these papers, together with my husband, these articles were carefully translated so they could be included as part of the history of anodizing.

Many great comprehensive papers go into more detail about the various theories for anodizing; that written by Diggle, Downie, and Goulding in 1968 covers in detail, between 1930 and 1968, many of the theories and concepts for compact (barrier-type) oxides and porous oxides. The paper by Poinern, Ali, and Fawcett in 2011 provides a review of past processes and updates the concepts to include additional insights into the use of AAO as templates for nano-technological applications; that by Lee and Park in 2014 provides a review of many of the classic theories for anodizing and covers some of the interesting current research. These papers are recommended for the details that exceed the scope of this book. Of course, the comprehensive, 2-volume set—*The Surface Treatment and Finishing of Aluminium and Its Alloys,* by Wernick, Pinner, and Sheasby, with six editions from 1956 to 2001—fills the gap between Diggle's paper and Poinern's review. Arthur Brace's *The Technology of Anodizing Aluminum* is a necessary reference book for all interested in historical yet applicable concepts and practices for anodizing; the most current version of Brace's book was published in 2003. A new book about the science of anodizing is therefore timely and necessary in order to—once and for all, definitely, and in clear terms—link the reactions that occur during anodizing to the metallurgy of the substrate. I believe that I have succeeded in fulfilling this need with *The Metallurgy of Anodizing Aluminum.*

Chapters 1 and 2 are not meant to be an entirely comprehensive history of aluminum and aluminum anodizing. What was desired was to present the interdisciplinary path to aluminum and anodizing science and technology in a coherent, progressive way in order to complement the following chapters, which focus on the reactions that occur during anodizing in every part of the electrochemical circuit to form the highly ordered nanoscale network that is the anodic aluminum oxide, including its function and application. Without the curiosity, invention, and courage of our scientific and technological forerunners, the discovery of aluminum, its natural passivity, the understanding the structure and properties of the aluminum oxide, and the anodizing process itself would perhaps not exist today. It is hoped that the background provided leads to continued insight into the science and technology of anodizing aluminum and its alloys.

Chapter 3 presents a basic overview of industrial anodizing that simplifies the operation by presenting the anodizing process as an electrical circuit comprised of a power supply, an electrolyte, a cathode, and the aluminum anode. Images of anodic oxides produced with different electrolytes by different process parameters as well as explanations for the various process steps that occur during anodizing are provided. In order to help build a virtual metallurgical picture of the anode, Chap. 4 provides a review of basic aluminum metallurgy that contains summaries of alloy chemistries, manufacturing techniques, and temper designations.

Chapters 5 through 7 are the nucleus of this book. In Chaps. 5 and 6, anodizing is presented as a corrosion process, and the relationship and characteristics of the passive aluminum oxide to the anodic oxide are developed, refining the concept of the Barrier layer. A new theory for anodizing is discussed, based on anodic oxidation as a corrosion process, and, in so doing, explains how the AAO structure nucleates and grows. A mathematic model for anodic oxide formation, based on the Tafel equation, is presented which provides a unifying thread tying initial surface oxidation together with porous oxide growth. This approach and model is substantially different from many of the theories for anodic oxide formation and presents my personal view of the subject, which continues to develop after over 35 years of academic, government, and industrial research in aluminum alloy metallurgy, surface science, and industrial metal finishing.

In Chap. 7, a more focused view is taken at the interface from which the oxide grows, and a clear description of how the elegance of anodic oxidation is confounded by the interfacial phenomena between the oxide and the substrate and how these phenomena impact anodic oxide growth are considered. The aluminum and aluminum alloy substrates, whose metallurgy and manufacturing processes have developed as virtual pictures in the reader's mind, can now be viewed as complex anodes for oxidation, developing a picture of how the anodic oxide grows by way of consuming the surface microstructure, from the atomic to the macroscopic level. The impact of the manufacturing process and therefore the base metal microstructure on the anodic oxide are presented in terms of how and why these effects occur, which brings insight to the development of methodologies for minimizing them.

The intimate relationship between the anodic oxide and the substrate microstructure was at the forefront of the process used to select the case studies presented in Chap. 8. Each case study has a background story as to why they were analyzed; some are actual industrial process failures and some are part of larger development programs that incorporate anodized aluminum as part of the engineering design. All of the samples provide actual insight into the effects of variations in the four basic components that impact the microstructure: alloy chemistry; composition (in other words, how the elements that comprise the alloy come together as intermetallic phases and compounds); thermal history; and lastly, deformation processing effects on the bulk and surface microstructure, which also impact the appearance and function of the anodic oxide. Basic details of the manufacturing processes—casting, extrusion, and forging—are presented in order to link the microstructural characteristics that they develop to what is developed at the surface by the anodizing process, and to the appearance and function of the anodic oxide. The information presented is original and new and has never been presented in its entirety. All of the images and photos are of actual industrial substrates and presented with the expressed consent of representatives from the companies for whom the work was accomplished.

In total, I hope that all who refer to this book, such as metallurgists, electrochemists, and other scientists and engineers who study and/or use the anodic oxide, metal finishers who employ and operate the anodizing process, and design engineers who may specify the anodic oxide as the finish of choice for aluminum components, find this informative and helpful. I am also hopeful that some of the seeds of concepts that are not completely developed will find purchase and growth in continued research.

Personally, over the last 35 years, I have noticed that no one starts his or her career by saying, "I want to be an anodizer." Instead, anodizing seems to find people who make it part of their career. There are fewer people for whom anodizing becomes a passion and an industry to which one feels devotion. I am one of them. For this reason alone, acknowledgment is due to all of the colleagues with whom I have worked and from whom I have learned and been inspired during my long career as a doctor of metallurgy with an emphasis in surface science. The various problems we have endeavored to solve together have always been a means for me to learn and to continue my anodizing and light metals research while directing and performing analysis.

It is impossible for me to acknowledge all of my colleagues individually, but I would like to especially acknowledge Mareike Krueger and Gerhard Thüringer of Impreglon Surface Engineering in Kaufbeuren, Germany; Bill Klaczinski of Anderson Shumaker, in Chicago, Illinois; David Parrett of the L. H. Thomson Company, in Macon, Georgia; Joel Peltier of IHC Corporation in Detroit, Michigan; and Josh Coaplen, who at the time was at Cane Creek Cyclery, in Ashton, North Carolina, for supporting the goals of this book by providing many of the samples analyzed and documented throughout. I am very fortunate to have had research support throughout the years, in various industries that utilize the anodic oxide for diverse applications, to help reinforce many of the scientific and engineering concepts for the process of anodizing aluminum presented in this book.

As I built my career around surface science as it relates to metal finishing, relationships have been constructed with amazing laboratories, whose scientists, microscopists, and technicians have provided the data that have led to the answers which solved various industrial problems and that continue to contribute to the building of the knowledge base that answers questions about the metallurgy of anodizing. Consequently, all of the metallographic and microscopic images that are presented throughout the book are of actual industrial samples, on common industrial alloys, rather than prepared in a research laboratory, on high purity substrates. I am very fortunate for having Robert Bulthuis, my longtime friend and *the best* metallographer of aluminum and aluminum alloys in SGS MSi Testing and Engineering, Inc., in Melrose Park, Illinois, and one of the best metallurgical analytical microscopists in the industry. I am grateful to SGS for granting Bob the time to provide the metallographic services and data for the samples of various wrought and cast alloys documented in this book. Over the years, I have used various laboratories for all types of microscopy, but for this book, Cerium Labs in Austin, Texas, deserves a special recognition for their support. I would like to especially acknowledge Tim Hossain, Clayton Fullwood, Jim Conner, and Lynette Ballast for providing TEM and SEM analysis and data as well as a wealth of brainstorming and valuable chemistry insights.

Among the many professionals involved in anodizing that I have met during my career, I would like to acknowledge the few kindred spirits for whom anodizing is also a passion. One such person is a dear colleague who was possibly the foremost authority on the technology of anodizing aluminum: Arthur Brace, Anodizing Consultant of Norwich, England, who passed away on April 5, 2017. I am grateful to Arthur for the conversations, discussions, and disagreements about the science and technology of anodizing. Fred Schaedel of Anodic Technical Services in Westminster, California, is also acknowledged for his passion and undying devotion to the anodizing industry. Fred continues to tinker with anodizing and is always excited to share and to teach what he learns about the anodic oxide and the anodizing process. Of my many colleagues at Apple, Inc., with whom I've had the pleasure to work, I would like to acknowledge Howard Bujtor for believing that I could contribute and make a difference in the amazing theater of consumer electronics. My friends Naja Tabrizian of the Danish Technical Institute, Aarhus, Denmark; Larry Chesterfield of Anodizing Technologies, Inc., Indianapolis, Indiana; Hans Sellge of Semano Industries, Inc., Hayward, California; and Walter DallaBarba of Italtecno in Modena, Italy, are acknowledged for being my resources over the years for production level and specialty anodizing, for reviewing and criticizing many conference papers and journal articles, and for our many technical collaborations. All colleagues mentioned have been generous with their time in providing industrial peer review for this book.

The late Moisey Lerner of the Sanford Process deserves special acknowledgement for his passion about the science and technology of anodizing aluminum. Moisey viewed the anodic oxide as an ionic solid and built an entire technology around the behavior of the oxide as a pin diode during anodizing. He said I was like the child in Hans Christian Andersen's "The Emperor's New Clothes" because I was never afraid to question the accepted theories in anodizing. Before he passed away, he trusted me in 2003 to give his last presentation on the electrical behavior of the anodic oxide during the anodizing process at the SUR/FIN Conference in Chicago, because he felt I was the one who best understood what he was trying to say about the structural behavior of the anodic oxide. I am honored to be able to cite his work in this book.

I would like to acknowledge my academic colleagues: Dale Barkey of the University of New Hampshire, USA; Mansour Soltanieh of the Iranian Institute of Science and Technology; and Marco Salerno of the Istituto Italiano di Tecnologia in Genova, Italy. Special thanks go to Dale, Mansour, and Marco for generously agreeing to be reviewers for this book. Dale, who I regard highly as a scholar of the anodizing process, brought me the initial insight regarding electrostriction theory and growth stress, which elucidated for me their important impact on the growth of the anodic oxide. I am thankful to Mansour for the opportunity to put the scientific and technical content of this book into a daylong course, which helped to solidify the concepts that make anodizing a metallurgical process, and to Marco for providing additional reviews, commentary, scientific insight and references. I am grateful to all of my academic colleagues that I have met and worked with over the years for their insight and comradeship and for presenting ideas that sometimes challenge my own, but always give me food for thought.

I am fortunate to have had the interest and support of so many important people in the aluminum industry: John Weritz, vice president of standards and technology of the Aluminum Association, for providing resources in the form of important people, reference books, articles, as well as insightful and valuable review for the history and basic metallurgy chapters. Dan Serafin of Alcoa is acknowledged for providing early patents about the anodizing process. I am grateful to both Michael Niedzinski of Constellium and Dean Malejan of Universal Alloy Corporation for providing samples for analysis and for discussing various manufacturing process issues that bring some of the many important challenges that come with manufacturing and anodizing complex alloys.

My dear friend and neighbor, Leora Douraghy, was essential in translating patents and technical papers from the late nineteenth century from their original French to English. My husband, Thomas Nussbaum, along with his moral support, also provided translation on many technical papers and patents from the original, and sometimes old German. It was wonderful to learn that these scientists predicted with excitement that the oxide finish on aluminum would be used for great things, for a great variety of applications.

I would like to acknowledge the Aluminum Anodizers Council, for providing the forum to present much of my research over the 15 years, and my friends and colleagues from the AAC, especially Greg Rajsky and Suzanne Spohr. Their technical expertise has brought a wealth of background information about the technology of anodizing that is presented in Chap. 3.

I would also like to acknowledge Springer Nature, specifically Ania Levinson, my first chief editor, for reaching out to ask me if I had ever considered writing a book about "Base Metal Microstructural Considerations for Anodizing Aluminum." It started this 3-year-old journey down the anodizing rabbit hole, which I had always thought I would take, but didn't know exactly when.

I don't know where to begin or exactly how to acknowledge and thank Joy Kaufman of JoyJoy Creations in Geneva, Illinois, who is my illustrator, graphic and layout artist, computer consultant, and colleague and dear friend. Since the beginning of this project, she has become the right side of my brain. Our discussions and my hand-waving descriptions have become the brilliant illustrations in my book. She is talented, creative, and has the brain of an engineer and scientist.

Many friends have supported me with encouragement and desire to read what I have to say. I would like to acknowledge Roger Alker and Britta Geiger for convincing me that, when it comes to understanding aluminum alloys and anodizing, I wasn't just another person who had "pet a pony and claimed to be a horse expert"; our conversation gave me the push to actually start writing. Special thanks go to Jan and Marita Siden, Peter and Ingrid Zingsheim, PJ Hardin, and Sowmya Laxminarayanan for insisting that this content be assembled and this book be written. My children, Andrew, Anthony, Robert, and Monica, and my grandchildren, Jack, Blaise, and Cecilia, who own a great part of my heart, also deserve acknowledgment for their acceptance and love while I have been absent, perhaps not bodily, but in mind and spirit, while pondering about what to write next.

Lastly, I would like to once again acknowledge my husband, Thomas Nussbaum, as my sounding board for ideas and interesting discoveries during my research; as my first-line reviewer, editor, and critic; and as a source for comfort and respite when I was simply exhausted from so much close work, by providing me *ein weitere blick*, both physically and mentally. He has always been a source of encouragement and business advice, from the moment he suggested I become an independent consultant, throughout the time traveling with me on long-term assignments, until now, during this book-writing time. I am grateful for his patience and pride throughout my career.

Bad Wörishofen, Germany Jude Mary Runge
Chicago, IL, USA
August, 2017

Contents

Chapter 1
A Brief History of Aluminum and Its Alloys

1.1 Introduction

About 8.2% of the earth's crust is composed of aluminum, making it one of the most abundant metals in the earth's crust; however, it is never found in nature as pure aluminum. Aluminum occurs in nature bound with other elements as compounds: as alum (potassium aluminum sulfate); feldspars (sodium aluminum silicate); micas and clayey earths (aluminum silicates); as well as bauxite, the common ore from which aluminum is extracted, in which it is bound as stable oxides in various ratios of gibbsite, boehmite, and diaspore [1]. The complex nature of aluminum-bearing minerals not only posed the scientific challenges to early researchers for reducing aluminum metal as an element; it ultimately posed interdisciplinary scientific and engineering challenges toward developing a viable industrial process for aluminum extraction, production, and manufacturing that were not met and solved until well after aluminum had been reduced in the laboratory. The stability of aluminum-bearing minerals and the associated scientific and engineering challenges therefore make aluminum and the processes surrounding it quite modern; however, the use of aluminum oxides and other mineral forms of aluminum and the development of the processes used to extract, reduce, and characterize them have an interesting history that dates back to ancient times.

The path to the discovery of aluminum, the development of its industrial extraction process and ultimately, the process to deliberately grow a functional, protective anodic oxide, corresponds with the path to modern science and technology. It originated in ancient times with the Urban Revolution, when the desire to find the composition and constituents of the earth's crust, how to extract them, and how to put them to use evolved into the development of specific skills or "crafts." Early craftsmen developed an amazing amount of technology; finding and extracting metals, alloying them, inventing and improving tools, even making machines to support the processes to make them. Beautiful objects were cast and wrought from gold, bronze, and iron.

J.M. Runge, *The Metallurgy of Anodizing Aluminum*,
https://doi.org/10.1007/978-3-319-72177-4_1

Throughout history, and even in modern times, methods to produce change were not always science based, and certainly in early history these processes could only be based on observation and experience. Craftsmanship and alchemy were the bases for engineering and scientific research and development that, without the exact knowledge of the engineers and scientists of the time, made use of the phenomena of thermodynamics and diffusion as they were developing the sciences of chemistry and physics. Alchemy, the early science of transmutation of material to produce a valuable change, provided some actual explanations for chemical and physical reactions. The steps to processes that yielded the desired changes to materials that produced the effects of increased strength, hardness, and enhanced appearance, as well as their specific recipes, were carefully developed and recorded. However, when no obvious explanation for the changes was found, magic was given the credit.

Consistent with modern events, the earliest technological and scientific developments became caught up with issues of wealth and power, which sometimes gave incentive and reward but also, at times, brought persecution and pain. Politics and religion significantly stilted scientific progress through the Middle Ages, but with the dawning of the Renaissance in the mid-to-late fourteenth century, the amazing pace of technological development was restored. The Age of Enlightenment was in full swing throughout Western Europe during the seventeenth and eighteenth centuries and gave birth to the Scientific Revolution. Modern chemistry and physics, together with modern metallurgy developed and grew, leaving their alchemic origins behind.

With the discovery of electricity and the invention and development of the battery at the end of the eighteenth century and the beginning of the nineteenth century, new sciences and related technology were born, of which electrochemistry and thermodynamics were particularly important. With the continuous discovery of new elements and developments to extract and refine them, in a natural curiosity to determine the properties of something new, the scientific community responded by characterizing them. Curiosity to learn the basis of all nature drove scientific research to understand the foundations of their material structure, how they react and how they come together to form compounds, giving birth to the new studies of crystallography, diffusion and modern atomic theory. Research and experimental results were often published or demonstrated to like-minded members of scientific societies as a means to garner affirmation and recognition from other colleagues, as well as to obtain additional research support from benefactors, which included national governments. Engineering and technological strides were made to support scientific discoveries, develop applications, and bring them to industry.

The discovery of aluminum in 1827 was the culmination of many results from the liberal scientific thought and experimental freedom that occurred during the Scientific Revolution. The "new metal," aluminum, was part of the intellectual challenge and excitement of the Scientific Revolution. From the time that the existence of aluminum was speculated, in 1789 by Antoine Lavoisier, the process to reduce aluminum from its mineral state was the source of scientific challenges and competition, as well as frustration and failure, until Friedrich Wöhler successfully reduced it almost 40 years later. Because of concurrent interdisciplinary research and development, less than one hundred years passed between the time when aluminum was first reduced and characterized until 1885 when the Hall–Héroult process was developed during the Industrial Revolution, which is the industrial process still in use today for the extraction and refining of aluminum.

The design and development of engineering applications for aluminum and its alloys track with the giant steps in science and technology that occurred during the Industrial Revolution in the late nineteenth and early twentieth centuries. This is clearly documented in developments in transportation, from the automotive to the railway and marine industries. However, no industry echoes the development of aluminum and its alloys more distinctly than the aircraft industry. Transportation, as the critical means to success in business and defense, meant that aluminum industrial research also tracked with specific national government interests, with distinct efforts being made within Western Europe, the United States, Japan, and Russia. Without a doubt, because of the amazing strength-to-weight ratio of aluminum alloys, together with their inherent corrosion resistance, the aluminum industry has grown with the aircraft and aerospace industries. Nevertheless, amazing achievements have likewise and simultaneously occurred in architecture, packaging, and electronics, as well as in every day household appliances. Aluminum and its alloys are ubiquitous, and in very many ways, mankind is richer for their presence on earth.

1.2 Ancient History

In order to understand why ancient times are germane to the path to aluminum, it is important to consider the nature of an element, and how it is bound. The idea that the simplest essential parts and principles of which anything can consist, exist in nature, is an ancient one. In 465 BC, the concept that all matter is composed of particles, called atoms, that are too small to be seen and cannot be split into further portions, was introduced by Leucippus and Democritus in ancient Greece. Their theory included that atoms are homogeneous and completely solid, with no internal structure; and that a void or empty space existed between atoms. Possibly most insightful is that Leucippus and Democritus theorized that atoms are different in their sizes, shapes, and weight, establishing "a permanent foundation of ultimate particles with which to build up everything we see" [2]. This insight might have been lost but for the fact that Aristotle vehemently disagreed with it, and argued against it by listing the details of Leucippus' and Democritus' atomic theory and refuting them citing the classical elements: earth, water, air, fire, and ether (or void) as the fundamental basis for the nature and complexity of all matter "in terms of simpler substances" [2]. In 350 BC, Aristotle defined an element as: "One of those bodies into which other bodies can decompose, and that itself is not capable of being divided into another" [3]. By disagreeing with the first atomic theory, although Aristotle garnered support for his theory that employed the classic elements, he preserved the tenets of the first known atomic theory in writing, making them available for future reference.

By the time Aristotle defined the nature of an element, several elements were already known, most of the naturally occurring. Of the nine elements first discovered in ancient times, *five naturally occur in the earth's crust*: gold, silver, copper, carbon, and sulfur are found commonly, in an unbound state, in nature. Anyone who was looking for these metals and substances, and was lucky, could find them and somehow extract them from the earth. As a result of the discovery of an element, or the ore of an element and the recognition of its value, mining and refining technologies were developed. Through experimentation and characterization carried out *during ancient times*, it was learned that no matter what was done in an attempt to change these metals, regardless of what alloys were produced and or other compounds formed, these processes could be reversed and the basic element, itself, would never change [4]. The remaining four elements: lead, tin, mercury, and iron were found to be more complex, and much credit is due the craftsmen and alchemists that determined the ways to extract them from their mineral states and to develop the technology and engineering for their many and varied applications for alloys with other metals and other materials such as wood, clay, common stones, and earths. The process of discovering and determining that a new material is an element was therefore something that was known and held in high esteem during ancient times, and therefore actively sought by early scientists. It is the development of the extraction process that is *the* important part of the path to the discovery of aluminum.

The mineral states of the four elements, that are not naturally occurring, all have rather low melting temperatures. This characteristic enabled the technology to extract them in ancient times, making it possible to obtain them as pure elements. Of the four, lead, tin, and mercury were by far the easiest to extract. Smelting, the industrial practice of extraction, was developed as a necessary process and industry, which consequently provided the basis for additional research and experimentation to find other elements. By mixing elements and other ores, the advantages of alloys were discovered and many were developed for a variety of daily engineering applications: the manufacture of agricultural and building tools as well as coarse machines, weapons and vessels, decorative objects, and, of course, coins. For example, tin extraction and use can be dated to the beginnings of the Bronze Age, around 3000 BC, in Persia, when it was observed that copper objects formed from polymetallic ores with different metal contents had different physical properties [5]. Two additional elements, arsenic and bismuth, were certainly known and utilized in ancient times, but it is not certain that they were actually ever recognized as elements. Knowledge of the element arsenic also goes back to ancient times; Dioscorides, a Greek physician in the court of the Roman Emperor Nero in the first century, described arsenic as an effective poison with a variety of applications. Arsenic was also used as an alloying element; similarly, bismuth was another element that was in use all over the known world during ancient times as an important alloying element in bronze [6].

Of all the known elements, iron proved to be the biggest technological challenge for the ancient world. Although iron can be found in a natural state in meteorites, such a condition is rare. Most of the iron in the earth's crust is found combined with oxygen as iron oxide minerals such as hematite (Fe_2O_3), magnetite (Fe_3O_4), and siderite ($FeCO_3$). Many igneous rocks also contain the sulfide minerals pyrrhotite and pentlandite [7]. The first iron production started in the Middle Bronze Age, samples of smelted iron from Asmar, Mesopotamia, and northern Syria were made sometime between 3000 and 2700 BC. Because of the higher temperature and atmosphere necessary to smelt pure iron, it took several centuries before iron displaced bronze as an early industrial metal.

In one word, the extraction of iron required *energy*. But how much energy, and how it needed to be applied was achieved over the next thousand years, and in about 1600–1200 BC, the process of iron extraction developed into a useable method such that iron was smelted in the Near East by the Hittites, in India, and in Africa by Zimbabweans. Beginning in 1600 BC, iron alloys were being smelted in China (Shang Dynasty) [8]. According to Pliny the Elder, in his *Naturalis Historia*, iron use was common in the Roman era. In fact, Pliny discusses the diffusion phenomenon of the "cementation" process, the diffusion of carbon in iron to yield steel [9]. That iron and steelmaking could be done with the technology available during ancient times is a truly awesome achievement.

The energy challenge was *the barrier* to the discovery of new elements well into modern times, which included aluminum. Aluminum is one of many elements found tightly bound in its mineral form. The ancient history of aluminum can be linked to alumen (alum), the mineral from which Sir Humphry Davy first took the name for the element aluminum in 1808 [10]. Alum is potassium aluminum disulfate: $KAl(SO_4)_2{\cdot}12H_2O$. Alum compounds form wherever clay or schists (mica, talc, chlorites) come in contact with sulfuric acid from wet volcanic soils and other sulfur-based minerals, and, in addition to aluminum, can contain other metals such as silicon, magnesium, calcium, and iron because of reactions with complex minerals such as hornblend or talc. Because of their volcanic nature, alum deposits are almost always located near the surface of the soil; they are easy to mine and can be stripped from the surface. In ancient times, and for the reason that Mediterranean soil is largely volcanic, alum is recorded as being used in the ancient civilizations of Mesopotamia, Sumeria, Egypt, Greece, and Rome. Clays would be mined and refined to obtain alum for applications such as: astringents, to coagulate blood; as an antiperspirant; as a fire retardant; as a preservative and pickling agent for foods and in dyes [11]. A treatise on alum, as a compound found naturally occurring in the earth, together with a variety of its medicinal uses can be found in "Alumen and the Several Varieties of It; Thirty Eight Remedies," Pliny the Elder, *Naturalis Historia*, Book 35, Chapter 52 [12].

Within his encyclopedic work, *Naturalis Historia*, The Natural History, written in 77–79 AD, Pliny mentions a strange, light, silvery metal "from plain clay," suggesting the possibility that metal with the characteristics of aluminum may have been discovered by accident and then deliberately and brutally forgotten over 2000 years ago: One day, a goldsmith in Rome was allowed to show the Emperor Tiberius, who was emperor from 14 AD to 37 AD, a dinner plate (sic—in another account, the dining piece is a goblet). The plate was very light, and almost as bright as silver. The goldsmith told the Emperor that only he, and the gods, knew how to produce this metal from clay. The Emperor became interested, but also concerned, because all of his treasures made from gold and silver would no longer be as valuable, should common people begin to produce metal from plain clay. Therefore, instead of giving the goldsmith praise and position the goldsmith expected, Tiberius ordered him beheaded [13].

Although the story is most likely a legend as it pertains to aluminum, it implies that other metals beside lead, tin, mercury, and iron may have been reduced from their mineral states quite early and even suggests the possibility that aluminum could have been discovered much earlier if technological development in the extraction and reduction of metals had been allowed to continue, yet clearly explains why that possibility disappeared in the Western World. Even today, when scientific achievement doesn't match the expectations of governmental or corporate benefactors, ideas get postponed and funding "cut off," along with the credit and acknowledgement properly due.

1.2.1 Alchemy

In considering the impact of ancient times on the science and engineering that led to the discovery of aluminum, the legend from *Naturalis Historia* not only presents the possibility of the discovery of aluminum, this story is also a great example of alchemy, the study and practice of the transmutation of baser substances into substances of great value, which was the basis for scientific experimentation and analytical thought throughout the ancient world. Alchemy was practiced from China to the Near East, and farther still to ancient Greece and Egypt; but it is the Egyptians who generally get the credit for developing the formal alchemic practices in the first century AD that were brought to Western Europe and practiced during the Middle Ages. Much of what was practiced as alchemy has bona fide technological and scientific value; it was the basis for art, trade, and occupation, culminating in craftsmanship and guilds, together with medical and scientific practice throughout the ancient world.

If one considers the Urban Revolution of 3000 BC, as the basis for all of science and technology during ancient times, it is arguably as significant as the Industrial Revolution of the nineteenth century. The Urban Revolution is concurrent with the Bronze Age and may indeed be the result of it. The technological advancements of textiles and writing; agriculture and irrigation; the invention of the wheel; and the metallurgy of gold, silver, copper, and bronze, were the dawn of the basic crafts of modern technology. From the point of view of wealth and power, metallurgical research had vast importance for nobility and defense purposes, and many wars were fought that enabled cultural mingling, which integrated and promoted further development of all technology. Science, as alchemy, and technology, as craftsmanship, were born during the Urban Revolution from the philosophy and technology of the Ancient Near East (Persia, Sumeria, Mesopotamia), India, and Greece [4].

Even before there was papyrus, codification for the terminology of science and technology, including: agriculture; engineering as the crafts of metalworking, architecture, and art; astronomy and medicine were reported on cuneiform and other permanent forms of ancient books (tablets). Ancient tablets connected the gods, metals, plants, and stars in a "universal sympathy." Most fascinating in the development of engineering and metallurgical processes are the actual recipes and procedures from ancient times that describe how to make alloys from metals. The accounts, taken from cuneiform codexes, report the technological strides that were made, and are quite detailed, including experimental data. Experimental results were determined by the empirical results of appearance and strength, and as experiments were performed that allowed ancient technologists to study and understand changes (transmutation), the ability to alloy metals was passed on as both useful, as a craft, and philosophical, as science; that is, it was also important to study for the sake of understanding alone! The various guilds of various craftsmen cast their body of knowledge into lists of chemical and physical tests that characterize materials by: outer form (shape), color, hardness, and effervescence in vinegar, and, insodoing determined how they were best applied. The lists are so comprehensive that scientists today can identify the various materials mentioned from the various listed test results and characteristics described found in ancient codices.

Alchemy embraced astrology, which embraced mathematics. At this time in history, early science as chemistry, physics, and metallurgy was budding and was 100% hands-on and observation based. The recognition of the importance of technological development in ancient communities is proven within agricultural records from the ancient Near East, in which it can be read that surplus harvests were given to craftsmen and early scientists so they could continue their experimentation and technical development work without having to worry about sustenance. As cultures expanded and overlapped due to wars, travel, and trade, technological information was shared, enabling the development and improvement of different technological methods for similar engineering processes. An example regarding cultural differences in metallurgical alloying and coating processes: Babylonian methods transmuted (alloyed or changed) metal through the embedment of the metal object to be changed within a supplementary coating of chemicals or other materials, which then penetrated the metal by way of diffusion through exposure to prolonged heating, thereby changing the appearance and properties of the metal object. Egyptian methods changed the appearance and properties of the object by roasting the metal object and projecting (subliming) chemicals onto the heated body, producing a coating that changed the appearance of the host object [4].

The Greeks, under Alexander the Great, took alchemic teaching from Mesopotamia in the sixth and fifth centuries BC during the time of the Greek occupation of Persia. Greek architects and scientists traveled and worked in what is today Iraq and Iran, and brought back new data and "strange beliefs." Four basic important books were written by Bolos Democritos of Mendes in 200 BC that developed from the integration of scientific knowledge from the Near East, and logically, in the Greek way, which "crystallized" Near East science into a clear-cut doctrine of four volumes that dealt with the making of gold; the making of silver; the making of gems; and the manufacture of the color purple [14]. Passages on the transmutation of matter, affected and indicated by the color changes produce by alloying, coating, and dyeing operations were combined with selected sections of technological "handbooks" and the recipes of craftsmen of their period. Experiments and data were systematically assembled that would allow the alchemist to study and understand the change, or "transmutation." These books were eventually brought to Alexandria and became the Universal Alchemic Codex, a melding of the scientific data, philosophy and technology from Egyptian, Jewish, Babylonian, and Persian sources.

The technological and scientific progress during ancient times is amazing. Unfortunately, in antiquity, scientific knowledge and experimental data were never collected to formulate results from which conclusions were drawn, as they are most of the time today from modern research data. Religion, philosophy, and science were one—every bit of scientific knowledge meant deeper knowledge of the Cosmos as a whole—a means to understand the Order of Creation and maybe, to master Nature. This sometimes had great value and was sometimes the stuff of very dangerous things, politically and religiously. It is reported that in fact, the Roman Emperor Diocletian ordered the destruction of the alchemic texts and codexes in the third century [4], with the intention of removing all people and things that contradicted and challenged his absolute power. This drove alchemic practices into secrecy, giving them the name "The Dark Arts." Unfortunately, the dark connotation followed alchemy, and the "dark practice" became associated with evil and demonic practices and was linked to occult, rather than religious, or more importantly for future researchers, scientific practice.

1.3 The Middle Ages

During the Middle Ages, from about the third to fourth centuries AD until the Renaissance began during the fifteenth century, scientific work of all types that was not solely for medicinal purposes was considered heresy. It was believed that anything that brought philosophical understanding or scientific reasons rather than a spiritual belief or religious faith to the origins of the earth was the work of the devil. Alchemy fell from favor when claims of magic were associated with it. Furthermore, the ruling nobility feared that those who knew the "Dark Arts" might gain power, which demonized the practice of science by those who actually knew. Many alchemists, who were learned men and women, and therefore ordained priests, nuns, and/or nobility, died at the hands of the Inquisition, or were killed by ruling nobility who viewed their scientific and technological advances dangerous to their ruling power. Many of the old alchemic codices had been destroyed, requiring the rediscovery and reinvention of quite a bit of science and technology. New ideas continued to come from the Near East, Arabia, and India, but were considered heresy if practiced in the known Western World. This is the basis for the veritable absence of the discovery of new elements prior to the Scientific Revolution.

Alchemic practices were only permitted to continue in medical research; and as an accepted area of study, were driven by the desire to discover the universal cure for disease and to discover a means to indefinitely prolong life. Consequently, most physical scientists were also medical doctors, well into the nineteenth century. Alchemy continued to be driven by the desire to transmute less valuable materials, most often lead and/or mercury into gold [4]. Research and experimentation was allowed by those in power not entirely to bring wealth, but to prolong the life of the wealthy. It was believed that the development of an ingestible gold would transfer the properties of this most indestructible metal to those who would eat it, ultimately, transmuting the individual. From this it can be concluded that if the Inquisition didn't kill the alchemist, lead and/or mercury poisoning did.

The lack of scientific progress during the Middle Ages was also due to an actual technical limit: the energy required to extract elements from their stable, complex mineral states was not available. This meant that alum remained an important mineral. Therefore, there is little recorded about chemistry as it relates to the discovery of elements during the Middle Ages. It is interesting to note that during the Middle Ages, Pliny's *Historia Naturalis* was still viewed as the main scientific reference throughout Western Europe.

Although little new science was discovered during the Middle Ages, technological improvements did occur because craftsmanship remained important. Alchemic experiments to transmute base metals to gold were the basis for the development of methods to reduce elements from their clays or earths, and an important basis for the science of extractive metallurgy during the Scientific Revolution and nuclear science today. There is evidence that two elements were reduced from their naturally occurring states during this time. It is reported that arsenic, as an element, was isolated in 1250 AD by Albertus Magnus in Switzerland and that bismuth was reportedly discovered around 1400 AD by an unknown alchemist [6]. Bronzes produced by the Incas in Peru around 1500 AD contain bismuth. These events enable some technological and scientific credit to the difficult times of scientific persecution that occurred during the Middle Ages. Additionally, Indian alchemists isolated zinc in the thirteenth century; in fact, alchemists called zinc oxide (ZnO) "philosopher's wool" because it looked like wool when collected on a condenser after burning zinc metal [15].

As the Middle Ages evolved into the Renaissance in the late fourteenth century, the beginnings of liberal thought and the development of new ideas are evident in the scientific theories and engineering inventions of Copernicus, Galileo, Leonardo da Vinci, Johannes Kepler, and Paracelsus. Fearing only for their souls with the threat of excommunication from the Catholic Church, instead of their heads, the rack, or other creative forms of medieval execution, scientists began to give up classical Greek ideas of alchemy. Rather than being associated with the occult, astrology and mathematics were accepted as sciences, and it was determined that physical science was indeed linked to medical science. This new intellectual freedom was the impetus for the Scientific Revolution.

1.4 The Scientific Revolution

From the beginning of the Scientific Revolution throughout Western Europe (mid seventeenth to mid nineteenth centuries), as chemistry, physics and related fields of study developed from alchemic practices during the mid-to-late eighteenth century, they became more like science and less like philosophy. It was accepted by most to study the physical and chemical properties of the earth's crust in a scientific context. During this time, the great physicist and alchemist Sir Isaac Newton revolutionized our concept of the universe with his new theories of gravity, motion, and classical mechanics. In order to continue the search to find the elemental basis for common minerals that comprised the earth's crust, experiments were carried out through chemical extraction under different conditions, by way of combustion, and with acid and alkaline solutions, work that was generally called "decomposition." Once a reaction product was derived, a great deal of experimentation would commence to characterize and therefore describe it.

New ideas that laid the foundations for modern chemistry and physics were abundant, and in Western Europe, scientific and technical ideas belonged to the *European Context of Shared Knowledge*, a rather unofficial but important philosophical society of all scientists in Western Europe during that time, whose recognition through sharing research brought particular credibility and practicality to work as a scientist and an inventor. Scientific and political rivalries developed within Western Europe, especially between France and England, which stimulated much experimentation and excitement when new results were reported when letters of communication were read in the company of other scientists. Consequently, scientific salons were assembled by elite societies during which experiments were performed, often very theatrically, and results presented in order to demonstrate and communicate new scientific results and garner the support of benefactors, thereby enabling further research. Philosophical societies were created at universities throughout Europe where the academics and scientists alike came to discuss experiments and new findings. The reports of these salons became the first regularly published scientific journals, The Philosophical Transactions of the Royal Society of London being among the most esteemed.

1.4.1 Birth of Crystallography and Development of Atomic Theory

Toward the end of the Renaissance, an alchemist recognized as one of the first modern chemists, Swiss-German Philip von Hohenheim (1493–1541, aka Paracelsus) postulated that minerals, the source for elements, were developed in the earth from "seeds" (sic) "the matrix element water, nourishes the seeds of minerals and metals, which grow into mature specimens in the earth…" [16]. The concept of the space and symmetry of the fundamental "seeds" that comprise natural bodies was

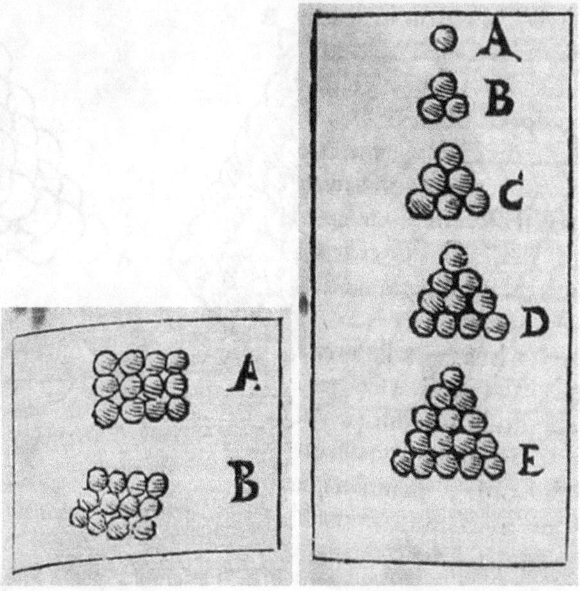

Fig. 1.1 Kepler's close-packed figures, Image 1 from "Strena Seu de Nive Sexangula" from the book exhibit "Crystallography: Defining the Shape of Our Modern World"—University of Illinois at Champaign Urbana, 2012. (http://www.scs.illinois.edu/xray_exhibit/books/kepler.php) [17]

expounded upon by Austrian astronomer, physicist, and mathematician, Johannes Kepler. In 1611, Kepler published a pamphlet "*Strena Seu de Nive Sexangula*" (sic) "On the Six-Sided Snowflake," in which he discusses the problem of why snowflakes are always six sided. Important to crystallography, Kepler proposed that ordered, symmetric shapes arise from the packing together of similar bodies, even though those bodies may not be symmetric, and included accounts of cubic and hexagonal close-packing of fundamental particles, an important concept of modern crystallographic studies today [17]. See Fig. 1.1.

During the mid-to-late seventeenth century, physical investigations regarding the structure of materials were being carried out with the newly invented compound microscope. In 1665, Robert Hooke (England) observed that "objects of regular figure that are so conspicuously various and curious…arise only from three or four several positions of globular particles…" which gave credence to Paracelsus' "seed theory." Hooke studied simple table salt, and showed, by way of drawing his observations, that it is composed of a texture of "*globules.*" Continuing on this work, in 1690, Christiaan Huygens, a Dutch philosopher and scientist, suggested through his work on optics and refraction, that materials comprise invisible equal particles, in fact, he called the assembly of particles, "crystals" and schematics were drawn to demonstrate how these particles were arranged [18]. See Fig. 1.2.

Huygens' crystal structures were constructed for the sake of the geometric structure alone and without regard for chemistry. This was due to the fact that he used the figures to describe how light traveled through crystals, not necessarily the

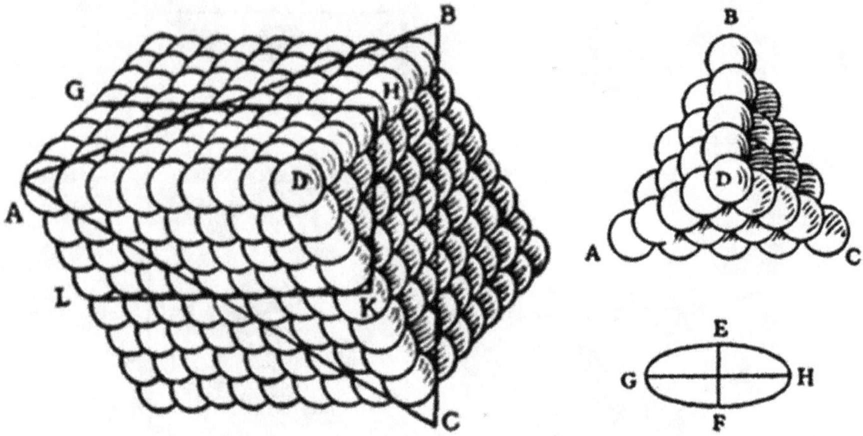

Fig. 1.2 Schematics of crystals from Huygen's Traite de la Lumieré. Optics and refraction of light were the source for early crystallographic research [19]. (http://www.gutenberg.org/files/14725/14725-h/)

nature of the crystals themselves, in particular "Icelandic Spar" (calcium carbonate from Iceland). Huygens' great assumption was that the crystals were regular in structure and packed orderly. By placing the condition of regular structure on matter as a mechanical system, in Cartesian coordinates, not only on matter, but on atmosphere as "*aether*," Huygens was able to show that the velocity of light, traveling in waves, is finite and uniform. Huygens theory was in contradiction of Sir Isaac Newton's theory, which treated light purely mathematically because he believed that light traveled in "corpuscles" or particles of the sun or stars from which they originated [19].

In 1735, in Sweden, continuing with the microscopic investigations on material structures, naturalist Carl Linnaeus (aka Carl von Linné: "*Minerals grow—plants and animals grow and live*") linked mineralogical observations to science through his studies of crystal morphology. In his *Systema Natura*, Linnaeus classified minerals by shape, number of crystal faces, and their observed chemical behavior. By linking crystals to salts through differentiation along the lines of acids and basis, a distinct connection between crystallography and chemistry was achieved. Ores were distinguished by how they reacted when burned. Reactions between crystals were observed to yield polyhedral molecules with similar structures as the parent reactants. According to Linnaeus, all crystalline minerals could be related morphologically, and therefore chemically. Minerals were identified within the system Lapidum, and classified into two groups, salts, *salia*, those that could be dissolved in various solvents, and *sulphura*, those that fumed or otherwise burned [20, 21]. Alumen was one of six salts: nitrum, natrum, borax, muria, *alumen,* and vitriolum. See Fig. 1.3.

Linnaeus characterized alumen as "*igne refractorum,*" meaning that it didn't burn when exposed to fire and was further characterized in various solvents such as

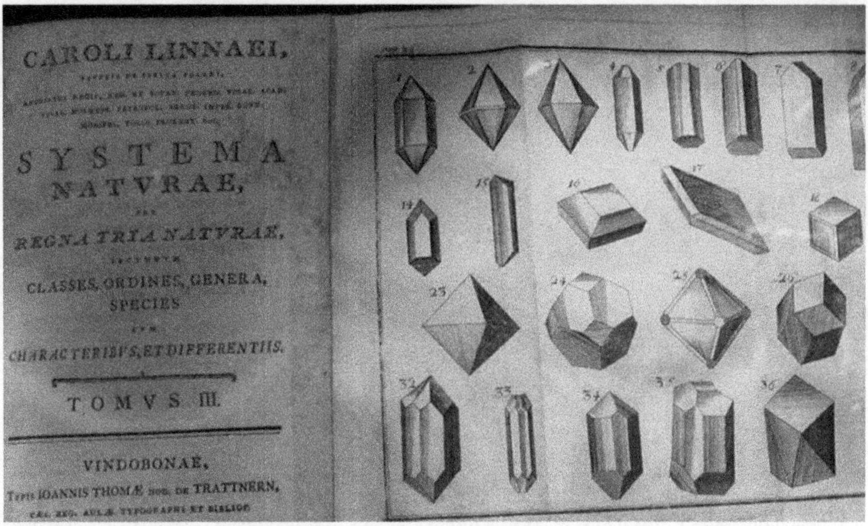

Fig. 1.3 Drawings of crystal structures observed by Carl Linnaeus in his Systema Natura, first published in 1735, represents structures described as rectangular, pentagonal, trigonal, octahedral, hexagonal, tetragonal, rhomboid, scalene, columnar, and pyramidal (from the third volume, published in 1770) [21]

various acids. Most interesting, alumen was described as having three phases: α, β, and γ, including sapphire and ruby as forms of alumen, as well as mixtures with other minerals such as marble and quartz. The α and β phases are both described as octahedral (although we know today that they are hexagonal close-packed) with α having a feathery appearance *(plumosum)* and β exhibiting surface transparency. Linnaeus observed no distinct form for γ alumen *(crudi non vidi).* It is clear from reviewing the comprehensive treatment for the various earths, minerals, and metals, that in spite of the treatment of mineralogy in biological terms much like Pliny's *Historia Naturalis*, Linnaeus' *Systema Natura* became an important resource for chemical and physical research during the Scientific Revolution, certainly referred to by the fathers of modern chemistry: Lavoisier, Berzelius, and Davy. (Linnaeus refers to some minerals as "fathers" of stones and some as "mothers," of which alumen is a father of stones, providing "seeds" that grow to form larger compounds; he also refers to alumen as "vegetable" as it is ubiquitous in soil) [22].

By 1784, the first molecular crystal structure theory was presented by French researcher, and Catholic priest, Abbé René Just Haüy. Haüy developed a two-stage molecular model for matter and crystal structures. The first model presented chemical compounds as polyhedral molecules built out of elementary constituent molecules. The second model is based upon fixed mineral species, which are characterized by fixed crystal form and constant chemical composition. From observations polyhedral fragments produced simply by dropping certain crystals on the floor Haüy concluded, from the appearance of the cleavage that they broke along consistent planes, producing the same appearance. Therefore, according to Haüy, the "crystal kingdom"

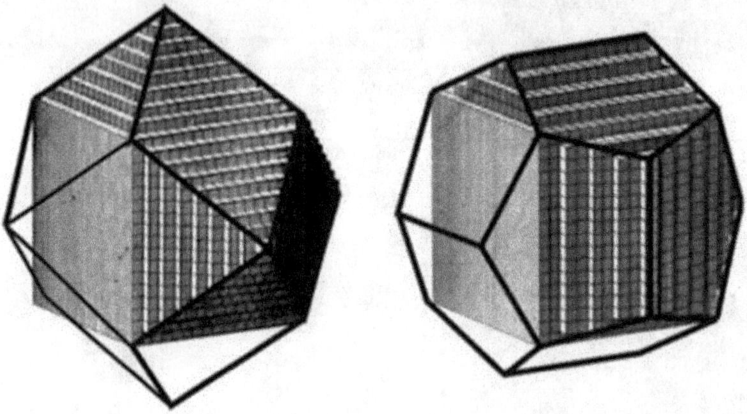

Fig. 1.4 Molecular structure models from Haüy's Traité de Minerologié. The figures show cubic integrated molecules rhomboid dodecahedron (12-sided structure comprises four-sided panels) and pentagon dodecahedron (12-sided structure comprises five-sided panels). The external shapes are added to the original to emphasize the external crystal "habit" developed by the individual "decrements" in each row of the crystal [25]

was constructed from three molecular forms: the tetrahedron, the triangular prism, and the rhombus. From these he constructed conceptual drawings and ideas of the "building bricks" of materials, which brought a new explanation to the structure of nature that was, albeit very, very small, yet tangible and sensible, elevating the science of matter from the spiritual to the more actual [23, 24]. Haüy's theory proved seminal yet controversial, providing the basis for other important scientific theories at both ends of a crystal structure: from the fundamental particle to the structures they form when they assemble. See Fig. 1.4.

Haüy's conclusions regarding crystal structure were criticized by British researcher, William H. Wollaston, who objected that a stable crystal could not result from such arrangements as the grouping of tetrahedral particles hanging together at their edges. Regarding Haüy's models as "precarious masonry," in 1812, Wollaston proposed alternative spherical crystal units, joined together as closely as possible by mutual attraction. His close-packed formations of triangularly arranged spheres imitated the commonly occurring crystal forms. From a rhomboid of spheres, tetrahedral groups of spheres could be removed, leaving an octahedron, accounting for the observed cleavage of rhomboidal fluorspar. Wollaston also constructed other forms from spheroids, earlier considered by Huygens. The most original part of his theory concerned the cubic form. He explained this in terms of two different kinds of sphere, which he referred to as "black and white balls," so arranged that each black ball was equidistant from all surrounding white balls; balls of the same type were also equidistant from each other. This produced a cube from two interpenetrating tetrahedra [26]. In the twentieth century, the lattice structure of sodium chloride was shown to be of this type; but in 1812 Wollaston's theory was only speculation. See Fig. 1.5.

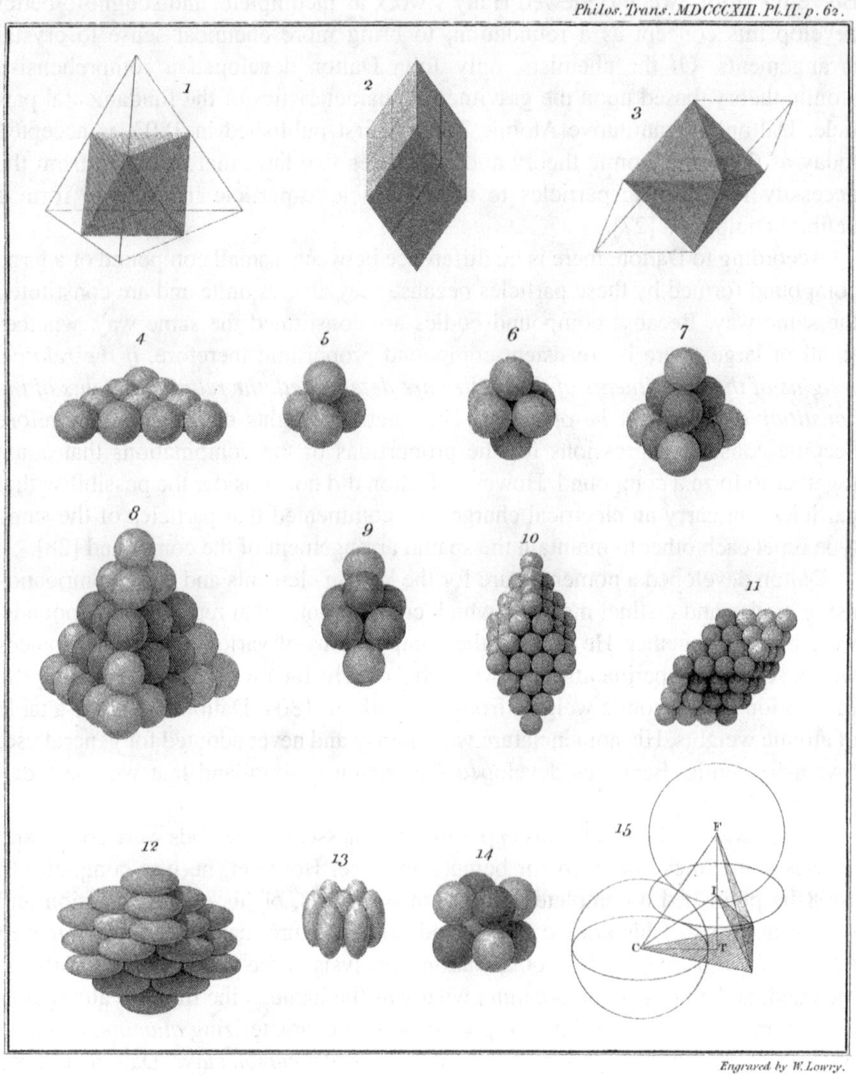

Fig. 1.5 Schematics of crystal structures comprising spheres from Wollaston's "On the Elementary Particles of Certain Crystals," 1813. (http://rstl.royalsocietypublishing.org/content/103/51.full.pdf+html) [26]

1.4.1.1 Development of Modern Atomic Theory

The idea of elementary constituent molecules was similar to the ancient Greek idea that all matter is composed of atoms that are too small to be seen and cannot be split into further portions. Separately and independently, chemists from different countries (Dalton from England, Higgins from Ireland, Richter from Germany, and

Berzelius from Sweden) viewed Haüy's work as incomplete, and sought to better develop this concept as a foundation, to bring more chemical sense to crystal arrangements. Of the chemists, only John Dalton developed a comprehensive atomic theory, based upon the gravimetric characteristics of the fundamental particle. Dalton's Quantitative Atomic Theory, first published in 1803, is accepted today as the basic atomic theory and comprises five laws that originate from the necessity for ultimate particles to unite particle-to-particle in order to form a definite compound [27].

According to Dalton, there is no difference between a small compound or a large compound formed by these particles because they always unite and are constituted the same way. Because compound bodies are constituted the same way, whether small or large, there is consistent compound proportion; therefore, *if the relative weights of the constituents of the bodies are determined, the relative weights of the constitutive atoms can be obtained.* The relative weights of the atoms therefore become constant expressions for the proportions of the combinations that come together to form a compound. However, Dalton did not consider the possibility that particles can carry an electrical charge, but commented that particles of the same type repel each other to maintain the spatial arrangement of the compound [28].

Dalton developed a nomenclature for the known elements and basic compounds using circles and distinct marks in which combinations, that represent compounds, were brought together. He verified the compositions of various known substances with a series of experiments from which his conclusions were drawn, enabling the calculation of the atomic weights from his work. In 1804, Dalton published a table of atomic weights. His nomenclature was clumsy and never adopted for general use; Swedish chemist Berzelius developed the chemical shorthand that we use today [27]. See Fig. 1.6.

Dalton was not known for his experimental finesse; his methods were coarse and he was sometimes criticized for being imprecise. However, he was complete. In 1808 he published a complete volume, in two parts, of his experimentation and results in terms of his atomic theory and nomenclature, called *A New System of Chemical Philosophy*. In his book, Dalton's analysis of the various known earths is included, and among them, *alumine*, what was the name at the time for alumina or aluminum oxide. Dalton admits to problems with characterizing *alumine*, identifying it as a "*fine white earth,*" the "*greatest portion of common clay.*" Dalton believed that alum was reducible to alumine atoms; however, because of its behavior when in contact with certain acids *("salts with acids are not 'crystallizable'")*, its weight could not easily be determined. Based upon his nomenclature, an "atom" of alumine was constituted of "*one atom of sulfate of potash in the center and four atoms of sulfate of alumine around it, forming a square*" [27]. Unfortunately, in his publication, Dalton did not include the nomenclature for neither sulfate of potash nor sulfate of alumine, making reproduction of the alumine "atom" impossible.

Dalton's characterization work garnered the praise of Sir Humphry Davy at the Royal Society of London. In Dalton's memoir, Davy is quoted as having said in 1809, "*the same proportions would follow from an application of Mr. Dalton's ingenious supposition*" [28].

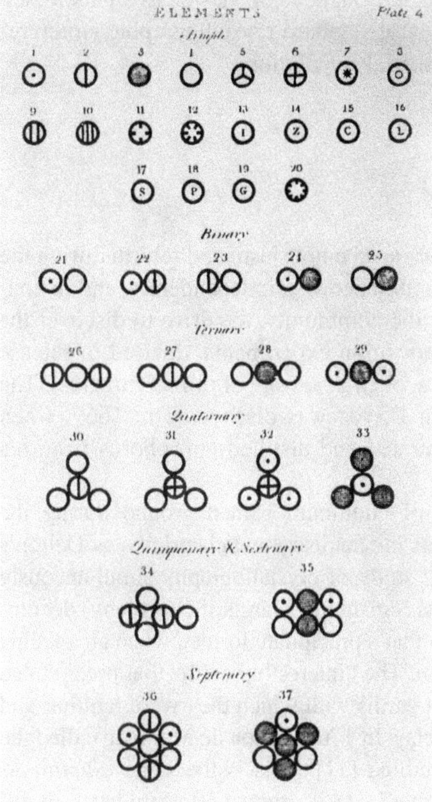

EXPLANATION OF THE PLATES. 219

PLATE IV. This plate contains the arbitrary marks or signs chosen to represent the several chemical elements or ultimate particles.

Fig.			Fig.		
1	Hydrog. its rel. weight	1	11	Strontites	46
2	Azote,	5	12	Barytes	68
3	Carbone or charcoal,	5	13	Iron	38
4	Oxygen,	7	14	Zinc	56
5	Phosphorus,	9	15	Copper	56
6	Sulphur,	13	16	Lead	95
7	Magnesia,	20	17	Silver	100
8	Lime,	23	18	Platina	100
9	Soda,	28	19	Gold	140
10	Potash,	42	20	Mercury	167

21. An atom of water or steam, composed of 1 of oxygen and 1 of hydrogen, retained in physical contact by a strong affinity, and supposed to be surrounded by a common atmosphere of heat; its relative weight = - - - - 8
22. An atom of ammonia, composed of 1 of azote and 1 of hydrogen - - - - - - - 6
23. An atom of nitrous gas, composed of 1 of azote and 1 of oxygen - - - - - - - 12
24. An atom of olefiant gas, composed of 1 of carbone and 1 of hydrogen - - - - - - - 6
25 An atom of carbonic oxide composed of 1 of carbone and 1 of oxygen - - - - - - 12
26. An atom of nitrous oxide, 2 azote + 1 oxygen - 17
27. An atom of nitric acid, 1 azote + 2 oxygen - - 19
28. An atom of carbonic acid, 1 carbone + 2 oxygen 19
29. An atom of carburetted hydrogen, 1 carbone + 2 hydrogen - - - - - - - - 7
30. An atom of oxynitric acid, 1 azote + 3 oxygen 26
31. An atom of sulphuric acid, 1 sulphur + 3 oxygen 34
32. An atom of sulphuretted hydrogen, 1 sulphur + 3 hydrogen - - - - - - - - 16
33. An atom of alcohol, 3 carbone + 1 hydrogen - 16
34. An atom of nitrous acid, 1 nitric acid + 1 nitrous gas - - - - - - - - 31
35. An atom of acetous acid, 2 carbone + 2 water - 26
36. An atom of nitrate of ammonia, 1 nitric acid + 1 ammonia + 1 water - - - - - - - 33
37. An atom of sugar, 1 alcohol + 1 carbonic acid - 35

Fig. 1.6 Table of Elements as Published in 1808 by John Dalton, using the Daltonian Atomic Nomenclature for selected elements and compounds. A legend for the various symbols on the left is presented in the table on the right [27]

Other great scientists of that time: Ampére, Avogadro, and Thénard to name a few, were interested in atomic theory and crystallography as the studies related to chemistry. Many papers were published, sharing opinions about the combinations of fundamental particles to make compounds, and the rule of definite proportion was a recurring theme. It was during this time that Avogadro used the concept of atoms and molecules to construct a table of atomic weights, which is useful for chemistry and physics yet today. Avogadro's number, 6.02×10^{23}, as the fixed number of anything that comprises a "mole," was one of the many constants that we still use in science today that came from investigations in atomic theory in the early nineteenth century [29].

By the 1830s, Haüy molecular crystal structure and Daltonian chemical atomic theory were combined to produce a view of the chemical molecule as the arrangement of atoms in space. All atoms were composed of point atoms with Daltonian gravimetric attributes located at solid crystallographic angles, producing polyhedral arrangement of atoms in space. Crystallography came to be viewed as an interdisciplinary study between chemistry and physics; standard crystallographic structures were identified and named during the Industrial Revolution.

1.4.2 *Birth of Modern Chemistry*

All facets of science and technology experienced enthusiasm and rebirth during the Scientific Revolution. During this time, simultaneously, independently, and in continuous competition throughout the scientific community, the drive to discover the next new element was underway. Decomposition experiments, devised to release metallic elements from their mineral states, or *calces*, were of primary interest. The first element of the Scientific Revolution Driven was discovered in 1669, when German alchemist Hennig Brand concentrated and distilled phosphorus from his own urine [30].

The long road toward the discovery of aluminum gained ground during the Scientific Revolution through developments in chemistry and electricity, as Dalton's new atomic theory gained acceptance and study of crystallography simultaneously progressed. In 1754, J. H. Pott and Andreas Sigismund Marggraf (Germany) decomposed alum into alumina by demonstrating that a precipitate formed when an alkaline solution was poured into a solution of alum. The "interesting earth" that precipitated was different from lime and chalk, distinct earths with which they were familiar and was found to be an ingredient in common clay. In 1761, Guton de Morveau called the "earth" *alumine*, which today we call alumina [11]. This is the same *alumine* as Dalton's "fine white earth," which shows that in 1808, almost 50 years later, aluminum remained trapped as stable *calces*. The lack of sufficient energy to produce a chemical reaction that would release the element from its mineral state was a known challenge, and, in the late eighteenth century a new method for decomposition studies was introduced using static electricity to decompose compounds to their metallic elements. Noteworthy, in the context of the determination of elements by way of electrolysis; Dutch scientist Martin Van Marum reduced (decomposed) zinc, antimony, and tin from their "calces" by way of electrolysis using static electricity in 1785 with his "Large Electricity Machine" [31].

1.4.2.1 Antoine Lavoisier

In France, from the years 1770 to 1794, through his rigorous treatment of chemistry as a science, Antoine Lavoisier was establishing himself as the Father of Modern Chemistry. By elevating the standards for experimentation and root cause explanation and demanding quantitative data in the form of weights and measures, Lavoisier helped to develop the metric system. In 1774, Antoine Lavoisier used the reaction of water steam with metallic iron inside an incandescent iron tube to produce hydrogen. Lavoisier's experiments led to the demonstration of the conservation of mass, a principle that was instrumental in changing chemistry from a qualitative science to a quantitative one, and his experiments in reaction chemistry led to the publication of a method and nomenclature for different oxidation states of various acids and alkali compounds. Lavoisier also wrote the first table of elements [32]. See Table 1.1.

In 1777, through his decomposition experiments, Lavoisier established sulfur as an element rather than a compound. To complement his work in decomposition, along with input from British chemist Joseph Priestly, Lavoisier studied the air released during combustion reactions and recognized it as a chemical substance, giving it the name oxygen in 1778. He identified oxygen as the substance that unites with metals to form their *calces* (basic compounds) and with combustible substances to form acids. Within 5 years, he identified and named hydrogen as a chemical substance (although another British chemist, Cavendish had isolated "inflammable air," which was actually hydrogen in 1766). Together with Pierre LaPlace, Lavoisier produced water from hydrogen and oxygen, proving by making water rather than decomposing it, that water was a compound, disproving what was considered one of the classic elements for more than 2000 years.

Lavoisier's work and discoveries were key in understanding the stability of different compounds, including alumina. Most importantly, Lavoisier was the first to recognize that alumina was an oxide of a metallic element. In his published table of the elements (See Table 1.1), within his publication *Traité Élémentaire de Chimie, ...* (Paris, France: Cuchet, 1789) in which sulfur, oxygen, and hydrogen are included, five "salifiable earths" (*substances simples salifiable terreuses*) are listed, which are ores that could be made to react with acids to produce salts (*salis* = salt, in Latin) [32]. The five salifiable earths are: *chaux* (calcium oxide), *magnésie* (magnesia, magnesium oxide), *baryte* (barium sulfate), *alumine* (alumina, aluminum oxide), and *silice* (silica, silicon dioxide). About these "elements," Lavoisier speculated (sic): "We are probably only acquainted as yet with a part of the metallic substances existing in nature, as all those which have a stronger affinity to oxygen than carbon possesses, are incapable, hitherto, of being reduced to a metallic state, and consequently, being only presented to our observation under the form of oxides, are confounded with earths...It is even possible that all the substances we call earths may be only metallic oxides, irreducible by any hitherto known process."—from page 218 of: Lavoisier with Robert Kerr, translator, *Elements of Chemistry, in a New Systematic Order, Containing all the Modern Discoveries*, 4th ed. (Edinburgh, Scotland: Mathew Carey, 1799) [34]. (The original passage appears in: Lavoisier, *Traité Élémentaire de Chimie, ...* (Paris, France: Cuchet, 1789), vol. 1, page 174) [33].

With *Traité Élémentaire de Chimie*, Lavoisier redefined the concept of an element; since ancient times, no metal, discoverable as a compound, was considered an element; furthermore, he clearly identifies alumina as one of the earths from which, by way of a process not yet known, a metallic element could be extracted. It is important to note that Lavoisier's work was done by producing chemical reactions involving heat and atmosphere; in other words, by combustion, without the aid of electricity. It is also necessary to remark that the list of ancient elements is based upon what we know today to be elements. During Lavoisier's time, experiments throughout the scientific community, while continuously finding something new, were reconfirming what had been discovered and lost from ancient times.

Table 1.1 Table of the elements according to Antoine Lavoisier, from *Traité Elémentaire de Chimie*, Cuchet, Paris, 1789, p. 192 [33]

	Noms nouveaux.	Noms anciens correfpondars.
	Lumière	Lumière.
Subfiances simples qui appartienment aux trois vignes & qu on prut regarder comme les dilema des corps.	Calorique	Chaleur. Principe de la chaleur. Fluide ignè. Feu. Marière de feu & de là chaleur.
	Oxgène	Air dèphlogiftiquè. Air empireal. Air vital. Bafe de l'air vital.
	Azote	Gaz phlogiftiquè. Mofete. Bafe de la mofete.
	Hydrogène	Gaz inflammable. Bafe de gaz inflammable
Subfiances fimples non metalliques oxidables & acidfiables.	Soulre	Soulre.
	Phofphore	Phofphore.
	Carbone	Charbon pur.
	Radical muriatique	Inconnu.
	Radical fluorique	Inconnu.
	Radical boracique	Inconnu.
Subfiances fimples mètalliques cridables & acidifiables.	Antimonie	Antimonie
	Argent	Argent.
	Arlenic	Arlenic.
	Bifmuth	Bifmuth.
	Cobolt	Cobolt.
	Cuivre	Cuivre.
	Etain	Etain.
	Fer	Fer.
	Manganèfe	Manganèfe.
	Mercure	Mercure.
	Molybdène	Molybdène.
	Nickel	Nickel.
	Or	Or.
	Platine	Platine.
	Plomb	Plomb.
	Tungftene	Tungftene.
	Zinc	Zinc.
Sufiances fimples falifiables terrenfes	Chaux	Terre calcaire, chaux.
	Magnèfie	Magnèfie, bafe du fel d'Epfom.
	Baryte	Barote, terre perfante.
	Alume	Arguile, terre de l'alun, bafe de l'alun.
	Silice	Terre filiceufe, terre vitrifiable.

Lavoisier's theories and experimental methods, which formed the basis of modern chemistry throughout Western Europe (publications of this book have been used as references from France, Holland, England, Italy, Sweden, Denmark, and Germany) as well as the British colonies that would become the United States, were brutally and literally cut short when he was guillotined during the French Revolution in 1794, becoming the second person related to the study of aluminum to lose his head. It is reported that, in the wake of Lavoisier's execution, Italian-French mathematician Joseph-Louis Lagrange commented, acknowledging Lavoisier's scientific contributions and stature: "*It took them only an instant to cut off that head, and a hundred years may not produce another like it*" [35].

Lavoisier's realization that certain metals have a stronger affinity to oxygen than carbon, therefore requiring other methods to reduce their compounds, inspired much research to find other methods for decomposition to discover other basic elements. Consequently, from the year 1770, until the successful extraction of aluminum metal in 1827, 40 elements were discovered by way of various decomposition processes throughout Western Europe.

1.4.3 Birth of Electrochemistry

1.4.3.1 Alessandro Volta

Alessandro Volta rocked the scientific world by introducing the first convenient electric battery, through a letter dated March of 1800 to the Royal Society of London, "*On the electricity excited by the mere contact of conducting substances of different kinds*" [36]. In his letter, Volta described work that he had commenced in 1797: "*A means to produce continuous electrical energy, based upon the chemical potential that is developed between dissimilar metals, specifically copper and zinc, separated by cardboard or felt pads soaked in sea water,*" showing them stacked in a "*Voltaic Pile*" connected in series and in parallel in order to provide an "*artificial source for electricity.*" It is claimed that chemist and future leader of the Royal Society of London, Sir Humphry Davy, said that Volta's work was "an alarm bell to experimenters all over Europe" [37].

Within a week of the disclosure of the "Voltaic Pile," as the battery was called, a rapid series of discoveries followed, including the electrical decomposition (electrolysis) of water into oxygen and hydrogen found by William Nicholson and Anthony Carlisle [37]. In Germany, Johann Wilhelm Ritter independently and simultaneously electrolyzed water. Additionally, Ritter collected and measured the amounts of hydrogen and oxygen produced in the reaction [38]. With the discovery of electrolysis, it was hoped that the new battery could provide the energy required to decompose calces that were too stable for combustion and static electricity. Research and experimentation continued in order to determine if the elements, which comprised Lavoisier's "*five salifiable earths,*" which included alumina, were within reach. See Fig. 1.7.

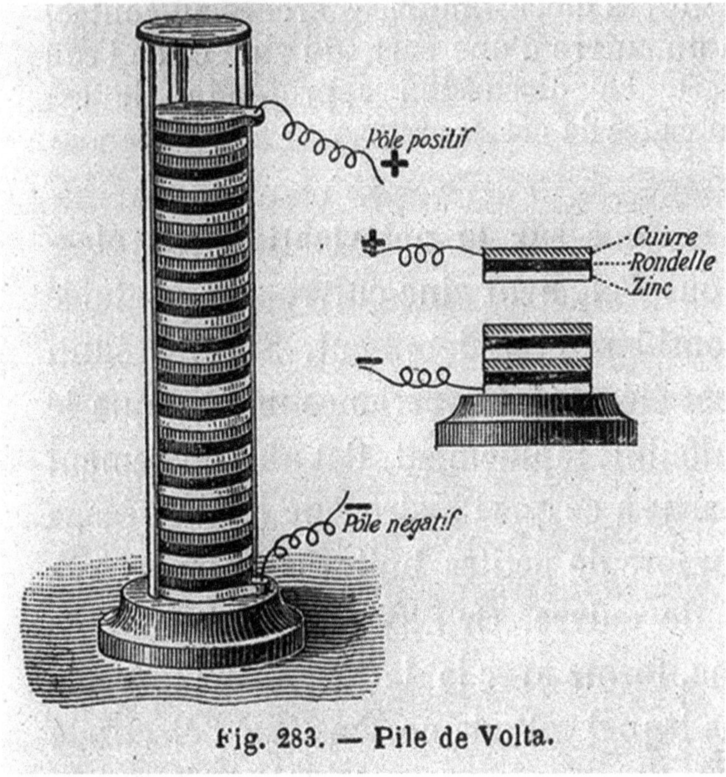

Fig. 1.7 Voltaic pile

The voltaic pile fulfilled the need for an electrical supply with increased constant voltage capability to enable the successful decomposition of minerals in order to yield new elements. Volta's fundamental experiments showed that in a closed circuit of electrical conductors, the electromotive forces would balance so that no current would flow; yet, if two conductors were placed in contact with one another through an electrolyte, current was observed to flow through the circuit. Volta considered that the electrolyte enabled the contact-electromotive force of the metallic junction to become effective, enabling the terms "*electromotive force*" *and* "*electric current,*" which are attributed to him. Volta's "*Contact Theory of Electrical Action*" formed one of the important fundamental concepts for what would become the science of electrochemistry. Volta observed: "*electricity and chemical affinity are different manifestations of the same force*" [36].

The second important concept that formed the basis for electrochemistry was the "*The Chemical Theory of Electromotive Force*". Shortly after the publication of Volta's letter in The Philosophical Transactions of the Royal Society of London, using the voltaic pile, in 1802, German philosopher and scientist Johann Wilhelm Ritter built an electrochemical cell with 50 copper discs separated by cardboard discs moistened by a salt solution that he used for electroplating gold metal onto silver [39]. Ritter observed that during his experiments with the voltaic pile, together with the electrical "action" brought about by contact with the battery, a chemical change was always observed, either in the electrolyte or at the electrodes. Because of these visual and significant changes, Ritter drew the conclusion that the chemical change, not electrical contact, should be regarded as the source for the electrical action.

Volta's Contact Theory and Ritter's Chemical Theory divided researchers within the new science of electrochemistry along this difference in basic principle. Those that had tried and failed to reduce elements from their tightly held mineral states until the invention of the battery, for example, Sir Humphry Davy and Jöns Jakob Berzelius, tended to hold with the Contact Theory. Their thinking was based on the success of isolating most of the elements discovered after 1800, by way of electrolysis, using the voltaic pile. Through his electrochemical experiments, and after reducing seven new elements in 1807 and 1808 with the use of the new battery, Davy echoed Volta by proposing that the tendency of one substance to react preferentially with another—its affinity—is electrical in nature. Through this understanding, Davy predicted the reduction of aluminum by way of a better battery—one that could produce energy sufficient to overcome aluminum's affinity with its mineral counterparts. However, the argument that this would never happen without a chemical reaction occurring first, drove much experimentation, publication, and discussion for the duration of the Scientific Revolution, into the Industrial Revolution of the mid-to-late nineteenth century and well into the twentieth century.

1.4.3.2 Sir Humphry Davy

British chemist Sir Humphry Davy, a follower of Lavoisier's work, continued decomposition studies that at the beginning were largely derivative of the Father of Modern Chemistry's, but with the development of the voltaic pile his achievements became diverse and colossal. Davy was a pioneer in the field of electrolysis using the voltaic pile to reduce common compounds and thus isolate many new elements. He went on to electrolyze molten salts and discovered several new metals, including sodium and potassium, highly reactive elements known as the alkali metals. Davy discovered both sodium and potassium in 1807, deriving sodium from caustic soda ($NaOH$) and potassium from caustic potash (KOH). Before the nineteenth century, no distinction had been made between potassium and sodium. Potassium was the first metal that was isolated by electrolysis using the voltaic pile [40].

In the treatise of his experiments, published as the Bakerian Lecture at the Royal Society of London, Davy states, "*Barytes and strontites have the strongest relations to the fixed alkalies of any of the earthy bodies; but there is a chain of resemblances, through lime, magnesia, glucina, alumina and silex. And by the agencies of batteries sufficiently strong, and by the application of proper circumstances, there is no small reason to hope, that even these refractory bodies will yield their elements to the methods of analysis by electrical attraction and repulsion*" [40]. Davy worked with electrolysis throughout his career and was first to isolate magnesium, calcium, strontium, and barium. By identifying aluminum as the metallic base for alum in 1808, Sir Humphry Davy echoed what Lavoisier had already announced in 1789; Davy's experiments to reduce pure aluminum from alumina were just as unsuccessful as Lavoisier's. In 1812, he wrote about alumina in his *Elements of Chemical Philosophy* "*This substance appears to contain a peculiar metal, but as yet Aluminum has not been obtained in a perfectly free state, though alloys of it with other metalline substances have been procured sufficiently distinct to indicate the probable nature of alumina*" [41]. In spite of his personal frustration, Davy's efforts were an inspiration for continued experimentation by many other scientists throughout Western Europe.

1.4.3.3 Jöns Jakob Berzelius

Swedish chemist Jöns Jakob Berzelius was an accomplished experimenter in the field of electrochemistry. He was a contemporary and rival of Sir Humphry Davy. Berzelius' contributions to the developing science of chemistry were significant. He became most famous in 1813 for his series of experiments that definitively established that the elements in inorganic substances are bound together in definite proportions by weight "*Chemical Law of Definite Proportion,*" which led to the understanding of stoichiometric relationships in chemical compounds. Using his experimental results, he determined the atomic weights of nearly all the elements then known. Berzelius invented a simple and logical system of symbols for the elements: H, O, C, Ca, Cl, and so forth—which is basically the same as the system we use today, except that the combining proportions of the atoms of elements in a compound were indicated as superscripts instead of subscripts. His nomenclature was much simpler and clearer to use than Dalton's, and it was quickly adopted throughout the scientific community. Berzelius' interest in all sorts of compounds led to his discovery of a number of new elements, including cerium, selenium, and thorium; his students discovered lithium, vanadium, and several rare earths [42]. See Table 1.2.

1.4.3.4 Michael Faraday

Michael Faraday, who began his work as a laboratory technician for Sir Humphry Davy in 1813, is attributed with the development and organization of the science of electrochemistry. He took copious notes while in the employ of Davy and published frequently. In doing so, Michael Faraday became the shining star in the European Science World of the nineteenth century. Even as Davy's successor as a member in the Royal Society, Faraday differed in opinion with his mentor in the Contact vs. Chemical Theories of Electromotive Force and was a supporter of Ritter's Chemical Theory of Electromotive Force. Faraday is credited with determining the connection between chemical action and the generation of electrical currents. All electrochemical methods of

Table 1.2 Comparison of early chemical nomenclature [43]

ELEMENT	ALCHEMICAL	DALTON'S	HASSENFRATZ & ADET'S	BERZELIUS'
Silver	☽	Ⓢ	Ⓐ	Ag
Gold	☉	Ⓖ	⊙	Au
Iron	♂	Ⓘ	Ⓕ	Fe
Copper	♀	Ⓒ	Ⓒ	Cu
Lead	♄	Ⓛ	Ⓟ	Pb
Tin	♃	Ⓣ	Ⓢ	Sn
Mercury	☿	◌	Ⓗ	Hg

corrosion monitoring are based on Faraday's electrochemical principles, and his laws that define the corrosion of metals in electrolytes as an electrochemical process are still in use today. Faraday also went on to become the Father of Electromagnetism [44].

First published in 1834, Faraday's Laws of Electrochemical Action [45] can be summarized as follows: (1) The Law of Definite Electrochemical Action: the amount of chemical change or decomposition is exactly proportional to the quantity of electricity that passes in solution; and, (2) The Law of Electrochemical Equivalents: the amounts of different substances deposited or dissolved by the same quantity of electricity are proportional to their chemical equivalent weights. In determining the connection between the electricity and chemistry, Faraday determined: *it is the current density (amps per unit area material) that impacts chemical change, and not charge alone*, which, unknown at the time, is a concept critical to the successful implementation of the aluminum anodizing process. In 1833, he worked out a new nomenclature for electrochemical phenomena based on Greek words, which is still in use today, for example, the words *ion*, *electrode*, and *electrolyte*.

1.4.3.5 Georg Simon Ohm

Around 1820, German academic Georg Simon Ohm began working with his own version of Volta's electrochemical cell, and determined that the amount of steady current through a material is directly proportional to the voltage across the material, for some fixed temperature:

$$I = V / R \tag{1.1}$$

Ohm had discovered the distribution of electromotive force in an electrical circuit and had established a definite relationship connecting resistance, electromotive force, and current strength. His 1827 publication "*Die Galvanische Kette: Mathematisch Bearbeitet*" (The Galvanic Circuit Investigated Mathematically) is an exhaustive mathematic proof of the seemingly simple relationship. Ohm presents his theory as one of contiguous action, a theory that opposed the concept of action at a distance. Ohm believed that the communication of electricity occurred between "*contiguous particles*" which is the term he developed to describe what would come to be known as electrons [47]. The resistance unit, Ohm (SI), is named for Georg Simon Ohm.

1.4.3.6 Andre Marie Ampére

Simultaneously, in France, Andre Marie Ampére had begun his work in electromagnetism. This is noteworthy in the relationship to Ohm's work as it applies to the movement of contiguous particles in order to communicate or conduct electricity. Ampére, continuing his early work in atomic theory and the law of definite proportion, began developing a mathematical and physical theory to understand the relationship between electricity and magnetism. In order to provide a physical understanding of the relationship between electricity and magnetism, Ampére theorized the existence of an "*electrodynamic molecule*" (the forerunner of the idea of the electron) that served as the fundamental component element for the motion of both electricity and magnetism [48].

Using the particle to explain current flow and magnetic repulsion, Ampére developed a physical account of electromagnetic phenomena that was both empirically demonstrable and mathematically predictive. He applied mathematics in generalizing physical laws from these experimental results, using the concept of a conductive particle that had also been proposed by Ohm. The most important of these was the principle that came to be called Ampére's Law, which states that the mutual action of two lengths of current-carrying wire is proportional to their lengths and to the intensities of their currents, which laid the foundation for the study of electrodynamics. Ampére also applied this same principle to magnetism. In 1827, the same year as Ohm published his laws for the communication of electricity in a galvanic circuit, Ampére published "*Mémoire sur la théorie mathématique des phénomènes électrodynamiques uniquement déduite de l'experience*" (Memoir on the Mathematical Theory of Electrodynamic Phenomena, Uniquely Deduced from Experience) [48, 49]. By coincidence, this was the same year that Friedrich Wöhler reduced aluminum.

1.5 Making a Better Battery

While much of the activity during the Scientific Revolution focused on the growth and development of new sciences through experimentation and communication of ideas, with the discovery of electricity and the invention of the voltaic pile, practical

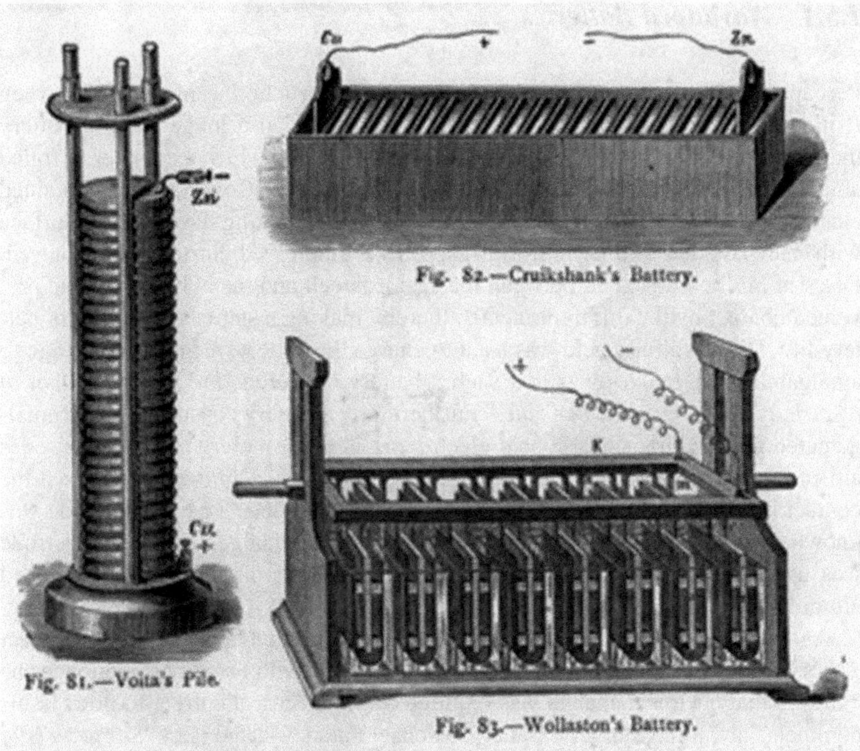

Fig. 1.8 Drawings of various early batteries, including those by Volta, Cruikshank, and Wollaston [46]

innovation was occurring in the development of an improved battery, as well as new methods to implement electrical measurements.

Rapid engineering developments that improved upon and exceeded the capability of the voltaic pile were made soon after Volta's battery was announced in England, first by William Cruikshank, in 1800 and by William Wollaston in 1810 (the same Wollaston who made his mark in optics and crystallography). Cruikshank's battery reoriented the position of the electrodes horizontally, in a trough of electrolyte, which improved battery performance by keeping the entire "pile" constantly immersed in electrolyte, but the lifetime of the battery was limited by the solubility of the zinc in the acid electrolyte. Wollaston, in 1812, developed a "better battery" with an innovation that incorporated removable zinc electrodes. Removing the zinc from the electrolyte prolonged the useful life of the electrodes and the acid electrolyte, providing the basis for improved electrical measurement systems that were easier to use and enabled more exact electrical measurements. Wollaston reduced the elements rhodium and palladium with his battery. He was so skilled at extraction and refining that he had developed a process to extract and refine platinum that was so proprietary, that he made a fortune for himself and his family by this alone. However, his method was not sufficient to reduce aluminum from its mineral state. The actual reduction of aluminum was almost 20 years away [46]. See Fig. 1.8.

1.5.1 Advanced Batteries

Perhaps the biggest improvement in battery construction came in 1830 when British electrical engineer William Sturgeon, developed a longer lasting battery that consisted of a single cylindrical cell of cast iron into which a cylinder of rolled amalgamated zinc, rather than common zinc, was placed. Common zinc contained elemental impurities, and it was found that by amalgamating the electrode surface with mercury, that a more uniform response to impressed current was achieved. Discs of millboard located between the cast iron cell and the cylinder of zinc prevented contact by the different metals, thereby making a stepwise increase in battery life. Dilute sulfuric acid was used to charge the battery. A major advantage to amalgamated electrodes was that such a battery delivered "*the full equivalent of electricity for the oxidation of zinc.*" Furthermore, a battery constructed with amalgamated zinc, with a sulfuric acid electrolyte, is active only when the electrodes are connected and ceases to act or be acted upon by the acid the moment that the contact is broken. The use of amalgamated zinc to produce a better electrode was known as early as 1826 (Sir Humphry Davy), and an amalgamated zinc electrode was used as part of the battery in Friedrich Wöhler's circuit that first reduced aluminum in 1827 [50].

As did most electrical engineers of that time, Sturgeon believed in the Contact Theory of Electromotive Force, which was not in line with Faraday's opinion, as he believed that chemical change was required for an electrical current to flow. In his comments about the Sturgeon Battery, Michael Faraday stated (sic) "*Sturgeon is of course unaware of the definite production of electricity by chemical action and is in fact quoting the experiment (with his new, improved battery) as the strongest argument AGAINST the Chemical Theory of Galvanism*" [51].

Beginning in 1830, with the encouragement of Michael Faraday, John Frederic Daniell, an advocate of Faraday's opinion that the Chemical Theory of Electrical Action, began electrochemical research to develop a better battery. By 1836, Daniell developed the Daniell Cell, a battery that resembled the batteries of today. The Daniell Cell consisted of a cylindrical copper vessel that served as the passive plate (cathode); within the outer copper vessel was a porous earthenware container or partition that held a zinc rod or active plate (anode). The space between the copper and the porous cup was filled with a solution of copper sulfate which was kept saturated by crystals of the salt lying on a perforated shelf. The porous cup was filled with dilute sulfuric acid. The porous earthenware kept the fluids from mixing without hindering the passage of current; it allowed ions to move through while the reaction of the cell was taking place. The sulfate of copper, in contact with the passive plate, served to take up hydrogen, thus preventing the buildup of hydrogen at the anode. The contents of the battery had to be dismantled when not used to stop the chemical reactions and conserve the metals. The amalgamated zinc rod (anode) had a binding screw. The top of the copper cylinder contained the other binding screw (cathode). This battery, which produced about 1.1 V, was ultimately used to power telegraphs, telephones, and even to ring doorbells [52]. See Fig. 1.9.

Fig. 1.9 A series of Daniell Cells. By grouping a number of batteries in series, more electrical energy could be realized in a circuit [50]. Wikipedia (https://commons.wikimedia.org/wiki/File:NMAH-Daniell_cell_batteries_1836.JPG)

1.5.2 The Wheatstone Bridge

Improvements in batteries that produced a standard voltage enabled the simultaneous development of electrical measurement devices. One of the most famous was invented in England by Samuel Hunter Christie in 1833, and improved by Sir Charles Wheatstone in 1843 and became known as the Wheatstone bridge. Based upon the principles of Ohm's Law, a Wheatstone bridge is an electrical circuit used to measure an unknown electrical resistance by balancing two legs of a bridge circuit, one leg of which includes the unknown component. When, the circuit was closed, a measureable electric current flowed, and when the circuit was opened, and "*electroscopic tension was developed.*" This tension was resistance to current flow. Resistance was defined as "the obstacle to the passage of electric current by the bodies through which it has to pass: it is the inverse of conductivity" [53]. The Wheatstone bridge was devised such that when materials of unknown resistance were placed in the circuit, the unknown resistance value could be calculated by way of proportions. Using resistors of the same element that weighed the same initially standardized the circuit. The Wheatstone bridge was an important electric tool utilized by early researchers to characterize the electrical properties of new elements, in order to refine and improve the Galvanic Series by way of their addition [54]. See Fig. 1.10.

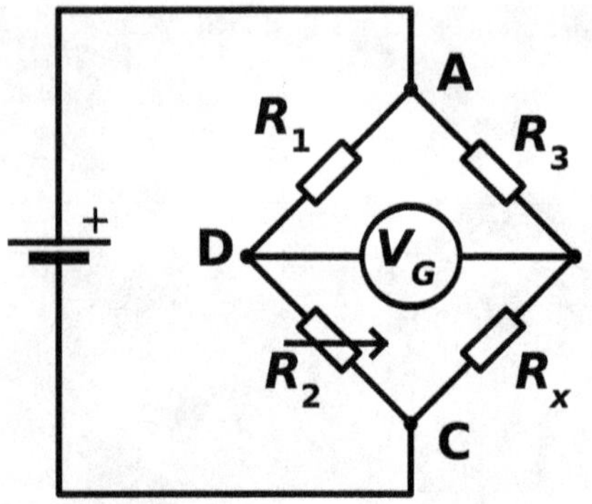

Fig. 1.10 Circuit diagram of the Wheatstone bridge. The unknown resistance R_x is to be measured; resistances R_1, R_2, and R_3 are known and R_2 is adjustable. If the measured voltage V_G is 0, then $R_2/R_1 = R_x/R_3$ [54]. (Wikipedia_Common license: https://en.wikipedia.org/wiki/Wheatstone_bridge#/media/File:Wheatstonebridge.svg)

1.6 Discovery of Aluminum

In 1808, while it seems the rest of Europe was also trying to discover the process to reduce aluminum from its mineral state, Berzelius attempted to decompose aluminum metal from aluminum fluoride using potassium amalgam, along with boron and silicon from their respective fluorides. In the case of extracting (decomposing) aluminum, Berzelius was not successful, most likely due to the high solubility of aluminum in caustic potassium that was produced in the reaction vessel during electrolysis. In fact, in his ignorance of the properties of aluminum metal, Berzelius treated the aluminum fluoride mass with water and potassium, dissolving everything. Possibly, Berzelius' most famous student was Friedrich Wöhler, who actually was the first to successfully reduce aluminum metal from aluminum chloride [55].

In 1821, Pierre Berthier discovered bauxite ore and is recognized as the first to learn that it contained alumina [56]. Bauxite is the most important naturally occurring ore of aluminum, as aluminum oxide, and contains additionally various oxides of silicon, iron, and titanium. Once bauxite could be reduced to aluminum oxide, a ready source for continued experimentation for aluminum extraction would be available. Also, the abundance of bauxite in France was an indication for the ready availability of aluminum metal, should a reduction process be developed, and accounts for the early success of for the industrial process for aluminum production in France. Extracting aluminum oxide from bauxite in order to provide a more readily "decomposable" mineral was a necessary step in the reduction of aluminum. The process was complex and required energy; this problem was not solved until 1888 when Austrian Karl Josef Bayer introduced the Bayer process to aid in the extraction of elemental aluminum [57].

The last recorded attempt to extract aluminum before Wöhler came in 1825, when Danish chemist Hans Christian Ørsted attempted to reduce aluminum from aluminum chloride, using potassium amalgam, a potassium-mercury mixture. He reported the development of an aluminum amalgam that in atmospheric air oxidized quickly and with distillation in mercury produced a metallic lump that he believed to be aluminum metal. He reported only that the reaction product, a metallic lump, had the color and shine of tin. Ørsted also reported his findings as interim, stating in fact that his attempt was not yet completed, and that it might take years to finish. With the encouragement of Ørsted, Friedrich Wöhler duplicated his experiments in 1827 and found that Ørsted's reaction had produced pure potassium. By conducting his own experiment, Wöhler finally reduced aluminum metal from anhydrous aluminum chloride by reacting it with metallic potassium in the vapor state and is given the credit for first isolating aluminum metal [55].

Following the successful reduction of aluminum, Wöhler continued his studies by characterizing aluminum metal, finding that it is "tin white" in color, a conductor of electricity, and that it fully oxidizes to alumina when heated in air. He further determined that the aluminum he extracted contained silicon, but this was not detected until the aluminum was heated until glowing. Wöhler also determined that aluminum is corrosion resistant, by showing that water didn't change the appearance of the metal, but with heating, hydrogen gas is evolved. In fact, Wöhler called the reaction oxidation. He reacted aluminum with sulfuric acid and determined that aluminum dissolves, evolving hydrogen gas; and when heated in sulfurous gas, that it dissolves quickly. Wöhler also determined that aluminum is soluble in caustic solutions such as potassium hydroxide and ammonium hydroxide. He also determined that aluminum could be converted in a stream of heated chlorine gas to aluminum chloride. Wöhler reacted aluminum with phosphorus, selenium, arsenic, tellurium, and antimony, finding that aluminum reacted to form alloys with all selected elements except antimony [58].

Wöhler continued his work in aluminum for the next 20 years, developing the discovery of aluminum from the few grams of aluminum powder into the production of aluminum samples large enough to stimulate interest in the development of an industrial process. Most notably, Wöhler determined the density of aluminum, discovering it's characteristic "lightness" and in 1845 reported the density to be 2.7 g/cm^3. Wöhler is also acknowledged as supporting Henri Sainte-Claire Deville in developing the double salt process for reducing aluminum, and most notably, developed the laboratory scale process for reducing aluminum from cryolite, which went on to be developed as the Hall–Héroult process [59]. In 1846, aluminum was considered a precious metal, with a price almost twice as expensive as gold, around $34 per troy ounce [60].

The discovery of the element of metallic aluminum, and the path taken to discover the process to reduce it, is the result of a long history of scientific and engineering development and truly a product of the *European Context of Shared Knowledge*. Great minds from France, Italy, England, Denmark, Sweden, and Germany communicated and collaborated across miles, most often by post to report success and failure while pushing for the same goal. Most of the researchers and scientists were academics. Aspects of chemistry and physics were combined following the discovery of electricity to develop the new science of electrochemistry, and each scientist's progress enabled continuation of the next one's work.

1.7 The Industrial Revolution

As scientific discoveries moved from research to develop engineering technology, national boundaries were sharply drawn, especially in the context of a defense industry. Production, alloy development, and the manufacturing processes adapted to and developed for aluminum became more nationally exclusive as political rivalries developed within Western Europe and the New World. All over the world, industry developed and boomed; by the mid-nineteenth century, the Age of Aluminum was dawning.

1.7.1 Crystallography in the Industrial Revolution

Structural chemistry with regard to atomic theory and crystallography continued as a basic interdisciplinary approach to understanding the properties of various chemical compounds, how and why they form, and consequently, how they decompose. Crystallography developed along chemical lines, organic and inorganic, with symmetry being one of the means to characterize types of crystal structures. In 1848, French physicist Auguste Bravais presented his Mathematic Theory of Crystal Symmetry, based upon the concept of the crystal lattice, as the basis to describe the atomic structure of solid materials. Bravais' lattice was an infinite array of points, which provided the same arrangement and orientation, regardless of the point of view, and related the external symmetry of a crystal to its internal symmetry, as embodied in the space lattice. The smallest unit produced by lattice points in a consistent arrangement was called a primitive cell, of which there were four types: simple cubic, body-centered cubic, face-centered cubic, and side-centered cubic. A crystal structure could similarly be comprised in terms of other symmetric structures that are larger than unit cells, and not unit cells: cubic, tetragonal, orthorhombic hexagonal monoclinic, triclinic, and trigonal. There are 14 basic Bravais lattice structures that are permutations of the basic lattice units. The Bravais lattice system proved a valuable basis for the characterization of all solid materials, whether metals, minerals, or semiconductors [61]. See Fig. 1.11.

British freelance scientist William Barlow was an independently wealthy genius whose hobby was crystallography. He was a member of the Royal Society of London. Barlow is noteworthy because his original view of the nature of crystalline matter was based upon Bravais' mathematical system of symmetry. In 1883, Barlow wrote "*The Probable Nature of the Internal Symmetry of Crystals*" [62], which began on the premise of Johannes Kepler that a crystal may be regarded as

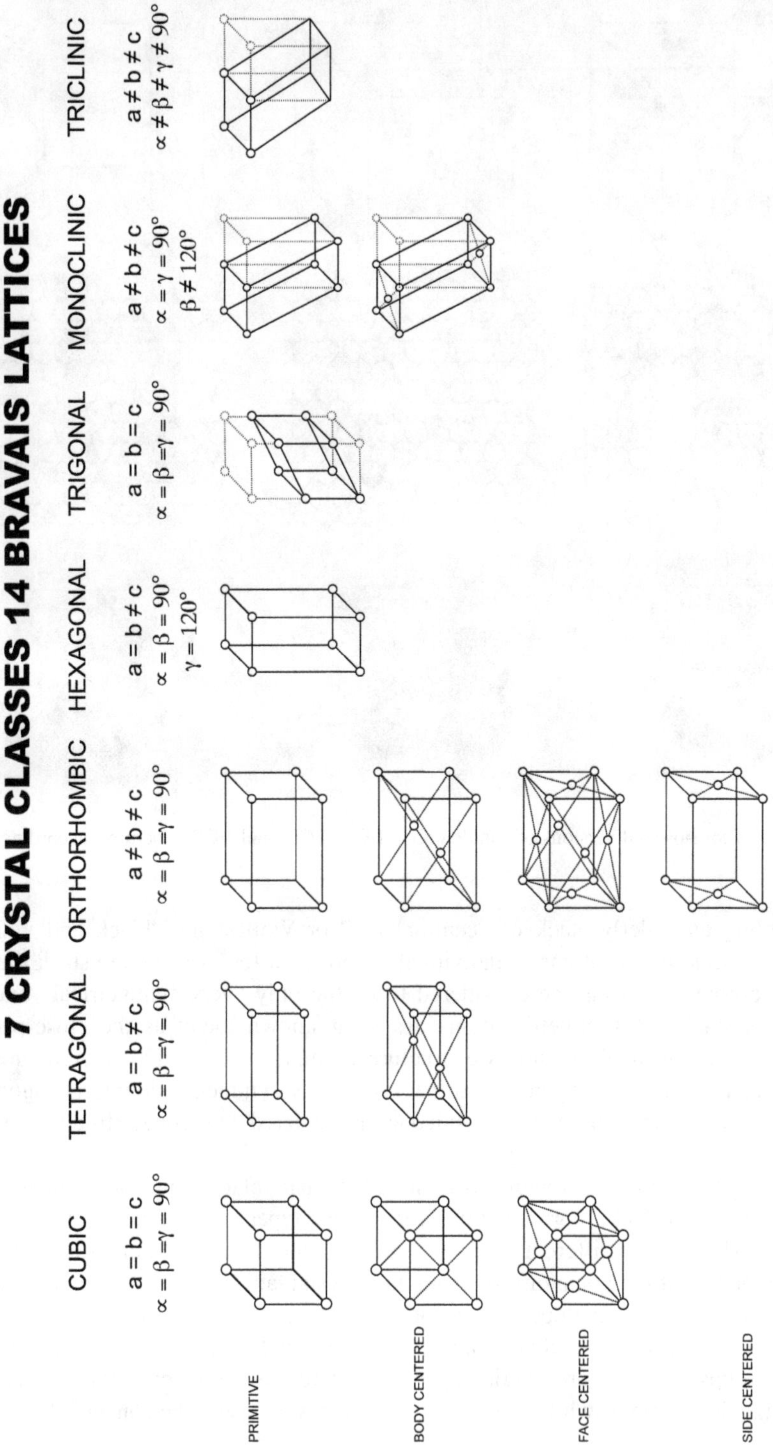

Fig. 1.11 The Bravais lattices: Of the 14 Bravais lattices, aluminum was found to be face-centered cubic [61]

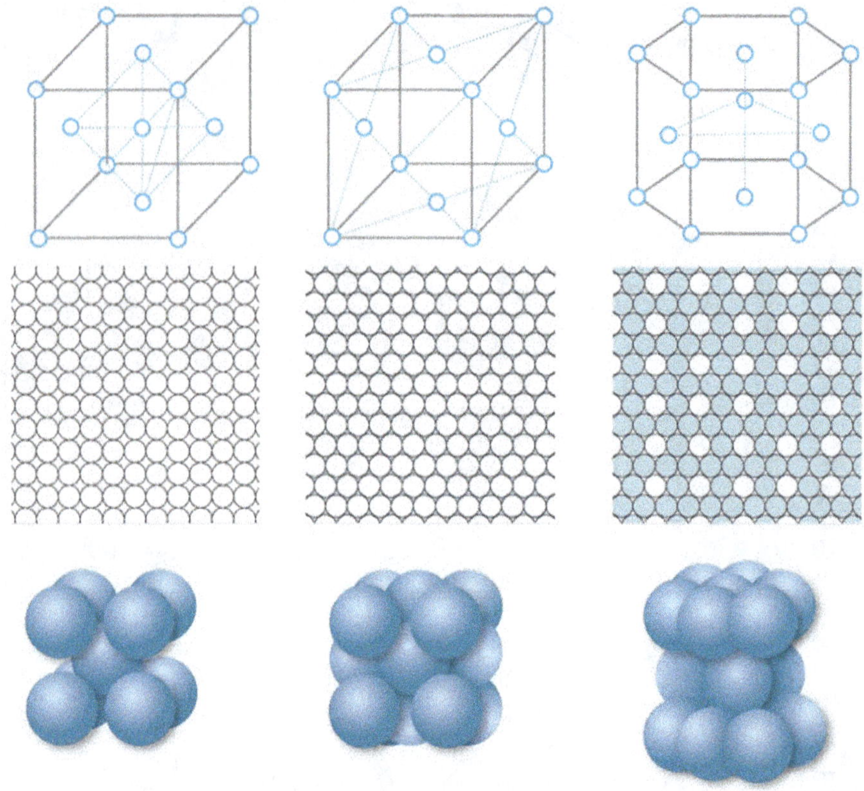

Fig. 1.12 Close-packed versions of simple cubic, BCC, FCC, and HCP structures, according to Barlow [61]

resembling an orderly stack of "cannonballs" or Wollaston's "black and white balls." By treating the atoms as individual spheres, Barlow was able to stack them in five arrangements that he considered to be the only "very symmetrical" ones. Barlow's stacks of symmetric atoms are what known today as the close-pack arrangements for the following five structures: simple cubic, body-centered cubic, face-centered cubic, hexagonal close-packed, and a symmetric double hexagonal structure that resembles a double-sided body-centered tetragonal structure. See Fig. 1.12.

The crystallographic structures presented the basic relationships between atoms in a solid as "perfect" representations for various types of materials. Although it could be shown how different types of structures could possibly bond together by sharing or substituting on atom-for-atom on a crystal lattice, the concept of a lattice defect was yet to be developed. Therefore, the effects of defects such as a vacancy (hole) or interstitial atom that could stretch the structure or the displacement of entire groups of atoms on a lattice by way of deformation, or even the effects of alloying with elements with different sizes of atoms had yet to be conceived.

With the discovery of X-rays by Wilhelm Roentgen in 1895 [63], contemporaries theorized that X-rays were composed of short electromagnetic waves, but the theory could not be proven because a diffraction grating with intervals small enough to measure the wavelength could not be constructed. In April of 1912, German physicist Max von Laue, a student of Max Planck, proposed sending X-rays through crystals, suggesting the regular symmetric structure of atoms as set forth by the Bravais lattices, which were known for 50 years at that time, would provide the intervals small enough for an atomic-level diffraction grating [64]. Experiments performed on copper sulfate by Laue's associate, Walter Friedrich, and student Paul Knipping demonstrated that irradiation of the crystal produced a regular pattern of dark points on a photographic film plate. These results proved two ideas: that X-rays indeed are electromagnetic waves rather than particles, and that crystals consisted of periodic arrangements of atoms.

Inspired by Laue's results, Australian-British father and son physicists William H. and William L. Bragg immediately investigated the phenomenon of X-ray diffraction and determined new methods for measuring the wavelengths of X-rays, developing the Bragg equation by the end of 1912 [65]. The Bragg equation enabled the calculation of the distances between atoms in a crystal lattice by measuring the intensities and direction of X-rays diffracted by the crystal, a great achievement. Soon, the determination of a crystal structure became a technical procedure that utilized the laboratory practice of X-ray diffraction and the well-recognized Bragg equation.

$$n\lambda = 2d\sin\theta \tag{1.2}$$

For which λ = the wavelength of the X-rays, d = the distance between two plates, and θ = the angle of reflection.

The crystal structures of many substances, including sodium chloride, diamond, and metallic copper were determined and calculated. These techniques were immediately employed in the characterization and development of new alloys and were significant in their application in the study of age hardening of aluminum alloys. By discovering the technique of X-ray diffraction and determining its usefulness and application, Laue and the Braggs provided the foundation of solid-state physics.

1.7.2 Industrial Scale Aluminum Production

The development of a large-scale industrial process to produce aluminum metal was no small undertaking. In about 1851, French university chemistry professor Henri Étienne Sainte-Claire Deville, based upon the works of Davy, Ørsted, and Wöhler, began his studies of aluminum, with the goal of developing an industrial process to reduce it from its natural state. Metallic potassium, the reducing agent that enabled the isolation of pure aluminum, had become less expensive and more accessible, which enabled large-scale experiments. The even more available and less expensive metallic sodium ultimately replaced potassium, which is considerably simpler and safer to use. Deville also developed a process to reduce the cost of sodium, and a method to replace the volatile aluminum chloride with a double salt of aluminum chloride and sodium chloride ($AlCl_3{\cdot}NaCl$). Within 3 years, in 1854, Deville developed a technique to obtain metallic aluminum by electrolysis from the double salt $AlCl_3{\cdot}NaCl$ the use of a carbon anode and a platinum cathode. Aluminum was deposited as a thin layer on the platinum cathode and was scraped off at regular intervals and remelted [66]. It is noteworthy to mention that Deville tried to obtain aluminum by electrolysis with other molten electrolytes, including cryolite, but the reactions could not go beyond the laboratory scale because it was impossible to obtain the necessary amounts of electric power on an industrial scale.

Deville's experiments gained the attention of the Paris Academy of Sciences in 1854, and beginning from that time, his work was funded by the government of Napoleon III of France. The excitement and anticipation for the possibilities offered by the new, lightweight, and strong metal, called "silver from clay" was such that bars of aluminum were exhibited at the *Exposition Universelle* (World's Fair) of 1855, in Paris. Napoleon, determined to use the new metal as lightweight yet strong armor for his soldiers, actually received some decorative military helmets cast from aluminum as well as an aluminum dinner set and some aluminum toys for the children of the Imperial Court. It is said that Napoleon held a banquet where the most honored guests were given aluminum utensils, while the others, held in less esteem, had to make do with gold.

In spite of the gains made in the industrial process to produce aluminum metal, in the mid 1850s it was still exceedingly difficult to produce, making pure aluminum more valuable than gold. The price of aluminum in 1855, at $32 per troy ounce, was only slightly less expensive than in 1846, while the price of gold in 1855 was significantly less expensive at $19 per troy ounce [67].

Deville was disillusioned by the ornamental and luxurious applications for aluminum; he is reported to have said: "*there is nothing more difficult than to make people use a new metal. Luxury items and ornaments cannot be the only sphere of its application. I hope the time will come when aluminum will serve to satisfy the needs of daily application*" [68]. Together with partners Charles and Alexandre Tissier, Deville built the first industrial scale aluminum refinery near Rouen (France) in 1856. He had yet to publish a formal book detailing his process when, in 1858, the brothers Tissier published their book, "*Ses Propriétés, Sa Fabrication et ses Applications par Aluminium*" taking Deville by surprise. Deville quickly published his own book "*de L'Aluminium*" in 1859 [69].

Possibly due to the abundance of Bauxite in-region, France was the first country in the world where aluminum production took place on an industrial scale. Henri Merle et Compagnie (later Pechiney and today's Constellium) [70] began producing aluminum metal in 1860 and in total, from 1856 to 1886, more than 200 tons of aluminum metal were produced by Deville's method. Excitement regarding the possibilities for aluminum was even echoed in literature. In 1865, French science fiction author Jules Verne, inspired by the exciting reports of the newly available metal, alludes to aluminum in his book *From the Earth to the Moon*: "*this fascinating metal possesses the whiteness of silver, the indestructibility of gold, the tenacity of*

Fig. 1.13 Cover of Jules Verne's book *From the Earth to the Moon*, published in 1865 [67]. The aluminum "projectile" is shown in this 1886 illustration from Henri du Montat's illustrated edition [65]. https://er.jsc.nasa.gov/seh/chapter23.htm

iron, the fusibility of copper, the lightness of glass. It is easily wrought, is very widely distributed, forming the base of most of the rocks, is three times lighter than iron and seems to have been created for the express purpose of furnishing us with the material for our projectile" [71]. The "projectile" was the rocket ship that first romanticized space travel as science fiction. In reality, because of Deville's industrial process for reducing aluminum, research to find engineering ideas and applications for aluminum literally "took off." See Fig. 1.13.

As with the initial extraction of aluminum from bauxite or other naturally occurring mineral state, the biggest hurdle to achieving lower cost aluminum production was the lack of a good power source. Even with the development of an advantageous electrochemical reaction, like Deville's Process, the power source needed to be sufficiently strong, sustainable, and economical. The growth of reliable, commercial electric power generators or dynamos, based upon Faraday's Law in the last third of the seventeenth century meant that reliable electrical power would be available wherever mechanical energy existed, and it first enabled the possibilities of an economical electrolytic process for aluminum. Improvements demonstrated by Zénobe Gramme in 1871 increased the dynamo's voltage and made the current more consistent and predictable. The Gramme dynamo was one of the first machines to generate commercial quantities of power for industry [72]. By 1886, the price for aluminum declined significantly, to \$15 per troy ounce [67].

Fig. 1.14 (Left) Photo of the Cap of the Washington Monument with the copper lightning rod "crown" being placed in a 1934 renovation to prevent lightening from blunting the tip. (Right) Replica of the 1884 tip of the Washington Monument, recast in 1984 to celebrate the 100-year anniversary [73]. (Citation: http://www.tms.org/pubs/journals/jom/9511/binczewski-9511.html Journal of the Mineral Metals and Materials Society)

In 1884, the capstone of the Washington Monument in Washington, DC, was cast from aluminum. It was the largest aluminum casting at the time, yet the piece is only 22.6 cm in height, 13.9 cm at its base, and weighed 2.85 kg. At the time, the price of aluminum was equal to the price of silver [73]. Within 2 years, in 1886, Charles Martin Hall, in the United States and Paul Héroult, in France, separately but simultaneously, determined the method to electrolyze aluminum metal by passing a direct current through aluminum oxide that was dissolved in cryolite (Na_3AlF_6), and the price of aluminum plummeted to a fraction of the cost. Two years later, Karl Bayer developed the process to extract aluminum oxide from bauxite, enabling a more stable production of alumina, rather than that of extracting from the double salt of aluminum and sodium (chloride). Five years later, the Deville double salt process of extracting aluminum metal was obsolete. In 1888, together with Alfred E. Hunt, Charles Hall founded the Pittsburgh Reduction Company, which is now known as the Aluminum Company of America, or Alcoa [74]. By 1891, aluminum production soared from a few tons per year to 300 tons per year, and the price fell to \$0.56 per troy ounce [67]. See Fig. 1.14.

In 1886, J.A. Price delivered an address to the Board of Trade in Scranton Pennsylvania, extolling the virtues of the new metal, stated: "*The enumeration of the properties of aluminum is as enchanting as the scenes of a fairy tale!*" In listing the various industries that aluminum could be applied, Price mentions household cookware (one of the first products from the Hall Process was a cast aluminum tea pot), architecture, transportation, construction, plumbing, communication and pneumatics, electrification, steam generation, and machine building. Aluminum has (sic) "*stood the egg on end! ...Cheap aluminum will revolutionize industry, travel, comfort and indulgence, transforming the present into a greater civilization*" [67]. And so it was: with the ability to economically produce large amounts of aluminum, the Aluminum Age began.

1.8 The Age of Aluminum: Alloy and Process Development

Remarkably, the claims and predictions of Mr. Price in 1886 were realized quickly. Early aluminum production approached 50 lb/day. At the time, aluminum and its alloys were most often selected as a replacement for other materials, mainly iron (steel) and copper alloys, in existing applications. The low density of aluminum drove engineering ambitions to develop light metal alloys with a high-strength-to-weight ratio, thereby increasing payloads and transporting farther with decreased fuel consumption. The three main driving forces for the development of aluminum and aluminum alloys were: electrification, transportation and defense. The development of generator plants for electricity by Thomas Edison enabled the technology and development of the Hall–Héroult process. By 1914, the Hall–Héroult Process had brought the cost of aluminum down to $0.18 per troy ounce [75].

The demand for products manufactured from the new metal was high, and the concurrent development of the science of aluminum metallurgy was rapid, leading to industrial success as well as some failures. For example, cast aluminum veneers were often bolted to iron-based frames to impart an expensive, decorative façade for some architectural applications. It is well known today that iron is active in the Galvanic Series with respect to aluminum, yet ignorance at the time led to the production of articles with reduced corrosion resistance; a situation aggravated by excitement, impatience, and the general lack of scientific knowledge and engineering experience with aluminum.

Unalloyed aluminum offered little strength for the applications developers were seeking to replace with it. The first alloys of aluminum were developed when the characteristics of unalloyed aluminum were inadequate to meet the needs of the end user, and the earliest approach to alloy development was strictly empirical. Prior to the Hall–Héroult process, in order to develop hardness for structural applications, the first cast aluminum products were initially alloyed with up to 6% copper in Europe and with up to 4% of nickel in the United States [76]. As aluminum metallurgy developed to keep pace with engineering design for the new metal, alloying elements were added and subtracted, records kept, and results evaluated by the end user; the motto was: "if a little is good, add more until something bad happens" [77]. In other words, if an alloying element was added and an improvement in a particular property was noted, more was added until other properties were adversely affected.

As the science of aluminum metallurgy developed, alloy researchers began to understand which alloying elements produced the desired changes in properties, yet didn't understand how and why the changes took place. Analytical methods and instrumentation were developed to characterize microstructure and properties, but the scientific community did not understand reasons for the effects that were observed usually for many years after the initial observation. The realization that microstructure controlled the properties of a component, and that microstructure was controlled by alloy composition and the casting, deformation, and heat treat processes preceded the ability to identify the particular microstructural features that controlled properties. Consequently, the connection between composition, microstructure, and properties was initially determined by trial and error.

The following brief sections on the history of aluminum alloy development are presented in the context of application. Detailed information pertaining to particular alloy types (cast or wrought), alloy designations, and microstructure development as it relates to a component surface is presented in Chap. 4.

1.8.1 Earliest Applications of Aluminum and Its Alloys

1.8.1.1 Development of Work Hardenable Alloys

The earliest aluminum alloys were developed to obtain acceptable mechanical properties, and manufacturing practices were based upon those for iron, copper, copper alloys, and steel. Technology for cast aluminum developed to produce net shape products and developed along with wrought processes in order to provide billet product for deformation processes such as rolling, forging, or extrusion. The first products that were produced in large quantities from aluminum by Alcoa were cast tea kettles and other cooking utensils, but this was only the beginning [74]. Automotive die-castings and ornamental architectural castings were produced in the late 1800s and early 1900s.

At first, to produce wrought product, aluminum cast billets were rolled and stamped. The first wrought alloy recorded, produced by way of the Hall–Héroult process, was the result of the inherent presence of silicon and iron in the aluminum as produced. This alloy was named Alcoa Alloy 2S in 1888 and was later given the designation 1100. It is the oldest commercial alloy in continuous use [74]. Alexander Dick developed the hot extrusion process for copper in 1894, which was further developed for aluminum. Aluminum alloys were being extruded for wire product and other simple shapes by 1902 [78].

One of the first applications for wrought aluminum was as foil. The earliest production of aluminum foil was in Switzerland in about 1903, by Industrialist Heinrich Alfred Gautschi, who employed the classical pack rolling method of reducing metal to foil thicknesses. Gautschi stacked a number of thin sheets of aluminum into a pack and rolled this between heavy iron cylinders heated internally by hot water. This was repeated each time with a progressively smaller gap between the iron cylinders, until the desired foil thicknesses were obtained. Gautschi and Jequier in Switzerland patented "*Papier Metallique,*" which came to be known as aluminum foil in 1905 [79]. Aluminum foil was produced and entered the US market in 1910; it was first used to wrap Life Savers Candies [80].

It was recognized almost immediately that aluminum as-produced became stronger with cold work. In the early 1900s, American Earl Blough of Alcoa was experimenting with additions of manganese to alloy 2S (today's 1100 alloy) in order to produce corrosion-resistant sheet for camera cases. Blough found that manganese additions increased the response to cold work, yielding improved hardness and strength. This alloy became the first 3000 series alloy, 3S, and is most similar to current day alloy 3003. In the 1920s, magnesium was added to alloy 3S to improve strength but increase malleability during cold rolling. This alloy was called alloy 4S and is most similar to the current day alloy composition 3004, the largest volume aluminum alloy in production due to its use as beverage can body stock [74].

1.8.1.2 Architectural Applications for Aluminum

Aluminum castings, mostly decorative, were some of the first applications for aluminum and were produced for architectural purposes. Cast aluminum was regarded as a valuable artistic accent because even at the reduced cost of the metal due to the Hall–Héroult process, it still substantially exceeded the cost of copper, brass, and bronze by about four times and was about 30 times the cost of steel [81]. Following the Washington Monument in 1884, decorative statues, capstones, steeples, and other cast or pressed sheet architectural accents were produced for landmark buildings throughout the world. From 1891 to 1895, notable early uses of aluminum were the elaborate cast elements for the Monadnock, Venetian (now demolished), and Isabella buildings in Chicago, Illinois, USA; the Anteros Statue in Piccadilly Circus in London, England; the Dome of the Chief Secretary's Building in Sydney, Australia: and the Dome and Roof of the Church of San Gioacchino in Rome, Italy. In most applications, cast elements were bolted to steel framework [82]. See Fig. 1.15.

Aluminum became the emblem of architectural style and modern design, a new wonder material capable of withstanding weathering and structural stresses. Prized for its light weight and white color, elaborate cast, forged and extruded aluminum elements were manufactured for the architectural features such as windows, spandrel panels, entrances, doors, and grilles of many buildings constructed during the Art Deco period of the 1920s and 1930s. In New York City, the Empire State Building is shaped like a rocket taking off, including four "fins" or winged buttresses made of cast aluminum; the façade of the main shaft of Rockefeller Center comprises Indiana limestone with aluminum spandrel panels. The Times Square Building in Rochester, New York, has cast aluminum "wings of progress" which define the city's skyline [83]. See Fig. 1.16.

Architectural development took a back seat to defense when it came to aluminum in the 1940s, but after World War II, the philosophy of "Place, Weight, and Time" became the mantra repeated during discussions about aluminum in the construction sector during the 1950s through the 1970s [84]. By applying aluminum using

Fig. 1.15 Examples of some of the first architectural applications of aluminum. (left) Cast aluminum stairwell at the Monadnock Building (1891) in Chicago (May need permission from: http://www.urbanremainschicago.com/news-and-events/2016/02/05/well-executed-monadnock-building-solid-aluminum-staircase-baluster-panel-recast-joins-bldg-51-museum/) (right) cast aluminum elevator enclosures at the Isabella Building (1892) in Chicago. (May need permission from http://www.urbanremainschicago.com/news-and-events/2017/03/26/isabella-buildings-pioneering-use-of-cast-aluminum-as-ornament/)

Fig. 1.16 The spire of Empire State Building is shaped like a rocket taking off, including four "fins" or winged buttresses made of cast aluminum. (Courtesy of Morgue File: by rosevita https://morguefile.com/terms#license)

Fig. 1.17 External applications for aluminum in architecture. The IBM Building in Chicago, Illinois, was finished in 1970, the year after Mies van der Rohe's death. It is the perfect example of the use of black anodized aluminum with tinted glass, which Mies and his colleagues utilized "to craft a uniform skin that lends the building an air of a simple imposing and impressive volume" [85]. Wikipedia by JCrocker (https://en.wikipedia.org/wiki/330_North_Wabash#/media/File:2004-09-02_1580x2800_chicago_IBM_building.jpg)

various methods, an architect could expand a space of a building, reduce the weight of any structure and speed up the construction process. These specific features turned out to be equally advantageous for avant-garde architects, businessmen, and metal manufacturers. In the early 1950s, Mies van der Rohe was the first to use a continuous aluminum and glass curtain wall for the Esplanade Apartments in the city of Chicago [85].

Aluminum is in use today to produce ceilings and walls, window frames, blinds, doors, stairs, roof covers, wall panels partitions, and HVAC systems and to construct houses and shopping centers, stadiums, and bridges. Furniture, household appliances, bathroom fixtures, and countless interior architectural accessories are manufactured from extruded and other wrought aluminum shapes. Aluminum provides efficient insulation and has low maintenance requirements. It is easily finished by way of anodizing to yield a durable, weather-resistant, beautiful appearance. See Fig. 1.17.

1.8.1.3 Electrical Applications for Aluminum

Applications for aluminum that were involved with bringing electric power on a commercial basis to industry and the general population in the late 1880s through the early 1900s were spearheaded by the likes of Thomas Edison and George Westinghouse. As smelter production increased and aluminum's price decreased, its competitive position versus copper improved. The electrification of the United States just after the turn of the century came at an ideal time for the aluminum industry to develop its first significant large-volume market. Aluminum electrical-conductor cable became a viable product, with J.W. Hoopes of Alcoa investigating ways to produce the new product by extruding lengths of aluminum wire in 1902. After abandoning the extrusion process, Hoopes developed alloys with sufficient conductivity but low strength. He solved the strength problem by reinforcing the soft aluminum wire with steel. This aluminum conductor steel-reinforced (ACSR) wire outperformed copper at a lower cost and withstood extremes in temperature. A patent for the product was granted in 1908. US business leaders and industrialists quickly recognized aluminum's superior properties and power transmission lines and elevated train electrical wiring as well as the framework for electrical towers were among the first to benefit from the advantages of aluminum and almost 500,000 km of aluminum conductor spanned the United States by 1929 [86].

1.8.2 Aluminum in Transportation: Ships, Automobiles, Trains, and Planes

1.8.2.1 Ships

In the transportation industry, the earliest applications for aluminum were in shipbuilding. Regardless of the high cost, the low density and strength of aluminum metal inspired thoughts of lower weight, increased freight loads and fuel savings, but most of all *speed*. Following the ambition of Napoleon III to outfit his armies with lightweight aluminum armor, the obvious defense application for aluminum came as navies all over the world became interested in aluminum. In 1894, the French navy placed an order for a torpedo boat, "*Le Foudre,*" with the British shipyard Yarrow & Co. The skin and frames were made from aluminum sheet 1–5 mm thick. In 1895, the same yard also constructed a 58 m (190 ft) long torpedo boat, the "*Sokol*" (Falcon), for the Russian navy. Powered by a 4000 HP engine, it attained the speed record for that time, of 32 knots [87].

The first steam launches, into which cast aluminum and pressed aluminum sheet components were incorporated, were built in Zurich, Switzerland. One of the first was called "*Le Mignon*" and was designed for Alfred Nobel and launched in 1891. It notably achieved 8 knots during speed trials on Lake Zurich. Partially made of aluminum, it confirmed the opportunity for using aluminum in shipbuilding. In 1893, in France, the first aluminum sailboat "*Le Vendenesse*" was constructed. The

Fig. 1.18 Photo of the yacht Defender (Copyright 1895 by J. S. Johnston, New York. Public domain: https://en.wikipedia.org/wiki/Defender_(1895_yacht)#/media/File:Defender(yacht).jpg) [83]

2 mm thick aluminum skin, riveted to steel frames, saved 40% on the weight of the hull. In regatta sailing, as an inshore racer on freshwater lakes and rivers, Le Vendenesse impressed William K. Vanderbilt. He and two other investors proceeded to have the skin of their America's Cup contender "*Defender*" constructed from aluminum. With such a weight savings, Defender won the America's Cup in 1895 virtually unopposed. The publishers of The Aluminum World, "*A journal for manufacturers of Aluminum and its Alloys, Dealers in Aluminum Goods, and for the Metal Industry in General, with Special Reference to the Reduction of Ores by Electricity,*" thrilled by the success of the Defender, praised her construction and heralded aluminum's future success in shipbuilding: "*…the noble metal hull was not strained. Not a seam gaped, not a plate loosened, and Defender lies at New Rochelle as trim and staunch as on the day she left the ways at Bristol. Now comes the day of the new metal!*" [88] See Fig. 1.18.

The performance of the Defender was inspiring and proved to be a hard lesson in the corrosion of aluminum. Within 4 years of its success in the America's Cup, she was broken up due to catastrophic corrosion. The construction of the Defender was a galvanic cell that operated in a seawater electrolyte. The aluminum-nickel alloy free-board plates, the part of the boat above the water line, came in contact with the manganese bronze yacht from the water line to the keel, under the water. The aluminum plates were riveted with steel rivets to a steel frame. Other aluminum boats had survived in freshwater, but Defender's aluminum hull became a "true battery" in the ocean.

Clearly, the drive to use aluminum in marine applications was because of its high strength-to-weight ratio, and by using aluminum, a weight savings of around 50% that of steel was achieved. However, the lack of understanding regarding the galvanic coupling of aluminum to steel or copper components in salt water, resulted in severe corrosion just a few weeks or at most several months after they were launched. The use of aluminum in naval construction was not pursued beyond 1900 owing to the fact that the service life of these vessels was generally very short. In fact, the superstructures of a number of ships of the US Navy that were already made of aluminum were rapidly replaced by steel [89]. Research efforts to develop aluminum alloys that could withstand the demands of the marine environment were successful, and by the 1920s ships were being constructed with aluminum alloys from aluminum-magnesium alloys which would become the 5000 series.

1.8.2.2 Continued Development of Work Hardenable Alloys: 5000 Series

Because of the low density of magnesium, detailed studies of aluminum-magnesium alloys of the 5000 family (or series) began before 1900; the development of industrial wrought alloys as we know them today was carried out in the period 1930–1960. The 5000 series alloys were developed to satisfy demands for sheet material with higher strength, good formability, high levels of corrosion resistance, and good weldability. Alloys 5052 (52S), 5056 (56S), and X54S (5154) were developed from 1932 to 1934, but problems with fabrication and temper stability delayed their widespread use. Coupled with magnesium shortages that occurred during World War II, the development of the 5000 series alloys was delayed until the 1950s. It is interesting to note that the goals of the 1930s continue to drive the needs of today. 5000 series alloy development is driven by the needs of the shipbuilding and other marine applications; armor plate and other defense applications; transportation in general (railroad and automotive), sheet for architectural applications, and many other industries [90]. See Fig. 1.19.

1.8.2.3 Automobiles

The first sports car featuring an aluminum body was presented at the Berlin International Motor Show in 1899. The first engine with aluminum parts was made 2 years later when in 1901 Karl Benz, of Benz and Company, which would become Mercedes Benz, presented a new car for the prestigious race in Nice (France) [91]. As with the yacht Defender, the lightweighting provided by aluminum added to the successful performance of the car, but difficulties in metalworking, lack of knowledge, and the high price of aluminum prohibited its use in early mass production. See Fig. 1.20.

Cast aluminum transmission housings and pistons have been commonly used in cars and trucks since the early 1900s. Aluminum and aluminum alloys appeared on the 1908 Bugatti, whose construction included extensive aluminum sand castings

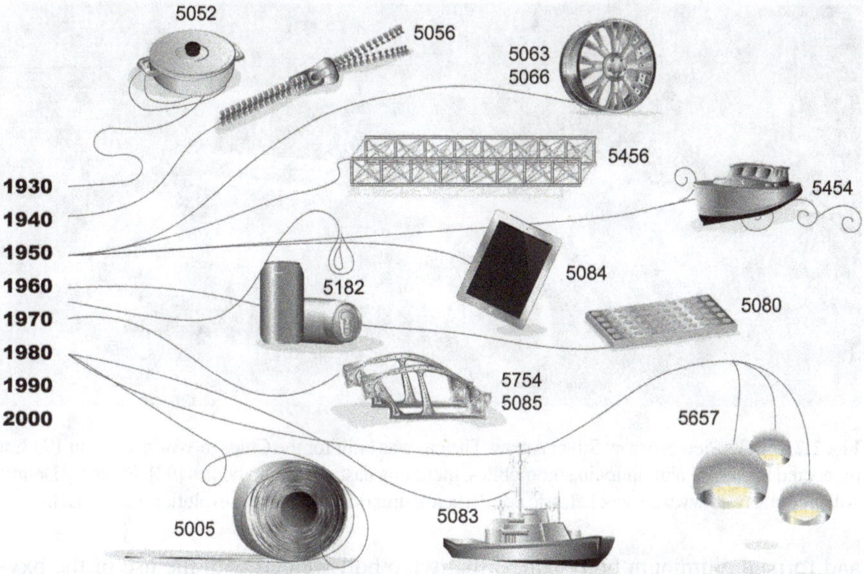

Fig. 1.19 Time line for development of important 5000 series alloys

Fig. 1.20 Image of "La Jamais Contente" (The "Never Satisfied"). In spite of its less-than-aerodynamic design, the vehicle was the first to exceed 100 km/h in 1899. The body alloy was called "Partinium," a laminated aluminum-tungsten-magnesium alloy [91]. (http://www.michelinchallengedesign.com/the-challenge-archives/2010-electrifying/2010-showcase-of-selected-entrants/la-jamais-contente-by-cie-generale-des-transports-automobiles-jenatzy-belgium-france/)

Fig. 1.21 1933 Pierce-Arrow Silver Arrow. The car was built for the Chicago World's Fair in 1933. It pioneered the use of aluminum in automobiles, including cast aluminum bodies [92]. Photo by Dennis Adler courtesy of www.heacockclassic.com/articles/streamliners-and-the-evolution-of-fastbacks/

and formed aluminum body panels that were butt-welded with the use of the oxy-fuel torch and handworked to a polished condition. Pierce-Arrow bodies were cast for several years—from 1912 until 1917 or 1918. The cast bodies were enormous sand castings and were often welded as part of the assembly process. After finish polishing, the cast product thickness was between 1/4 and 1/8 in. thick, and the bodies were of good appearance, durable, and structurally sound. The six-cylinder engines produced by Pierce-Arrow were cast in pairs and secured to aluminum crankcases. Europeans in particular were very fond of aluminum and, during the 1920s and 1930s, made quite artistic automotive bodies using the material. Beautiful limited production custom and racing coachwork for automobiles of the 1930s exhibited extremely fine aluminum metal artwork that reached its pinnacle during this period. Today, cast products, primarily engine and other under-the-hood components, make up more than half of the aluminum used in cars [92]. See Fig. 1.21.

Lightweighting, driven by aluminum's extraordinary high-strength-to-weight ratio, continues to be an important requirement in the automotive industry today. The 5000 series alloys are popular candidates to fill the rapidly growing need for automotive structural sheet. In these applications, strength and corrosion resistance are needed but formability is also an important requirement. Today, 6000 series extrusions are in use as part of automotive and lightweight truck framework. Anodizing is becoming the finish of choice for bonding applications where tolerances are tight and standard fasteners cannot be used [93].

1.8.2.4 Trains

Aluminum and aluminum alloys became key in the development of the railway industry worldwide by fulfilling the economic demands of the transportation industry through the exploitation of lightweighting, which enabled fuel savings and the

Fig. 1.22 (left) Photo of Japanese aluminum railroad engine bodies, (right) the Acela Express Train (courtesy Amtrak). 300 (Left) and 700 Series Shinkansen at Tokyo Shinkansen Station; Tokyo, Japan. Photographed by Richard Brown (https://commons.wikimedia.org/wiki/Commons:GNU_Free_Documentation_License)

delivery of increased payloads faster and over greater distances. Possibly the earliest application of aluminum in a railway application was in high-speed railway cars. In 1894, the New York, New Haven and Harford Railroad implemented lightweight car seats that had aluminum frames. Freight trains were outfitted in 1931 with all aluminum hopper cars for dry product such as seeds and coal. In 1933, main-line passenger cars constructed from aluminum were on display at the World's Fair in Chicago [94].

Aluminum and its alloys as plate, sheet and extrusions, have found application today as shipping containers and freight cars; in light rail and inner city commuter trains, metros, and underground trains; as well as in express, intercity, and passenger trains. Aluminum is used in the advanced high-speed trains such as the TGV (France and Czech Republic), Acela (USA), Transrapid (Germany), and even the Maglev (China) in which aircraft construction, and therefore aluminum construction, is consistently adopted. See Fig. 1.22.

Aluminum alloys chosen for railcars are selected based on the ease of fabrication, mechanical strength, weldability, and corrosion resistance. The most common alloys selected are from the 5000 series (aluminum-magnesium) alloys and the 6000 series (aluminum-magnesium-silicon alloys). 7000 series (aluminum-zinc-magnesium-copper) alloys have also been used, but to a lesser extent. A great advantage for aluminum in freight car applications is that they retain a high salvage value at the end of their useful life. A recent published example is that several 100-ton hopper cars initially built around 1967 were recently sold for almost 90% of their original manufactured cost!

Fig. 1.23 Photo of the Wright Brothers' first aircraft. Wikipedia. The library of congress. John T. Daniels—This image is available from the United States Library of Congress'. Prints and Photographs division under the digital ID ppprs.00626

1.8.2.5 Planes

The history of aviation is closely related to aluminum and the history of developing aluminum alloys. The more durable and reliable aluminum became, the higher, farther, and safer airplanes flew. As for all applications, before it became the essential and most important material for aircraft manufacturers, aluminum navigated a long route from pure metal to high strength alloy, and replaced wood, steel, and copper alloys. Aluminum and its alloys have become indispensable for the aerospace and aircraft industries in particular for its use as a lightweight structural material for flight applications.

Aluminum found early industrial uses in aircraft engines, such as the one built in 1903 by the Wright brothers to power their first biplane. In order to attempt powered flight, the Wrights calculated they needed an engine that produced at least 8 horsepower and weighed no more than 200 lb (91 kg). No such engine was available for purchase; therefore, it was decided they would make their own. The Wrights were advised that they could save weight if they cast the engine block from aluminum. Although this was a soft metal, alloys had recently been developed that were much stronger—both Benz and Daimler in Germany were successfully making engines with aluminum blocks. The Wrights decided to cast the block from an alloy of 92% aluminum and 8% copper, known as P12. The completed engine weighed 180 lb and developed 12 horsepower at 1025 revolutions per minute [95]. Without aluminum as part of the engine, the flight of the first power-driven aircraft could not have taken place at that time. See Fig. 1.23.

Fig. 1.24 The Apollo Command Module employed an aluminum honeycomb inner "skin." Courtesy of NASA. https://www.hq.nasa.gov/alsj/a17/AS17-145-22261HR.jpg

Defense was, and continues to be, one of the biggest driving forces for aluminum alloy development as well as for the aerospace and aircraft industries. World War I pushed the demand for material to meet the needs of new technologies and lightweight aluminum became essential in aircraft design and development. During WWII, the aluminum production soared. US aircraft production (July 1940–August 1945) totaled an astounding 296,000 planes and related aircraft. After the war, the technology used for warplanes transitioned into commercial aircraft. More than half were made primarily from aluminum. From there, alloys were used to construct early rockets. The body casing and oxide fuel lines of the Avantgarde and Titan rockets used for launching the first American rockets into orbit, including Gemini manned flights, were made from 2000 series aluminum alloys [96]. Today, many of the components for the NASA space shuttle and international space station are made from 2000 series aluminum alloys which include the important aluminum-lithium alloys. See Fig. 1.24.

1.8.2.6 Heat Treatable Alloy Development

In the early twentieth century, possibly the most important breakthrough in the development of aluminum alloys, which impacted all industries that use aluminum, occurred in Germany. In a clearly defense-driven application for aluminum alloys, in 1903, Alfred Wilm, was tasked by the German War Munitions factory of Berlin to find an aluminum alloy that had the characteristics of brass [97]. This began a frustrating 2-year-long investigation involving the strengthening of Al-Cu alloys by heat treatment.

Wilm proceeded with his investigations and experiments following the rationale for iron and steel alloy development, assuming that quenching from a high temperature would produce hardness. However, he quickly determined that Al-Cu alloys remained soft rather than hard after quenching from a high temperature, contrary to the behavior of carbon steels, and began experimenting with Al–Cu–Mn alloys. Using this composition, the necessary strength to compete with brass was almost achieved, but hardness remained too low. The happy ending to the frustrating story proceeds as follows: together with his technician, Wilm's most famous iteration occurred on a Saturday morning: 0.5% Mg was added to the alloy melt; sheet was rolled from the cast product. The sheet was heated to 520 °C and quenched. Hardness measurements were recorded and were frustratingly low. Wilm then departed for a weekend sailing trip. Upon returning on the following Monday, hardness measurements were repeated on the same sheet, and Wilm was astounded to learn that the hardness values measured were significantly higher than those measured on the previous Saturday. The calibration of the hardness tester was checked and measurements showed that hardness increased over the next 4 days before remaining constant. Not knowing the cause, Wilm was the first person to recognize the phenomenon of age hardening.

The composition of Wilm's alloy was 3.5–5.5% Cu with less than 1% Mn and Mg. Wilm was awarded a patent, and by 1908 experimental work had advanced to commercial production. The alloy was given the name "Duralumin," a combination of the names Durener, the name of the plant in Germany that licensed the alloy, and aluminum. The name "Duralumin" was copyrighted in 1909. The alloy went on to becoming the most important aluminum alloy of the early twentieth century: The Vickers Company in England took out a license for duralumin in 1911 and produced the British Vicker Mayfly rigid airship; Alcoa (USA), also in 1911, produced laboratory heats of Al–Cu–Mn–Mg alloys and commercialized its own version known as 17S, an alloy of Al–4Cu–0.6Mg–0.5Si–0.6Mn. The use of duralumin in early aviation applications continued into World War I: Zeppelins were retrofitted with duralumin components; most importantly, the first all-metal airplanes were manufactured utilizing corrugated duralumin as the skin, the most successful was the Junkers J7, known as the "Flying Tank." As much as 750 metric tons were produced per year during this time. Duralumin is still produced today throughout the world and is registered by the Aluminum Association in the United States as alloy 2017 [97].

Wilm's discovery stimulated research to develop other aluminum alloy systems that would age harden. The empirical approach to alloy development continued

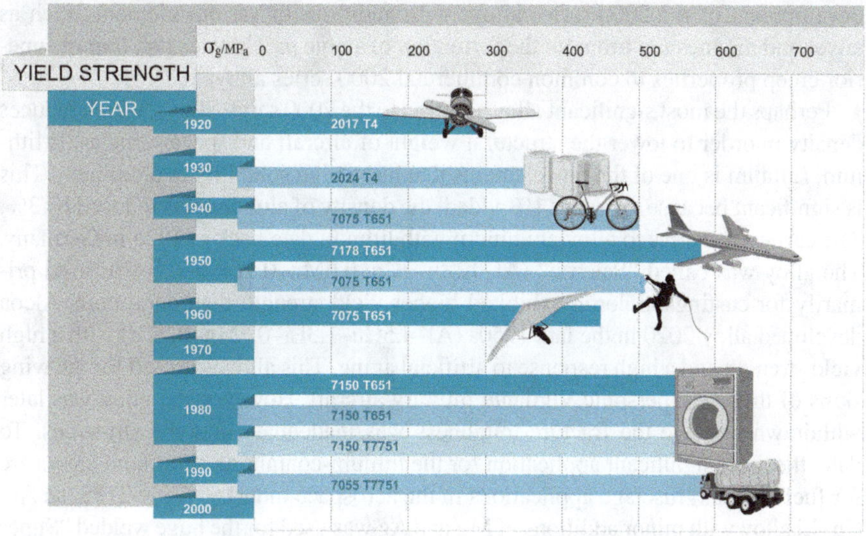

Fig. 1.25 Age hardening alloy time line for some common precipitation hardenable alloys

while analysis methods improved for understanding of the metallurgical phenomena, which produced the desired characteristics of the new alloys. Three series of cast and wrought alloys emerged that are based on the following systems: Al–Cu–Mg (2000 series); the Al–Mg–Si (6000 series); and Al–Zn–Mg–(Cu) (7000 series). Alloys have been developed with increasingly higher strength and continue to be used in aircraft and aerospace applications. These improvements have come from the development of new alloy systems, modifications to compositions within particular systems, and from the use of a range of multi-stage aging treatments (tempering heat treatments). See Fig. 1.25.

1.8.2.7 2000 Series Alloy Development

With duralumin as a springboard, and the early demands of the aircraft industry as a driver, more and stronger Al–Cu–Mg alloys were developed, which became today's 2000 series alloy designation from the Aluminum Association. Wrought alloy 2024 (Al–4.3Cu–1.5Mg–0.6Mn) was developed with yield strength 20% higher than duralumin (alloy 2017). When used in the T3 temper, 2024 has a high ratio of tensile to yield strength, which promotes damage tolerance, a plus when used as armor. Alloy 2024 was used in the clad aluminum condition for the commercial DC-3 airplane and has been used for the fuselage of most passenger aircraft built since the mid 1930s. Alloys based on the Al-Cu system have the advantage of superior creep strength at elevated temperatures; alloy 2618 was used for the skin and much of the structure of the high-speed Concorde aircraft. Alloy 2219 is weldable and has been used for fuel tanks in several space vehicles. Continued development of the 2000 series alloys with additions of various elements such as silver and magnesium promote the formation of stable precipitates that impart superior creep properties to common commercial 2000 series alloys [97].

Perhaps the most significant alloy addition to the 2000 series alloys, which reduces density in order to lower the structural weight of aircraft and space vehicles, is lithium. Lithium is one of the few elements that has a high solubility in aluminum. This is significant because for every 1% added, the density of aluminum is reduced by 3%. The earliest attempts to alloy aluminum with lithium date back to 1924 in Germany. The alloy was called "Scleron" (Al–12Zn–3Cu–0.6Mn–0.1Li) and it was used primarily for castings. Scleron exhibited higher yield strength than duralumin. Alcoa developed alloy 2020 in the late 1950s (Al–4.5Cu–1.3Li–0.5Mn–0.2Cd) with a high yield strength and a high response to artificial aging. This alloy was used for the wing skins of the US supersonic Vigilante military aircraft. However, the alloy was later withdrawn because the fracture toughness was inadequate in some situations. To date, the most significant application for the lithium-containing aluminum alloys are for fuel tank and fuselage applications in the aerospace industry. Alloy 2195, an Al–Cu–Li alloy with minor additions of Mg and Ag was used for the huge welded "Super Light Weight" external launch tank of the US Space Shuttle [97]. See Fig. 1.26.

1.8.2.8 6000 Series Alloy Development

The early twentieth century experiments to achieve an age hardening response with binary Al–Mg and Al–Si alloys failed; however, attempts to heat treat ternary Al–Mg–Si alloys were successful. The low alloy content develops a homogeneous microstructure whose hardness and strength is based upon the formation of Mg_2Si precipitates during aging. The first age hardenable Al–Mg–Si alloy was introduced in the United States in 1921. The composition was Al–0.5Mg–1Si (present day alloy 6051). Although not as strong as duralumin, it was stronger than alloy 3003, and as easy to deform as alloy 1100. Alloy 6051 also has a much higher corrosion resistance than duralumin. In spite of the susceptibility to intergranular corrosion, alloy 6051 found use as forgings, extrusions, and tubing; therefore, in the T4 temper, it was mainly used in the manufacture of furniture because of the unique combination of strength, corrosion resistance, and the ability to be fabricated. An improved version of 6051, 6151, was introduced in 1928. It developed higher yield strength along with higher elongation values, due to variations in iron and silicon, and since it was the first variation, it was called 6151. Another variation, 6351, exhibited better finishing characteristics than either 6051 or 6151 [98].

Many wrought Al–Mg–Si alloys have been developed since then that have become the most widely used general-purpose alloys for extrusions. The additions of Mg and Si are made in either balanced amounts such that quasi-binary Al-Mg_2Si alloys would form (Mg:Si 1.73:1) or with excess Si above that needed to form Mg_2Si [99]. In 1935, alloy 6061 was introduced with 1% Mg and 0.6% Si, in the ratio to form Mg_2Si. It contains alloying additions of 0.25% Cu and 0.04–0.35% Cr. It has set the standard for a lightweight, economical, medium-strength, and

Fig. 1.26 Aluminum-lithium alloy 2195 was used for the external tank of the Space Shuttle (Courtesy NASA). (https://www.jpl.nasa.gov/edu/teach/activity/street-math-space-shuttle-style/)

corrosion-resistant alloy for general-purpose structural applications. Alloy 6063 was introduced in 1944 and became the most widely used extrusion alloy. It contains nominally 0.67% Mg and 0.4% Si, but more custom versions are made or this alloy than any other. Alloy 6063 can be extruded readily with low pressures at low temperatures and is quenchable at the press by forced air cooling; it can be easily finished by almost any process. Most of the aluminum extrusions in the building and construction industry are versions of 6063. See Fig. 1.27.

Alloys with higher copper have been developed to increase yield strength, but have reduced corrosion resistance. 6000 series forgings and extrusions are used for decorative applications because the cleanliness of the microstructure yields a surface that is readily anodized and dyed, producing beautiful components for a variety of applications.

Al–Mg–Si alloys, with high levels of silicon, also have important applications as castings. These alloys are in the 300 series for aluminum castings. One such alloy, alloy 356 (Al–7%Si–0.3%Mg) has a hypoeutectic microstructure in which the α-aluminum phase is hardened by Mg_2Si precipitates. The alloy is widely used for automotive die-castings in under-the-hood applications such as engine blocks or pistons.

Fig. 1.27 The Copenhagen Opera House with its anodized façade is a striking example of the beauty of anodized aluminum in architectural applications. (Wikipedia by Julian Herzog https://commons.wikimedia.org/wiki/File:Copenhagen_Opera_House_2014_04.jpg)

1.8.2.9 7000 Series Alloy Development

In 1923, German researchers determined that some ternary Al–Zn–Mg alloys showed a greater response to age hardening than any other of the aluminum alloy compositions investigated at that time. Immediately, the potential of these alloys for aircraft materials was recognized, but the adoption was delayed because it was proven that the alloys were very susceptible to stress corrosion cracking (SCC). Much research was done in the United States, Europe, and Japan in the late 1920s and 1930s to understand the phenomenon and to develop commercial alloys.

Defense again proved the biggest driver for aluminum alloy development. In 1938, an alloy known as Extra Super Duralumin (ESD) was developed in Japan that reduced the weight of the Zero fighter aircraft. When this successful innovation was revealed in 1940 to the Allies by way of chemical analysis of a crashed Japanese aircraft at the beginning of World War II, similar alloys were quickly used for the manufacture of military aircraft in the United States and England. The major result was the modification of the base composition of Alcoa alloy X74S (5.2% Zn, 2.1% Mg, 1.5% Cu, and 0.4% Mn), an alloy which had shown susceptibility to SCC with plastic deformation.

Following the analysis of the downed Japanese Zero, X74S was modified with small additions of high melting temperature elements and alloys containing 0.2–0.35% Cr were developed that exhibited high resistance to SCC. Higher strength demands were met by increasing the primary alloy constituents, and alloy 75S was developed and introduced as a commercial alloy in 1943 [97]. 75S became the best-known 7000 series alloy, 7075. Alloy 7075-T6 sheet was used as the skin and stringer members of the upper wing in the B-29 Super Fortress bomber. The reason

Fig. 1.28 Photo of a 7000 series anodized bicycle frame with anodized accessories. Photo credit: Thomas Nussbaum

why the chromium addition developed resistance to SCC was not known, but key alloy properties of high strength and SCC resistance were achieved.

Alloy 7075 was widely adopted for post-war commercial passenger and military aircraft, including the Boeing 707 and the ill-fated British Comet in the late 1940s and 1950s. The Comet failed numerous times due to design flaws: sharp corners at the windows and at other high stress places in the fuselage exhibited fatigue cracks that opened to cause several crashes because of cyclic fatigue. The windows, that were engineered to be glued and riveted, had been punch riveted only. Unlike drill riveting, the imperfect nature of the hole created by punch riveting may cause the start of fatigue cracks around the rivet [100]. Investigations to determine the root cause of these tragic failures resulted in analysis protocol that are still in use today.

The continuous demand for materials with increased strength-to-weight ratios has led to the development of the entire group of alloys designated as the 7000 series. These alloys have been adopted for additional applications with this requirement and also include alloys that exhibit resistance to high cycle fatigue. In the 1960s, copper was removed as a primary constituent to develop alloys that were both weldable and resistant to SCC (alloys 7005 and 7004). Although lower in strength when compared to 7075-T6, these alloys exhibit comparable strength to 6061-T6 and 2219-T8. Demands in the 1970s for an economical, lightweight material for bright-finished auto bumper face plates led to the development of alloys 7016 and 7029. These alloys contain moderate copper (0.5–1.0%) to optimize the response to mechanical finishing to produce a reflective surface [97]. In addition to sheet product for aircraft and aerospace applications, the 7000 series alloys are used as extrusions and forgings in automotive and sport applications, such as off-road vehicles, firearms, and bicycles. See Fig. 1.28.

1.9 Summary: History of Aluminum

The nature of elemental aluminum, tightly bound as an oxide in a mineral state, made the path from ancient times to its discovery and extraction in the nineteenth century, difficult, frustrating, painstaking, and sometimes even dangerous. The numerous researchers who attempted to "reduce aluminum from its calces" all deserve credit for its ultimate discovery: from (chronologically and in order), Antoine Lavoisier in the late eighteenth century, to Sir Humphry Davy, Jöns Jakob Berzelius, Hans Christian Ørsted until 1827, when enough electrical energy was present and the correct method was employed by Friedrich Wöhler to finally extract aluminum metal. The path was indirect, and all involved were personally connected with one another as scientific researchers and academics bound by the ethics established by way of the European Context of Shared Knowledge.

When one considers the origins of science and technology, and how the sciences of chemistry and physics developed, and what new elements were discovered along the way, it is astounding. Without the concurrent development of other sciences and technologies, specifically, electrochemistry and electrical engineering, it could be that civilization would still be waiting for the Aluminum Age to arrive; however, the science and development that brought us the Aluminum Age is a sign of the times that were the Scientific and Industrial Revolutions. For extracting aluminum in quantities that could be considered an industrial process, Henri Étienne Sainte-Claire Deville is acknowledged for developing the state-of-the-art process for 1854; and even though Deville and Wöhler knew that aluminum could be extracted in cryolite, it wasn't until there was sufficient electrical power was available that Charles Hall and Paul Héroult could develop the industrial process in 1886 that we still use today. Karl Bayer deserves acknowledgement for making the Hall–Héroult process more efficient in 1888, by determining how to extract aluminum oxide from bauxite. From 1854 to 1888, aluminum's status changed from rare and precious to that of an industrial commodity, guaranteeing its place in industry worldwide for new artistic ideas and engineering applications.

The industrial production of the "new metal" proved advantageous for politicians, businessmen, and metal manufacturers. Aluminum alloy development began when the properties of unalloyed aluminum were insufficient to meet the requirements for end-use applications. Defense, transportation, and electrification were the first industries to drive aluminum alloy development. Most of the time aluminum alloys would substitute existing materials, such as iron, steel, copper, and copper alloys; and all of the time, the approach to alloy development was strictly empirical. Nevertheless, alloys, both cast and wrought have been developed according to applications such that 1000 series alloys are primarily used as sheet and foil product, and 2000 series wrought alloys and 200 series cast alloys are used for high strength applications in the defense, aerospace, and aircraft industries. 3000 series alloys are produced as plate and sheet for higher strength applications such as cookware and beverage cans; while 300 series castings are the workhorses of the aluminum casting industry. 5000 series wrought alloys and 500 series castings are primarily utilized for automotive and marine applications in which strength and corrosion resistance are of primary importance. 6000 series alloys have been developed for structural applications in architecture and in machine building, with extrusions of various standard and custom compositions fulfilling a variety of needs. Lastly, the 7000 series alloys fill the need for high strength, fatigue-resistant applications that have, in addition to the defense, aircraft, and aerospace industries, have a niche in sporting goods.

New aluminum alloys are developed to fulfill new requirements every day. Advanced analytical equipment, computer software, and test methods to characterize and develop materials with specific properties are available today in the twenty-first century and are routinely utilized to aid in this process. Although the process of alloy development has improved as aluminum metallurgy has grown and developed, and although the relationship between structure and properties is better known today than it has ever been, these methods can only succeed in predicting how the alloy microstructure might develop and react in the intended service environment for an application, serving as a guideline to alloy developers. Consequently, alloy development has advanced to an "enlightened empiricism," from the time when the philosophy for alloy development was alchemic, and made to produce the suitable "color of gold," or the more practical of the twentieth century: "if a little is good, add more until something bad happens."

Today, the science of metallurgical thermodynamics, together with mechanical design engineering and computer technology, can determine the initial iteration for an alloy chemistry that can yield the desired characteristics to meet the application target for a particular design application much closer than ever before. Nevertheless, the heuristic approach of trial and error to achieve specific properties by way of alloy composition tweaks and heat treatment modifications must still be undertaken until the actual target is reached. A more detailed approach to the characteristics of aluminum and its alloys from composition to microstructure development, as they apply to the metallurgy of the surface is presented in Chap. 4.

References

1. Davis, J. R. (Ed.). (1993). *ASM specialty handbook: Aluminum and aluminum alloys* (pp. 3–18). Metals Park, OH: ASM International.
2. *The history of the atom: The ancient Greeks*. AAAS Science Net Links. Retrieved from sciencenetlinks.com/lessons/history-atom-ancient-greeks (2017)
3. Partington, J. R. (1937). *A short history of chemistry* (1st ed.). New York: Dover.
4. Forbes, R. J. (1953). On the origin of alchemy. *Chymia, 4*, 1 – 11, JSTOR, Retrieved from www.jstor.org/stable/27757160
5. Harms, W. (1994, January 6). Bronze Age source of tin discovered. *The University of Chicago Chronicle, 13*(9).
6. Doyle, D. (2009). Notoriety to respectability: A short history of arsenic prior to its present day use in haematology. *British Journal of Haematology, 145*, 309–317.
7. Klein, C., Hurlbut, C. S., Jr., & Dana, J. D. (1985). *Manual of minerology* (20th ed.pp. 278–279). New York: Wiley.

8. Rostoker, W., & Bronson, B. (1990). *Pre-industrial iron: Its technology and ethnology* (Archeomaterials monograph). University Press.
9. Mehrer, H. (2007). *Diffusion in solids, fundamentals, methods, materials, diffusion-controlled processes*. New York: Springer.
10. Davy, H. (1808). The Bakerian Lecture on some new phenomena of chemical changes produced by electricity, particularly the decomposition of the fixed alkalies, and the exhibition of the new substances which constitute their bases; and on the general nature of alkaline bodies. *Philosophical Transactions of the Royal Society of London, 98*, 1–44. Part 1, Bulmer and Co. printers, London.
11. Davis, K. (2010). Material review: Alumina (Al_2O_3). *School of Doctoral Studies (European Union) Journal, 1*, 109–114.
12. Pliny the Elder. *Alumen and the several varieties of it, thirty-eight remedies*. Naturalis Historia, Book 35, Chapter 52 (1991).
13. Skrabec, Q. (2006). *The metallurgic age: The Victorian flowering of invention and industrial science, Ch. 8, Aluminum – Victorian Gold* (pp. 120–129). Jefferson, NC: McFarland.
14. Linden S. J. (Ed.) (2003). *The Alchemy reader: From Hermes Trismegistus to Isaac Newton, Introduction*. Cambridge: Cambridge University Press.
15. Oldroyd, D. (1974). Some neo-platonic and stoic influences on mineralogy in the sixteenth and seventeenth centuries. In A. Debus (Ed.), *Alchemy and early modern chemistry* (p. 220). Huddersfield: Jeremy Mills Publishing for the Society for the History of Alchemy and Chemistry.
16. Pagel, W. (1982). *Paracelsus: An introduction to philosophical medicine in the era of the renaissance, "The philosophy of Paracelsus" and "the elements" and the "three principles" (sulfur, salt and mercury): General considerations* (2nd Rev. ed., pp. 50–62 and 82–98). Basel: Karger.
17. Ball, P. (2011). In retrospect: On the six-cornered snowflake. *Nature, 480*, 455.
18. Huygens, C. (1962, November 8). *Traité de la Lumiere, Leyden, Pierre van der Aa* (Parke-Bernet Galleries, Trans.). (Original work published 1690).
19. Shapiro, A. E. (1989). Huygens' 'Traite de la Lumiere' and Newton's 'opticks': Pursuing and eschewing hypotheses. *Notes and Records of the Royal Society of London, 43*(2), 223–247. Science and Civilization under William and Mary. JSTOR. Retrieved from www.jstor.org/stable/531384.
20. Bressan, D. (2016, June). How biology pioneer Carl Linnaeus once tried to classify minerals. *Forbes*.
21. Linnaeus, C. (Carl von Linné). (1770, January 1). *Systema Naturae per Regna Tria Naturae, Secundum Classes, Ordines, Genera, Species cum Characteribus, Differentiis, Synonymis, Locis* (Vol. 3, pp. 12–35). Holmiae: Impensis, direct, Laurentii Salvii.
22. Linné, C. (Sir Charles Linné aka Linnaeus). (1806). *A General System of the Mineral Kingdom, systematically divided into its several classes, orders, genera, species and varieties with their habitations, manners, economy, structure, and peculiarities* (translated and augmented with a biography of Carl Linné by William Turton M.D., fellow of the Linnean Society, London, printed for Lackington, Allen and Co., pp. 209–226).
23. Haüy, par M. L'Abbé. (1822). *Traité de Minérologie* (2nd ed., Vol. 4, reviewed, corrected and augmented). Bachelier, Libraire, Successeur de Madame V. Coucier, Quai des Augustins.
24. Haüy, Par M. L'Abbé. (1822). *Traité de Cristallographie, d'une application des principes de cette science a la determination des espêces minerals, including a novel method to measure the form of crystals in projection* (Vol. 2). Bachelier et Huzaed, gendres et successeurs de Madame V. Coucier, Libraire pour les sciences, Rue du Jarunet, Saint-Andre-des-Arcs.
25. Authier, A. (2013). *Early days of crystallography*. Oxford, UK: Oxford University Press.
26. Wollaston, W. H. (1813). The Bakerian Lecture: On the elementary particles of certain crystals. *Philosophical Transactions of the Royal Society of London, 103*, 51–63.
27. Dalton, J. (1808). *A new system of chemical philosophy, part one* (Manchester: Printed pp. 209–218, by S. Russell, R. Bickerstaff, Eds., Strand, London).

28. Smith, R. A. (1856). *Memoir of John Dalton and history of the atomic theory up to his time* (pp. 230–246). London: H. Bailliere.
29. Wisniak, J. (2004). André-Marie Ampére. The chemical side. *Educación Química, 15*(2), 166–176.
30. Greenberg, A. (2007). *From Alchemy to chemistry in picture and story* (pp. 225–228). New Jersey: Wiley.
31. *The supplement (1803 edition) to Encyclopedia Britannica 3rd edition,* (1797). (Vol. 1, p. 225).
32. Donovan, A. (1996). *Antoine Lavoisier: Science, administration and revolution.* Cambridge: Cambridge University Press.
33. LaVoisier, A. (1789). *Traité Élémentaire de Chemie* (Vol. 1, p. 174). Paris: Cuchet.
34. Kerr, R. (1799). *Elements of chemistry, in a new systematic order, containing all the modern discoveris* (4th ed.). Mathew Carey: Edinburgh, Scotland.
35. Donovan, A. (2017, June 22). *Antoine-Laurent Lavoisier.* Encyclopedia Britannica, Inc. Retrieved from www.britannica.com/biography/Antoine-Laurent-Lavoisier
36. Volta, A. (1800, June 26). On the electricity excited by the mere contact of conducting substances of different kinds. [Letter to Rt. Hon. Sir Joseph Banks and read at the Royal Society of London], pp. 403–431.
37. Desmond, K. (2016). *Innovators in battery technology: Profiles of 95 influential electrochemists, Davy, H. (1778–1829) The "Great Battery"* (p. 53). Jefferson, NC: McFarland.
38. Russel, C. (2003). *Enterprise and electrolysis.* London: The Royal Society of Chemistry.
39. Wetzels, W. D. (1968). *Johann Wilhelm Ritter: Physik im Wirkungsfeld der Deutschen Romantik.* Doctoral thesis, Princeton.
40. Davy, H. (1840). On some new phenomena of chemical changes produced by electricity, particularly the decomposition of the fixed alkalies, and exhibition of the new substances which constitute their bases; and on the general nature of alkaline bodies, first presented to the Royal Society in November, 1807. In J. Davy (Ed.), *The collected works of Sir Humphry Davy* (Vol. V, pp. 57–99). Bakerian Lectures and Miscellaneous Papers from 1806 to 1815. Cornhill, London: Smith, Elder.
41. Davy, H. (1812). *Of metals; their primary combination with other undecompounded bodies, and with each other, Section 7. Aluminum, from elements of chemical philosophy, Part 1* (Vol. 1), Division 5, London. (Printed for J. Johnson and Co., St. Paul's Church-Yard, p. 355).
42. Sutton, M. (2008, November). A clash of symbols. Historical profile of Jöns Jakob Berzelius. *Chemistry World*, 56–60.
43. Childs, P. E., (1998). From hydrogen to meitnerium: Naming the chemical elements, Section 2.5, Berzelius and the agreement on symbols. In K. J. Thurlow (Ed.), *Chemical nomenclature*, Table 2.1 (p. 36).
44. Faraday, M. (1834). Experimental researches in electricity. Eighth Series. *Philosophical Magazine*, paragraph 918, 126. London: Taylor and Francis.
45. King, W. J. (1968). *The development of electrical technology in the 19th century: Part 1. The electrochemical cell and the electromagnet. Bulletin 228: Contributions from the Museum of History and Technology* (pp. 231–271). Cambridge: Harvard University.
46. Niaudet, A. (1890). Elementary treatise on electric batteries (L. M. Fishback, Trans. in French, 6th ed., pp. 13–15). New York: Wiley.
47. Ohm, G. S. (1827). *Die Galvansiche Kette, Mathematisch bearbetitet.* Berlin: T. H. Riemann.
48. Blondel, C., & Wolff, B. (2013, January). *Ampére lays the foundation of electrodynamics (September 1820–January 1821), ampere.cnrs.fr* (French version, March 2009) (English translation).
49. Ampére, A.-M. (1826). *Mémoire sur la théorie mathématique des phénomènes électrodynamiques uniquement déduite de l'experience* (Memoir on the mathematical theory of electrodynamic phenomena, uniquely deduced from experience), Paris.
50. King, W. J. (1968). *The development of electrical technology in the 19th century: Part 1. The electrochemical cell and the electromagnet. Bulletin 228: Contributions from the Museum of History and Technology* (p. 241). Cambridge: Harvard University.

51. Faraday, M. (1839). *Experimental researches in electricity.* (Reprinted from *Philosophical Transactions of 1831–1838*, p. 304, paragraph 999, London: Taylor and Francis).
52. Desmond, K. (2016). *Innovators in battery technology: Profiles of 95 influential electrochemists, Daniell, John Frederic (1790–1845), The constant cell* (pp. 50–53). Jefferson, NC: McFarland.
53. Wheatstone, C. (1879). The scientific papers, as published by the Physical Society of London. *An account of several new Instruments and processes for determining the constants of a voltaic circuit* (pp. 97–132). London: Taylor and Francis.
54. Bowers, B. (2001). *Sir Charles Wheatstone FRS: 1802-1875* (pp. 101–114). London: The Institution of Electrical Engineers.
55. Minet, A. (1905). *The production of aluminium and its industrial use* (L. Waldo, Trans, 1st ed.). New York: Wiley.
56. Berthier, P. (1821). Analyse de l'alumine hydraté des Beaux, department des Bouches-du-Rhóne. *Annales des Mines, 1st series, 6*, 531–534.
57. The Aluminum Association. (2017). *Alumina refining, the bayer process*. Retrieved from info@aluminum.org
58. Wöhler, F. (1827). XI. Ueber das Aluminium. *Annal d. Physik, B.87*, St. 1, J. St.9, 146–161.
59. Wisniak, J. (2004). Henri Étienne Sainte-Claire Deville: A physician turned metallurgist. *Journal of Materials Engineering and Performance, 13*(2), 117–128.
60. Price, J. A. (1886, March). Aluminum. *Scientific American Supplement, 21*(351), Logo 8471, text 8482-3.
61. Authier, A. (2013). *The birth and rise of the space-lattice concept, Section 12.11: A. Bravais "Systèmes forms par des point distribués régulièrement sur un plan ou dans l'espace, 1848", Chapter 12, The early days of crystallography*. Oxford: Oxford University Press.
62. Barlow, W. (1883). The probable nature of the internal symmetry of crystals. *Nature, 29*, 205–207.
63. Assmus, A. (1995). Early history of X-rays. *Beam line. SLAC National Accelerator Laboratory, Stanford, CA*, 10–24.
64. Eckert, M. (2012). Max von Laue and the discovery of X-ray diffraction in 1912. *Annalen der Physik, Berlin, 524*(5), A83–A85.
65. Bragg, W. L. (1922, September 6). The diffraction of X-rays by crystals. *Nobel Lecture*.
66. Morachevskii, A. G. (2006). Henri Étienne Sainte-Claire Deville (To 150th anniversary of the development of the first industrial method for production of aluminum). *Russian Journal of Applied Chemistry, 79*(10), 1731–1735.
67. Polmear, I. (2006). *Production of aluminum, Section 1.2. Chapter 1, Light alloys, Traditional alloys to nanocrystals* (4th ed.). Butterworth-Heinemann, an imprint of Elsevier, Oxford, UK.
68. The history of the aluminium industry. Retrieved from Aluminiumleader.com
69. Deville, H. *de L'Aluminium.* Taschenbuch (Replica of the book originally published in 1859).
70. Constellium. (2017). *Who we are. Company timeline*. Retrieved June 15, 2017, from constellium.com
71. Verne, J. (1886). *From the earth to the moon, Chapter 7, The hymn of the cannon-ball* (John W, American Translation). New York: Lovell.
72. King, W. J. (1963). *The development of electrical technology in the 19th century: Part 3. The early arc light and generator. Bulletin 228: Contributions from the Museum of History and Technology* (pp. 334–406). Cambridge: Harvard University.
73. Binczewski, G. (1995). The point of a monument: A history of the aluminum cap of the Washington Monument. *Journal of Management, 47*(11), 20–25.
74. Alcoa. (2017). *Who we are/History/Timeline*. Retrieved from alcoa.com. Copyright 2017.
75. Plunkert, P., & Jones, T. (1999). *Metal prices in the United States through 1998* (A compilation published by the US Department of the Interior, US Geological Survey, pp. 1–4), Reston, VA: US Government Printing Office.
76. UC Rusal. (2017). *All about aluminium, history*. Retrieved from Aluminiumleader.com.

77. Staley, J. T. (1989). *Aluminum alloys—Contemporary research and applications, Treatise on materials science and technology, "Overview" in part 1: Historical background* (Vol. 31, p. 28). San Diego, CA: Academic Press.
78. Lake, E.F. (1913). *The extrusion process, The extrusion of metals* (pp. 3–27). New York: The Industrial Press, Publishers of Machinery.
79. Gautschi, & Jequier. (1905, March 25). *Papier Metallique* (Swiss Patent No. 33290).
80. The Aluminum Association. (2017). Foil and packaging. Retrieved from www.aluminum.org
81. Smith, G.D. (2003). *From monopoly to competition: The transformations of Alcoa, 1888–1986, Chapter 3: Building a big business* (pp. 77–131). Cambridge: Cambridge University Press.
82. Sennott, R. S. (Ed.) (2004). *Encyclopedia of twentieth century architecture* (A–F, aluminum, Vol. 1, p. 38). New York: Taylor and Francis.
83. Sheller, M. (2014). *Aluminum dreams, the making of light modernity*. Cambridge, MA: Massachusetts Institute of Technology Press.
84. Jester, T. C. (Ed.). (2014). *Twentieth century building materials, history and conservation*. Los Angeles, CA: Getty Conservation Institute.
85. La Trace, A. J. (2015, March 27). *Less is more: Mapping Mies van der Rohe's career in Chicago*. Retreived from chicago.curbed.com
86. Sanders, R. E. (2001). Technology innovation in aluminum products. *Journal of Management, 53*(2), 21–25.
87. Alcan Marine. Aluminium, the marine metal. *Historical Review*, 10–13 (2005).
88. The Aluminum World. (1895, October). *The Aluminum Yacht Won! 2*(1), 1–2.
89. Hobson, R. P. (1897). Notes on the yacht defender and the use of aluminum in marine construction. In Proceedings of the US Naval Institute (pp. 523–562).
90. Sanders, R. E., Hollinshead, P. A., & Simielli, E. A. (2004). Industrial development of non-heat treatable aluminum alloys. *Metals Forum, 28*, 53–64.
91. Genta, G., Morello, L., Cavallino, F., & Filtri, L. (2014). *The motor car: Past, present and future, Ch. 4 "Powertrain", section 4.41 electric cars* (p. 144). New York: Springer.
92. Vaughan, D. (2011, July). *Pierce arrow*. Retrieved from Conceptcarz.com
93. Brown, L. (2015). Exploring the opportunity for anodizing in automotive bonding. In *Proceedings of the Aluminum Anodizers Council, San Diego*.
94. Skillingberg, M. (2007). Aluminum applications in the rail industry. *Light Metal Age, 65*(5), 1–5.
95. Skrabec, Q. (2017). *Aluminum in America: A history, Ch. 7. Metallurgical wars and monopoly* (p. 68). Jefferson, NC: McFarland.
96. Skrabec, Q. (2017). *Aluminum in America: A history, Ch. 13: The war and the new aluminum industry* (pp. 134–147). Jefferson, NC: McFarland.
97. Polmear, I. (2004). Aluminium alloys—A century of age hardening. *Materials Forum, 28*, 1–14.
98. Kaufman, J. G. (2008). *Properties of aluminum alloys: Fatigue data and the effects of temperature, product form and processing*. Materials Park, OH: ASM International.
99. Mondolfo, L. F. (1976). *Aluminum alloys, structure and properties, Chapter 4-2: Aluminum-silicon and aluminum-magneisum silicide alloys* (pp. 759–806). London: Butterworths.
100. Simons, G. (2013). *Comet! The world's first jet airliner, Ch. 8: Disasters—Investigation—Inquiry* (pp. 125–167). Barnsley: Pen and Sword Aviation.

Chapter 2
A Brief History of Anodizing Aluminum

2.1 Introduction

Since ancient times, surface treatment methods have been used to change the appearance and surface properties of metal and other materials for a variety of reasons. Coatings, specifically, have always been primarily used for the purpose of decoration, and whether for personal, religious, or governmental/tribal objects or articles, they were and still are used for identification and other communication purposes. In keeping with alchemic practices, coatings almost always changed the surface of the material to make it appear to be of higher value. As the methods for various coating processes developed, the discovery of the actual engineering properties of the coating itself soon found application. With these discoveries, the purpose of surface treatment exceeded decoration, and soon coatings were used for corrosion protection and wear resistance. As the forerunner to electrochemical processes, a background in early coating technology is a definite part of the history of anodizing.

The history of anodizing aluminum goes beyond the history of coating technology. The overlap of the interdisciplinary concepts that made the reduction of aluminum possible is similar for the development of anodizing. With the discovery of aluminum in the early nineteenth century, understanding the electrochemical circuit that yielded the aluminum developed into studies to understand the mechanisms of what happens in each circuit element. New areas of study: physical chemistry and solid-state physics, with consideration for new and developing studies in atomic theory, crystallography, diffusion, and thermodynamics were the major areas of research utilized to develop new sources for electrical power (batteries), the materials that comprised the electrodes, the solution chemistry of the electrolyte and ultimately, the properties of the oxide that forms naturally, and with electrochemical processing, on aluminum.

These studies lead to the identification of the characteristic of some metals, which lends itself to the formation of a stable surface oxide when exposed to various

J.M. Runge, *The Metallurgy of Anodizing Aluminum*,
https://doi.org/10.1007/978-3-319-72177-4_2

chemical and electrochemical treatments, also known as the condition of being passive. With the discovery and improvement of the battery, it soon followed that the passive quality of aluminum could be externally controlled and that the anodic oxide had specific engineering qualities, which led to the development of engineering processes that deliberately anodically oxidized aluminum for different types of applications. Beginning with the determination that the anodic oxide functioned as a semiconductor at the end of the nineteenth century, to the determination of additional engineering properties of corrosion protection and wear resistance in the early twentieth century, the process of electrochemical oxidation of aluminum came to be called anodizing by the mid-twentieth century, well after the process had already developed into an industry. Today, the anodic oxide finish is as ubiquitous as aluminum products that it enhances, used in a variety of applications that not only exploits the characteristics of the anodic aluminum oxide as an integral part of the component, but also exploits its unique structure as a stand-alone entity.

In this chapter, the chronological events in science and technology that lead to the development of anodizing industry are presented. After a review of the developments of the related important sciences that contributed to the discovery of the anodic oxide on aluminum, its characterization and development of applications for it, a review of important research and development specific to anodizing theory is provided that spans the late nineteenth and twentieth centuries together with some of the important research that is being carried out today. Rather than providing a comprehensive list of names and dates, the work of a few of the most important scientists that contributed to the birth of the anodizing industry and the course of its theoretical development were selected and summarily discussed. Key figures in modern anodizing are also presented. Should the reader like a more comprehensive treatment, there are reviews available by Diggle [1]; Thompson [2]; Lee and Park [3]; and Poinern, Ali, and Fawcett [4] that can be used and are referenced in this chapter as well.

2.2 Ancient Coating Technology

Proven by many ancient texts and actual archeological finds, the various ancient communities and cultures possessed a vast lore of properties of metals and alloys, methods to refine and test them, and they actually manufactured some of them. At the same time, various coating processes were also developed to modify the decorative appearance and/or engineering performance of many different types of objects, for both practical and purely esthetic applications.

Clearly, the art and science of physical change produced by way of alchemic practices was much more than the magical hocus pocus we typically think of when we think of alchemy. Many coating methods were employed as a means to "transmute" metal surfaces in order to change them to appear more like gold. Beautiful inlays of different metals and materials as well as other decorative metalwork with various colors were produced with pure gold, gold and silver alloys, and various brasses, bronzes, and other copper alloys.

Coatings were produced by a variety of methods: cladding with gold and silver foil, staining with organic matter, etching of base gold alloys to remove the base and reveal the gold, surface leaching, enrichment, pickling, and gilding were methods for producing gold finishes of different colors to achieve an appearance that was linked to a gold color with a particular magical power or thought to bring luck or fortune. For example, rose gold, a color to which a particular large amount of magic was attributed, was produced in ancient Egypt by placing a translucent coating of iron oxide over a gold object by dipping it into an iron salt solution and heating it. Other coating methods were used to change the surface appearance of common stones, making them appear like gems. Stones were heated in solutions of acids or bases; entire objects were embedded in chemical solutions and roasted. Enameling processes were developed that created the turquoise look of lapis lazulis and the lustrous shine and color of copper and bronze [5]. Much like the anodic oxide on aluminum today, ancient coating processes were decorated with gold and valuable stones more economical but no less valuable in appearance; to ancient nobility, skilled alchemists and craftsmen were "worth their weight in gold."

Within the realm of Ancient Alchemic Coating Technology, a unique device was found in an antiquities museum in Iraq in 1938, called the Parthian Galvanic Cell, which dates to the first century BC or the first century AD. The cell consists of a pottery jug in which a hollow cylindrical tube is made of rolled and soldered copper sheet. The base of the tube is closed at the bottom with a copper disc that is fixed to the copper tube and to the ceramic jug with a layer of asphalt. An iron rod is suspended from an upper asphalt plug down the central axis of the tube; as assembled, both the iron and copper components projected above the surface of the upper asphalt plug. Several of these cells have been found in other places as well, with variations in the nature of the central electrode—some have a fixed electrode, and some have a removable electrode [6]. The features of the assembly are that of an electrochemical wet cell, driven by the potential difference between the copper tube and the iron rod when the cell is filled with acid, very similar to the voltaic cell, invented by Alessandro Volta in the late eighteenth century (See Sect. 1.3.3). See Fig. 2.1.

As a unique therapeutic method and ultimately a means for coating metal, the galvanic current produced by the Parthian cell was most likely discovered by accident by way of the feeling produced by the electric potential difference when eating salty vinegar-containing food from a copper bowl with an iron spoon. The charge would have been similar to that produced by an electric eel. Eels and other "electrified fish" were kept alive in captivity in ancient Greece and Rome for medicinal purposes; they were used to produce a palpable electrical charge to the extremities of the human body, immersed in electrified solutions in order to help with the analgesic treatment of gout, and other forms of neuropathy. However, such a galvanic cell may have had additional technical uses in chemistry and metallurgy, for the purpose of mineral extraction and/or for electroless or immersion metal plating processes, and the discovery and ultimate production of such a visual change by way of galvanic coating almost certainly would have been an important alchemic achievement.

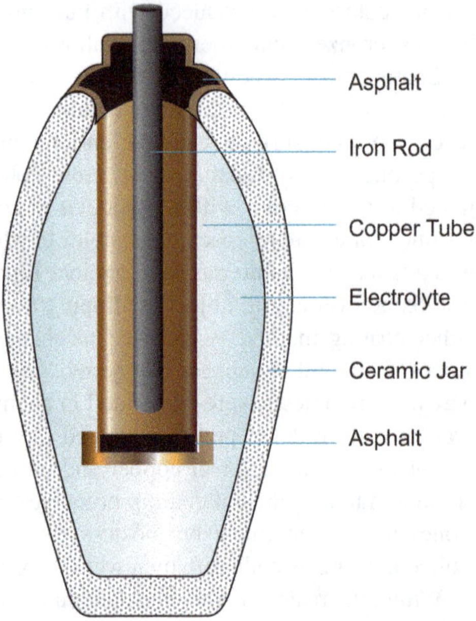

Fig. 2.1 Schematic of the Parthian Galvanic Cell, a jar with the central cathode. Galvanic current was generated by the potential difference between the copper tube and the iron rod which acted as electrodes in the acid electrolyte [6]

Nevertheless, exactly how the Parthian Galvanic Cell was definitely applied is still a question. Found in the Near East, a great distance from a great body of water, these cells might have produced the relief from pain (numbness from electrical stimulation) that ailments such as rheumatism and gout cause, simply by way of the potential difference produced between copper and iron in an acid electrolyte, removing the need for living electrified fish inland. From the perspective of coating technology and plating processes, voltage measurements taken from models built of the Parthian cell are insufficient to support direct electroplating, but several in series support electroless or immersion coating processes, just by way of the naturally occurring potential difference between copper and iron.

When considering an electrolyte, acid solutions were not uncommon in ancient times, and the strongest acid solution commonly available was acetic acid derived from spoiled wine; and, significantly, copper is soluble in acetic acid. In the first century AD, both Dioscorides, the Greek physician in Nero's court and Pliny, report the use of copper in solution with sulfuric acid (as green vitriol) as well as the electroless deposition of copper on iron. It is therefore clear, "transmuting" an iron coin to copper using electrolytes of copper acetic acid was possible, which supports the conclusion that the Parthian cell may have been used for electrochemical coating purposes. However, based upon the age of the Parthian cell, it would have been exclusively for the plating of copper; the cell could not have been applied to electroplate or even immersion plate gold because an acid capable of digesting gold, or even silver would be required to do so. Mineral acids, such as hydrochloric acid and nitric acid were not discovered until about 800 AD by the important Persian alchemist, Jabir ibn Hayyan (Geber) [7]. In spite of this, it is certain that the technology existed for electrochemical processes during ancient times that may have been used for the express purpose of immersion coating iron with copper.

With new discoveries, warnings to keep knowledge sacred to minimize the growth of power and influence were wielded. This was not necessary to protect trade secrets or intellectual property of a craft or guild, but to keep the practitioners of the science of alchemy safe. This quote, from the Middle Ages, is attributed to the seventh century AD medical texts: "*Let him that knoweth, show him that knoweth, but he that knoweth shall not show him that knoweth not!*" [5]. Similar warnings are also reported *on tablets* (cuneiform) from 1700 BC Ancient alchemic codification ends with the Greek Zosimos of Panopolis who, in the third and fourth centuries, summarized the entire alchemic doctrine and literature into four books: "Keys of the Craft"; "Keys of Magic"; "The Book of Elements"; and "The Book to All the Wise of the Craft." Already, in Zosimos' writings, a strong religious overtone is present, promising, "Salvation can be obtained by the Great Work" [8].

Because of the limitations placed upon alchemic practices by the Church during the Middle Ages, science and technology in general did not advance a great deal beyond what was already developed in ancient times. In fact, the argument can be made that with the destruction of many of the alchemic codices in the west under the rule of Roman Emperor Diocletian in the third century AD, some technology was lost and later rediscovered or reinvented. Coating processes did not advance beyond the already known methods for gilding objects with silver and gold, and nothing is found in the literature that discusses immersion coating of dissimilar metals in Western Europe.

2.3 The Scientific Revolution

The curiosity and experimentation that originated during the Renaissance were the foundation of the basic sciences of chemistry and physics during the Scientific Revolution. As multiple, concurrent, yet sometimes, divergent paths of research developed, the sciences themselves were partitioned into areas of focus: the drive to understand the *nature* of the earth and all constituent materials resulted in the sciences of metallurgy, chemistry, and physics; the drive to understand the *structure* of all materials and *how they react* resulted in the sciences of physical chemistry and with the discovery of electricity and electrochemistry. With time and development, chemistry divided along inorganic and organic lines, and by using atomic theory, explanations developed not only for electrochemical decomposition, but also for how structures come together to form compounds. Atomic theory and crystallography developed well beyond its Daltonian and Haüyistic roots.

As electrochemistry progressed during the Scientific Revolution, with the advent of the study and comparison of how materials behave in electrochemical reactions, whether they be equilibrium, under the effect of *chemical action*, or in reactions driven externally, under the effect of *electrical contact*, connections between the physical structures of different materials and how they behave as part of an electrochemical circuit were studied in great detail. In 1810, Berzelius hypothesized that materials comprised differently charged particles after observing how charge moves through an electrolyte solution, migrating from the anode to the cathode during the reactions in an electrochemical experiment. Insodoing, Berzelius developed the first electrochemical theory, which was called Dualism. "*By arranging the atoms in the order of their electrical affinities, one forms an electrochemical system, which is more suitable than any other arrangement to give an idea of chemistry*" [9]. With this guiding law, Berzelius was able to interpret chemical reactions, explain the properties of compounds, and classify them.

New studies continued to develop during the Scientific Revolution: crystallography along with studies in diffusion, offered explanations for the structure of all matter and explained the movement of the "contiguous particles" that comprised their structures, including electric current and the atmosphere. Studies to determine how charged particles move through electrodes and the electrolyte in electrochemical circuits and how they produce change in resistance measurements by way of differences in electromotive force between different metals and materials were the foundation that enabled Ohm to develop the theory and the mathematic laws that the communication of electricity occurred between "contiguous particles" [10]. The relationship between voltage, current, and resistance was the basis for: the engineering of new and better batteries; the research into the electrical behavior of metals and metal compounds, and combinations thereof; and the discovery of distinct material characteristics of conductor, insulator, and ultimately, semiconductor.

The energy required for various reactions was accounted for by way of thermodynamic calculations. Possibly, the single most controversial discovery regarding metal surfaces, surface treatment, and coating technology was the discovery of passive behavior in the late eighteenth century. The nature of passive behavior, as a development of a physical change in the metal or simply the development of an oxide layer, became as controversial as the fundamental argument of electrochemistry (chemical action vs. electrical contact). Engineering strides were made and drove the Scientific Revolution into the Industrial Revolution. Engineering was innovative and important, enabling the elevation of scientific discoveries to industrial reality. Many of the methods that were used are equally relevant today for a variety of sciences, materials, and applications; how they apply to aluminum is an interesting and important part of these discoveries and developments.

2.3.1 Important Research in Physical Behavior of Metals

2.3.1.1 Diffusion

Diffusion is an extremely important concept when considering the basis for the mechanism of anodic oxidation of aluminum. During anodizing, aluminum and alloy ions are dragged with electrons as part of the applied current to the substrate surface where they react with electrolyzed ions from the electrolyte to form anodic alumina. With the applied current, the simultaneous diffusion and mass transport of alloying elements, atomic level defects, and intermetallic particles confound the oxidation reaction by creating all kinds of electrochemical and physical phenomena. Therefore, some of the history of diffusion is presented briefly to see how it evolved from a theory to an accepted physical process. Diffusion is defined as the movement or transport of material (mass) from a high to a low concentration gradient. An important distinguishing feature of diffusion is that there is no change in the bulk where the diffusion takes place, only mixing or mass transport within the enclosed system.

The first known demonstration of solid-state diffusion is published in the *Naturalis Historia* in which Pliny describes the *cementation process*, the diffusion of carbon in iron to make steel. The first record of modern diffusion experiments during the Scientific Revolution occurred in the seventeenth century when Robert Boyle reported the progressive penetration of zinc into a copper coin. As one can read in *Naturalis Historia*, diffusion processes were used for coating applications as far back as ancient times; Boyle's study was done much like those described in the ancient codices [11, 12].

By the late eighteenth and early nineteenth centuries, with the new regard for the composition of solids in terms of atomic particles, the idea that atoms have distinct mass and come together with specific proportion to form compounds implied that there was movement involved to achieve the "coupling" of elements and molecular structures. Studies to understand the movement or "diffusion" of atoms and molecules began in 1827, the same year that Friedrich Wöhler successfully reduced aluminum metal for the first time. Scottish botanist, Robert Brown, reported that he observed in the field of view under the optical microscope when small pollen particles were dispersed in water, they moved erratically. Observing the same motion of inorganic particles in another experiment, Brown drew the conclusion that the movement was not "life related," that is, it was a phenomenon typical for small particles in a suspension, even suggesting that molecules could possibly play a role in the motion. Although Brown was not able to determine the mechanism for the phenomenon, the effect itself has come to be called "Brownian Motion" and is regarded as one of the cornerstones for diffusion [13].

Scottish chemist Thomas Graham, one of the fathers of physical chemistry, initiated a quantitative study of diffusion in gases in 1828. His work and publications on this subject were published over the period of time from 1828 to 1833 and are probably the first formal studies in diffusion. Graham's studies, which became Graham's Law of Diffusion, can be summarized: Gases of different nature, when brought into contact, do not arrange themselves according to density, with the heaviest undermost and the lighter uppermost, but they spontaneously DIFFUSE, mutually and equally through each other, and so remain in the "*intimate state of mixture*" for any length of time. The diffusion or spontaneous intermixture of two gases is effected by an interchange in position of indefinitely minute volumes of the gases, which volumes are not of equal magnitude, being, in the case of each gas, inversely proportional to the square root of the gas. Graham extended his studies from gases to the diffusion of salts in liquids, and to the uptake of hydrogen in metals during electrochemical reactions [14]. The work was so carefully executed that other scientists were inspired to continue diffusion studies, some connecting diffusion with electrochemistry.

2.3.1.2 Physical Chemistry

In what can be called a rebirth of alchemy came with the development studies devoted to organic chemistry. The need to transmute or *synthesize* materials in order to find replacements for resources that were being exhausted, such as wood and charcoal; and new sources for medicines, dyes, and cosmetics, fell on the shoulders of scientific researchers. This was especially true in Western Europe, where forests were being depleted in order to obtain charcoal for fuel. Coke, from coal, was the ready substitute, fueling lamps and other important industrial processes. All of the noted European scientists that were involved in the race to reduce aluminum were also involved in research of organic chemistry: Davy, Berzelius, Faraday, and Wöhler. In another example of Friedrich Wöhler's process expertise that exceeded his teacher, Jöns Berzelius, Wöhler, in 1828, just 1 year after being the first to successfully reduce aluminum from aluminum chloride, demonstrated that natural organic chemicals could be synthesized from inorganic chemicals, when he synthesized urea by combining cyanic acid and ammonium [15]. This was the beginning of modern organic chemistry, and the connection to alchemy didn't escape chemists of the time. In fact, organic chemists, because of the extensive work that was done with the syntheses of various compounds found in coal tar, were often called *Schwarzbrenner*, literally translated, *Black Burners*—another reason why alchemy and chemistry could be called "dark arts."

Michael Faraday, shortly before Wöhler's synthesis of urea, discovered that benzene was present in coal tar. It was similarly determined that aniline is a derivative of coal tar, too, as is naphthalene (moth balls). Pharmacology was born from these reactions when it was discovered that quinine, the "miracle drug" for treating malaria, could be synthesized from aniline. The toxicity of these reactions, sometimes not realized, significantly shortened the lives of these researchers; they often suffered terrible illnesses, headaches, and several committed suicide. But with the pain and suffering that can be associated with alchemy at any time in history, a great deal of learning was also accomplished. Possibly, the first fine chemicals that were derived by these reactions were dyes. Certainly, with respect to anodizing, the development of organic dyes was a great achievement, but this would not be realized for another century [16].

Of greatest importance, in the realm of organic and physical chemistry, was the understanding of the structure of organic molecules and the reactions that took place between the structures during the synthesis of new materials. In 1837, French chemist Auguste Laurent theorized that there was an intimate relationship between crystal form and atomic-molecular arrangement within a crystal, which governed its chemical properties and intermolecular relations. Laurent's work focused on organic compounds, in particular with naphthalenes, and his work theorized a *fundamental unit cell with a nucleus*, which could be a single atom, to which *free radical* groups could attach themselves. Laurent conceived that two bodies could share the same atoms, united in the same proportion (as isomers), with atoms arranged in

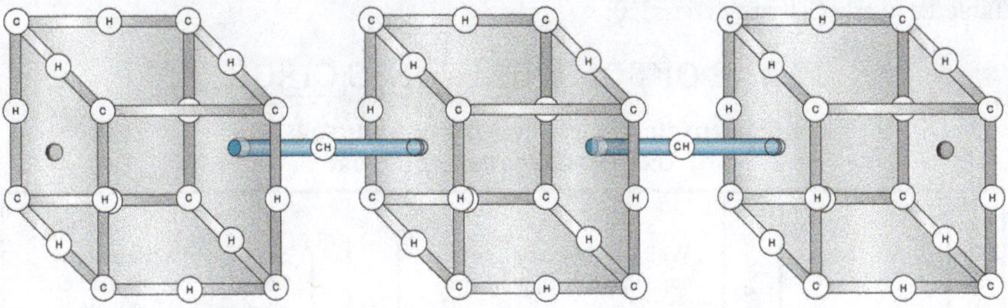

Fig. 2.2 An example schematic for Laurent's nucleus theory. Atomic structures were illustrated as cubes with functional groups connecting them at specific locations

the same manner (as isomorphs) but as long as the order, that is, the relative arrangement of the atoms was different, the product of the reaction produced by sharing, would be different. Similar arrangements of atoms in a molecule were responsible for similar properties in compounds, while different arrangements of the same atoms produced different properties [17].

The above stated theory that came to be called Laurent's *Nucleus Theory*, for which he was given the name the Father of Modern Organic Chemistry, is not electrochemical. It is based on chlorine substitutions, which showed that chlorine ions could substitute for hydrogen ions in a reaction, even though chlorine is a negative ion and hydrogen is a positive ion. The theory is based on Laurent's background in crystallography and states that all compounds have a fundamental group in common, or adopting from crystallography, a primitive form or core *nucleus*. The nuclei do not decompose by heat or alkali, they maintain shape, they have the same properties, and the same molecular arrangement. Families of similar chemical substances all share a common nuclear "radical," modified among the members of a family by substitutions of the hydrogen atoms of the molecule by atoms (or atomic groups) of other elements, for example, between aluminum hydroxide radicals in the anodic oxide, or in a conversion coating and acid counterions in the electrolyte or a polymer modifier (See Chap. 3). Laurent's theory is pure chemistry, neither inorganic nor organic, and is critical in understanding interfacial phenomena, in particular, bonding reactions. Laurent's theory challenged Berzelius' Dualism Theory, but even Berzelius conceded by recognizing the Nucleus Theory accounted for differences in the structure of compounds as isomers and isomorphs as well as the basis for differences in their properties [18]. See Fig. 2.2.

2.4 Applied Electrochemistry: Impact on Surface Science and Corrosion

At the beginning of the nineteenth century, the voltaic pile became the foundation for electrochemical research and development. Batteries were not only used for decomposition experiments, to discover new elements, they were used to drive

Table 2.1 Early Galvanic Seriess [20]

TABLE OF SOME GALVANIC CIRCLES,

COMPOSED OF TWO PERFECT CONDUCTORS AND ONE IMPERFECT CONDUCTOR.

More oxydable substances.	Less oxydable substances.	Oxydating fluids.
Zinc	With gold, charcoal, silver, copper, tin, iron, mercury.	Solutions of nitric acid in water, of muriatic acid and sulphuric acid, &e. Water holding in solution oxygen, atmospheric air, &e.
Iron	—— gold, charcoal, silver, copper, tin.	
Tin	—— gold, silver charcoal	
Lead	—— gold, silver.	Solution of nitrate of silver and mercury, nitric acid, acetous acid. Nitric acid.
Copper	—— gold, sliver.	
Silver	—— gold.	

characterization experiments for new elements, as well as "known" metallic elements such as iron. As more new elements were isolated by various decomposition methods at the end of the eighteenth and the beginning of the nineteenth centuries, experiments on the "new metals" were immediately performed to identify their unique properties. Discernment and measurement of material characteristics were enabled by noting consistencies in the electrical behavior of the electrochemical circuit, as well as the chemical reactions within the electrolyte and at the electrodes.

Characterization enabled the tabulation of important data such as specific reactivity in water and selected acid and alkaline solutions. Electrical conductivity measurements were made at various temperatures and in different atmospheres. As a result of characterization studies, Volta found that the various known metals could be arranged in a series, each of which was positive with respect to those preceding it and negative with respect to those following it. The reference list is called the Galvanic Series, named for Luigi Galvani whose work in early electrochemistry, using dissimilar metals to generate electric current to make frog's legs twitch, preceded Volta's [19]. An early Galvanic Series reported by Sir Humphry Davy in 1801 [20], produced from experiments on circuits driven by a voltaic pile, is presented in Table 2.1.

As electrochemical experimentation proceeded, engineering characterization of the various chemical reactions with new elements were studied and the elements were characterized by how the reactions proceeded in electrical circuits that employed various electrolytes, electrodes, atmospheres, and other materials such as glass, leather, cardboard, and tar, resuslting in new applications for the electrochemical circuit. As a result, modern applications for electrochemical phenomena were developed. Electroplating began in the early 1800s with Volta; he successfully deposited lead, copper, silver, and zinc onto gold and iron substrates. Sir Humphry Davy also used the rules of galvanic action for corrosion prevention. In 1824, Davy developed a method to protect the copper hull of the ship *Sammarang* with cast iron sacrificial anodes [21]. While cathodic protection had been used since ancient times to prevent corrosion, this is the first time in recorded history that it is acknowledged as an electrochemical method. It is also noteworthy that Napoleon I supported and sponsored Volta's research in galvanic action in order to gain knowledge and derive the benefits of two major electrochemical phenomena that have become important industries: electroplating and corrosion protection. Reports of experiments and their results successfully communicated information about galvanic processes throughout Europe, the New World, and in Russia.

The idea that charged particles comprise chemical compounds, which is attributed to Berzelius in 1810, was one of the most important conclusions from electrochemical decomposition experiments that was not a newly discovered element. Berzelius derived this concept from the understanding of the operation of the electrochemical circuit: compounds were decomposed by an electrical current and released elements that were electrodeposited at the poles in an electrolytic cell; atoms were charged and chemical combination resulted from the mutual neutralization of opposite charges. Observations of galvanic cells determined that metal was usually dissolved or oxidized at the anode and reduced at the cathode. In 1819, this trend led Louis Jacques Thénard to draw the conclusion that corrosion of metal was an electrochemical process [22].

Faraday, in 1833, concluded with Berzelius' hypothesis, naming the particles *ions*, thereby connecting chemical action and the generation of electrical currents in what would become his *Laws of Electrochemical Action* [23]. His experiments to identify the electrical behavior of metal compounds led to the discovery that the resistance of specimens of silver sulfide decreases when they are heated, which is actually an example of semiconductor behavior. This discovery demonstrated that some materials behave differently, even contrary to the behavior of metallic substances such as copper, which develops an insulating oxide with increased temperature, thereby increasing resistance to the flow of electricity, decreasing conductivity. The identification of electrical properties of materials enabled the development of better sources for electrical power, as well as helped to identify the role of ions and the fundamental structure of materials. In the time leading up to the Industrial Revolution, the innovative use of different combinations of materials led to the development of batteries that are very similar to those in use today.

The concept of particles with complementary charges comprising chemical compounds (Berzelius) to contiguous particles communicating electricity (Ohm) and electrodynamic molecules providing the vehicle for electricity and magnetism (Ampére) was taken to the rest of the electrochemical circuit by German physicist Gustav Theodor Fechner. From 1834 to 1839, Fechner worked with Ohm's Law and electrochemical circuits to show that it could also be applied to electrolyte solutions. In his experiments, Fechner encountered resistance to current flow at the interface between the electrolyte and the electrode, identifying it as charge-transfer resistance (*Übergangswiederstand*) at the metal electrode-electrolyte interface, an important concept to consider when surface modification of the electrode is the purpose of the circuit, and of great importance when deliberately growing an oxide (anodizing), especially in terms of hydrogen overpotential in the case of concentration polarization [19, 24].

In 1826, Italian chemist Leopoldo Nobili obtained, through electrochemical oxidation in a voltaic cell, interference colors on metal substrates, called "*Metallocromia*." Nobili's work, achieved with platinum electrodes using a lead acetate electrolyte, produced variations in color on a platinum anode due to the deposition of lead oxide layer from the lead acetate electrolyte. The interference coloring effect occurred because the oxide layer was progressively thinner as the distance between the anode and the contact tip of the cathode increased. In discovering this new phenomenon, which he called "*electrochemical appearances*," Nobili devoted himself to study them not only for scientific reasons, but also for (sic) "*the advantages that this new technique of coloring metals may lend to the arts*" [25]. His treatise on the anodizing and coloring of platinum was published in 1830 at the height of experimentation with the voltaic cell.

Specific to electrochemical concepts that dealt with the metallurgical quality of the electrode in a galvanic circuit, Swiss physicist Auguste-Arthur de la Rive, in 1830, determined the impact of alloying elements and other impurities in metals by showing differences in the rate of corrosion on electrodes of the same base metal, but of different purities, in the same electrolyte (dilute sulfuric acid). De la Rive determined the accelerated corrosion observed on electrodes that were not pure was the result of an electrical reaction between the parent metal (zinc) and (sic) "*portions of copper, lead, cadmium, or other metals*" within the zinc. All of the impurities, being in contact with the zinc, form small but very active voltaic circles (cells) which cause great destruction of the zinc and evolution of hydrogen, apparently on the zinc, *but really upon the surface of these incidental metals* [26].

This result was very important at the time because many new elements were not pure as reduced; and specific to industrially produced zinc, was often contaminated by iron. By relating differences in chemical sensitivity to changes in the electrical condition of the electrode, and by showing that the changes are due to the effects of impurities (alloying elements), de la Rive showed the significance of chemical potential differences across the surfaces of electrodes that would otherwise be considered uniform. He also determined that the communication of electricity was hindered in proportion to the amount of alloy loading as well as on the progressive deterioration of the battery performance, thus showing that alloy complexity has a great effect on the speed and effects of the oxidation reaction. De la Rive was a key figure in the development of industrial scale electroplating, viewing it as an engineering means to protect a surface by preventing corrosion, not only a means to decorate, as it had been since ancient times.

2.5 Passive Behavior of Metals

Perhaps the most important milestone along the path to the anodizing process came before aluminum was discovered with the discovery of passive behavior in the late eighteenth century. The ensuing studies that proceeded to characterize and control it in the early nineteenth century still continue today. The determination that anodic polarization could develop passivity in many metals became one of the methods to characterize the electrochemical properties of existing industrial metals, such as iron, and newly discovered metallic elements, such as aluminum (after 1827). Soon after the discovery that aluminum displayed passive behavior with anodic polarization, useful properties of the oxide were determined, and the development path for anodizing bifurcated along science and engineering lines, with scientists striving to understand the mechanism for passive behavior, as well as the chemistry and structure of the oxide layer while engineers characterized the oxide, determined engineering applications for it, and developed a process to produce it on an industrial scale. Indeed, these two paths exist today in the anodizing industry, and although intimately related, they are often disconnected, as science and engineering regretfully often are. In this section, a review is presented of some of the scientific research undertaken to understand passivity throughout the nineteenth century, as part of the development of anodic oxidation from a means to characterize metallic elements into an important industry for finishing aluminum as the anodizing industry is presented, as well as some of the various past and current theories that center on how the unique structure of the anodic aluminum oxide nucleates and grows.

2.5.1 *James Keir*

In 1790 James Keir, a Scottish scientist, who rejected the theories of Antoine Lavoisier because they were not spiritually grounded, observed that iron, after treatment with concentrated nitric acid, had lost the capability to precipitate silver from a silver nitrate solution, and no longer reacted with dilute nitric acid. Keir also noted that when the surface of the treated iron was scratched, or brought into contact with an active metal, it again became active [27]. He is acknowledged as first reporting the phenomenon of what would be called "passive behavior" of metals. Keir did not call the behavior passive. It is important to note that there was no external electrical contact to actuate the reaction between the iron and the nitric acid; the behavior was initiated by way of the electromotive force stimulated by contact with an electrolyte.

2.5.2 *Gustav Fechner*

With the concurrent rapid discovery of new metals, experimental methods in electrochemistry and the acceptance of natural science, Keir's report fell into relative oblivion until 1826 when it was rediscovered by Wetzlar and republished by Gustav Fechner in 1828 [28]. Fechner and Sir John Herschel published the first results explaining passivity on the basis of electrical phenomena. Herschel claimed to have done his work with iron in 1825, but did not publish it until 1833 in the *Annales de chimie et de physique* (French for *Annals of Chemistry and Physics*). In his report, Herschel concludes: (sic) "*Once iron preparé resists nitric acid of the concentration in which it had been treated to be preparé, or even more dilute, which shows that these phenomena are due, not to a lack of solvent for the iron salt, but rather to a certain permanent electrical condition of the surface of the metal.*" Like Keir, Herschel did not call the observed resistant condition passive, using the French terms *preparé* and *indifferént*. Michael Faraday confirmed Herschel's work in July 1836 with a letter to the London and Edinburgh Philosophical Magazine and Journal of Science [29].

During this time, in 1827, Friedrich Wöhler successfully reduced the first aluminum metal. As part of his characterization studies, Wöhler determined and reported that heating aluminum in air reduced its shine and that it appeared darker in color; he found that aluminum does not oxidize in water, sulfuric acid, and nitric acid at room temperature, but that it oxidized in boiling water, and hot sulfuric acid. He found that it was attacked by dilute hydrochloric acid and sulfuric acid. In his extensive report, although the aluminum exhibited variations in oxidative behavior, he did not report the new metal as exhibiting passive behavior. This is because the terms "*passive*" and "*active*" as they pertain to the electrochemical reactivity of a metal were not classified as a characteristic until 1836. It took another 20 years to classify aluminum as passive, and another 20 years to understand how the anodic oxide could be utilized and applied in an engineering application.

2.5.3 Christian Friedrich Schönbein

Michael Faraday, as the premier expert in electrochemistry of his time, was often consulted by contemporaries for his opinion on their results obtained in chemical and electrochemical experiments. He hosted his own periodical and salon for electrochemical scientific discussion, called the Faraday Lectures, that are still presented today. Consequently, there is a great deal of communication in the form of letters and articles available for review today that provides great insight about the development of the new sciences.

The property of a metal to change its surface reactivity due to exposure to an acid electrolyte after anodic polarization was published in May of 1836 in The Philosophical Transactions of the Royal Society of London by editor Michael Faraday after German Scientist Christian Friedrich Schöbein had sent him a letter describing "*facts, …that seem to me not only to be new, but at the same time deserving the attention of chemical philosophers*." The letter was published under the title "*On a peculiar voltaic condition of Iron, by Professor Schœnbein, of Bâle*," (*fr.* Basel, Switzerland) [30]. In this letter, Schönbein calls this property, for the first time, "*passive*" and also introduced the term "*active*." Schönbein described the metal as passive if it was only slightly or not at all attacked either when acting as an anode in an electric circuit, or when treated with chemical reagents. It was described as active when, under the same conditions, it was vigorously attacked. These experiments are extremely significant because they are the first recorded electrochemical studies regarding passive behavior of metal.

In communication with the joint editor of the Philosophical Magazine, Faraday pointed out that Schönbein's experiments afforded an additional proof that "*voltaic electricity is due to chemical action, and not to contact*," and went on to explain that the passive behavior of iron in nitric acid, observed by Schönbein, was (sic) "*the surface of the metal is oxidized or else the superficial particles of the metal are in such relation to the oxygen of the electrolyte as to be equivalent to oxidation, meaning by that, not an actual oxidation but a relation*" [31]. In stating "*the cause of the phenomena is due to a relation of the superficial particles of the iron to oxygen*," Faraday meant that the film of oxide that formed on the surface of the metal during the anodic treatment had the property of resisting dissolution by nitric acid of a *particular strength*. It is noteworthy that Faraday is referring to the reaction between superficial particles that comprise the iron with oxygen, acknowledging perhaps, the fundamental atomic structure of the metal. The three tenets of Faraday's conclusion [32], which state that the formation of an oxide layer is the source for passive behavior of materials, are as follows:

1. All known passivity phenomena are oxidation processes.
2. Iron coated with oxide is insoluble in acids.
3. The passive condition readily disappears on polishing the metal.

Schönbein immediately disagreed with Faraday, citing his experiments that showed the passive iron remained passive through a variety of conditions, and although he couldn't explain the reason for it, he believed the development of the passive condition was more complex than the formation of an oxide layer alone. In fact, he believed that passivity was brought about by a change in the metal itself. In keeping with the European context of shared knowledge, and in order to get another opinion, Schönbein contacted Swedish chemist Jöns Jakob Berzelius, of the same circle of researchers who attempted unsuccessfully to extract aluminum metal.

In the exchange of letters between Schönbein and Berzelius [33, 34], Berzelius agrees with Schönbein, that Faraday had not suggested an acceptable theory for the passive behavior of iron in nitric acid, citing Schönbein's experiments show a "*change in electrical condition*" that was also observed by the French scientist, de la Rive. Auguste de la Rive had published that corrosion was an electrochemical phenomenon. Berzelius goes on to suggest the various chemical states of iron (valences), under which the oxidizable characteristics change, and suggests that these may be the source for passivation, suggesting as well that electrical contact with the anode may indeed produce chemical change. Berzelius goes on to cite Volta "*that electricity and chemical affinity are different manifestations of the same force*."

Schönbein could not admit to Berzelius' opinion, that the Contact Theory of Electrical Action was the source for passive behavior, because he believed as Faraday, in the Chemical Theory of Electromotive Force, in other words, that chemical change was the source of electrical current. Nevertheless, Berzelius encouraged Schönbein to continue his studies with passive behavior stating, "*it is conceivable that you may be able to discover a method for making iron in any quantity retain this condition for any length of time: such a discovery would be of the highest value*."

Insodoing, Berzelius foresaw the value of an industry based on corrosion protection by the controlled formation of a stable oxide layer, such as anodizing.

2.6 The Industrial Revolution

With the continued development of the science and technology of electricity and electrical processes, the means by which scientific experimentation and engineering measurement were achieved became steadily more sophisticated and understanding grew in direct proportion. Mathematic modeling of physical processes enabled new relationships between chemistry and physics, and atomic theory flourished as the foundation for all matter. Clearly, as there is today, a cyclic interdependent progress between scientific learning and related technology developed during the Industrial Revolution. Understanding how electricity flows through each component of a circuit became a source for new scientific research and engineering applications that were sometimes purely electrical in function, and the effects of "*electrical action*" were found to have benefits that have engineering applications on their own. This was the intention behind learning about and developing an understanding of how passive oxide layers form on metals that were developed in electrochemical characterization experiments, such as those performed by Heinrich Buff on aluminum. In review, as one considers each part of an electrochemical circuit, it is remarkable how much was discovered and explained throughout the rest of the nineteenth century that is germane to anodic oxidation and the anodizing process.

In 1855, the technology for aluminum extraction was maturing and ambitions for engineering possibilities for the new metal were rapidly developing. Henri Sainte-Claire Deville developed the double salt process of chemical extraction of aluminum, resulting in a rapid increase in production in France, and production of aluminum worldwide was at the highest rate ever, exceeding 200 tons over the following 30 years. As a new element, the scientific community was eager to establish the physical and chemical properties of aluminum in order to determine its practicality as an engineering material. Electrochemical characterization experiments that determined active and passive behavior of aluminum, extracted at the laboratory level, also became important for characterizing the properties of aluminum that was produced on the industrial level. In addition to changing mechanical properties, the effects of impurities and alloying elements were found to change passive behavior, as well as the condition of the passive layer. Passive behavior was determined to be a key characteristic for aluminum surfaces, and research focusing on the electrical properties of the oxide layer formed by way of electrochemical treatment enabled revolutionary engineering applications for the passive oxide, formed by way of anodic polarization.

2.6.1 *Advances in Physical Science*

2.6.1.1 Thermodynamics

From 1840 to 1843, English physicist James Prescott Joule, famous for his studies regarding the nature of heat and its relationship to mechanical work, did his first work in electrochemistry. Through experimentation, Joule determined a physical law that expressed the relationship between the heat generated and current flowing through a conductor with a complex experimental system that incorporated both voltaic and electrolysis cells. Using this system, Joule also compared the electrical energy of "*galvanic elements*" with the heat evolved by dissolving the metals in acids and in other aqueous salts. Heat measurements were accomplished with a calorimeter. The energy values developed by oxidation (dissolution) of the metal electrodes were used by Joule to theorize that electrical energy is equivalent to chemical energy. In this way, Joule's work brought the diverging electrochemical theories into a more complementary point of view, as well as supporting what would later become the First Law of Thermodynamics: *Energy in a system is constant that can be transformed from one form to another, but it can neither be created or destroyed* [35].

Understanding the energetic nature of physical and chemical reactions became especially critical in the design of industrial processes. Thermodynamics, the studies of electrolytes in solution, electrochemical kinetics, and the reactions that occurred during extractive processes, foundry and deformation processes were studied with the intent of meeting the energy requirements on a large scale. These studies were first rigorously developed as part of the Industrial Revolution from 1860 to 1880s. During this time, American chemist Josiah Willard Gibbs first began measuring and cataloging the thermodynamic heat of various reactions, in an accounting system that would be called Gibbs free energy [36]. The application of the principles of thermodynamics and diffusion as they pertain to anodizing aluminum will be developed in Chaps. 5–7.

Perhaps the most famous scientist of this time was Scottish physics genius, James Clerk Maxwell. Maxwell is considered as important to the development of modern physics in the nineteenth century as Sir Isaac Newton was in the eighteenth century, and Albert Einstein in the twentieth century. He is known for unifying Faraday's Laws of Electricity and Magnetism with Ampére's Law, describing elegantly, how electric and magnetic fields propagate, interact, and how they are influenced by objects in space. Maxwell was the first to determine that the speed of propagation of electromagnetic waves was the same as the speed of light, concluding that electromagnetic waves and visible light are identical [37]. The importance of electromagnetism to the background for electrochemical reactions is that propagation phenomena explain the nature of the Helmholtz's electrical double layer on the surface of an electrode and the behavior of the double layer when the electrode is polarized, concepts critical to the structural order of the anodic aluminum oxide.

2.6.1.2 Diffusion

Adolf Fick

In 1855, German mathematician, physicist, and medical doctor, 26-year-old Adolf Fick, while working as an anatomy demonstrator in Zürich, proposed his law for diffusion. He used Graham's research, stating his goal as "the development of a fundamental law, for the operation of diffusion in a single element of space." He asserted a deep analogy between diffusion and conduction of heat, similar to Fourier's Law for Heat Conduction (1822) or electricity, creating a logic and proof method that is similar to Ohm's Law for Electric Current (1827). Fick's First Law of Diffusion is: The molar flux due to diffusion is proportional to the concentration gradient [11, 38].

$$J = -D_i x \frac{dc_i}{dx} \tag{2.1}$$

In which J = molar flux = number of particles per unit area per second
D_i = diffusivity of i, which is a constant for dilute solutions of i in a specific medium
x = point in space
c_i = concentration of i

The foundation of Fick's First Law is the concept of diffusivity, or how one substance moves through another in a consistent way. Fick identified diffusivity as a constant "dependent upon the nature of the substance." If one considers flux to be the change in concentration, c_i; with respect to time, t; over a distance, x, Fick's Second Law can be calculated as the second derivative of his First Law.

$$\frac{\partial c_i}{\partial t} = D_i \frac{\partial^2 c_i}{\partial x^2} \tag{2.2}$$

Fick utilized his Second Law to model Graham's diffusion experiments by placing his own geometric boundary conditions on the equation. The first was to use concentrated salt water in a cylinder, placed into a bath of pure water; the second was to use a conical funnel. Both conditions produced a concentration gradient over which the salt concentration vs. depth (length of the cylinder) over time could be measured gravimetrically by way of the density of the solution in the cylinder. The concept of diffusivity, which defines the diffusion constant, D, was proven when Fick used it to measure the diffusivity of salt in water.

For the rest of the nineteenth century, Fick's Laws for diffusion were approached phenomenologically, that is, accepted in terms of how Fick, himself, conceived diffusion, although his concepts were based upon Graham's observations of diffusion in gases, Ohm's Law for the movement of contiguous particles in an electrical circuit, and Fourier's equations for heat transfer. The distance and time over which diffusion could be modeled were set by boundary conditions and mathematic solutions calculated by Josef Stefan and Franz Neumann in 1879. One of the most important boundary conditions, the so-called *Thin Film Solution*, that of a thin film of atoms, M, deposited at a fixed initial point ($x = 0$) and allowed to diffuse, is very important to tracer studies of a unique diffusing metal or even radioactive tracer in experiments for a variety of reasons. Tracer experiments are extremely useful for studying the mechanism of the development and growth of the highly ordered anodic aluminum oxide structure [39, 40].

$$c(x,t) = \frac{M}{\sqrt{\pi Dt}} \exp\left(-\frac{x^2}{4Dt}\right) \tag{2.3}$$

Sir William Chapman Roberts-Austen

An important researcher in solid-state diffusion, Sir William Roberts-Austen, was devoted to the subject having been the assistant to Graham at the Royal Mint. Austen performed a great many diffusion and other metallurgical studies that included the effects of impurities in precious metal alloys, most importantly the determination of the nonmagnetic phase of the iron-carbon alloy that bears his name, Austenite. Mr. Austen was the first chemist and assayer at the Royal Mint from 1869 to 1902. In 1896, he presented and published in the proceedings of the Royal Society, *the Bakerian Lecture on the Diffusion in Metals, Part I: Diffusion of Molten Metals, and Part II: Diffusion of Solid Metals* [41]. The paper is a detailed description of the experimental work in solid-state diffusion documenting the movement of precious metals in liquid lead and in liquid bismuth, hosts selected because of their low melting temperature. From precise measurements obtained by using the assayer's balance at the mint, Austen used Stefan's solutions to Fick's Second Law to calculate the diffusivities of gold, platinum, rhodium in liquid lead; gold, silver, and lead in liquid tin; and gold in liquid bismuth. Austen's measurements for the diffusion were carried out at a fixed temperature of 492 °C.

Austen also performed interesting studies to determine the diffusion of gold into solid lead by fusing thin plates of gold onto lead rods. The diffusion of gold into lead, which, because of the limited solubility of gold in lead is not easy to trace metallographically, and the corresponding calculation for the diffusivity, D, at the two temperatures the experiments were performed (251 and 200 °C), are found to be close to those determined with radioactive tracers. It is curious and noteworthy that even though Austen was by nature and employment a very precise man, he neglected to include any reference to the effects of temperature on the diffusion process, even though he carefully measured and maintained the temperatures at which the experiments were performed, and recorded the time over which diffusion was allowed to proceed. It can be that it wasn't reported because there is no

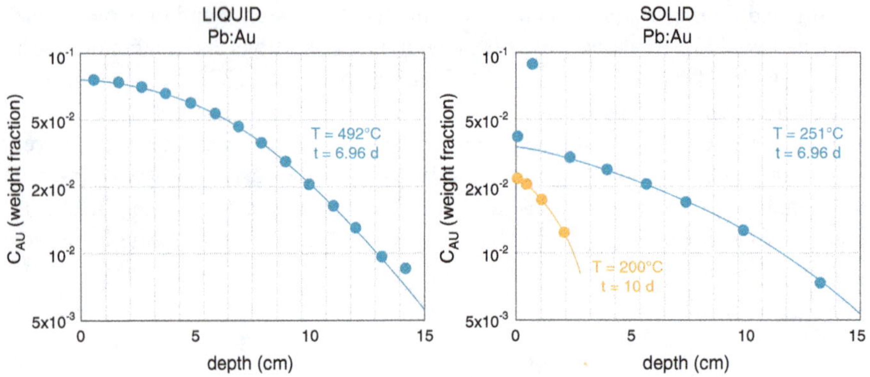

Fig. 2.3 Diffusion profiles of gold in liquid lead (left) and gold in solid lead (right) reconstructed by Mehrer et al. from Austen's tabulated data presented to the Royal Society in 1896 [39]

temperature variable in Fick's Laws. Figure 2.3 presents the diffusion profiles that were computer generated from Austen's data by Mehrer [39].

Svante Arrhenius

It is Swedish physical chemist Svante Arrhenius who put the important parameter of temperature into William Roberts-Austen's studies of the solid-state diffusion by developing the temperature dependent form of the rate coefficient for chemical reactions. The Arrhenius Equation is one of the greatest contributions to diffusion studies.

$$k = A\exp\left(-\frac{E_a}{RT}\right) \tag{2.4}$$

In this relationship, k is the rate coefficient for the kinetics of the reaction, R is the gas constant, and A is the pre-exponential factor. Two important thermodynamic factors, E_a, the activation energy and most importantly, T, as the temperature of the reaction are critical factors that are considerations for all sorts of metallurgical reactions from the solidification of metals and their alloys, to heat treatment effects on microstructure, to the temperature dependence of the rate of solid-state diffusion reactions. Danish ionist Jacobus Hendricus van't Hoff is also credited with publishing a similar relationship, proposed on the relationships between equilibrium constants and rate constants for the forward and reverse reactions of several chemical reactions [42].

2.7 Atomic Theory

In spite of the clear, actual physical relationships that could be drawn between Fick's phenomenological equations and actual diffusion, no one formally addressed the phenomenon of Brownian motion until the end of the nineteenth century. (Recall Brown's observation of the random motion of pollen in liquid was first published in 1827, the same year Wöhler successfully reduced aluminum.) In 1889, French chemist Georges Gouy reinvestigated Brownian motion by studying different kinds of particles in various types of fluids. In addition to confirming the phenomenon of Brownian motion, Gouy showed that particulate motion is independent of external forces and that it is constrained in more viscous liquids [43].

It is interesting to note that well into the early twentieth century, that while chemists embraced atomic theory, physicists avoided the nature of matter by embracing its mathematic representation, taking atomic theory for granted as a convenient means to explain the math, or rejecting it completely, because, although the concept of an atom as the basis for all matter made explanations convenient, the existence of an atom had never been actually physically proven. In the November 1904 issue of the periodical Electrochemical Industry, an editorial written by E.F. Roeber, Ph.D., "On the Reality of Atoms and Ions," was published in which Roeber stated that "*it is doubtful that (sic) atoms can be proclaimed as realities*," and "*The conceptions of atoms and ions have proved to be an enormously useful working hypothesis and this fact represents their merit in physical science. Whether atoms or ions really exist, is rather a question of metaphysics*" [44].

2.7.1 Albert Einstein, Marian Smoluchowski, and Jean Perrin

The erratic motions of particles without a clear trajectory, by definition, have no velocity. This is because there is no clear distance over which the speed of the particle can be determined, a critical aspect to calculating diffusivity. Without a frame of reference, for amorphous bodies or the crystal lattice for solid bodies, no explanation for Brownian motion was published until 1905, when Albert Einstein published in Leipzig, in "Der Annalen der Physik"; "*Die von der Molecular kinetischen Theorie der Wärme forderte Bewegung von in ruhenden Flüssigkeiten suspendierten Teilchen*" [45], "On the kinetics of molecular movement, with thermal stimulation, of particles suspended in static solutions." The frame of reference turned out to be the nature of the fluid, represented by its viscosity, η. Einstein's thermodynamic approach to Brownian motion, related the diffusivity of a particle with known radius, r, to the viscosity of the fluid in which it is suspended, η, by addressing the constraint between the particle and the liquid by including the Stokes frictional force, $6\pi\eta r$. The resulting equation is known as the Stokes–Einstein equation, in which R_g = the ideal gas constant, T = temperature, and N_A = Avogadro's number.

$$D = \frac{R_g T}{N_A} * \frac{1}{6\pi\eta r} \tag{2.5}$$

In relating the constraint between the particles to a resultant frictional force that causes motion, Einstein suggested that molecules of water were actually "pushing" the particles of pollen around. In order to address the velocity of particles that appear to randomly move while suspended in a liquid, Einstein derived an equation that relates the mean-square displacement of particles of radius r to the diffusion coefficient, *D*, and time, *t*. In which *R* is the total displacement of each particle in a suspension, during time *t*, which consists of many individual displacements that are equal to the radii of the individual particles.

$$\langle R^2 \rangle = 6Dt \tag{2.6}$$

This portion of the equation enables the relationship between the kinetics of the reaction with the thermodynamics and is called the Einstein–Smoluchowski relation for Polish physicist Marian von Smoluchowski who addressed the kinetics of Brownian motion by suggesting collisions between particles in a closed system, rather than the thermodynamics alone. Smoluchowski's paper was published in 1906, after Einstein's thermodynamic treatment, but his statistical approach to molecular physics with respect to Brownian motion originated in 1901 [46].

The idea that Brownian motion could be the result of inter-particle collisions was the first suggestion of the active role that atoms play in the structure and properties of all matter. However, this concept was necessary to prove in order to convince the entire physics community of the existence of actual atoms. Believing that energy, not matter, was the chief component of the universe, German physical chemist Friedrich Ostwald was opposed to atomic theory. According to Ostwald, atoms should be viewed as convenient placeholders in physical and chemical reactions, and definitely not reality. His opposition was so vehement that he forbade his colleagues at the university in Leipzig, where he was professor of chemistry, to teach atomic theory or even chemical or physical concepts in terms of atoms.

In the time before his conversion, Ostwald's anti-atomic theory philosophy proved particularly frustrating for Austrian physicist Ludwig Boltzmann, who was forbidden to teach his kinetic theory of gases in terms of atomic theory in 1900, because it was based on the statistical mechanics derived from atomic mass, charge, and structure. Boltzmann taught statistical mechanics from the perspective that the physical properties of matter: viscosity, thermal conductivity, and diffusion are determined by the properties of atoms. Together with Boltzmann, many notable chemists and physicists also disagreed with Ostwald: German colleague Hermann von Helmholtz, Scottish physicist James Clerk Maxwell, who suggested the concept of self-diffusion in gases and on whose equations Boltzmann founded his statistical treatment of gases; and American, J. Willard Gibbs, noted founder of thermodynamics, all accepted that atoms are the fundamental components of all matter [39].

Ostwald was finally convinced to accept the actual presence of atoms in 1909 after French physicist Jean Baptiste Perrin published his elegant proof of the Einstein–Smoluchowski theory for Brownian motion, which explained the incessant motion of small particles by fluctuations through by calculating the dependence of the mean-square displacement (motion or fluctuations) on the small particle (atomic) radius using carefully prepared monodispersed solutions. By observing and measuring the positions of particles at specific time intervals, without the confusion that develops by using solutions with dissimilar particles, Perrin showed agreement with the mean-square displacement by the Einstein–Smoluchowski relationship [47].

Solid-state physics was born in 1912 when Max Von Laue developed the technique of X-ray diffraction. However, the crystals analyzed by way of the developed by Laue were perfectly ordered and could not explain all solid-state diffusion. At that time, it was not known that atomic defects are necessary to bring about diffusion in crystals. Russian physicist Yakov Il'ich Frenkel first introduced the concept of disorder in crystals in 1926. Frenkel's defects were created by the displacement of atoms from their lattice positions, which produce a vacancy at the original lattice site as well as a corresponding interstitial defect, between lattice sites. The defect created does not change the chemical properties of the crystal, and only involves the migration of ions, preserving both volume and mass, and therefore density [48].

Frenkel's theories were refined shortly thereafter through the collaboration of German physical chemist Carl Wagner and German physicist Walter H. Schottky. Wagner is known as the Father of Solid-State Chemistry, best recognized for his work on oxidation rate theory; counterdiffusion of ions, defect chemistry, and their effects on thermodynamics, electrical conductivity, and diffusion. Schottky is best known for his theory of rectification behavior at the metal-semiconductor interface, which led to the development of the Schottky diode. This principle is key in the understanding of the reactions that occur at the aluminum-anodic oxide interface during anodizing by providing the explanation for conduction through the oxide structure that enables oxide growth.

A Schottky defect is defined in two ways: for a nonionic crystal, much like crystals defined by Bravais, this first definition simply defines the defect as a vacancy on the crystal lattice. The second definition explains defects in semiconductors and other ionic crystals, such as sodium chloride; in which oppositely charged ions leave their lattice sites, creating vacancies. The vacancies are formed stoichiometrically, maintaining the neutral charge of the solid [49]. See Fig. 2.4.

Together with Wagner and Heinrich Ulich, Schottky authored in 1929 an important textbook of thermodynamics that presents lessons in cyclic physical processes and chemical changes and chemical equilibrium, *Thermodynamic*. The book is dedicated to Berlin University professors Max Planck and Walther Nernst [50].

In the 1940s, American metallurgist Ernest Kirkendall showed that not all atoms diffuse at the same rate, by way of his experiments in solid-state diffusion utilizing diffusion couples comprised brass (Cu-Zn alloy) bars that were electroplated with pure copper [51]. Molybdenum wire markers were placed at the surface of the brass prior to electroplating. The diffusion couples were annealed and the markers

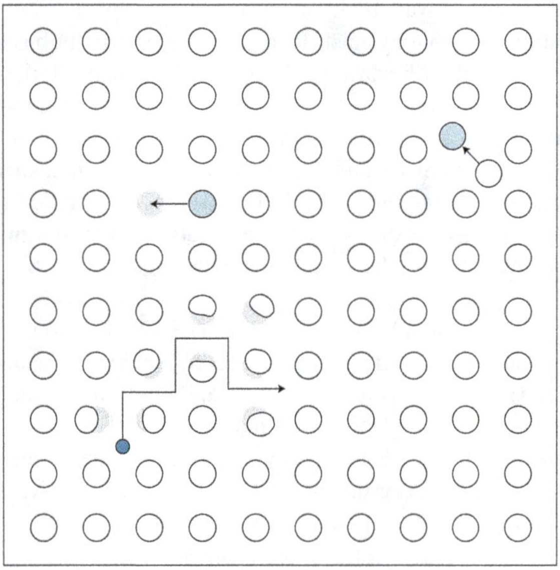

Fig. 2.4 Diagram for a crystal lattice with defects. The defect at the top left is representative of a Bravais-type vacancy. The defect at the top right is an interstitial atom of the same type of atom as the lattice, similar to those that occur on ionic crystals; the defect at the bottom left is a foreign interstitial atom that causes distortion on the lattice by moving through the structure

migrated inward as the zinc atoms migrated outward, faster than the copper atoms diffused inward, toward the brass. In some cases, with higher temperature annealing, zinc diffusion was so rapid that voids were left behind in the brass. The effect of the movement of the molybdenum wires is called the Kirkendall effect and the voids that form are called Kirkendall voids. These effects can be seen in other alloy systems, including aluminum-zinc alloys, in which zinc diffuses more rapidly than aluminum. See Fig. 2.5.

The basis for diffusion in solids is the movement of atoms, ions, or molecules on a lattice. The link between diffusivity and the mean-square displacement of particles with respect to the medium in which they diffuse, as proposed by the Einstein–Smoluchowski relationship, is therefore valid for crystals as well. The effects of constraint in diffusion transcend viscosity, as in a liquid, to any type of resistance, as with the atomic packing on a crystal lattice, in which the lattice restricts the positions and migration paths of atoms. Constraint, and therefore resistance, becomes more significant as alloy complexity increases. Consequently, if one considers the diffusion stimulated by an electrical current, the resistance to the flow of electrons can also be considered as affecting diffusion, because diffusion will be biased in the polarization direction and will increase with alloy complexity in the electrode, and with changing ionic concentration in the electrolyte.

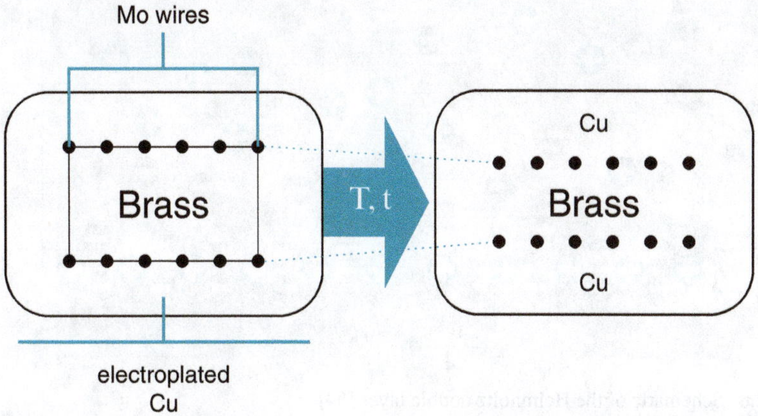

Fig. 2.5 Schematic for the Kirkendall effect. The inward migration of the molybdenum wire documented the rapid, outward solid-state diffusion of zinc toward the surface of the diffusion couple. Had the rate of diffusion between the copper and zinc been the same, the markers would not have moved

In this respect, the Stokes–Einstein equation can be used to refine the concept of diffusivity, and therefore can be used to explain the movement of particles that are not electrons in an electrochemical circuit.

2.8 Electrochemical Behavior of Materials

In 1853, Johann Wilhelm Hittorf discovered concentration charges that were produced at the electrodes during electrolysis, which lead to the determination of the first laws to explain ionic transport theory in electrolytic solutions. Hittorf derived a method for the measurement of the fraction of charge carried by different migrating ions in an electrolyte, arriving at the conclusion that the loss of concentration around any electrode is proportional to the speed of the ion moving away from it. He found that different types of ions move with different speeds under the same potential gradient. Consequently, cations and anions carry different amounts of electricity during electrolysis. The fraction of charge carried by different ions was characterized by a specific *transport number*, which have proven to be significant both technically and fundamentally when considering the conductivity of an electrolyte, but also in the characterization of electrode materials. The migration of oppositely charged ions from the anode to the cathode of an electrochemical circuit was important to the industry of electroplating, by understanding the ratio of different types of ions, and how their respective behavior in solution changes when other ions are present. In his publication "*On*

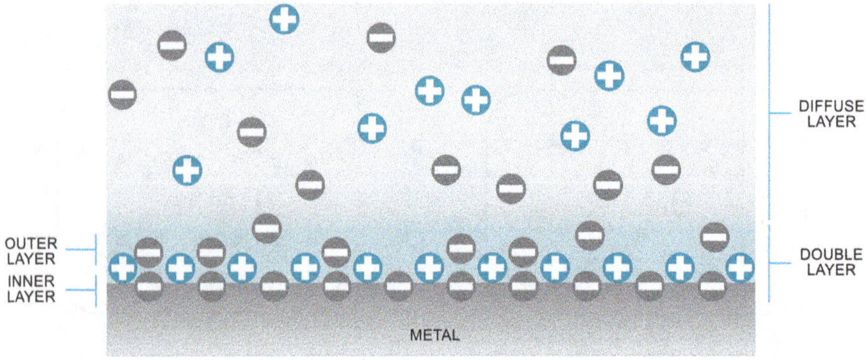

Fig. 2.6 Schematic of the Helmholtz double layer [54]

the Migration of Ions during Electrolysis," Hittorf discusses polarization as "*the first action of current,*" which consists of bringing the "*particles of the body*" to be decomposed into such a position that the cation of each molecule is turned to the cathode and the anion to the anode [52]. Hittorf's findings were important for understanding how changes in electrode materials affected the efficiency of an electrochemical process. Furthermore, Hittorf's findings are germane to anodizing aluminum: polarization of the aluminum electrode is critical to the order of the self-assembled anodic oxide.

More understanding of electrical contact and chemical relationships in electrochemical reactions were elucidated by Hermann von Helmholtz in 1853 with his realization that when an *electronic* conductor is brought in contact with a solid or liquid *ionic* conductor (electrolyte), a common boundary (interface) among the two phases appears; charged electrodes immersed in electrolytic solutions repel the co-ions of the charge while attracting counterions to their surfaces; in this way, layers of opposite polarity form at the interface between electrode and electrolyte. In 1853, Helmholtz showed that an electrical double layer is essentially a molecular dielectric and stores charge electrostatically. Below the electrolyte's decomposition voltage, the stored charge is linearly dependent on the applied voltage. This determination was critical to developing an understanding of interfacial reactions such as electroplating and oxidation (anodizing), as well as the development of the passive layer on aluminum. Together with physicists and chemists who supported atomic theory especially in terms of thermodynamic reactions, Helmholtz believed that the heat and work of chemical reactions could be explained in terms of the mass, thermal conductivity, and atomic structure of the reacting compounds. Helmholtz included all matter by claiming, "*mechanics, heat, light, electricity, and magnetism are manifestations of energy*" [53]. See Fig. 2.6.

2.8.1 First Report of Anodic Oxidation of Aluminum

In 1855, Sir Charles Wheatstone initiated the research for characterizing aluminum such that it could be incorporated into the Galvanic Series. Wheatstone was remeasuring and tabulating the characteristic potentials of some of the known elements; specifically silver, iron, and copper, in various solutions by taking measurements utilizing the famous Wheatstone bridge in order to compare them with what would be the newly determined characteristic potential for aluminum. Based upon his observations, Wheatstone concluded that aluminum could not be assigned a fixed place in the electromotive series because the condition of aluminum changed, depending upon the electrolyte used and with what metal it was combined in the circuit [62].

A contemporary of Wöhler and Deville, in 1857, German physicist Heinrich Buff took this work over from Wheatstone and issued a report of the experiments he performed to determine the electrical behavior of aluminum on samples obtained from Friedrich Wöhler, who had produced his aluminum wires by way of laboratory-scale reduction in cryolite, and industrially produced aluminum that was commercially obtained in Paris presumably by way of Deville's industrial process. By repeating Wheatstone's experiments, Buff declared that aluminum exhibits passive behavior, confirming work done in 1855 by Heeren in Hannover, Germany. Buff identified passive behavior as follows, (sic) "*a metal is passive when, under the effects of a solution in which it is immersed, it becomes coated with a soluble oxide layer; or, additionally, when polarized in an acid electrolyte, that even though there is no visible oxide layer, the metal resists further attack.*" Buff's experiments, a continuation of Wheatstone's, characterized silver, copper, iron, and additionally characterized the two aluminum samples from Paris and Wöhler. Zinc was used as the counter-electrode in the Wheatstone bridge; nitric acid and dilute sulfuric acid were used in various combinations as electrolytes. The results of his experiments are provided in Table 2.2.

Table 2.2 Comparison of observed (measured) electrical resistance, with silver having the highest conductivity

Wire Composition	Specific Gravity	Resistance of 1 m Wire Length 1 g weight	1 mm thickness	Same length and thickness
Silver Maximum conductivity	10.420	1.884	0.2303	100.00
Silver Chemically pure, soft, pliable	10.489	1.976	0.2399	104.16
Copper Chemically pure, very soft	8.933	1.717	0.2452	106.46
Copper High iron, very soft	8.729	2.104	0.3071	133.34
Copper Unchanged, hard, but pliable		2.143	0.3131	135.94
Aluminum From Cryolite, almost pure, soft and flexible	2.670	0.962	0.4598	199.63
Aluminum Industrially produced, brittle	2.664	1.019	0.4858	210.92
Iron Low elasticity, flexible wire	7.740	9.471	1.5587	676.71

From Buff, Über das electrische Verhalten des Aluminiums, Annalen der Chemie und Pharmacie, 1857

Most remarkably, Buff observed that when aluminum was polarized as the anode, that (sic) "*the metal became coated immediately with a dark gray coating that tended toward black in appearance, and, as coverage became more complete, the effects of the acid were disturbed*," he also wrote, "*as the metal coated itself, the coating was well adherent, and became thicker the longer it remained in the current path…it had the property of being a poor conductor*." When placed in an alkaline solution he noted, "*the aluminum becomes once again active, and the effects of the current are seen in the dissolution*." Based upon Buff's knowledge of the presence of silicon as a contaminant in both samples, and based upon the knowledge of the color of elemental silicon, he identified the coating as a silicon layer, not an oxide layer. However, his description of the layer and its characteristics are based upon observation only, and it can be concluded that what Buff saw was most likely an anodic oxide affected by the inherent coloring of alloying elements; therefore, Buff's paper, published in 1857, appears to be the first paper on the process of anodic oxidation of aluminum.

Buff's findings, published in *Annalen der Chemie und Pharmacie,* are also interesting regarding the electrical characteristics and mechanical performance of aluminum alloys. In his paper, he comments on the purity of his aluminum samples provided by his colleagues for his experiments. He had received aluminum bars from the Paris factory, which most likely ran Deville's double salt reduction process. These bars were drawn down to brittle wires. Buff comments that the purity of these samples was inferior to sheet and wire samples he had obtained from Wöhler, who had reduced the aluminum metal from cryolite by way of a laboratory, not industrial, process. Consequently, due to the scale of the process, but also by way of the method, the samples provided by Wöhler were remarkably purer than those received from Paris, with the only noted contaminant being silicon. Wöhler's samples weren't as brittle as the commercially prepared samples, which support the conclusion that the samples from Paris were contaminated with other elements, which Buff surmised was primarily iron. In this publication, Buff clearly reports that the technology to reduce aluminum from a process that uses cryolite as an electrolyte was possible in 1857, which must have been a process similar to the Hall–Héroult Process, albeit on an experimental basis. This is noteworthy as the paper is published 29 years before the Hall–Héroult process revolutionized the industrial production of aluminum, suggesting that although possible, energy sources sufficient to reliably run the aluminum reduction process with a cryolite electrolyte on a large-scale industrial basis were simply not available in 1857.

In 1866, German chemist Wilhelm von Beetz, by performing his own experiments, proved that the layer formed on the surface of the aluminum wires in Buff's experiments was not silicon, but an oxide of aluminum. Beetz ran three experiments: (1) he measured the changes in weight of the aluminum anode before and after polarizing the wire as an anode and running the current to produce the layer; (2) he measured the aluminum dissolved into the dilute sulfuric acid electrolyte; and (3) he collected the gas liberated at the anode in order to determine the oxygen content. These findings were published in "*Annalen der Physik und Chemie*" in 1877 [63].

2.9 The Ionists

From about 1872, the electrical behavior of solutions became a focal point for studies carried out by a number of scientists who came to be known as ionists. It is noteworthy that the scientists comprising this group were mainly organic chemists, yet studied and pioneered modern physical chemistry as well as furthered modern electrochemistry by way of their thermodynamic treatment of solutions based upon ionic dissociation in electrolytes [55]. In studying the structure and properties of reacting materials as well as the mechanisms by which reactions occur, physical chemistry finds use and application regardless the nature of the materials reacting. These studies incorporated thermodynamics, diffusion, and crystallography.

2.9.1 Wilhelm Kohlrausch

German physicist Friedrich Wilhelm Georg Kohlrausch, the founder of the ionists, observed and measured the dependence of electrolyte conductivity on dilution, noting that the electrolyte has a definite and constant amount of resistance that depends upon the type of ion and its concentration. In 1874, Kohlrausch published the details of his studies, as well as the *Law of Independent Migration of Ions* "at infinite dilution where the ionization of all electrolytes is complete, and all inter-ionic effects are absent, each ion migrates independently of its counterion and makes a definite contribution towards the total equivalent (molar) conductance of the electrolyte which depends only on the nature of the ion itself, and not on its counterion, with which it is associated in solution" [56]. Kohlrausch's most notable students at the Technical University of Hannover and various colleagues from 1870 to 1875 were Swede Svante August Arrhenius, and Germans Walther Hermann Nernst and Friedrich Wilhelm Ostwald. In 1902, Kohlrausch was the doctoral advisor for Adolf Güntherschulze, who became a key figure in developing the foundation for the theoretical treatment of anodic oxide formation.

2.9.2 Friedrich Ostwald

A colleague of Kohlrausch, Russian-German Friedrich Wilhelm Ostwald is one of the principle founders of physical chemistry and is known for his investigations of the fundamental principles governing chemical kinetics (rates of reactions) and equilibria. He is best known for his Electrochemical Dilution Law (1888), which enables the relationship between conductivity and electrolyte concentrations using the activities of the acid ions in solution. By applying the law of mass action to the dissociation of weak electrolytes, Ostwald showed that a dynamic equilibrium exists between the ions in solution and undissociated molecules. He is also the author of the concept of *Ostwald ripening*, the thermodynamically driven spontaneous process that occurs between particles of different sizes in order to reduce the overall energy of a solution: the larger particles grow at the expense of smaller particles [57]. In the year 1887, he founded the first journal of physical chemistry *Die Zeitschrift für physikalische Chemie*.

2.9.3 Svante Arrhenius

A student of both Ostwald and Kohlrausch, Swedish physical chemist Svante August Arrhenius' doctoral thesis was titled "*Investigations of the galvanic conductivity of electrolytes*." The novelty of his topic, keeping in mind Kohlrausch's Law of Independent Migration of Ions, was that some molecules could be split into ions when in aqueous solution, a significant concept to comprehend because it is the ionic composition in solution that governs the conductivity of the electrolyte. Arrhenius' *Theory of Electrolytic Dissociation* [58], for which he won the Nobel Prize in 1903, was the outcome of his investigations, and is based upon five basic concepts that follow:

1. The electrolytes in aqueous solution form charged particles called ions. The positively charged particles are called cations and the negatively charged particles are called anions.
2. A dynamic equilibrium is present between undissociated molecules and dissociated ions of the electrolyte in aqueous solution (Ostwald's Electrochemical Dilution Law).
3. Ions move freely under the influence of the applied electric field. Cations move toward the negative electrode (cathode) and anions move toward the positive electrode (anode).
4. The number of cations and anions in the solution of an electrolyte may be different. However, the total charge carried by cations is equal to the total charge carried by anions. The solution as a whole is electrically neutral.
5. The extent of dissociation is different for different electrolytes and depends on the nature of the ions, their concentration, and the temperature of the solution.

In closing his Nobel Lecture, Arrhenius stated, "*This theory has also shown us that atoms or groups of atoms, charged with electricity, play a highly significant part in the world of chemistry.*"

2.9.4 Walther Nernst

German Walther Hermann Nernst, a colleague of Kohlrausch, made fundamental contributions to electrochemistry and thermodynamics, complementing the ionist's solution theory with solid-state chemistry. Nernst was the first to treat the thermodynamic equilibrium for an electrode, assuming equilibrium between the electrode (metal) and the solution phases, showing that these relationships can be used to calculate the electrode potential from the composition of the contacting phases. Based upon Fick's First Law of Diffusion, Nernst considered the effects of electrostatic forces and concentration due to osmotic pressure and chemical potential of different particles in solution, producing a means to understand the kinetics of electrochemical reactions. The result was the most important equation developed for electrochemistry: the Nernst equation, published in 1888, relates the reduction potential of a chemical species to the standard electrode potential, the temperature of the reaction, and the activities (approximate concentrations) [59]. The basic form of the Nernst equation is given by:

$$E_{\mathrm{Cell}} = \frac{\Delta G}{nF} \tag{2.7}$$

In which ΔG is the Gibbs energy change of the cell reaction, n is the charge number of the electrochemical cell reaction, and F is the Faraday constant.

For an electrochemical oxidation reaction, the Nernst equation becomes more complex. The electrochemical potential for an oxidation-reduction chemical reaction of a metal, M, with activities a_{O} and a_{R} written as:

$$M_{\mathrm{O}} + ne^{-} \Leftrightarrow M_{\mathrm{R}} \tag{2.8}$$

can be expressed as:

$$E = E^{\mathrm{o}} + \frac{2.303RT}{nF}\log\frac{a_{\mathrm{o}}}{a_{\mathrm{R}}} \tag{2.9}$$

In which E^0 is the standard oxidation potential for the oxidation–reduction couple.

It is this form of the Nernst equation that Julius Tafel used to derive his famous Tafel equation in the early twentieth century, which can be used to determine the rates of corrosion reactions. The Tafel equation is an excellent model for anodic oxidation that enables the explanation and understanding of the effects of interfacial phenomena on the kinetics of the anodizing reaction, in terms of the anodizing process parameters.

2.9.5 *Julius Tafel and Tafel Behavior*

Swiss organic chemist turned electrochemist Julius Tafel, in earning the credit for his work in electrode processes, was given the title "Father of Modern Electrochemistry" [60]. He was active from 1885 to 1910, beginning his career as a student of Nobel Prize winner Emil Fischer, who pioneered work in carbohydrate chemistry; among some of his classmates and contemporaries were Wilhelm Kohlrausch, Wilhelm Röntgen, and Adolf Fick. From about 1893, Tafel turned from working in pure organic chemistry to physical chemistry, while working on reduction reactions with strychnine and amino compounds. In 1896, Tafel worked with Ostwald at the University of Leipzig, and began his work in electrolysis, moving into a period of electrochemical study that made an indelible mark on surface finishing science.

Through his investigations of electrode processes, Tafel proposed the mechanism of catalytic hydrogen atom combination with the electrode as the rate-determining step in electrolytic hydrogen evolution, which came to be called *concentration polarization* [61]. In working with circuits to develop organometallic reactions, Tafel noted that lead, cadmium, copper, and silver electrodes exhibited active and passive "states," depending upon the polarization history of the anode and the process parameters of the reaction: the access of the electrolyte to the electrode (design of the cell), concentration of the electrolyte, agitation, and time. By comparing different types of electrodes, Tafel determined that metals exhibit different characteristic polarizations and that some metals attained maximum polarization more slowly, at a given current density.

Tafel stressed the overriding importance of current density in electrochemical reactions because he found that electrode surface reactions are rate controlled, unlike thermodynamic reactions that are equilibrium controlled. His work showed that irreversible chemical reactions could be studied systematically, and that catalysis takes place with respect to the type of electrolyte. Tafel identified differences in the cathodic overpotential for different metals, by the rate of hydrogen gas generated at the surface of the electrode. Metal electrodes with the smallest overpotential accelerate most strongly the rate of hydrogen evolution; for example, at high current densities, there is a stiffing effect developed by the evolution of hydrogen gas at the electrode. Through the learnings from these studies, Tafel determined that overpotential phenomena reside in the surface layer on the electrode, which is related to the surface polarization.

The Tafel equation represents the total polarization of the electrochemical cell, which is a sum of two types of polarization: surface polarization of the electrode and concentration polarization, and reflects the hydrogen gas evolution at the surface of the electrode that is necessary to complete the electrochemical circuit.

$$N = \eta_{\text{surface}} + \eta_{\text{concentration}} \tag{2.10}$$

$$N = \beta_{\text{a}} \log\left(\frac{i_{\text{a}}}{i_0}\right) + \frac{2.3 * RT}{nF} \log\left(1 - \frac{i_{\text{a}}}{i_{\text{L}}}\right) \tag{2.11}$$

In this equation, the applied current density for the circuit is represented by i_{a}; i_{o} is the exchange current density, the thermodynamic quantity for the current density as it crosses from the electrode into the electrolyte, it is controlled by the material comprising the electrode. The limiting current density, i_{L}, represents the rate with which hydrogen escapes from the electrode to complete the circuit. β_a is a constant that is electrode material dependent; usually about 0.1 V and T is the heat of reaction.

The Tafel equation is an excellent model for all surface reactions that enables the understanding of the effects of the electrode on the kinetics of the electrochemical reaction. The Tafel equation regards the reactions that occur at the surface of the electrode as well as the reaction that occur in the electrolyte to complete the corrosion circuit. The function of the Tafel equation as a model for anodizing aluminum is introduced in Chap. 5 and fully developed in Chaps. 6 and 7.

2.10 Continued Research in Passive Behavior

Research regarding the passive behavior for all known elements continued the debate about the Contact Theory and the Chemical Theory for chemical reactions through the end of the Industrial Revolution and well into the early twentieth century. Faraday's theory for passivity remained the accepted concept for passive behavior from the time of his communication with Schönbein, in 1836, until 1898 when Johann Wilhelm Hittorf determined, through characterization studies, that chromium, too, exhibited passive behavior and produced additional understanding of passive behavior in terms of valency, claiming that, as Schönbein had concluded, passivity is the result of an actual change in the metal. Hittorf's conclusion bolstered the electron theory of materials. Later, German chemist Hans Finkelstein supported Hittorf's theory with the claim that passive behavior is linked to the "modifications," or differently charged ions of a metal that has different valences. He assumed that all ionic states for the metal were present in the electrode in certain proportion, depending upon the temperature of the reaction, and affect passive behavior depending upon the concentration or the proportion of the various valences at the surface during the reaction [64].

Early characterization studies showed conclusively that aluminum exhibited passive behavior, and that although insulating, the passive layer exhibited "valve" behavior, allowing current to flow in only one direction, away from the anodically polarized aluminum electrode (as found by Eugene Ducretet in 1875) [65]. Aluminum was found to have no allotropes, and inasmuch as Hittorf determined that valency has an impact on passive behavior, American chemical engineer and chemistry professor William Roy Mott determined the passive behavior exhibited by aluminum is different from the passive behavior of iron, chromium, and silicon, all of which have allotropes [66]. This finding also supported Schönbein's conclusion that passive behavior was much more than the formation of an oxide layer and was related to a change in the metal itself. However, while these learnings developed an interesting additional explanation for passivity, the new explanation supplemented Faraday's theory that passive behavior was solely due to the formation of an oxide layer, rather than replacing it.

By the end of the nineteenth century, there were three main theories for passive behavior. The first was Faraday's oxide theory, and the second was Hittorf's valency theory. The third was from German electrochemist, Max Julius LeBlanc and was called the Reaction Velocity Theory [34]. LeBlanc's theory of passive behavior was not well received. It was based upon observations that led him to theorize that metals in the passive condition send their ions of the lowest valency only very slowly into solution. He believed that the ionization of the metal is associated with *chemical change* and when these changes are slow, passivity phenomena occur. In this way, the nature of the chemical change governs the rate at which any electrical change, such as the passage of current due to passive layer breakdown, would occur.

Recalling Ritter's Chemical Theory of Electromotive Force, many hypotheses for the possible various chemical reactions that could be occurring at the surface electrode to elicit passivity were proposed, some of the various hypotheses that complemented LeBlanc's Reaction Velocity Theory were important to the characterization of oxide films. For example, Finkelstein, in questioning the presence of a coating on a passive electrode, claimed it must conduct electricity "*like a metal*"; German-Polish chemist Fritz Haber discovered that many oxides are porous and that the mechanical hindrance (resistance) to the escape of metal ions from the electrode increased as the oxide thickness increased. German physical chemist Karl Fredenhagen determined that the electrolyte acid concentration had a bearing on the formation of a passive layer. Electrolytic reactions were explained as oxygen "*deposition*" at the anode surface and hydrogen "*deposition*" at the cathode surface. When optical measurements showed no change in the surface, but the electrode exhibited passive behavior, the theory that passivity is a change in the condition of the metal, not the growth of an oxide was supported. All of the hypotheses showed that Faraday's Oxide Theory was incomplete, but had credibility, and set the stage for continued research into the nature of passive behavior and oxide structure, much of which focused on aluminum [64].

The relationships between chemical reactions, how electric current flows in an electrochemical circuit, and the consequent changes to the electrode were the critical point for understanding and challenge of the studies on passive behavior. The role of an electron, a rather new charged particle in the scope of atomic theory, on the structure of the materials in the chemical reaction was not fully understood in the early twentieth century, as Perrin's diffusion experiments that convinced the western scientific world about the existence of atoms weren't published until 1909. In 1916, Bennett and Burnham published a cautionary definition for passivity in their paper "*On the Passivity of Metals*" in the Transactions of the American Electrochemical Society. They wrote: passivity is "*a condition of a metal when it is exposed to oxidizing agents, whereby its surface becomes more noble and its solubility is decreased*" [29].

2.11 Birth of the Anodizing Industry

During the period of time immediately following the development of the Hall–Héroult process, the science and technology of aluminum bloomed. Studies continued in order to determine the nature and characteristics of the passive film, and as the unique quality of aluminum to build a robust anodic oxide was discovered to have interesting properties, the first engineering applications for anodized aluminum were developed. The first applications for the anodic oxide on aluminum were as a part of electrical apparatus that used the anodic oxide as a capacitor and diode in electrical circuits. Research to understand the nature and cause for the formation of the anodic oxide followed.

World War I was the first proving ground for aluminum and its alloys in mass production, and most importantly, for the rapidly growing aircraft industry. The aluminum industry grew with the aircraft industry, and applications for the anodic oxide as a corrosion barrier and a base for paint grew simultaneously. Many patents for anodizing and coating processes were published in the 1920s for the technology developed during the war. Research to determine the mechanism and structure of the anodic oxide occurred simultaneously, and many of the names on patents for engineering applications are on research papers published in professional journals, such as the Electrochemical Society and other academic proceedings.

2.11.1 Early Applications for Anodic Aluminum Oxide

2.11.1.1 Eugene Ducretet

French scientific instrument manufacturer, Eugene Ducretet is credited for first determining that the anodic oxide, formed by electrochemical dissolution, has engineering value. In 1875, he wrote "*Note Concerning a Liquid Rheotome with Constant Direction Based on a Novel Property of Aluminium*" in which he reports the invention of an electrochemical cell in which he placed aluminum as the positive electrode in a cell with dilute acid in water as an electrolyte to grow an "isolating coating." The coating allowed for the passage of current in one direction, but resisted the electric current flowing from the other direction. The effect produced by the coating was produced instantaneously, constant and long lasting, "*no matter how numerous the inversion of the battery's current via the rheotome*." He is credited with assigning the term "*valve metal*" to this behavior. Ducretet intended to use this property of aluminum to construct liquid circuit breakers. Limitations with Ducretet's invention were found when the coating delaminated, at a low dielectric strength of about 20 V. Nevertheless, Ducretet was excited about the anodic oxide. He wrote: "*This novel property of aluminum when electrified positively is capable of being used in ways (applications) that practice will be able to sanction.*" Ducretet

Fig. 2.7 Pollak's Rectifier, there are four condensators in series. Such a rectifier provided current in one direction of a pulsating value as each condensator charged and discharged when the AC current was applied

discusses applications for many novel devices, including telegraphic apparatus and a teletypewriter [65].

2.11.1.2 Charles Pollak

No specific electrolyte, other than one that is slightly acidic, is described for Ducretet's process; however, in 1896, German inventor Charles Pollak patented an electric liquid capacitor with aluminum electrodes in which he cites Ducretet's work. He describes development of anodizing processes with different types of electrolytes, both neutral and alkaline, which contain salts of aluminum that yield oxides with higher dielectric strength. These may be the first published statements that suggest changing the properties of the anodic oxide by changing the electrolyte, apparently building dense, barrier-layer type anodic oxides [70]. For a DC condensator, only one of the electrodes needed to be aluminum, the other could be any arbitrarily selected conductor, so long as the anode was always aluminum. For an AC condensator, both electrodes had to be aluminum, so that as the current alternated, the oxide would build on both electrodes, charging and discharging with operation of the condensator.

By 1901, Pollak was using condensators in series to make an electrolytic rectifier. A quite in-depth paper on subject was published in 1902 by American electrical engineers C. F. Burgess and Carl Hambuechen, in the first Transactions of the American Electrochemical Society. "*The Electrolytic Rectifier*" was presented as a means to transform an alternating into a direct current, to alleviate "disadvantages attendant upon existing commercial means from alternating current lines." The main disadvantage mentioned was uncontrolled sparking that occurred with the commonly used rectifiers. It is noteworthy that the first applications for the anodic oxide would be electrical since the discovery of the passive behavior of aluminum was the result of characterization in electric circuits [67]. See Fig. 2.7.

2.11.1.3 Early Patents

As research continued investigating the properties of the anodic oxide on aluminum from the perspective of passive behavior, industrial research proceeded to find commercial applications for anodized aluminum. The earliest patent found, that is not related to electrical applications (Charles Pollak updated his 1896 invention of the condensator in 1901), was granted in 1913 in the United States to Mershon for anodized cookware and other heat transfer applications such as cast radiators [68].

One of the most famous patents for anodizing aluminum was published in 1923 by noted British researchers in passive behavior, Guy Dunstan Bengough and John McArthur Stuart. The patent deals with the process for chromic acid anodizing, rather specifically for German patented alloy duralumin (most like aluminum-copper alloy 2017) and other commercially available alloys. It is noteworthy that these alloys were utilized in aircraft flown during World War I. The application is for corrosion protection, especially for increased resistance to pitting in seawater, implying new marine applications for anodized aluminum. The process for anodizing is presented in detail, citing the electrolyte concentration, times, and formation voltage. It is interesting to note that no seal process is included in the patent; however, it is mentioned that the finish should be painted or treated with grease or lanoline to enhance the corrosion resistance [69]. It is noteworthy that their studies in the development of surface treatments for the corrosion protection of aluminum, Bengough and Stuart were a part of their greater work in long-term corrosion protection. Also in 1923, Bengough determined that the differential aeration currents set up by oxygen concentration as pO_2 are only one of many factors determining the potential existing at different points on a metallic surface.

In 1925, Fulton Flick assigned his invention for the process of anodizing aluminum and aluminum alloys to Alcoa. His process utilizes electrolytes of ammonia, ammonium sulfide, and ammonium polysulfide to produce dense, adherent, and absorbant coatings to protect against corrosion and to provide electrical characteristics of insulation and dielectric applications. This patent is the first that mentions the application of the anodic oxide in wear applications, in order to prevent the seizing of bearings, and the first that mentions dyeing of the finish to produce a color. It also mentions alloy effects on the color of the oxide, "*copper produces a greenish tint.*" With Flick's process, one could anodize and dye simultaneously or anodize first and dye in a separate post-process operation. It is interesting to note again, that no seal is mentioned as part of the anodizing process in Flick's patent [70]. Later in that same year, Aladar Pacz assigned his invention for an immersion coating to produce an "*adherent, hard, non-metallic permanent coating*" on aluminum, the nature of which may vary as an oxide or hydroxide, capable of being painted, varnished, enameled, or even electroplated. This process was achieved in a hot ammonia electrolyte [71].

In 1927, British researcher Charles High Roberts Gower patented an anodizing process for cast and wrought aluminum and aluminum alloy products that employ a sulfuric acid electrolyte. This patent details much about the dyeing of the anodic oxide, including specific dye chemistry and the colors that can be produced, mentioning repeated process exposure to various solutions, with the anode polarized to obtain specific colors. Like Flick, Gower includes the effects of alloy chemistry on the color of the oxide. Post-treatment for the oxide is either a layer of paint or the surface can be oiled to optimize corrosion resistance. Typical applications cited in this patent are for components that require "*resistance to seawater and similar liquids*" [72]. As of 1927, no mention of hydrothermal sealing is made.

Possibly, the first handbook for the surface treatment of aluminum and its alloys was published in Germany in 1935. In *Das Handbuch der Anorganische Chemie: Oberflächenbehandlung des Aluminiums und seiner Legierungen*, (The Handbook of Inorganic Chemistry: The Surface Treatment of Aluminum and Its Alloys) the practice of the electrochemical treatment of aluminum, both electrolytic oxidation and electroplating, is presented in a methodical, step-by-step manner [73]. Not only are sources across all countries in Western Europe and the United States cited, but also, for the first time in western literature, sources from Japan are cited that go back to 1924. In this highly detailed resource book, comments are inserted about the challenges, success, and failure of different processes. For example, regarding the Bengough and Stuart process for chromic acid anodizing, it is mentioned: chromic acid anodizing of alloys containing more than 5% copper cannot be achieved because the withstand voltage of the oxide is exceeded at 30 V. Notably, among the many cited articles, there are two by Simon Wernick going back to 1931. Mr. Wernick is one of the authors of the famous two-volume series *The Surface Treatment of Aluminium and its Alloys*, by Wernick, Pinner, and Sheasby [74].

In considering all known processes for anodizing (the section of the handbook on electroplating was not perused), each step of the process, including pre- and post-treatment, and each process parameter is presented and discussed with the most up-to-date, as of 1935, considerations. Anodizing processes are categorized by electrolyte: chromic acid, sulfuric acid, oxalic acid, alkaline electrolytes, and neutral electrolytes; various additives are considered (especially at high acid concentration); tables are presented with the parameters of electrolyte concentration, bath temperature, formation voltage, current density, and their corresponding oxide properties. Sealing by way of hydrothermal methods, including by steam, is presented, citing Setoh and Miyata from a 1930 publication. Engineering properties of the oxide layer are discussed in terms of the anodizing process, as well as the steps needed to achieve them that were published in the current literature of the time. For example, the properties of density, hardness, corrosion and abrasion resistance, adhesive strength, to name a few, are mentioned. External influences on anodic oxide quality such as moisture, pressure, and temperature; as well as the influence of the base material, process parameters, and specific test methods are presented in a systematic way, and fully cited to enable reference. An example of some of the valuable details regarding the anodizing process provided in the Handbook is provided in Table 2.3.

Table 2.3 Process parameters for sulfuric acid anodizing with corresponding oxide characteristics

H_2SO_4 concentration (weight %)	Solution temperature (°C)	Voltage (V)	Current density (A/dm²)	Characteristic of the oxide	
5–12	20–35°	20–30	–	High and consistent hardness, even gray color	1
10–20 where appropriate with additions	Usual temperature	4–6	–	Corrosion and high-wear resistance	2
15–35	Usual temperature	6–20	–	High-wear resistance, highly absorbent for colors and oils, etc. on aluminum with natural color; on duralumin and on alloys containing Sn and Si, gray color	3
35–60 Addition of glycerin	20–30°	2–9	0.3–1	Heat resistant, highly absorbent for colors, oils, waxes, half transparent, off-gray color; on Si containing alloys almost black, on Cu containing alloys, yellow to white	4
60–77 Most appropriate 65%, addition of 1 vol. glycerin in 15 vol. solution	20–30° Most appropriate 25°	Up to 12	0.6–1	Colorless or whitish; especially suitable for coloring; electrical isolation, heat resistant, high-wear resistance, high corrosion resistance, increases by sealing with oils and lacquers	5
90–100	80–90°	Starting at 90 increasing up to 500	0.2–6	Very dense and corrosion resistant; after polishing has a porcelain appearance. A glass sheet at the end of the oxidation is applied to the anodized sheet and connects firmly with it	6

Literature: (1) M. Tosterud. übertr. an Aluminum Colors Incorporated (A.P. 1900472 [1931]) (2) Aluminum Colors Incorporated (D.P. 600387 [1928]). (3) Ch H. R. Gower, übertr. an Aluminum Colors Incorporated (A.P. 1869058 [1921]). (4) Aluminum Colors Incorporated übertr. von B. Ritzinger Haueisen (E.P. 403560 [1933]). (5) H. Bengston. übertr. an Aluminum Colors Incorporated (A.P. 1869041 [1920]. 1869042 [1930]), Aluminum Colors Incorporated (E.P. 378521 [1931]; F.P. 718144 [1931]). (6) Comp. Nationale de Matières Colorantes & Manufactures de Products Chimiques du Nord Rèunis, Estblissments Kuhlmann (F.P. 783166 [1934])

2.12 Foundations of Theory for Mechanism and Structure of Anodic Oxide

With the turn of the twentieth century, academic, defense, and industrial research brought about technological advances that encompassed each part of the electrochemical circuit used for characterization of all types of materials, extraction processes, and various other electrochemical cells. Research focused on understanding chemical reactions in various types of electrolytes as well as in the electrode (metal) itself. Laboratory batteries became more sophisticated, evolving from the original voltaic pile to gigantic dynamos capable of producing enough energy to run industrial level processes, which culminated in the establishment of the commercial electrical energy companies. Physics research boomed as importance was assigned to understanding how electricity flowed through different materials, together with the changes that occurred in materials due to the flow of electricity, resulting in the development of new analytical instruments and application-oriented products that actually ran on electricity. The new tools increased the ability to observe and measure, which enabled more in-depth analysis of various phenomena, including the structure of the anodic oxide on aluminum and its alloys.

2.12.1 William Roy Mott

In 1904, in the periodical "*Electrochemical Industry*," published by the Electrochemical Publishing Company of New York, American chemical engineer William Roy Mott authored four important and practical articles detailing his studies regarding the reactions that occur on aluminum anodes, the application of anodized aluminum as a means to mitigate aluminum corrosion, as well as a comprehensive list and technical history of the scientists and technicians who have worked on aluminum electrodes since the discovery of aluminum by Wöhler in 1827. In his first paper from this periodical, published in April of 1904, "*The Corrosion of Aluminum and its Prevention*," Mott discusses that a film forms on an aluminum anode in certain solutions, for example, sodium acid phosphate. The film greatly reduces corrosion of the aluminum in acid solutions and doesn't really inhibit corrosion when in a caustic solution. Measurements taken by Mott showed that the protective action of the film is directly proportional to the voltage at which the film is formed. Engineering applications for the film formed on aluminum anodes were recommended as follows: for use in contact with food, such as jars and cups; for personal use, on combs and buttons; for furniture handles and electric lines that are exposed to sea coasts. Mr. Mott's paper is the first in which a new application for the anodic oxide, that as a corrosion barrier, is suggested [66].

Mott's other papers published in July, September, and November of 1904 refer to the application of anodic aluminum oxide in electrical applications and relate to the studies of Burgess and Hambuechen from 1902 that were published in the first Transactions of the American Electrochemical Society [75]. All of the research presented by Burgess and Hambuechen was proceeding at the University of Wisconsin, where Mr. Burgess was an assistant professor in electrochemistry and Mr. Mott was a professor of chemistry (Mr. Hambuechen was a student of Mr. Burgess). These papers are fascinating in that they further characterize the nature of the anodic oxide in terms of its thickness, its composition, and how the oxide grows, by way of substrate consumption. In "*Dimensions of Films on Aluminum Anodes*," published in July 1904 [76], Mott carries out this study because anode film thickness is important in electrolytic rectifier and condenser applications due to the fact that thickness of the film constitutes the essential capacity of the condenser. In this study, Mott determines that the gain in thickness due to film formation is not the sum of the thickness of the films on both sides of the electrode, but only a fraction of this—inasmuch as some of the aluminum goes into the formation of the film; concluding, "*the formation of the film is accompanied by more or less corrosion of the anode*." He also concludes that the film is not a true oxide, but depending on the electrolyte varies between $Al(OH)_3$ and $AlPO_{4-}$ (recall that Mott's electrolyte was sodium acid phosphate). At the conclusion of this article, Mott recommends further study of anodes from other metals to learn the "*real*" mechanical and electrical forces that act on molecules and ions at the moment of electrolytic separation.

Through the review of Mott's articles, it is clear that the amount of literature research was extremely comprehensive. He mentions several scientists from Western Europe, Russia, and the United States important to the development of the understanding of the electrochemical behavior of aluminum when polarized as the anode, as well as the nature of the anodic film as an oxide, as well as brief commentaries on selected work, what the particular opinion was to the constitutive makeup of the anodic film and why other work may have been significant. In reporting that his anodic films sometimes exhibited interference coloring, Mr. Mott comments on Russian chemist Slouginoff, who made hard, dense films in sulfate solution that could withstand abrasion with emery paper, who believed that the anodic film was an oxide, a hydrate, or other aluminum compound [77].

In his article from November 1904, "*Colloidal Precipitation upon Aluminum Anodes*" [78], Mott considers the composition of the film as well as the formation of the anodic film in terms of colloidal precipitation of aluminum hydroxide upon the aluminum anode from various electrolyte solutions of chromates, phosphates, tartrates, citrates, etc. According to Mott, the hydroxide reacts with acid radicals from the electrolyte to produce dense, insulating films on the aluminum anode. Electrolytes containing acid radicals of high valence give high critical voltages and films of great specific resistance and high dielectric strength; therefore, thinner films are produced on the anode. The films described are those that are utilized for their valve properties, mostly on condensators and in rectifiers that comprise condensators with aluminum electrodes; however, Mott's first paper in the series on the theme of the reactions that occur at aluminum anodes, from April 1904, cites another application for the film, that of corrosion protection. The editor, E. F. Roeber, Ph.D., after summarizing the article at the beginning of the issue states: "*Mr. Mott's paper appears to give for the first time an explanation of the complex phenomena in the anode film, and since this explanation is also able to predict what to expect from new electrolytes, with respect to the properties of the film formed on the aluminum anode, Mr. Mott's paper deserves the greatest interest of all interested in this subject which is of equal importance scientifically and practically*" [79].

2.12.2 Adolf Güntherschulze

German experimental physicist Günther Adolf Eugen August Güntherschulze, who completed his doctoral degree under Wilhelm Kohlrausch in 1902, over the period of time from about 1903 to 1914 worked as a research scientist at the German National Laboratory, the Physikalisch-Technische Reichsanstalt (PTR) in Berlin. From the beginning of his time at PTR, Güntherschulze ran a variety of experiments with a simple electrochemical cell, characterizing a variety of metal anodes, in a variety of electrolytes in order to develop better electrical rectifiers and capacitors for applications in radios and transmitters. In 1906, he published his treatise, "*Über das Verhalten von Aluminiumanoden*" (On the Behavior of Aluminum Anodes) in Annalen der Physik [80]. In 1907, he published studies on tantalum, and in 1909, on niobium. All of the papers followed the same analysis and reporting in order to answer the question: how does the specific metal perform in the specific circuit, under various electrical and electrolyte conditions?

Over the course of the work performed, Güntherschulze refined his theory of films formed on metal anodes and classified them. In his last paper, "*Die elektrolytische Ventilwirkung des Niobs und eine Klassifizierung des Verhaltens electrolytischer Anoden*" (The electrolytic Valve-effect of Niobium and the Classification of the behavior of the electrolytic anode) published in Die Annalen der Physik in 1909 [81], his theory that provide a description for both a passive film and an anodic oxide was fully developed. This description and explanation for anodic oxide formation, with only minor modifications, became the basis for most theories of anodic oxidation throughout the twentieth century, and his work is still considered important by most familiar with anodizing processes today. Güntherschulze's work continued into the 1930s, which included studies of ionic transport in oxide structures on a variety of metal electrodes, including radioactive materials. His contributions to understanding the oxide on aluminum electrodes from an application standpoint involved further structural characterization with various electrolytes are cited in "*Das Handbuch der Anorganische Chemie: Oberflächenbehandlung des Aluminiums und seiner Legierungen*" [73].

Definition of a passive layer, according to Güntherschulze: a passive layer is conductive and protects the surface of the electrode from further attack by the counterion (anion) in the electrolyte. The layer is consistent and adherent. Because the counterion doesn't diffuse through the passive layer, it remains very thin, at a molecular thickness. The film isn't visible, and the metal appears "noble." Examples of some of the metals that form passive layers are iron, nickel, and chromium, in oxygenated salt solutions.

Definition of an anodic oxide, according to Güntherschulze: the anodic oxide is not conductive. If the voltage of the electrochemical circuit is sufficient, anions from the electrolyte mechanically penetrate the forming layer, producing very fine pores that are "tube-like." The tube-like structure enables continued growth, so long as the current is applied. The current is pushed through the tube-like vias in the structure, producing the valve effect, caused by differences in resistance experienced by anions and cations in the fine pores. An extremely thin layer is formed at the base of the solid porous layer, which maintains the valve effect as growth continues. Gas is developed in the pores as the layer builds. There is a fixed ratio that is constant for this type of film, which relates the layer thickness and the electrolyte. This ratio enables relationships that govern the speed of oxide formation, the sensitivity to current interruption, the maximum voltage the oxide can withstand, and "other characteristics."

In the 1916 Transactions of the American Electrochemical Society, Bennett and Burnham published "*The Passive State of Metals*," in which the history of passive theory as well as current research is summarized. In this paper, the authors describe the passive oxide on aluminum; the theory presented is clearly derived from Güntherschulze and cites his work. The characteristics of the "passive film" on aluminum are described (sic) "*the film is unquestionably a non-conducting one, hence it is probably not continuous. A film of gas serves to give the aluminum electrode 'capacity.' Aluminum is passive in air, and remains so in solutions, which do not attack the oxide covering. It is supposed that this film consists of a very thin non-porous layer of oxide, covered by a porous one, with gas occluded in the pores of the outer film. The outer film may slowly dissolve, while the inner one becomes porous and is replaced by a new one formed by the oxygen liberated, breaking through the original film*" [28].

Güntherschulze's explanation for the mechanism by which the anodic oxide formed became the basis from which other researchers studied and developed their own theories. Clearly, this description and theory for anodic oxide formation on aluminum as a dual structure of a thin, nonporous layer that later came to be called the "*barrier layer,*" with a porous outer layer, persisted through the twentieth century, and is similar to some of the theories proposed today. The next sections present some of the most important theories and their authors.

2.12.3 Setoh and Miyata

In 1932, S. Setoh and A. Miyata published, "*Researches on the Anodic Film of Aluminium. II. Anodic Behaviours of Aluminium in Aqueous Solutions of Oxalic Acid*" [82]. Setoh's oxide, which is developed in oxalic acid electrolytes under strict measurement DC conditions by way of constant current and constant voltage processes consistently comprises two layers, one visible and one invisible, the so-called active layer. The layers are described as having a close mutual relation between them. The "active layer" comprises mainly Al_2O_3 and is present on the surface of the anode. It is very thin and is more or less porous. The source of the pores is "uniform destruction" of the active layer all over the surface by the attack of the anions from the electrolyte. Setoh is specific that destruction of the layer is NOT due to dielectric breakdown at weak points in the oxide. Oxygen gas in the pores "violently" combines with exposed aluminum at the interface, at the places where the active layer is perforated by anion attack, which is the mechanism by which the aluminum continuously forms a new active layer.

Setoh's paper is in full support of the theory and dual-structure model of Güntherschulze and rejects the colloidal precipitation theory proposed by Mott from 1904. Setoh comments about the various theories of anodic oxidation on aluminum, stating, "*there is much obscurity about the formation mechanism of anodic film and the so-called active layer, and no conclusive theory that is convincible to everybody has (not) been obtained.*"

2.12.4 Burgers, Claasen, and Zernike

In this 1932 paper, supported by Philip's "Gloeilampenfabrieken" (Light bulb or Lamp factory) in Einhoven, the Netherlands, "*Über die chemische Natur der Oxydschichten, welche sich bei anodischer Polarisation auf den Metallen Aluminium, Zirkon, Titan and Tantal bilden*" (On the chemical nature of Oxide Layers, which develop by way of anodic Polarization on Aluminum, Zirconium, Titanium and Tantalum), a comprehensive study was carried out by Burgers to determine the chemical and crystallographic nature of the compact or barrier-type anodic oxide formed on electrodes from each metal [83]. Changes to the process parameters, specifically voltage and temperature were investigated to determine, if any, structural development in the oxide layer. In this study, a borax–boric acid electrolyte was employed, and hydrated alumina as Al_2O_3 was formed on the aluminum anode. The structure was found, via X-ray diffraction, to be cubic γ-Al_2O_3. It was found that the crystallinity depended upon the formation voltage and electrolyte concentration and determined that oxide "films" produced at high current density (80 mA/cm^2) up to 100 V in 4% oxalic acid were amorphous, but with heat treatment, they became crystalline cubic γ-Al_2O_3.

Van Geel and Schelen repeated Burgers' study at Philips in 1957. Anodic aluminum oxide was used to preserve the surface of high purity aluminum reflectors used in lamps manufactured by Philips. The paper, "*Some Properties of Oxide Layers Produced on Aluminium by Electrolytic Oxidation*," describes a similar procedure in which the various properties of density, thickness, and dielectric constant of the oxide are measured; analysis to determine the crystallographic nature of the oxide was also performed [84]. The results of this study showed that the oxides produced with a borax–boric acid electrolyte were amorphous, but heat treating (oxidizing) at 100 °C, produced a mixed amorphous-crystalline γ'-Al_2O_3 structure. By sealing the amorphous structures in boiling water, through the production of boehmite, some crystalline γ'-Al_2O_3 was also produced.

2.12.5 Schumann

Winfried Otto Schumann was full professor for theoretical electrical engineering at the Electrophysical Laboratory at the Technical Hochschule of Munich from 1924 until 1961. His area of research was the electrical breakdown stress of gases with an emphasis on discharge phenomena in highly ionized gases (plasma) and on the coincident wave propagation within the gas [85]. During the 1930s, in parallel with the Güntherschulze's work on rectification and the properties of anodic polarization and oxide growth on valve metals, Schumann advised at least two students who published important work on the mechanism and properties of anodic oxides on aluminum. The students, Theodor Rummel and Werner Baumann, published their work on the reactions that occur on the surface of aluminum electrodes, with respect to their application as electrodes for capacitors and condensers within rectifiers; however, theories for the mechanisms for two types of anodic oxides were studied in detail and published as part of their engineering research, as a report or *Mitteilung* from the Electrophysical Laboratory. The papers are rich in technical detail, are the first to provide metallographic analysis of the anodic oxide, and include a wealth of information on processes for anodic oxidation that involve a variety of techniques that are unique even by today's standards: plasma electrolytic oxidation; pulse anodizing, and AC anodizing.

2.12.6 Rummel

In 1936, Theodor Rummel published "*Über Wachstum und Aufbau elektrolytisch erzeugter Aluminiumoxydschichten*" (On the growth and development of electrolytic Aluminum Oxide Layers), in which he discusses the tuning of the dielectric properties of the anodic aluminum oxide, depending upon the amount of bound water, as hydroxide in the oxide layer [86]. Two types of layers are discussed: the thin oxide type grown for the express purpose to operate as part of a rectifier; and thicker oxides grown in acid electrolytes such as oxalic, chromic, and sulfuric acids; Rummel's favorite electrolyte was reported as a mixture of oxalic and succinic acids.

According to Rummel, the mechanism for the formation of a thin oxide layer proceeds with the application of the formation voltage, and, as a thin, compact "barrier" layer forms with low dielectric strength, and pores form in the oxide by way of voltage breakthrough. New oxide forms underneath the porous layer where breakthrough has occurred and the cycle repeats itself until the resistance of the building layer exceeds the formation voltage and the rest of the voltage isn't strong enough to continue to break through; at this time, the growth is finished. This theory assumes that transport of current happens in the pores by way of gas discharge through the oxygen in the pores that forms due to the electrolytic reaction. At the end of the process, with total insulation, a glow effect may occur at the oxide interface with the anode, and the heat causes hot corrosion of the anode. This process is similar to plasma electrolytic oxidation, which is used today to produce compact oxides for a variety of applications, from electrical components to parts that require a high level of impact and wear resistance, such as weaving shuttles.

Rummel claimed that the growth of the thicker oxide layers is initially similar to that of the thin layer. But when the electrolyte dissolves the forming oxide layer, the oxide becomes hydroxide, and the semiconductor properties of the hydroxide pre-

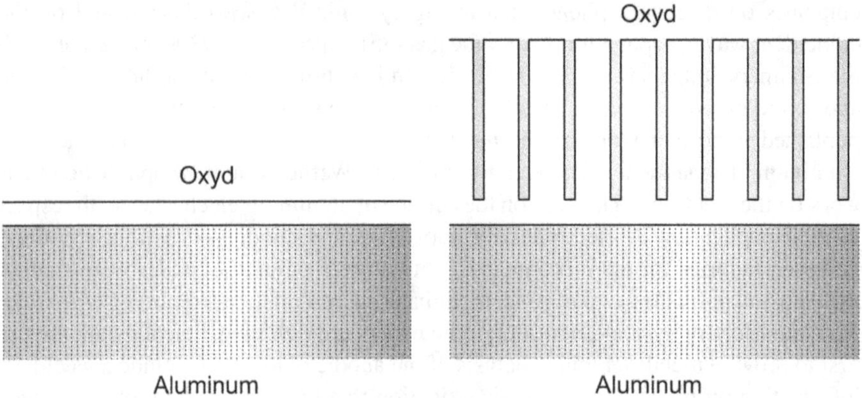

Fig. 2.8 Schematics presented as figures 28 (left) and 29 in Rummel's paper. It was possibly the first schematic representation of the porous oxide layer [86]

vent the barrier layer from forming. Electrolyte penetrates the pores and, rather than a gas buildup that requires discharge to enable continued oxide growth, oxide growth continues across the initial oxide layer by way of the semiconductor action of the hydroxide. The measured resistance of the thicker layers was much lower than that of the thin barrier-type oxide layer. Figure 2.8 shows the schematics of the two types of oxide layers investigated by Rummel. They are interesting representations based upon his laboratory observations during oxide growth and from metallographic examination.

In characterizing the anodic oxides, Rummel experimented with masks and dyes, showing conclusively that the oxide forms at the interface with the anode and consumes the aluminum as it grows. He showed the arrangement, population, and size of the pores in the oxide depended upon the formation voltage, and grew perpendicular to the anode surface, in the direction of the electric field. In his investigations, Rummel determined that the barrier layer exhibited holes due to imperfections in the anode and studied anodizing around a corner. In investigating the phenomenon that the pore would fill with gas as the electrolytic reaction proceeded, Rummel determined that pulse anodizing helps to overcome these effects; that by introducing a pulse, the gas would become displaced by electrolyte penetration in the pores.

2.12.7 Baumann

In 1938, the second important paper from W.O. Schumann's group at the Technical University of Munich was published regarding the nature and mechanism for the formation of anodic aluminum oxide. In his paper "*Entstehung und Struktur elektrolytisch erzeugter Aluminiumoxydschichten*" (The Origins and Structure of electrolytically produced Aluminum Oxide Layers) author, Werner Baumann, takes a new approach to the development of the three types of oxide layer, the barrier type, the thick type, and the reflective type, all which helps to elucidate phenomena that were previously not investigated [87].

Baumann is the only early researcher who approached the oxidation reaction at the atomic level, explaining how the molecular structure of the electrolyte interacts with the structure of the aluminum atomic lattice to form a continuous thin oxide that varies in resistance effects across the surface because of lattice defects in the anode. The oxide that grows is proportional to the amount of current passed. At the oxide-electrolyte interface, a gas layer develops in which the anion radical can react with the outer surface of the thin oxide layer. When this forms, the oxide layer is simultaneously under high pressure, high electric field strength, and consequently high temperature, the condition is especially interesting for the transport of electricity, following Faraday's First Law of Electrolysis. When the anode is completely covered by oxide, voltage breakdown occurs, which causes the resistance to decrease across the oxide, but more oxide forms at the points of breakdown as hemispherical particles, which become the "foot" of the pore. The continuous growth of hemispherical particles through the oxidation-breakdown process, finds a balance with respect to each mechanism, this balance establishes the order of the oxide structure, as well as the location and size of the pores.

The size and distribution of the pores depends upon the anodizing voltage and stand in relationship with one another and by that determine their dimensions and actual number. The electrolyte penetrates the pore space and the reaction occurs by virtue of ionic gas that develops at the base of each pore, which keeps the oxide growing into the anode. It is the first time ever that the anodic oxide is characterized as a semiconductor. Baumann proposes that all anodic oxides are developed basically by way of the same process, but that electrolyte type governs the characteristics that differentiate the three basic types. The paper contains many interesting optical photomicrographs that show the columnar structure of the AAO. Rummel cited Baumann in his 1936 paper; the two were most likely colleagues, as researchers that worked for W. O. Schumann, or classmates who were advised by him. Professor Schumann's research in dielectric breakdown and how electricity travels through liquid and gaseous electrolytes lead him to determine the formulae for the global electromagnetic resonances, generated and excited by lightning discharges in the cavity formed by the earth's surface and the ionosphere, which are called today the Schumann resonances. These resonances have definite ballistic significance for defense purposes and were calculated while Schumann was working in the United States; however, the Schumann resonances also have meteorological significance.

2.12.8 Edwards and Keller

Junius Edwards and Fred Keller of Alcoa discuss the structure and mechanism for anodizing and sealing, in order to explain alumilite [95] process in their paper, "*Formation of Anodic Coatings on Aluminum*," which was presented in the Transactions of the Electrochemical Society (Pittsburgh section) from 1940 [88]. The authors describe two types of anodic oxide, barrier type and porous, with the structure of the barrier type, formed in boric acid and borax, being as presented in 1932 by Burgers et al.; that is, crystalline gamma-type alumina. Porous oxides are described as soluble in the electrolyte, especially in dilute sulfuric, oxalic, and chromic acid electrolytes. And these oxides "grow in thickness as long as a suitable electric potential is applied." The pore, according the authors, is the vehicle that enables current flow since the electrolyte must penetrate to the metal surface in order to establish continued growth. The oxide, therefore, according Edwards and Keller, is an insulator during the growth process.

Edwards and Keller cite Rummel and Baumann in their presentation of metallographic images of the anodic oxide that confirm the columnar nature of the oxide structure, claiming that the solvent action of the electrolyte increases the pore volume. This publication was the first in which the ratio of aluminum consumption with respect to changes in the dimensions of the anode due to oxide growth is presented as 70% substrate consumption and 30% oxide thickness exceeding the original surface of the anode. Additionally, Edwards and Keller stated that alloy effects have a definite effect on oxide growth as they increase dissolution by the electrolyte, and, as more dissolution occurs, more oxide growth proceeds, in "electrochemical equivalent to the amount of current flowing."

2.12.9 Wernick

In the 1947 book, *Electrochemical Polishing and Bright Plating of Metals*, in Chap. 5, "Electrolytic Polishing of Aluminium" [89], Simon Wernick presents a rather comprehensive summary for the anodizing process at the time that includes his theory for anodic oxidation. In this chapter, Wernick actually refers to the industrial process of electrochemical oxidation of aluminum as *ANODIZING*. As an introduction, Wernick comments that that the anodic treatment of aluminum has progressed, as a means to protect the aluminum surface electrolytically by anodic oxidation such that its "*proneness to corrosion under humid, especially marine conditions, has been fully corrected by the effective provision of an artificially produced oxide layer on the aluminum anode surface*." As the book deals with the theme of electrochemical polishing, which is an anodic treatment, Wernick also points out that "*anodizing has become synonymous with the surface treatment of aluminum for its protection*," and that anodic oxidation has "*trespassed*" into the field of electrolytic polishing.

Much like Setoh and Miyata, Wernick recognizes the complexities of the anodizing process, but claims that insight is gained to the mechanism of anodic oxidation by considering his theory, thus introducing a concept for the formation of the anodic aluminum oxide, which describes anodic oxidation as the formation of a stable dispersion of colloidal particles (peptization) that form and impress themselves on the anode due to the dissolution of the anode by the electrolyte forming small hydroxide particles. Wernick's basic concept is similar to William Roy Mott's fundamental theory for the formation of barrier layer oxides from 1904, but he embellishes the nature of the establishment of the oxide layer with the stabilization of the particles by way of a peptization process that forms a "*peptized gel*" that is like a close net-like sponge over the whole anode surface. As a consequence of its formation, electro-osmosis occurs through the sponge, resulting in the expulsion of the electrolyte; and with alternating current, the charge on the gel varies with the periodicity, and the electrolyte is correspondingly alternately adsorbed and ejected from the structure.

In discussing the physical structure of the oxide, in which Wernick cites the work of Edwards and Keller, the oxide film is described as consisting of two layers, the upper layer that has no structure, and the lower layer that consists of alternating light and dark bands normal to the base metal. Near the oxide-metal interface, Wernick acknowledges the presence of a darker band along which there is a "delicate tracery," this being the part of the oxide most recently formed. According to the data presented by Edwards and Keller, Wernick identifies the dark bands in the lower layer as pores, which connect to the base metal, and as a result, enable film growth; the oxide layer is composed of "*walls of oxide surrounding the pores*." The oxide structure near the surface of the film is altered by the solvent action of the electrolyte; partial solution of the walls of the oxide "tubes" (a term used by Edwards and Keller) produces a sponge-like structure at the top. In this publication, Wernick doesn't offer a source for the nature of the pore; however, in the handbook that would become *the* reference book for anodizing aluminum, *The Surface Treatment of Aluminum and its Alloys*, first published in 1954, Wernick claims that electric field-assisted dissolution is the nature of the pores in the oxide layer [74].

In true handbook style, Wernick proceeds by discussing the factors affecting the film properties and how they can be controlled by the anodizing process parameters, even discussing the effects of additives, such as glycerin, to the electrolyte. By adjusting the process parameters, Wernick claims the anodizer can produce differences in the oxide film that affect: appearance, compactness, hardness, corrosion resistance, electrical insulation, and adsorptive qualities; such that from one type of an electrolyte it is possible to obtain a "*glass-hard, compact, brittle but non-absorptive oxide film*," and from another, a "*medium-hard, relatively spongy, quite ductile, and highly absorptive film*." He goes on to point out that additives tend to produce the latter type of oxide; a lustrous film is, on the contrary produced without additives. With the correct process parameters of electrolyte composition (a sulfuric acid electrolyte with a concentration of 10%), temperature of the electrolyte (that can be set by external cooling and/or air agitation, to about 75–80 °F), anodizing time to produce the "*desired effect*" (usually 5–10 min), and current density (for a sulfuric acid electrolyte) of about 8–10 A/ft^2, a "*closely grained, glass-hard layer*" can be produced. Sealing is also addressed as part of the anodizing process to increase the corrosion resistance and bring about a more or less "*stainless*" aluminum surface, either in boiling water or in a boiling solution of potassium dichromate, for a time of 30 min.

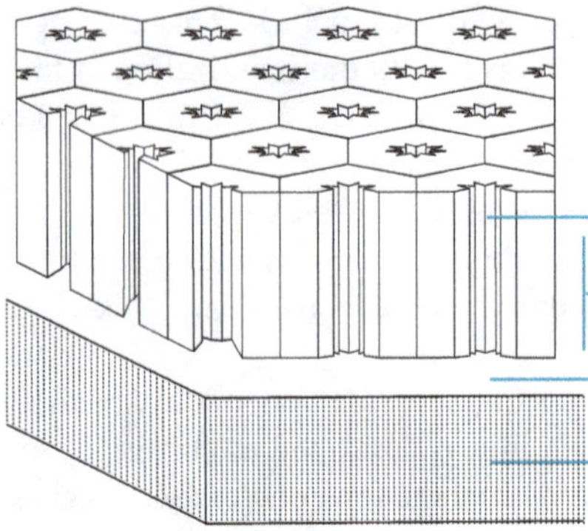

Fig. 2.9 Based upon the description of the anodic oxide provided by Edwards and Keller in their 1940 paper, this schematic of the structure of the aluminum anodic oxide layer is useful in showing an illustration of the structure developed according to the observations carried out by early researchers and is current with pre-1953 research. This schematic was obsolete at the time of its publication in 1976 by J.A. von Fraunhofer in his book *Basic Metal Finishing* [90]. Research by Keller, Hunter, and Robinson in 1953 and O'Sullivan and Wood in 1970 had already superseded the idea that the pores must contact the aluminum base material in order to facilitate oxide growth [91, 92]

2.12.10 Summary of Early 20th Century Anodizing Research

With the review of Wernick's summary of the anodizing process, it is quite clear that by the end of World War II, the Western World, and most likely Japan and Russia as well, had achieved a rather stable and mature industrial anodizing process. Early research and investigation into the mechanism for the formation of the anodic aluminum oxide was carried out primarily by measuring the electrical response to anodizing process parameters during oxidation, observing and characterizing the resultant oxide layer by way of various electrical tests (dielectric breakdown, changes in interfacial resistance) and microscopic examination. With the discovery of X-ray diffraction (as performed by Burgers et al.) and the development of metallographic analysis techniques (performed by Rummel; Baumann; and Edwards and Keller), the atomic structure, hardness, and microstructure were elucidated. Three basic theories were developed during this time.

Early theories for the formation of the anodic aluminum oxide are summarized as follows:

1. Colloidal Gel Theory: Aluminum hydroxide colloids precipitate from the electrolyte as it reacts with the electrode. The colloids are uniformly dispersed over the electrode surface by way of peptization and consolidate to form a compact solid layer with high dielectric strength. These layers are typically referred to as *barrier layer oxides*. Some researchers also believe that the colloidal gel can develop into the porous columnar structure. *W.R. Mott; Wernick*
2. Two-Layer Structure Theory: Nonporous barrier layer forms on electrode; no specific mechanism is proposed. As the oxide layer grows, pores form by anions

from electrolyte "mechanically penetrating" the barrier layer. As the current is applied, it pushes through the pores in the direction of the current bias, producing a "valve effect." The pores, which connect to the base metal, enable film growth; the oxide layer is composed of "*walls of oxide surrounding the pores.*" The oxide structure near the surface of the film is altered by the solvent action of the electrolyte; partial solution of the walls of the oxide "tubes" produces a sponge-like structure at the top. See Fig. 2.9. *Güntherschulze; Setoh and Miyata; Edwards and Keller*

3. Dielectric Breakdown Theory: Mechanism is the same for barrier layer oxide structures and two layer oxide structures; differences are produced by different electrolytes. Barrier layer oxide structures nucleate on the electrode and grow, merging together; forming a rather continuous layer. When the layer is consolidated, the layer randomly breaks down where its strength is lowest, causing more oxide to form. A competition between breakdown and conduction on the oxide structure causes pores to develop in the oxide structure. Oxides produced in electrolytes that dissolve the oxide tend to be porous. The oxide formed in this reaction is γ' Al_2O_3. *Rummel; Baumann; Burgers, et al.*

In the early theories that provided the foundations for the current theories for anodic oxidation, there is no proposed strict mechanism for the order of the oxide; however, beginning with Baumann, it is proposed that the size and distribution of the pores depends upon the anodizing voltage, and that the pores develop in relationship to one another by way of a continuous current flow with respect to the voltage, which determines their dimensions and actual number. Of all the early researchers, including Keller, Hunter, and Robinson, and O'Sullivan and Wood, who developed possibly the most influential theories for anodic oxide formation in the mid-twentieth century, Baumann is the only one who approaches the oxidation reaction at the atomic level and proposes that it is influenced by defects in the electrode microstructure.

2.13 Mid-century Research on Mechanism and Structure of Anodic Oxide

In the second half of the twentieth century, with the development of new analytical tools and methods, the ability to observe the anodic aluminum oxide reached levels of detail that brought new insight to the mechanism for its growth and development. Actually being able to *see* the oxide and to determine and measure its composition when processed in different electrolytes with different electrical parameters also brought additional understanding about and explanations for its engineering performance.

The effects of the anode microstructure on the formation of the electrolytic oxide were not explored; rather, understanding the mechanism of porous oxide development and growth was a primary research goal. While the three basic theories provided the foundation for continued research, new applications in aircraft and the new and fascinating aerospace industry drove the need for a more complete theory, grounded in basic physical sciences, as well as industrial processes that yielded consistently high quality anodic oxides. New analytical methods that yielded more detailed information about the oxide structure were carried out that utilized radioactive isotopes as markers while tracing the diffusion aluminum and oxygen during the anodizing process. In reviewing the great deal of interesting scientific research published on aluminum anodizing during the 1950s and 1960s, one can only conclude that scientific research during the mid-twentieth century must have been very stimulating and satisfying.

2.13.1 Keller, Hunter, and Robinson

Based upon their research performed on polystyrene impressions and replicas of AAO layers in the newly developed scanning electron microscope (SEM), in 1953, Fred Keller of Alcoa (of authors Edwards and Keller, from the 1940 publication cited in the previous section), together with M.S. Hunter and D.L. Robinson, wrote the first of many decisive papers from various authors about the mechanism for the growth and development of the anodic aluminum oxide, "*Structural Features of Oxide Coatings on Aluminum*" [91]. This paper is seminal regarding a detailed description of how the oxide forms, with corresponding SEM images to support the conclusions, and draws conclusions with respect to how the oxide can be engineered to yield specific dimensions by changing the electrolyte and forming voltage, relating dimensions to characteristics and behavior of the oxide with respect to engineering applications.

Important characteristics for barrier-type oxides are presented initially, such as the quality of the coating is a function of the substrate alloy, with current leakage occurring at intermetallic particles in the oxide, decreasing the ability of the finish to function as a diode. The barrier coating thickness is characterized as a function of the applied voltage. What is far more interesting and carried a much bigger impact are the images from the SEM, which were formed by way of replication. These images enabled the conclusions about the development and growth of porous oxide. See Fig. 2.10.

The theory begins with a simple statement that "*the barrier layer forms and solvent action of the electrolyte develops pores.*" Initially, current and voltage excursions are observed, but when they level out, oxide growth proceeds. With the solvent action of the electrolyte, thickness decreases at a single point, and current flows to these thinned areas to repair damage, with a corresponding increase in temperature and continued dissolution, the aluminum is consumed and the oxide forms a cell. Attack is described as random, not ordered, and since voltage and current fields are spherical in nature, each advancing cell front into the aluminum is likewise spherical.

Based upon calculations for the breakdown voltage of anodic aluminum oxide, the "*barrier thickness*" was determined to grow at a rate of less than 14 Angstroms

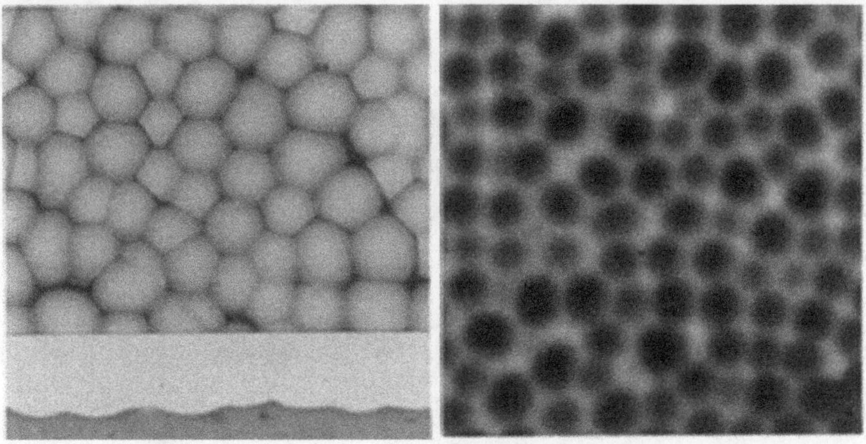

Fig. 2.10 SEM images of an anodic oxide grown in a phosphoric acid electrolyte at 120 V. Images are reported at ×35,000. The image on the left is a replica of the base of an anodic oxide, prepared by way of aluminum surface replication with "formvar replicas" after the oxide had been stripped away. This revealed in rather accurate detail, the scalloped bases for the columns comprising the oxide. The top surface of the oxide (right) was replicated using a mercury amalgamation technique that at the time was unique. Both processes disclosed much structural information about the anodic aluminum oxide, but also distorted some of the features. It was not until coating methods were developed that enabled surface conductivity within the SEM that more precise imaging was achieved

per volt. As the oxide forms, it has the shape of a cylindrical cell, having a roughly hemispherical end and a central cylindrical pore. According to Keller et al., the cells form singularly, not as a network, such that pillars of metal remain between the oxide columns and eventually oxidize, developing a layer of "*close-packed cells of oxide, each has the shape of a hexagonal prism. Each prism has a spherical shaped end and contains a centrally located pore, star-shaped in cross section.*" This remained the model of choice for anodizing aluminum until 1970, when O'Sullivan and Wood published their model; nevertheless, the KHR model is still referred to today. The schematic for the Keller, Hunter, and Robinson (KHR) oxide is presented as Fig. 2.11.

Anodic oxides were prepared from phosphoric acid, oxalic acid, sulfuric acid, and chromic acid electrolytes. Trends in structural details such as pore size and population, cell wall thickness, barrier layer thickness (the oxide thickness at the base of the pore) were measured and related to the formation voltage. It is interesting to note that the KHR investigation found no correlation to formation voltage and pore diameter, and attributed the measured differences to the different electrolytes. In terms of engineering control, in this paper KHR connects the relationship of the electrolyte and the formation voltage to the engineering characteristics and performance of the oxide as they impact hardness and resistance to wear, corrosion resistance and the reactions that occur during sealing. See Fig. 2.12.

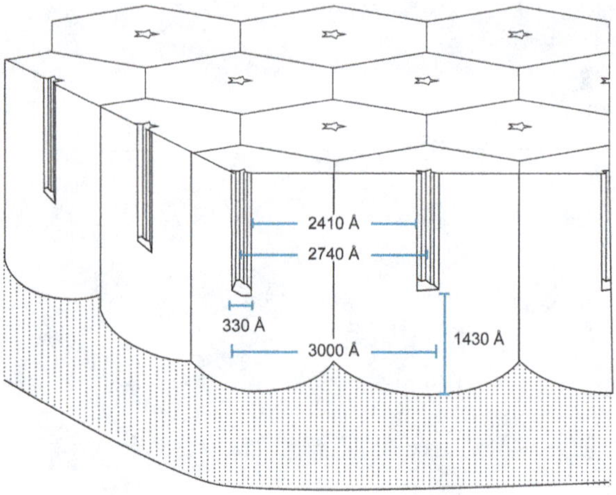

Fig. 2.11 Model of the anodic aluminum oxide according to Keller, Hunter, and Robinson, 1953; the updated schematic, showing the proposed characteristics of a dense barrier layer and star-shaped pores form by way of dissolution is drawn to scale, according to the dimensions published in the article

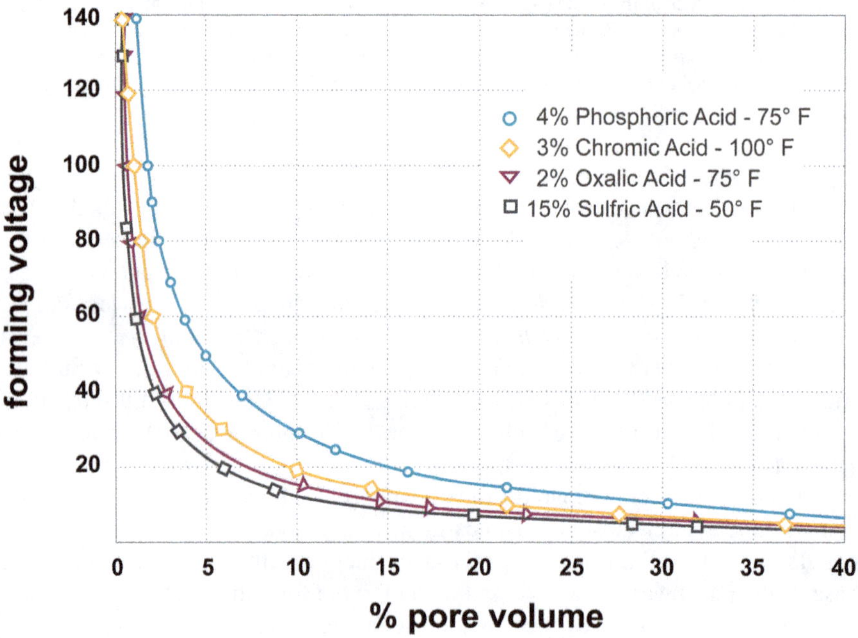

Fig. 2.12 Relationship between pore volume and forming voltage for oxides produced in different electrolytes [91]

2.13.2 *Wernick, Pinner, and Sheasby*

In 1954, the first of six editions of *The Surface Treatment and Finishing of Aluminum and Its Alloys* was published. The book was well received as an authoritative text for all steps in the chemical treatment for aluminum, which covered anodizing, conversion coating, and other methods for coating aluminum substrates. Most of the research collected and presented were the results from research on anodic oxidation of aluminum and its alloys that continues today at the University of Manchester, Science and Technology in Manchester, England. The handbook is still considered as one of the most important reference guides today for the surface finishing, especially anodizing, of aluminum [74].

2.13.3 *Hoar and Mott*

In considering the dual layer theory for anodic oxide formation, in 1959, English corrosion scientist and electrochemist Thomas Percy "Sam" Hoar published "*A Mechanism for the Formation of Porous Anodic Oxide Films on Aluminium*" together with English physicist Sir Nevill Francis Mott, who, during a lifetime rich in professional achievement, was awarded the Nobel Prize in 1977 for his work on the electronic structure of disordered systems, especially amorphous semiconductors, which is by coincidence, the nature of anodic aluminum. In this brief article, the importance of counterdiffusion through the developing "barrier" layer is presented as necessary in order for the oxide to grow [93]. Hoar and Mott present a diffusion model for the movement of oxygen, from the electrolyte, and aluminum, from the anode. See Fig. 2.13.

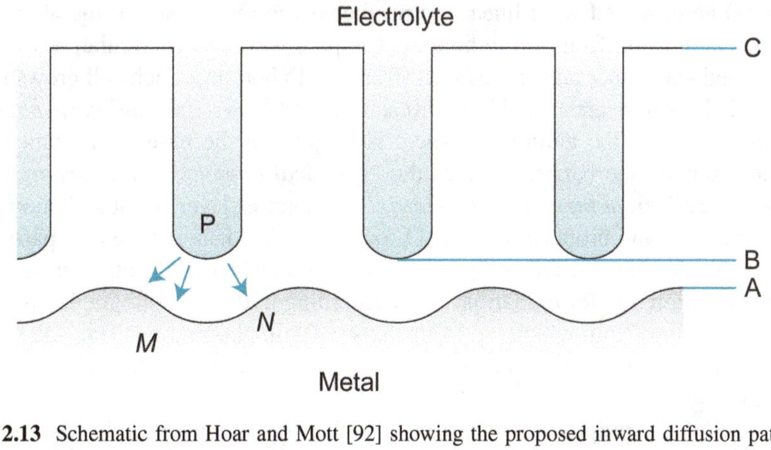

Fig. 2.13 Schematic from Hoar and Mott [92] showing the proposed inward diffusion path for oxygen across the "compact layer" A–B. Oxygen diffuses from the base of the pore, P, according to the path PM, PN. At the same time, aluminum ions, Al^{3+}, move from area MN, through the compact layer, to "neutralize" oxygen ions at the oxide-electrolyte interface, much as with the formation of barrier layer films formed in neutral electrolytes

Hoar and Mott propose that some oxide grows from the oxide-electrolyte interface, and that the temperature of the reaction is sufficient to continually dissolve the forming oxide; however, as a counter-reaction, aluminum ions that go into solution from area MN also react with OH^- ions (that form from the decomposing acid) to form oxide at the metal-oxide interface. The competing reactions are complicated by differences in electrical field strength from area *P*, at the bottom of the pore, and in area *MN* at the metal-oxide interface. In presenting this theory, the idea of merging cells for oxide growth becomes clear, as does the possibility for the actual displacement, or flow, of oxide as it develops. This diffusion phenomenon was a key consideration for O'Sullivan and Wood, in considering the development of their theory for the mechanism of anodic oxide development and growth.

2.13.4 Akahori

In 1961, with the new data that showed the anodic oxide comprised individual columnar cells with rounded bottoms, the origin of the total structure from the so-called barrier layer was questioned. In considering the formation of a barrier layer, Japanese researcher, Hiroshi Akahori found the nature of a thin, nonconductive layer, that breaks down and reforms, suggested by Rummel, or the micropore reaching the aluminum through the oxide to the aluminum matrix by way of anionic action, suggested by Güntherschulze or Setoh and Miyata, difficult to accept. Akahori's observations of AAO growth with different electrolytes under different conditions yielded important characteristic information about the anodic oxide, and an interesting theory for its formation were published in "*Electron Microscopic Study of Growing Mechanism of Aluminium Anodic Oxide Film*" [94].

Akahori's observations in the SEM contradicted the models proposed by Baumann and Keller, Hunter, and Robinson. The oxide "unit cells" that comprised the AAO he observed were linear (perpendicular) to the surface of the aluminum matrix and uniform from top to bottom. The pore was always circular, never star-shaped and was rather uniform as well (from top to bottom). Each cell grows independently from its nearest neighbor, and each unit cell grows, forming "*hemispherical concaves*" toward the aluminum. There is no pore at the base of the unit cells. Akahori defined the barrier layer as the "*spherical boundaries that separate the oxide unit cells from the aluminum matrix.*" The barrier layer is not stoichiometric Al_2O_3, but is some proportion of Al_mO_n (m:$n > 2{:}3$). Akahori noted no particular relationship between the anisotropy of the Al matrix and the AAO although he states that the reaction rate for oxidation varies according to the crystallographic orientation of the grains from which the oxide grows. These observations lead Akahori to theorize that AAO grows as the result of a reaction between molten aluminum and oxygen.

A part of the proposed mechanism for Akahori's theory is indeed similar to Baumann's, that electrolysis of the electrolyte produces oxygen ions, and a gas layer forms at the surface of the anode. Heat generated, coupled with the abundance of oxygen ions forms a thin oxide or "barrier layer." The thickness of the gas layer depends upon the formation voltage, and this thickness determines the curvature of the base of each unit cell's barrier layer. Unique to Akahori's theory: the ionic current produced by electrolysis, coupled with the electric field "impacts" the interface with the oxide where the electric field is strongest and *the aluminum matrix locally melts.* The molten surface of the aluminum is deformed by the pressure of the gas layer, and oxidation of the aluminum occurs by the ionic conduction of oxygen in the gas layer. This action creates a convective action that continually "blows" away oxide at the base of each cell, forming the central pore and maintains cell growth, because the pore is filled with electrolyte.

2.13.5 Csokan

Paul Csokan's work with the Elypovist microscope throughout the 1960s, a combination of an optical microscope with an electrolyzing cell, brought further insight into the formation of the anodic oxide through the documentation of oxide nuclei under nonstandard, dilute anodizing conditions. Csokan's experiments were limited by actual anodizing conditions, specifically electrolyte concentration, and by the magnification capabilities of his microscope. Nevertheless, the images obtained are impressive and provide a great deal of insight regarding how the anodic oxide originates at discrete nuclei. The details of Csokan's work are presented in *Advances in Corrosion Science and Technology* edited by Fontana et al. [95]

By documenting the formation of anodic oxides under controlled, nonstandard conditions in real time, Csokan strived to show the morphological phenomena of how the anodic oxide formed *during anodizing*. His work shows conclusively that anodic oxide films on aluminum in a variety of electrolytes comprise discrete nuclei that grow in three dimensions. These findings support the classification/characterization of the passive film as a "network former" and connect the formation of the passive film to the anodic oxide film. By documenting oxide nucleation under conditions that are not typical for growing industrial anodic finishes, it was also shown that oxide nucleation favors higher energy centers; therefore, many nuclei are located along surface texture developed by deformation, or at edges of intermetallic compounds that intersect the surface of alloy substrates; to contrast, on homogeneous surfaces, fewer nuclei form, and particles tend to be larger and slower growing. See Fig. 2.14.

Csokan's work brought him to the conclusion that when anodic oxidation is started at a higher temperature with higher cell voltage, the number of nuclei is greater and their distribution is more uniform than for oxides formed with lower energy levels. Nuclei appear not only on crystallographic or microstructural defects, but also what he classified as randomly, producing what appeared to be uniform coverage of oxide nuclei over the surface. Through his work, he showed that the anodizing parameters of electrolyte type and concentration, temperature, and time determine the characteristics of the AAO.

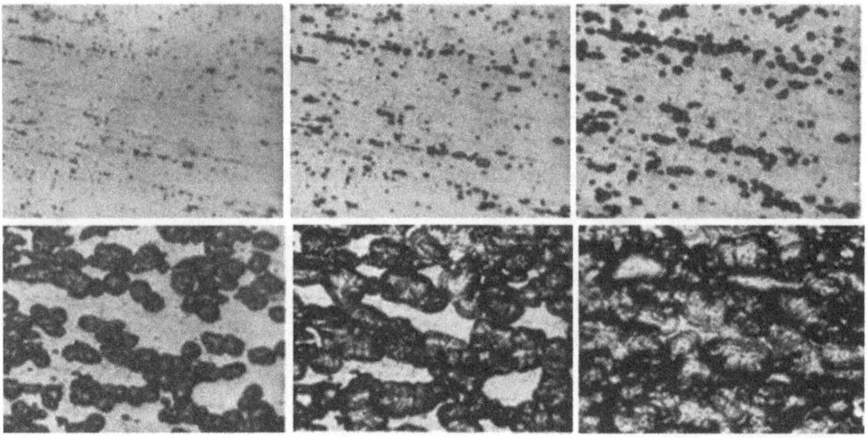

Fig. 2.14 Progressive oxide development over a period of ~10 s in a 5% sulfuric acid electrolyte with 36–38 V at ±1 °C. Nuclei are reported to appear in the first half-second of exposure after switching on the current [95]

2.13.6 Murphy and Michelson

In their paper, published in the proceedings of the Conference on Anodising Aluminium at the University of Nottingham in 1961, J. F. Murphy and C. E. Michelson suggest that the anodic oxide forms due to the consolidation of a colloidal gel [96]. Their model brings to light interesting ideas of how the structure of the oxide is sufficient to enable electrolytic conduction through itself during the oxidation reaction, accounting for the ability of the barrier layer to sustain oxide growth. The proposed structure is described as a triple layer of colloidal gel. The colloids are identified as "*submicro crystallites*." The composition of each of the three layers was suggested as follows: (1) The first layer, adjacent to the substrate, that comprises a compact, anhydrous form of alumina is nonuniform in structure and composition yet hydration can occur in order to enable conduction between the substrate and the oxide; (2) The second layer, identified as an "intermediate transition region," in which the conversion of the inner anhydrous region to the outer region of high hydration occurs; and (3) The third region, adjacent to the electrolyte and consisting of hydrated oxide, in which, according to Murphy and Michelson, the precursors to the pores occur. Within the framework of this concept, the pores in the oxide do not penetrate the oxide to the barrier layer and have less of a significant role than the pores do in the Keller, Hunter, and Robinson model.

Similar to Wernick's theory for anodic oxide formation from 1947, Murphy and Michelson seem to have explained the origins for barrier layer formation, without much consideration of the pores in the oxide layer. Rather than Wernick's elegant description for the formation of an "oxide net" over the surface of an anode by way of peptization of the colloidal particles, the Murphy and Michelson model presents more complex detail, describing the electric field strength as sufficient to promote the ion exchange of hydroxide ions and water molecules in the direction of the barrier layer, while the field causes migration of *protons*, not cations, from the barrier layer. The movement of anions into the structure is considered to occur through the oxide on a hydrogen-bonded network in which the anion progresses by moving from one lattice unit to the next, making and breaking hydrogen bonds as it proceeds. The opposing actions explain the observed dependence of the barrier layer thickness on the applied formation voltage.

Diggle [1] draws an interesting parallel to a model similar to Murphy and Michelson's *Colloidal Gel Model*, by citing Russian researcher Bogoyavlenskii in whose theory the anodic oxide layer is viewed as consolidation of micelles, referred to as *nonons*, of $Al(OH)_3$, in gel-like columns, oriented vertically from the aluminum surface, in the direction of the electric field. Pores separate the columns, along which additional material is transported for continued growth. The continuity of the barrier layer is established at the end of the anodizing reaction, when the current is removed, and the nonons fill in the bases of the pores.

2.13.7 *Davies, Domeij, Pringle, and Brown*

The authors of this fascinating piece of work on the migration of ions across the base of the pores in the anodic oxide were from the chemistry branch at the Chalk River Nuclear Laboratories in Ontario, Canada. The paper, published in 1965, "*The Migration of Metal and Oxygen during Anodic Film Formation*" [97], present the studies by Davies et al., utilizing beta-ray spectroscopy and radioactive tracer (Xe^{125}) techniques to determine the diffusion of metal and oxygen ions across various types of metal oxides formed by way of electrochemical processes (anodizing). The metals investigated: aluminum, niobium, tantalum, tungsten, zirconium, and hafnium exhibited differences in ionic migration across the anodic oxides; Al, Nb, Ta, and W displayed both metal and oxygen migration, Hf and ZR displayed only oxygen migration.

For the metals other than aluminum, less than 1% of the total anode oxidized went into solution, but depending upon the current density, up to 40% of the total aluminum anode dissolved. This study showed that the aluminum anodic oxide grows only from the electrode-oxide interface, and that no significant oxide grows on the surface of the oxide at the interface with the electrolyte, although tracer anions were adsorbed. It was also noted that not the entire amount of aluminum ions that diffused from the metal-oxide interface and were dissolved in the electrolyte actually became oxide; in fact, most of the Al^{3+} ions migrated through the oxide layer and went into solution with the electrolyte did not build additional oxide at the interface with that of the electrolyte at all. This was an important conclusion because this result contradicted the building of oxide from the oxide-electrolyte interface by way of colloidal deposition. Based upon the amount of radioactive tracer migration measured, the transport numbers for the Al^{3+} ions and O^{2-} ions were calculated to be 0.72 and 0.28, respectively.

2.13.8 O'Sullivan and Wood

With the introduction of the transmission electron microscope (TEM), O'Sullivan and Wood deepened the dive into understanding the mechanism for anodic oxide formation. Their theory, published in 1970 in "*The Morphology and Mechanism of formation of Porous Anodic Films on Aluminium*," explains the development of the pore by the "merging of thickening oxide regions" on the anode surface, which is similar to Baumann's theory from 1938. According to O'Sullivan and Wood, oxide regions are related to the substructure of the substrate, and the current is concentrated in thinner areas across the oxidized surface. Pores develop randomly and grow in diameter and number until a steady-state morphology is reached that becomes the "*barrier layer*." The steady-state barrier layer thickness, the cell diameter and pore diameter are all directly proportional to the formation voltage of the anodizing circuit. With continued application of the voltage, an equilibrium develops between oxide formation in the barrier layer and thermally enhanced electric field-assisted dissolution at the pore bases, which determines the cell and pore sizes by way of the geometric order established by the spatial relationship between nearest neighbor cells, the so-called *simple geometrical mechanism* [92].

The comprehensive nature of the work and the detail of the images published by O'Sullivan and Wood in this important paper presented the formation of the anodic aluminum oxide in a very comprehensive way that had never been accomplished before, which provided documentation and insight for the formation of the structure of the anodic aluminum oxide that led to the schematic used today when considering the general structure of the anodic oxide. The TEM images were produced by way of replication, rather than directly, from actual anodic aluminum oxide. Nevertheless, this work reinforced and refined the conclusions that were arrived at by way of observation by many of the their predecessors. See Figs. 2.15 and 2.16.

Prior to this study, or possibly concurrently, an extremely interesting study was carried out by O'Sullivan, Hockey, and Wood, "*Infra-Red Spectroscopic Study of Anodic Alumina Films*" [98], in which the structure and composition of anodic oxides produced in a sulfuric acid electrolyte were examined by way of infrared spectroscopy and electron microscopy on freshly prepared anodic oxides, those hydrothermally sealed at 100 °C and then, unique to this study, oxides prepared by way of isotopic exchange techniques using heavy water vapor (deuterium) to impregnate the structure following evacuation. The intent of the investigation was to elucidate some of the mechanism of hydrothermal sealing, while gaining new insight to the structure of the anodic aluminum oxide; "*By correlation of these results with visual and structural evidence, a relatively detailed morphological, structural and analytical model is proposed for sealed and unsealed films.*"

The results of this investigation determined that the material within the pores of oxides produced in sulfuric acid consists of fine microcrystallites of anhydrous alumina joined by intercrystallites of molecular water, hydroxyl groups, and acid anion, which is similar to the conclusions drawn by Murphy and Michelson. Additionally, however, O'Sullivan and Wood determined that the aluminum oxide comprising the columnar structure exhibits ionic properties with some covalent

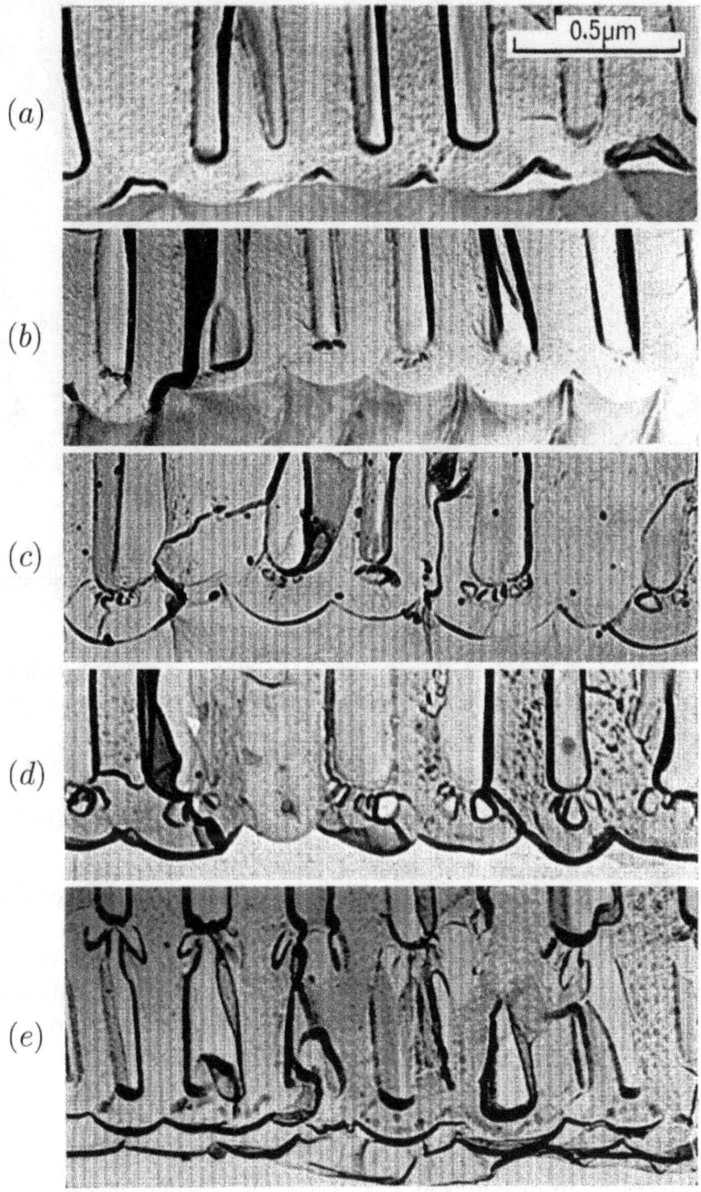

Fig. 2.15 Sample TEM images of the anodic oxide provided in O'Sullivan and Wood. Distortion of the replicas certainly was introduced during sample preparation; nevertheless, this technique elucidated features of the anodic aluminum oxide never seen before. These micrographs represent sections of oxides at various times after a sudden voltage drop from 115 to 85 V in 0.4 M H_3PO_4 at 25 °C: (**a**) 1 min; (**b**) 5 min; (**c**) 7 min; (**d**) 9 min; and (**e**) 13 min after the voltage drop. Note the discontinuities and ultimate thinning at the pore bases [92]

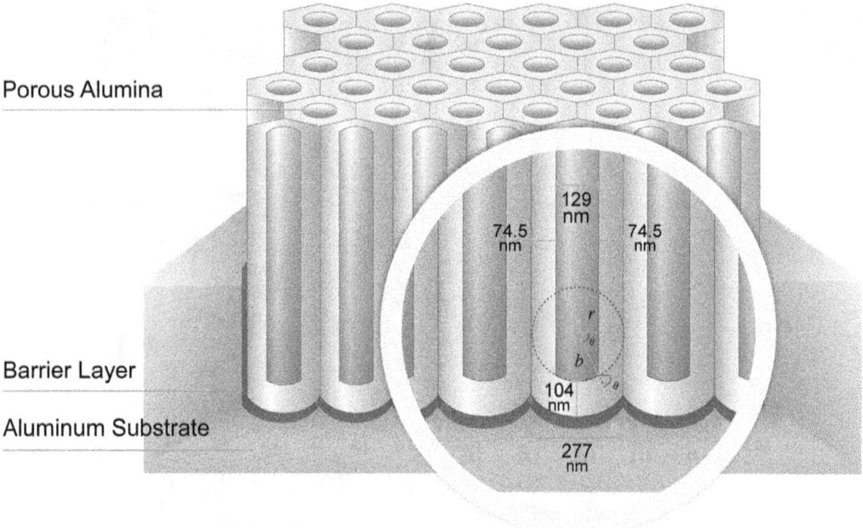

Fig. 2.16 Schematic for the general structure of anodic aluminum oxide presented to scale according to the measurements determined by O'Sullivan and Wood

character; and, depending upon the anodizing conditions, it was concluded that the likelihood for short-range crystalline order within the oxide was probable.

2.13.9 Summary of Mid-20th Century Anodizing Research

By reviewing and comparing the various theories developed during the mid-twentieth century, one notices that there is a trend in development and detail with the development of analytical equipment such as metallographic analysis, scanning electron microscopy, and transmission electron microscopy. According to these theories, the formation of the initial oxide layer has three possible origins, two of which depend on electrolyte composition, which may develop distinctly different types of oxides, either by way of the formation of a colloidal gel (*Murphy and Michelson*) or by way of the random formation of oxide regions that merge together, *The Merging Cell Theory*. The third possible origin for the structure of the anodic oxide was proposed by *Akahori* and suggests that the aluminum at the interface with the oxide is molten at the time the structure forms as individual *unit cells*.

The prevalent theory for anodic aluminum oxide formation that continued through the rest of the twentieth century and is still sometimes taught today can be called *The Merging Cell Theory*, in which individual oxide cells randomly form and grow on the electrode. As the cells merge, they come together to form a layer, in which pores develop, where the layer is thinnest and the electric field is of highest strength. The electrolyte enhances the pore diameter by dissolving the oxide layer in the area of the pore. Keller, Hunter, and Robinson first developed this theory in 1953. In 1970, O'Sullivan and Wood refined it by showing the interrelationships between barrier layer thickness, pore diameter, and cell diameter, and that they are proportional to the formation voltage. Csokan showed that oxide particles formed preferentially but randomly and that the pore formation occurs by way of electric field-assisted dissolution.

Interesting basic science research also took place in the mid-twentieth century when Hoar and Mott reasoned that ion exchange should be occurring in order for anodic oxide growth to occur and was supported by radioactive tracer studies performed by Davies et al. This study supported The Merging Cell Theory, and confirmed the work of Rummel and Baumann, who reported the semiconductor properties of the anodic oxide. The insulator quality of the anodic aluminum oxide, once formed, made it tough to accept this semiconductor quality of the oxide while it was growing, and versions of The Merging Cell Theory emerged that resemble the older, Dual Layer Theory. It is also important to mention that while this exciting scientific research was happening, interesting engineering processes and commercial anodized products were being developed throughout the anodizing industry, especially for the aerospace and aircraft industries.

2.14 Modern Research on Mechanism and Structure of Anodic Oxide

With the continued development of analytical tools and instrumentation, more detailed research and analysis to determine the mechanism, structure, and properties of the anodic oxide has been accomplished through the end of the twentieth century and continues today, aiding in explanations for the mechanism of oxide formation, its order and functionality for a variety of applications. Since the publication of O'Sullivan and Wood's model for the formation of the anodic oxide, the idea of merging cells as the source for the cellular oxide network has evolved into several different theories, not one can be considered 100% correct and yet each theory can be considered relevant.

In addition to the actual papers referenced in this section, there are comprehensive review papers about anodic oxidation that go into more detail about these and other theories not presented by: Diggle, Downie, and Goulding in 1968 [1], which covers many of the theories and concepts for compact (barrier-type) oxides and porous oxides in detail for the time between 1930 and 1968; Poinern, Ali, and Fawcett in 2011 [4], which in addition to providing a review of past processes, updates the concepts to include additional insights into the use of AAO as templates for nanotechnological applications; and Lee and Park in 2014 [3], which provides a review of many of the classical theories for anodizing, and covers some of the interesting current research. These papers are recommended reading for the details that exceed the scope of this book. Of course, the comprehensive two-volume set, *The*

Surface Treatment of Aluminium and its Alloys, by Wernick, Pinner, and Sheasby [74], with six editions from 1954 to 2001, fills the time gap between Diggle's paper and the reviews by Poinern and Lee. Arthur Brace's *The Technology of Anodizing Aluminium* [99] is a necessary reference book for all interested in historical yet applicable concepts and practices for anodizing; the most current version of Brace's book was published in 2003.

By providing a chronological summary of some more modern research and interesting conclusions regarding the mechanism for anodic aluminum oxidation, the next section highlights some of the more well-known researchers who have made an indelible mark on the understanding of anodizing aluminum, but also provide some that are less famous, and nevertheless important and relevant to the study of the structure and properties of the anodic aluminum oxide. Before beginning, it must be mentioned that in every bibliography for every paper written on the subject of anodic oxidation of aluminum since about 1975, there is a citation that includes professor of corrosion science and engineering, George E. Thompson, of the University of Manchester Institute of Science and Technology (UMIST). Thompson's research in anodic oxidation spans the time covered in this section, which, in addition to striving to arrive at a model for anodic oxidation, covers practical engineering aspects by investigating the reactions that occur when anodizing various aluminum alloys, while developing new analytical methods for investigating AAO. The wealth of information from UMIST, regarding the science and engineering of anodizing aluminum and its alloys, has been, since before the time of O'Sullivan and Wood, and continues to be, of tremendous value to the understanding and development of the aluminum surface finishing industry, for which we are very grateful.

2.14.1 Advanced Analysis of Anodic Oxide I

In the 1970s and 1980s, clearly, the major advances in understanding the mechanism for anodic oxide formation were due to the advances in the means to prepare and analyze it. In all branches of ultra-microscopy from scanning electron microscopes (SEM) to transmission electron microscopes (TEM), techniques developed that enabled imaging of actual anodic oxides, rather than replicas. Certainly, by today's standards, the images are coarse, and damage was introduced by ultramicrotomy techniques used to prepare ultra-thin sections for imaging. Nevertheless, advances in microscopy proved key in garnering new information about the structure and composition of the anodic oxide.

The research conundrum for the aluminum anodic oxide in the 1970s that continued into the twenty-first century comprised the myriad of questions surrounding the function of the barrier layer during porous oxide growth. Although it was certain that the oxidation reaction occurred at the surface of the aluminum anode, the questions regarding the mechanism were: what is the nature of ion migration across the barrier layer? Did the oxide grow exclusively from the interface with the electrode, or was there deposition of colloidal particles of aluminum hydroxide on the surface of the growing oxide from the electrolyte, too? While it was assumed that single ions could migrate through the oxide when under the electric field, the actual nature of ion migration was uncertain. How much transport was actually through the alumina "crystallites" within the pore, and how much was through the intercrystallite or "cell wall" regions? Understanding ion migration was crucial to determining how new oxide was continually formed. Was it deposited, if so, where? Why did incipient pores that apparently develop on the barrier layer surface convert into larger, steady-state pores? Throughout the end of the twentieth century and into the twenty-first century, efforts continued to explain the various aspects of anodic oxide formation, especially when discussing the elegance of the growth of the apparently self-ordered porous anodic oxides.

2.14.2 Thompson

In 1978, Thompson, Furneaux, Goode, and Wood published "*Porous Anodic Film Formation on Aluminium Substrates in Phosphoric Acid*" [100], in which TEM images of the interface with the base metal and oxide are presented that show, in rather high detail, the progressive formation of the oxide. These are beautiful images that present a sequence of growth over 3 min, at a constant current density of 5 mA/ cm^2 and a potential response of 120 V. The best images, with the least amount of distortion are of the thinnest oxides, taken earliest in the anodizing sequence in which no tearing of the oxide structure through microtoming is apparent. Some of these oxides clearly show the presence of minor "*protuberances*" of oxide, coinciding with "*ridges*" in the aluminum substrate document the formation of individual oxide cells as minor cells develop and are consumed by continued oxidation until steady-state growth of the primary cells occur. See Fig. 2.17.

Within 3 years, Thompson and Wood published "*Porous anodic film formation on aluminium*" [101], in which improved TEM imaging was enabled by way of ion-beam milling. This technique introduced far less damage to the samples during preparation, and enabled even thinner sections to be imaged, which increased the level of detail one could observe with the microscope. Unfortunately, no actual images are available in this paper, but the result of the enhanced details disclosed with improved sample preparation techniques, an interfacial band comprised relatively pure alumina at the base of the cell and at the cell walls. A compositional gradient, most likely determined by way of Rutherford backscattering spectrographic (RBS) analysis, was determined to be present from the interface with the substrate to the surface of the pore showing varying amounts of adsorbed counter-ion, depending upon the electrolyte composition. This method was discussed in more detail in a 1983 publication, in which RBS and other nuclear methods were used to determine the density and composition of barrier-type anodic oxides [102].

In 1986, Thompson, Xu, Skeldon, Shimizu, Han, and Wood published "*Anodic oxidation of aluminium*" [103], in which ionic transport as part of pore origination

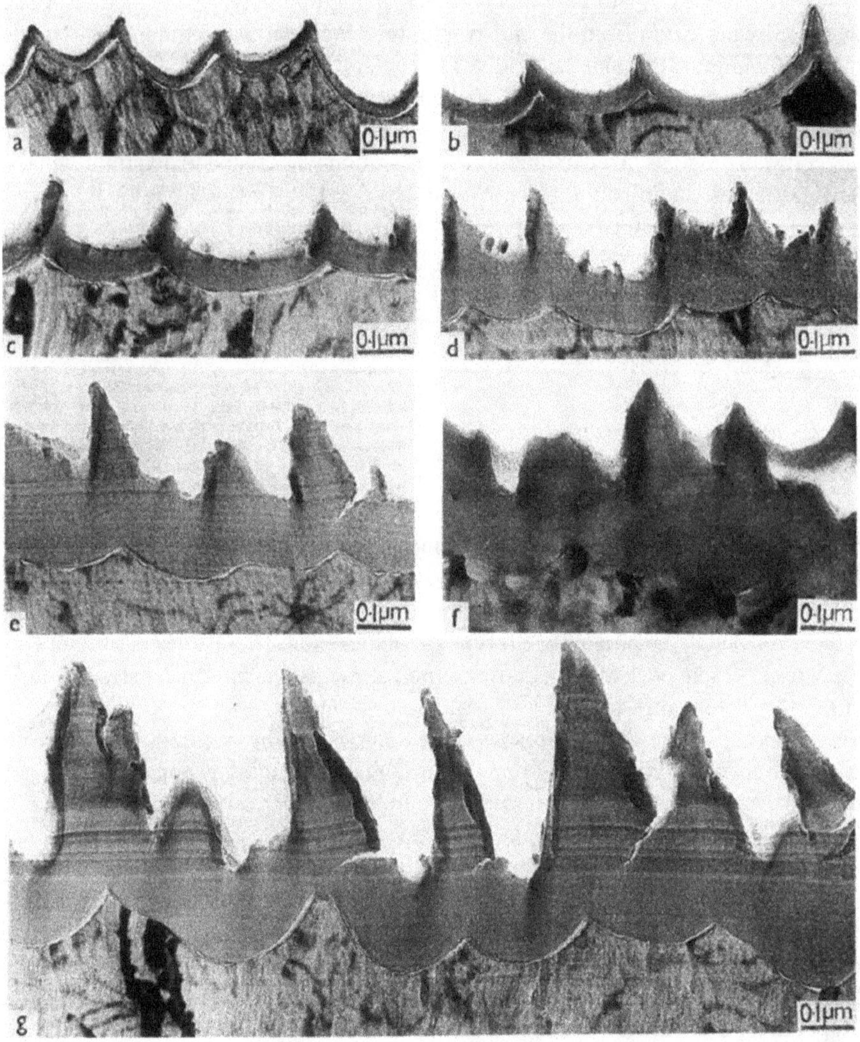

Fig. 2.17 TEM micrographs of an ultramicrotomed anodic oxide. The oxide was formed at 5 mA/cm^2 in 0.4 M H_3PO_4 at 25 °C for the following times: (**a**) 15 s; (**b**) 30 s; (**c**) 60 s; (**d**) 90 s; (**e**) 120 s; (**f**) 150 s; (**g**) 180 s. Note how the bases of each pore correspond to a scallop at the interface with the aluminum substrate. Raster marks, from the electron beam are clear in some of the images, as are unidirectional tears at the bases of the cell walls. Nevertheless, these images were of actual oxides and provided the opportunity to develop new insights for the mechanism of anodic oxidation

is revisited and studied using more advanced techniques. Inert marker layers and tracers are used to enable the determination of the location of solid-film growth (from the metal-oxide interface) and electrolyte ion effects. Pore initiation was suggested by way of penetration paths established by the ingress of O^{2-} and OH^-

anions from the electrolyte through the forming barrier layer, faster or in greater amounts than Al^{3+} cations can egress (re. Güntherschulze; Davies et al.); major pores ultimately develop and sustained by way of field-assisted dissolution (re. O'Sullivan and Wood). However, this paper is unique because it is the first to suggest the development of electrostriction stress at various locations in the film section, beneath and closely adjacent to the pores. It also comments about the effects of the substrate alloy, and even process contamination, that the oxide growth rate and structure can be altered.

2.14.3 *Kniep, Lamperter, and Steeb*

In 1989, citing the interest in the use of anodic alumina for application as a stand-alone membrane, but also the common industrial uses for anodized aluminum, Rudiger Kniep, Peter Lamperter and Siegfried Steeb, published a rather extensive but concise analysis of the structure and composition of the anodic oxide in "*Structure of Anodic Oxide Coatings on Aluminum*" [104], in which freshly prepared porous anodic oxide coatings were investigated by way of a variety of techniques. Analytical methods that are mentioned as having been implemented to obtain structural data are: X-ray diffraction (XRD) as angular dispersive X-ray diffraction (ADXD) and energy dispersive X-ray diffraction (EDXD); and Infra-red (IR) analysis of dried porous anodic oxide, by citing work by Kiss et al. that determined the aluminum atoms in the structure had coordination numbers of four (tetrahedral) and six (octahedral). Structural information was also determined by way of neutron diffraction at the Institut Laue-Langevin. Methods implemented to obtain compositional data are: X-ray fluorescence (XFA), ion chromatography (IC), energy dispersive X-ray spectroscopy (EDS), and thermal gravimetric analysis (TG).

Chemical analysis by way of XFA (Al and S), IC (SO_4^{2-}), EDS (Al and S), and TG (H_2O) determined a gross formula for anodic alumina: ($5Al_2O_3 \cdot SO_3 \cdot 0.66H_2O$) $\cdot$ $1.33H_2O$. This is somewhat different from what is published in Wernick, Pinner, and Sheasby [74] in which Scott is cited as determining the composition as: Al_2O_3, 72%; H_2O, 15%; SO_3, 13%. The difference is in the amount of water in the formula given by Kniep et al. accounts for constitutive water as OH^- groups (within the brackets) and the rest of the water as molecular water, not bound within the structure or adsorbed on the pore walls. How the water is present in the structure is an important part of understanding the structure anodic aluminum in total because constitutive water is released from the structure only at temperatures above 470 °C and accounts for hydrogen bonding in the oxide structure, while molecular water evaporates and releases over a much lower temperature range, from 80 to 250 °C.

By performing and comparing careful analysis by way of ADXD and EDXD, the amorphous nature of the anodic oxide was confirmed; however, neutron diffraction showed that the coordination number of oxygen atoms around aluminum is 4, which was determined by evaluating the data representing the Al–O nearest neighbors as well as the density of aluminum. This result is of most importance regarding

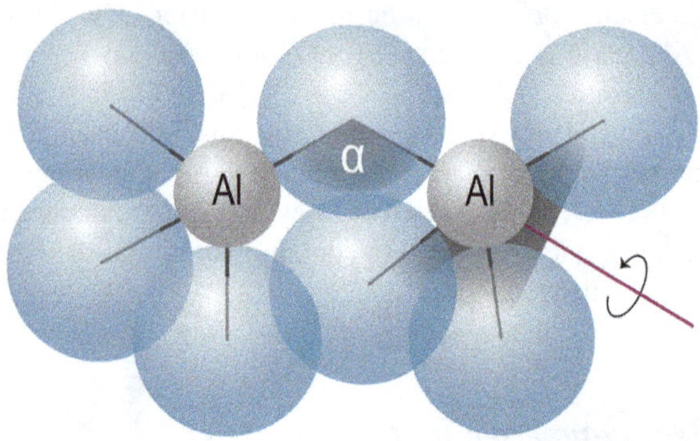

Fig. 2.18 Neutron diffraction done at the Institut Laue-Langevin in Grenoble, France by Kniep, Lamparter, and Steeb in 1989 [113] determined each cell of the anodic oxide comprising AlO_{4-} tetrahedra with short-range order (periodicity of <5 units). The tetrahedra are the basic building blocks that comprise the internal network of the individual cells, which make up the nanoscale network of the anodic oxide

the presence of short-range order within the structure. Bonding lengths were determined to be smaller than that of boehmite or conundrum, which makes sense because the anodic oxide is amorphous and an unsealed film will have shorter bonding lengths than that of the boehmite that forms with sealing. The structure of short-range alumina was described as two AlO_{4-} tetrahedral that share a common oxygen corner, forming an irregular three-dimensional structure controlled by the angle at the shared corner. The bond lengths reported correspond to those reported by Kubicki and Apitz [105]. See Fig. 2.18.

2.14.4 Anodic Aluminum Oxide and Nanotechnology

Beginning in the 1980s, nanoscale materials and their engineering possibilities were an interesting research topic, especially in the areas of synthetic membranes for filtration, bioreactors, tissue culture, analytical devices including sensors and supports for active materials. Nanotechnology was bourgeoning, and already anodic oxidation was being investigated as a means to produce unique and complex structures on a variety of metals, including aluminum.

In 1980, electrochemist J. P. S. Pringle of the chemistry branch at the Chalk River Nuclear Laboratories in Ontario, Canada, published his work in the simultaneous anodic oxidation of superimposed dissimilar metals, in particular niobium and tantalum, to make complex structures that can be utilized for patterning and other purposes, including that of the development of an inert marker, especially dur-

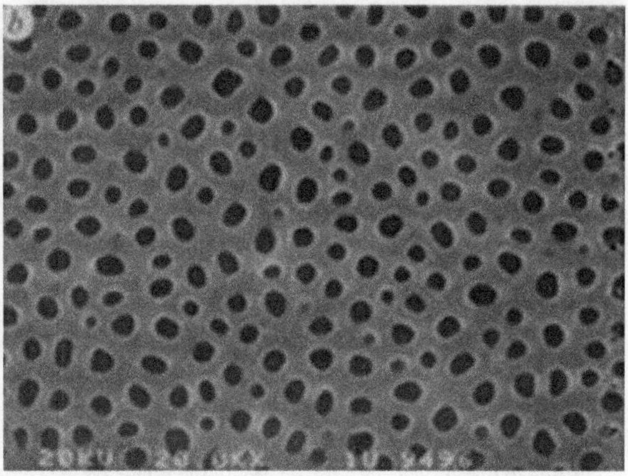

Fig. 2.19 Transverse image or plan view of an anodic aluminum oxide membrane

ing the anodizing process [106, 107]. Complex meshes of tantalum oxide with niobium oxide-filled holes were thus manufactured, and using the TaO mesh as a marker, the transport numbers of metal and oxygen could be measured. In this way, some of the first nanoscale sensors were developed.

2.14.4.1 Furneaux, Rigby, and Davidson

To contrast with the complex processes used for making some nanoscale devices, the ease of making highly ordered structures of anodic alumina with controlled dimensions was the basis for the work performed by the research group at Alcan Aluminium at the Banbury Laboratory in Oxfordshire, UK, led by corrosion scientist Robin Furneaux, to develop a method by which precision anodic alumina could be formed, with a narrow pore size distribution, high pore density, and a reliable thinness. While these attributes were easy to tune with the anodizing process parameters, a disadvantage to anodic aluminum oxide was that a portion of the aluminum anode remained attached to the oxide. In 1989, Furneaux et al. [108] presented a method by which anodic aluminum oxides, with very ordered structures could be detached by steadily reducing the anodizing voltage, thereby enabling separation of the film as a porous membrane. This technology went on to be called *Anopore*© and was the basis for a spin-off company from Alcan called *Anotec Separations, Ltd.* Anopore membranes are still available for sale today and represent possibly the first use of anodic alumina as a stand-alone product. See Fig. 2.19.

Fig. 2.20 Nickel nanowires produced in an anodic aluminum oxide template. In the SEM image on the left, the template is only partially removed, not the wires are grouped in small domains, according to the grain size of the anode. In the image on the right, the template is entirely removed. The wires have a diameter of 60 nm [110]

In 1995, Hideki Masuda and Kenji Fukuda, of the Department of Industrial Chemistry at Tokyo Metropolitan University, by producing highly ordered nanohole arrays in platinum and gold by way of a two-step replication process of the anodic oxide [109], revolutionized the approach to applications for the anodized finish as a stand-alone layer. Masuda et al. produced nearly ideal anodic oxide templates using "long period anodization," using anodized high purity aluminum (99.99% pure) substrates, and a 0.3 M oxalic acid electrolyte at a constant voltage (40 V). Longer anodizing times produced larger defect-free regions (domains) across the substrate to use as a template. The columns in the ideal areas are unidirectional, perpendicular to the substrate, without branching and have a high aspect ratio, perfect for nanowires. See Fig. 2.20.

With the excitement surrounding nanotechnology, naturally the scientific community turned to the anodic aluminum oxide because of its inherent nanoscale porous structure. Over the last 20 years, many universities and research institutions have been working with anodic alumina to develop processes and new applications. The anodic aluminum oxide has taken on an entirely new role as a self-assembling template for nanowires, nanotubes, and nanoindenters for medical, computer, and aerospace industries. AAO has become a platform for hydrogen and thermal sensors. The need for long range, highly ordered AAO structures has limited the process of producing them to ultra-pure aluminum substrates such as sputtered layers or foils, creating the research niche for finding appropriate alloys for such processes. Alloys of aluminum are rarely employed to produce AAO of this quality; to contrast, in industry, it is rare to encounter components that require high purity aluminum because of its limited engineering properties.

2.14.5 Advanced Analysis of Anodic Oxide II

The morphology of the oxide as a network that comprises columnar cells, each containing a central pore, was generally accepted in the 1990s. The myriad of techniques available for the analysis and characterization of anodic oxides provided more details about the structure than ever before. Nevertheless, the mechanism for anodic aluminum awaited clarification.

Vasily Parkhutik developed a mathematic model of the oxide based upon the existing theories, "*Theoretical modeling of porous oxide growth on aluminium,*" which was published in 1992 [111]. Parkhutik based his model upon the "*elementary processes involved in porous oxide growth*": (a) field-assisted dissolution of the oxide at the pore base; (b) ion transport across the barrier layer; (c) growth at both the metal-oxide interface and the oxide-electrolyte interface. In considering the lateral distribution of the electric field along the base of the oxide layer, Parkhutik made an interesting assumption that the lateral distribution of the electric field was inhomogeneous, with the maximum electric field concentrated at the pore center and decreasing toward the pore walls, which enabled modeling of the three-dimensional electric field distribution through the aluminum-oxide-electrolyte system as applied to steady-state pore growth.

Georgios Patermarakis developed several mathematical models that were used to describe the growth mechanism and transport phenomena inside the pores for anodic alumina. The various process parameters of temperature, current density electrolyte concentration, and agitation were considered with regard to the kinetics of porous oxide formation, even to the extent of predicting the shape of the pore. According to the shape of the pore, other, more detailed process details were shown to have an impact on oxide growth, such as the electrolyte concentration along the length of the pore, and differentiating between the rate of anion transport produced by field-assisted pore base oxide electrochemical dissolution and purely chemical dissolution of the pore wall, taking into account the electrolyte concentration within the pore itself. These models were written and presented in several papers published from 1995 to 2001 [112–115].

In 1997, George Thompson published "*Porous anodic alumina: fabrication, characterization and applications*" [2], in which he provides a review of then-current formation theories for both barrier layer and porous films on aluminum, together with a summary of analysis tools and instruments that were also then-currently available for studying the anodic oxide. In reviewing barrier layer oxides, Thompson reiterates what is already commonly known about ion migration and states that the mechanism awaits clarification; field-assisted dissolution of the pore base at the oxide-electrolyte interface balanced against film growth at the metal-oxide interface was still considered to be the source for stability of the cellular morphology of the anodic oxide. However, in describing the use of inert markers and tracers that enable monitoring of the oxide growth and extrapolation of the effects of growth conditions such that processes can be developed to control oxide morphology, Thompson provided a preview of exciting research to come.

2.14.5.1 New Theories for Mechanism of Anodic Oxide Formation

Runge

In 2000, the author presented a theory for anodic oxide formation based upon lateral growth of the oxide from preferential nuclei on the anode surface in the paper "*Anodic Oxide Film Formation: Relating Mechanism to Composition and Structure*" [116]. The approach to the oxidation reaction was that of a corrosion reaction. In this article, it was proposed that the source for the pore was the mechanical stress between impinging nuclei that reconstructed the anode surface and the electric field effects on the inside surface of the pore. The field effects functioned to drive ion exchange along the inside surface of the pore, keeping the column erect through the anodizing process. The theory was based upon observations made by way of TEM analysis, XPS, and FT-IR analysis at the synchrotron at Brookhaven National Laboratory during the author's doctoral research at the University of Illinois at Chicago, and was presented at the Aluminum Surface Science and Technology (ASST) Conference in May of 2000 at UMIST.

Lerner

In 2003, Moisey Lerner wrote "Is the barrier layer responsible for the cellular structure of aluminum oxide film anodized in sulfuric acid?" in which he presented his theory for the progressive formation of the anodic oxide as it functions as a p-i-n diode [117]. The basis for his theory was an effort to clarify the function of the barrier layer, whose foundation was laid by the Keller, Hunter, and Robinson Theory in 1953. Lerner found that it was strange that the barrier layer did not function as a "barrier" to high current, citing the industrial example of hard anodizing in which the direct current can reach 100–200 A/ft^2. The key concept for understanding Lerner's approach was the acceptance that the anodic oxide as formed has interesting rectifying properties, and that the "barrier layer," that is at the base of the anodic oxide, exhibits unilateral conductivity, in other words, it functions as a diode. The electric field forces in the anodizing circuit enable the diffusion of oxygen through the finish as it forms, and an oxygen gradient exists from the electrolyte, where the oxygen content in the oxide layer is highest, the so-called p-layer, to the n-layer, adjacent to the anode, where the oxygen content is lowest. A counter-gradient of aluminum exists from the aluminum substrate to the finish surface. Lerner's Diode Model can be applied to barrier layer anodizing, which was first reported by Ducretet in 1875, as well as to porous anodic oxide formation.

According to Lerner, three processes take place during anodizing: the first process was described as uniform dissolution that produces a p-i-n diode that is pore free. This accounts for a passive layer formation (equilibrium oxidation of aluminum) as well as barrier-type oxides formed in weak acids such as boric acid. The second process was described as nonuniform dissolution, in which consumption of the anode is centered at a spot of initial nucleation, where n-type conduction domi-

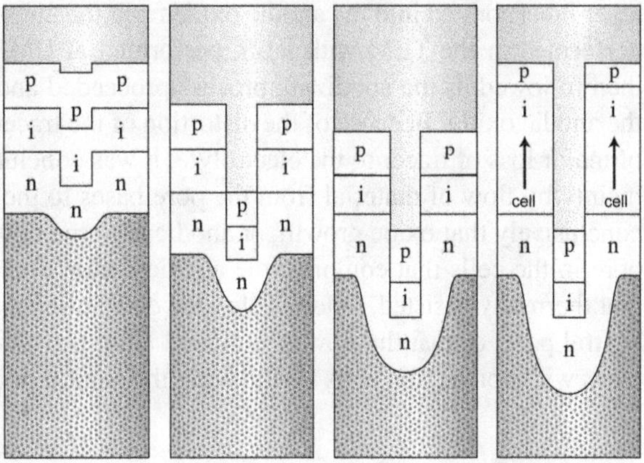

Fig. 2.21 Formation of an oxide cell according to Lerner's second scenario of nonuniform dissolution during anodizing [117]

nates. Dissolution is the consumption of the aluminum by oxide, best described as corrosion of the anode, not the consumption of the oxide by electrolyte to form a pore. Electric field effects are minimized at tangential points surrounding each nucleus, where p-type conduction dominates. The i-layer is an intermediate layer that constantly changes between the n and p layers as the oxide grows and develops, pumping aluminum ions outward from the substrate and oxygen ions inward, from the electrolyte. The third process that can take place during anodizing occurs when imposing an AC current over a DC current. The resultant AC current through the semiconductor (anodic oxide) layer is higher than the Faraday current formed in the finish, suggesting that the imposition of AC current can effectively reduce the required current density to form an equivalent anodic oxide with DC current. See Fig. 2.21.

Garcia-Vergara

In 2006, *Garcia-Vergara et al.* published similar findings, in the Proceedings of the Royal Society, "*Mechanical instability and pore generation in anodic alumina*" [118]. TEM, RBS, and nuclear reaction analysis (NRA) were employed to show that the growth of the anodic film at the barrier region is accompanied by displacement of the comprising material (oxide). Displacement was determined to be the result of growth stresses and the plasticity of the barrier layer that arises from ionic transport processes. The reported result was that the porosity of the porous anodic aluminum oxide is essentially of mechanical origin, initiated by the instability of the film under growth stresses, providing no new material developed at the film-electrolyte interface. In "*A flow model of porous anodic film growth on aluminium*" [119] and "*A Tracer Study of Porous Anodic Alumina*" [39], both published in 2006, Garcia-Vergara and Skeldon et al. present the result of studies that employ a tungsten tracer layer incorporated into the anodic oxide from the aluminum substrate. Analysis was performed in the TEM with RBS, performed at UMIST. The tungsten tracer was then followed as the anodizing process proceeded and was found to move through the anodic oxide. Because of the distortion of the tracer distribution and the absence of major loss of tracer to the electrolyte, it was concluded that the pores are created mainly by flow of material from the pore bases to the cell walls. This study shows conclusively that oxide growth, or anodic alumina flow, is the source for the central pore in the cells that comprise the anodic oxide. This is in contrast to the theories that thermally assisted, field-accelerated dissolution by the electrolyte produces the central pore, or that the flow of colloidal alumina particles comprises an oxide network with convection cells establishing the central pore. See Fig. 2.22.

Hauser and Hebert

In 2009, *Hauser and Hebert* presented a further developed new theory for porous anodic oxide formation when they published "*The role of viscous flow of oxide in the growth of self-ordered porous anodic alumina films*" [120]. The authors suggest that the ordering of porous anodic aluminum oxide is driven by a mechanism in which the oxide motion at the metal-film interface is regulated by the combination of ionic migration in the oxide and stress-driven interface diffusion of metal atoms.

Paschanka

Another interesting theory, with apparent foundations in the earlier work of Murphy and Michelson [95], as well as the work of Bogoyavlenskii work reported by Diggle [1], was developed by Mikhail Paschanka and published in 2013 [121]. Paschanka presents anodic oxide growth as a cooperative motion of particles responsible for the self-organizational process of AAO development and growth. The theory correlates pore formation and size to the dynamic flow of a colloidal oxide, inferring that AAO originates as alumina slurry. Convection cells, driven by ion exchange, are uniformly spaced by anodic polarization. The convection cells that form across the colloidal anodic oxide layer plow out the individual oxide columns, forming uniformly spaced pores throughout. The reaction proceeds to the interface with the aluminum anode, and the bases of each pore form at the end of the oxidation process by way of colloids falling back to fill in the bottom formed by convection. Paschanka's model describes the formation of a porous oxide, not a network of individual oxide cells; however, in a second paper, the dimensions of the features comprising the oxide are shown to be electrolyte dependent can be calculated in terms of anodizing process parameters [122].

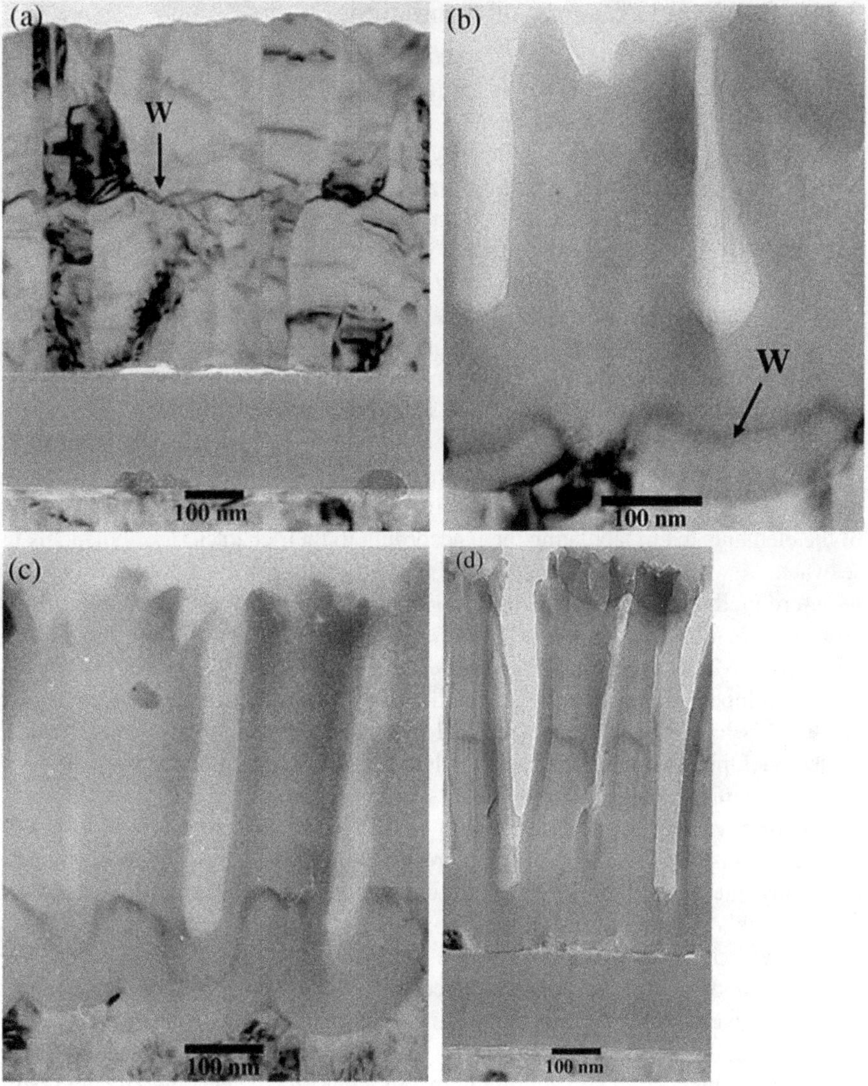

Fig. 2.22 TEM images documenting the movement of the tungsten tracer through the anodic aluminum oxide from the time it was included as a layer in the aluminum substrate, (**a**); to various times during the anodic oxide growth process at (**b**) 180 s; (**c**) 240 s; (**d**) 350 s. The anodizing conditions were constant current at 5 mA/cm^2 in 0.4 M H_3PO_4 at 20 °C [39]

2.15 Summary: History of Anodizing

As with most important discoveries, an indirect path led to the discovery of anodizing aluminum as a means to develop a functional oxide. The interdisciplinary origins of anodizing encompass the physical sciences that were born in ancient times with alchemy and developed during the Scientific and Industrial Revolutions of the eighteenth and nineteenth centuries. Physics, chemistry, electrochemistry, and the interrelated sciences of crystallography, corrosion, thermodynamics, and diffusion, all play a role in understanding what happens during anodic oxidation.

After aluminum was successfully reduced in 1827, the existing methods for electrochemical characterization for new elements set the stage for the rapid concurrent development of aluminum and its alloys with the anodizing process. Through the determination and understanding of new electrochemical principles while carrying out characterization studies on new elements, it was found that many chemical reactions could be externally controlled by way of an electrical power supply. This discovery opened the door to new engineering opportunities: electrical characterization of the elements led to tabulating them according to their characteristic potentials in seawater in what became the Galvanic Series, enabling new understanding and therefore methods for corrosion prevention and protection that were implemented as a modern approach to extend the useful life of engineered components and assemblies. Modern electroplating was developed; industrial level electrolysis cells were developed to extract metals. Relationships between chemical reactions and electrical behavior were recognized and in some cases, could be externally controlled, yielding measurements that enabled the characterization of various types of materials as insulators, conductors, even semiconductors.

The discovery of the phenomenon of passive behavior created a philosophical debate that lasted more than 100 years, resulting in tremendous scientific research and many interesting theories that stem from the original concept of the nature of certain metals to spontaneously form an oxide layer. The growth of the anodic oxide layer on aluminum that serves to insulate electrically provides decorative corrosion protection and/or wear resistance is based upon the characteristic of aluminum to form a passive oxide layer. It is in studying the passive behavior that the deliberate growth of oxide layers on reactive metals was discovered. Heinrich Buff first documented this phenomenon on aluminum in 1857, but Eugene Ducretet did not document practical applications for the oxide until 1875.

The first applications for anodizing were electrical, employing the capacitive properties of the oxide in condensers and rectifiers. The application springs directly from the discovery of the nature of aluminum to form an oxide when polarized as an anode. Therefore, the first anodic oxides were barrier-type oxides; thin, compact, and very sensitive to voltage breakdown, when the surface of the anode was totally reconstructed by oxide. The electrolytes were not necessarily acid solutions, but were often solutions of mineral salts that were either pH neutral or alkaline. Other applications were soon found for the oxide, and anodizing was performed to provide corrosion protection, especially in seawater. The ability to form thicker oxides using acid electrolytes was discovered soon after the process for growing barrier-type oxide layers. Anodizing without the possibility of dielectric breakdown drove investigations to determine the properties of the thicker oxides. It was soon determined that the anodic oxides are porous, which made interesting developments for decoration and lubrication, as it was found that the oxide absorbed dyes, soaps, and oils. These developments expanded applications for the anodized finish to encompass decoration, wear, and abrasion resistance.

While the practical applications for the anodized finish were being investigated wherever aluminum and its alloys were being applied (household goods, aircraft and automotive industries in particular), scientists were studying the oxide to understand the nature of its formation and growth. There are as many ideas and theories for the mechanism and formation of the electrochemical formation for the anodic aluminum oxide as there are for passive layer formation, and they continue to develop as research proceeds not only to understand anodic oxidation, but to design and development strive to find new applications for it. In 1932, Setoh wrote (sic) "The anodic aluminum oxide has been extensively studied, and even so, the mechanism of anodic oxidation of aluminum is still not fully established." Even today, this statement is repeated throughout anodizing literature.

The evolution of concepts for anodic oxidation is in parallel with the development of analytical tools and instrumentation, and this is reflected in the various conceptual drawings of the AAO. Early conceptual schematics present the anodized aluminum oxide as having two basic components. The first, directly adjacent to the aluminum substrate, is commonly called *the barrier layer*. The second component is *the porous oxide*, which is distinct, but not separate from the barrier layer. From analysis and more detailed data that are available surrounding the structure of the anodic oxide, we know today that the oxide is a network of nanoscale electrochemical cells, each with their own structure and properties that are inextricably linked to the structure and properties of the electrode from which it grows.

Most existing theories for anodic oxide formation suggest the barrier layer forms simultaneously with the porous oxide and that the porous oxide is the result of field-assisted dissolution by the electrolyte. It is further assumed that the characteristic pores in the anodic oxide structure are the result of oxide breakdown at defect sites in the substrate, with geometry being the reason for its order. Csokan suggested that AAO originates at preferential sites on the substrate as oxide nuclei, growing to form a barrier layer and concluded with field-assisted dissolution as the source for the pores. Through their tracer studies, Skeldon, Garcia-Vergera et al, support alumina flow as the source for porous oxide formation, rather than field-assisted dissolution by the electrolyte, but do not address the origins of the oxide. Paschanka presents a novel point of view in which the source of the porous oxide is Rayleigh–Bénard convection cells developing at colloidal oxide particles, which are equally spaced due to variations in pH near the surface. However, in regarding the process of initial oxide formation as the "deposition of colloidal alumina films," rather than the growth and development of an oxide layer, the key role of the substrate as the source for porous oxide formation is overlooked.

The theory for anodic oxide nucleation and growth described in the coming chapters of this book shares some of the tenets of other theories for anodic oxide, especially those that propose anodic oxide growth by way of surface reconstruction (Csokan) and alumina flow (Houser and Hebert, Garcia-Vergara et al.). Important details to the theory presented in the coming chapters are: 1) Polarization is the overriding electrochemical parameter, key to setting the order for the anodic oxide, and 2) electrostriction is the source for oxide movement as it is constrained by the electric field to form the central pore. Ion exchange facilitated by convection in order to maintain the central pore over its entire length is a key concept for anodic oxide development and growth the author shares with Paschanka. However, the theory presented in the following chapters differs from and enhances existing theories as it explains anodic oxidation on aluminum in total, by addressing oxide nucleation and growth as a function of substrate polarization and progressive surface reconstruction by way of ordered aluminum corrosion, and most importantly, considers the impact of the substrate microstructure on the total process of anodic oxidation.

References

1. Diggle, J., Downie, T., & Goulding, C. (1968). Anodic oxide films on aluminum. *Corrosion Science, 8*(12), 857–920.
2. Thompson, G. E. (1997). Porous anodic alumina: Fabrication, characterization and applications. *Thin Solid Films, 297*, 192–201.
3. Lee, W., & Park, S.-J. (2014). Porous anodic aluminum oxide: Anodization and templated synthesis of functional nanostructures. *Chemical Reviews, 114*, 7487–7556.
4. Poinern, G. E., Ali, N., & Fawcett, D. (2011). Progress in nano-engineered anodic aluminum oxide membrane development. *Materials, 4*, 487–526.
5. Forbes, R. J. (1953). On the origin of Alchemy. *Chymia, 4*, 1–11.
6. Keyser, P. T. (1993). The purpose of the Parthian galvanic cells: A First century A.D. battery used for analgesia. *Journal of Near Eastern Sudies, 52*(2), 81–98.
7. Karpenko, V. (1992). The chemistry and metallurgy of transmutation. *Ambix, 39*, 47–62.
8. El Khadem, H. S. (1996). A translation of a Zosimos' text in an Arabic alchemy book. *Journal of the Washington Academy of Sciences, 84*(3), 168–178.
9. Levere, T. H. (2001). *Berzelius and Laurent: A question of method. Chapter 8, "The rise of organic chemistry" of transforming matter: A history of chemistry from alchemy to the Buckyball* (pp. 100–106). Baltimore: Johns Hopkins University Press.
10. Ohm, G. S. (1827). *Die Galvanische Kette*. Berlin: T.H. Riemann.
11. Mehrer, H., & Stolwijk, N. (2009). *Heroes and highlights in the history of diffusion*. diffusion-fundamentals.org. *11*(1), 1–32.
12. Boyle, R. (1743). The experimental history of colours. *The works of the Honourable Robert Boyle* (Vol. 1, with a forward and biography of Mr. Boyle by Thomas Birch, 1743, Annotation IV, p 780 (on the production of brass through the diffusion of zink in copper, from about 1640)).
13. Brown, R. (1829). *A brief account of microscopical observations made in the months of June, July and August 1827, on the particles contained in the pollen of plants; and on the general existence of active molecules in organic and inorganic bodies*, 28 July 1829, pp 465–486.
14. Wisniak, J. (2013). Thomas Graham I. Contributions to thermodynamics, chemistry and the occlusion of gases. *Para Quitarle El Polvo, Educación Química, 24*(3), 316–325.

15. Wöhler, F. (1828). Ueber künstliche Bildung des Harnstoffs. *Annalen der Physik und Chemie, 12*, 253–256.
16. *Presentation: The evolution of formulas and structure in organic chemistry in the 19th century*. ursula.chem.yale.edu.
17. Wisniak, J. (2009). *Auguste Laurent. Radical and radicalss* (pp. 166–175). Educación Química, Abril des: Para Quitarle El Polvo.
18. Brooke, J. H. (1975). Laurent, Gerhardt, and the philosophy of chemistry. *Historical Studies in the Physical Sciences, 6*, 405–429.
19. Langmuir, I. (1916). The relation between contact potentials and electrochemical action. *Transactions of the Twenty-Ninth General Meeting of the American Electrochemical Society, 29*, 125–182.
20. Davy, H. (1839). Outlines of a view of Galvanism. Journals of the Royal Institution, Vol. i., 1802. In Dr. Thomas Young and Sir Humphry Day, editors, *The collected works of Sir Humphry Davy*, Vol. II. Early miscellaneous papers (p. 193), edited by his brother, Dr. John Davy. London: Smith, Elder and Cornwall.
21. Davy, H. (1824/1825). On the corrosion of copper sheeting by seawater, and on methods of preventing this effect, and on their application to ships of war and other ships. *Philosophical Transactions of the Royal Society, 114*, 151–246; *115*, 328–346.
22. Groysman, A. (2010). Humanitarian aspects of corrosion science and technology, Chap. 6. In *Corrosion for everybody* (pp. 252–253). New York: Springer.
23. Faraday, M. (1922). Electricity of the voltaic pile and On the source of power in the Voltaic pile, Chapters 6 and 7. In E. Rhys (Ed.), *Experimental researches in electricity* (pp. 172–312). London: Everyman's Library.
24. Fechner, G. T. (1831). *Massbestimmungen über die galvanische Kette* (pp. 242–244). Leipzig: F.A. Brockhaus.
25. Oleari, C. (2013). Leopoldo Nobili from Chapter 9 Color in optical coatings. In A. Piegari & F. Flory (Eds.), *Optical thin films and coatings from materials to applications* (p. 397). New Delhi: Woodhead Publishing.
26. DelaRive, A. (1858). A treatise on electricity, in theory and practice, Chap. II. In *Chemical applications* (Vol. III, pp. 512–566). London: Longman, Brown, Green, Longmans & Roberts.
27. Keir, J. (1790). Experiments and observations on the dissolution of metals in acids and their precipitations, with an account of the new compound acid menstruum, useful in some technical operations of parting metals. Part II, Section II, On the precipitation of silver from nitrous acid by iron. On the alterations which iron or its surface undergoes by the action of a solution of silver in nitrous acid, or of a pure concentrated nitrous acid. *Philosopical Transactions, LXXX*, 374–384.
28. Bennett, C. W., & Burnham, W. S. (1916). The passive state of metals. *Transaction of the Twenty-Ninth General Meeting of the American Electrochemical Society, 29*, 217–269.
29. Faraday, M. (1836). Letter to Mr. Brayley on some former Researches relative to the peculiar voltaic condition of Iron reobserved by Professor Schoenbein, supplementary to a letter to Mr. Phillips, in the Last Number. *The London and Edinburgh Philosophical Magazine and Journal of Science, IX*, 122–123.
30. Schoenbein, C. F. (1899). Letter to Michael Faraday, published in the Phil. Mag. S.3 vol.9 (1836) p.53 under the title: On a peculiar condition of Iron, by Professor Schoenbein, of Bâle. In G. Kahlbaum & F. Darbishire (Eds.), *The letters of Faraday and Schoenbein, 1836 – 1862* (p. 1). London: Williams and Norgate.
31. Faraday, M. (1922). On a peculiar voltaic condition of iron (Schoenbein) and On a peculiar voltaic condition of iron (Faraday). In E. Rhys (Ed.), *Experimental researches in electricity* (pp. 317–332). London: Everyman's Library.
32. Senter, G. (1914). A general discussion on the passivity of metals. *Transactions of the Faraday Society IX*, Part 3, pp. 203–213.

33. Schönbein, C. F. (1900). Letter to J. J. Berzelius, dated 22 April 1836. In G. W. A. Kahlbaum (Ed.), *The letters of John Jakob Berzelius and Christian Friedrich Schönbein 1836 – 1847* (pp. 18–24). London: Williams and Norgate. Bibliolife, www.bibliolife.com/opensource.
34. Berzelius, J. J. (1900). Letter to C.F. Schönbein, dated 4 May 1837. In G. W. A. Kahlbaum (Ed.), *The letters of John Jakob Berzelius and Christian Friedrich Schönbein 1836 – 1847* (pp. 24–27). London: Williams and Norgate. Bibliolife, www.bibliolife.com/opensource.
35. Cropper, W. H. (1988). James Joule's work in electrochemistry and the emergence of the first law of thermodynamics. *Historical Studies in the Physical and Biological Sciences, 19*(1), 1–15.
36. Wolfram, S. (2002). Notes for Ch. 9, Fundametal physics, Section: Irreversibility and the second law of thermodynamics. In *A new kind of science* (p. 1019). Champaign, IL: Wolfram Mediz.
37. Maxwell, J. C. (1865). A dynamical theory of the electromagnetic field. *Philosophical Transactions of the Royal Society of London, 155*, 450–512.
38. Shewmon, P. (2016). Atomic theory of diffusion. In *Diffusion in solids* (2nd ed., pp. 53–96). Berlin: Springer.
39. Skeldon, P., Thompson, G. E., Garcia-Vergera, S. J., Iglesias-Rubianes, L., & Blanco-Pinzon, C. E. (2006). A tracer study of porous anodic alumina. *Electrochemical and Solid State Letters, 9*(11), B47–B51.
40. Mehrer, H. (2007). History and bibliography of diffusion, Chap. 1. In *Diffusion in solids: Fundamentals, methods, materials, diffusion-controlled processes* (pp. 1–16). Berlin: Springer.
41. Roberts-Austen, W. C. (1896). Bakerian Lecture. On the diffusion of metals. *Philosophical Transactions of the Royal Society of London, 187*, 383–415.
42. Arrhenius, S. (2012). In A. Bard, G. Inzelt, & F. Scholz (Eds.), *Electrochemical dictionary*. New York: Springer. (pp. 45–46).
43. Pais, A. (2005). The reality of molecules, section 5d. Eleven days later: Brownian motion, from Ch. 5 "Statistical Physics". In *Subtle is the lord: The science and life of Albert Einstein* (pp. 93–104). Oxford: Oxford Universtiy Press.
44. Roeber, E. F. (1904). On the reality of atoms and ions. *Electrochemical Industry, 11*(11), 436–437.
45. Einstein, A. (1905). Die von der Molekularkinetischen Theorie den Wärme Gefordete Bewegung von in ruhenden Flüssigkeiten Suspendierten Teilchen. *Ann Physik, 17*, 549–560.
46. Góra, P. (2006). *The theory of Brownian motion: A hundred years' anniversary* (pp. 52–57). Cracow, Poland: Marian Smoluchowski Institute of Physics, Jagellonian University.
47. Newburgh, R., Peidle, J., & Rueckner, W. (2006). Einstein, Perrin, and the reality of atoms: 1905 revisited. *American Journal of Physics, 74*(6), 478–481.
48. Frenkel, J. (1926). Über die Wärmebewegung in festen and flüssigen Körper. *Zeitschrift für Physik, 35*(8–9), 652–699.
49. Welker, H. (1996). Obituary, Walter Schottky. *Physics Today, 29*, 63–64.
50. Wagner, S. (1929). Vorwart. In *Thermodynamik: die Lehre von den Kreisprozessen den physikalischen und chemischen veränderungen un gleichgewichten*. Berlin: Verlag von Julius Springer.
51. Nakajima, H. (1997). The discovery and acceptance of the Kirkendall effect: The result of a short research career. *JOM, 49*(6), 15–19.
52. Hittorf, W. (1899). On the migration of ions during electrolysis. In Memoirs by H. Faraday, F. Kohlrausch, & H. M. Goodwin (Eds.), *Harper's scientific memoirs, VII. The fundamental laws of electrolytic conduction* (pp. 49–79). New York: Harper and Brothers.
53. Helmholtz, H. (1853). Ueber einige Gesetze der Vertheilung elektrischer Ströme in körperlichen Leitern mit Anwendung auf die Thierisch-elektrischen Versuche. *Annalen der Physik*, 211–233.
54. Lewerena, H. J. (2013). On the structure of the Helmholtz layer and its implications on electrode kinetics. *ECS Transactions, 50*(52), 3–20.

55. Arabatzis, T., & Gavroglu, K. (2005). Physical chemistry, Chap. 6. In C. A. Russel & G. K. Roberts (Eds.), *Chemical history: Reviews of the recent literature* (pp. 135–143). Cambridge: RSC publishing.
56. Kohlrausch, F. (1899). On the conductivity of electrolytes dissolved in water in relation to the migration of their components. In Memoirs by H. Faraday, F. Kohlrausch, & H. M. Goodwin (Eds.), *Harper's scientific Memoirs, VII. The fundamental laws of electrolytic conduction* (pp. 85–92). New York: Harper and Brothers.
57. Pickover, C. (2008). Kohlrausch's laws of conductivity, 1874 and 1875, from Section 1800 – 1900. In *Archimedes to hawking: Laws of science and the great minds behind them* (p. 387). Oxford: Oxford University Press.
58. Arrhenius, S. (1903). Development of the theory of electrolytic dissociation. *Nobel Lecture*, December 11, 1903.
59. Nernst, W. (1888). Zur Kinetik der in Lösung befindlichen Körper. Erste Abhandlung. Theorie der Diffusion. *Zeitschrift für Physikalische Chemie, 2*(9), 613–637.
60. Tafel, J. (1905). Über die Polarisation bei kathodischer Wasserstoffentwicklung. *Zeitschrift für Physikalische Chemie, 50*, 641–712.
61. Müller, K. (1969). Who was Tafel? *Journal of the Research Institute for Catalysis, Hokkaido University, 17*(1), 54–75.
62. Buff, H. (1857). Ueber das electrische Verhalten des Aluminiums. *Annalen der Chemice und Pharmacie, CII, Bandes drittes Heft* (pp. 265–284).
63. Beetz, W. (1866). Ueber Wasserstoff-Entwicklung an der Anode. *Annalen der Physik, 203*(1), 45–57.
64. Müller, W. J. (1931). On the passivity of metals. *Tranactions of the Faraday Society, London, 18*, 737–751.
65. Ducretet, M. E. (1875). Note sur un Rhéotome Liquide a Direction Constate, Fondé sur une Propriété nouvelle de L'Aluminium. *J Phys Théor, 4*, 84–85.
66. Mott, W. R. (1904). The corrosion of aluminium and its prevention. *Electrochemical Industry, II*(4), 129–130.
67. Pollak, C. (1897). *German Patent: Elektrischer Flüssigkeitskondensator mit Aluminiumelektroden*, No. 92564. Granted May 19, 1897.
68. *Mershon patent*, 1913.
69. Bengough, G. D., & Sutart, J. M. (1924). *British Patent: Improved process of protecting surfaces of aluminium or aluminium alloys*, No. 19,838/23. Accepted No. 3, 1924.
70. Flick, F. B. (1925). *American Patent: Coating aluminum articles*, No. 1,526,127. Granted Feb 10, 1925.
71. Pacz, A. (1925). *American Patent: Coated aluminum articles and process and means for producing same*, No. 1,551,613. Granted September 1, 1925.
72. Gower, C. H. R. (1928). *British Patent: An improved process for providing a resistant coating upon the surfaces of aluminium and aluminium alloys*, No. 27,880/27. Accepted May 24, 1928.
73. Zentrale, et al. (1935). *Gmelins Handbuch der Anorganischen Chemie: Aluminium*. Oberflächenbehandlung von Aluminium und Aluminiumlegierungen (pp. 451–534). Aluminium Tashenbuch, Herausgeber: Aluminium–Berlin.
74. Wernick, S., Pinner, R., & Sheasby, P. G. (2001). The surface treatment and finishing of aluminum and its alloys (6th ed., Vol. 1). Middlesex, England: ASM International, Finishing Publications.
75. Burgess, C. F., & Hambuechen, C. (1902). The electrolytic rectifier. *The Transaction of the Inaugural Meeting of the American Electrochemical Society, Philadelphia, 4*, 147–163.
76. Mott, W. R. (1904). Dimensions of the films on aluminium anodes. *Electrochemical Industry, II*(7), 268–271.
77. Mott, W. R. (1904). The electrical properties of the films that form on aluminium anodes. *Electrochemical Industry, II*(9), 352–355.

78. Mott, W. R. (1904). Colloidial precipitation upon aluminium anodes. *Electrochemical Industry, II*(11), 444–447.
79. Roeber, E. F. (Ed.). (1904). The film on the aluminium anode. *Electrochemcial Industry, II*(11), 436.
80. Guntherschulze, A. (1906). Über das Verhalten von Aluminiumanoden. *Annalen der Physik,* 929–954.
81. Guntherschulze, A. (1908). Die elektrolytische Ventilwirkung des Niobs und eine Klassifizierung des Verhaltens elektrolytishcer anoden. *Annalen der Physik, 330*(4), 775–782.
82. Setoh, S., & Miyata, A. (1932). Researches on the anodic film of aluminium, II. Anodic behaviours of aluminium in aqueous solutions of oxalic acid. In *Sci. Pap. I.P.C.R.*, Vol. 19, No. 397, pp. 237–291.
83. Burgers, W. G., Claasen, A., & Zernike, J. (1932). *Über die chemische Natur der Oxydschichten, welche sich bei anodischer Polarisation auf den Metallen Aluminium, Zirkon, Titan and Tantal bilden* (pp. 593–603). Eindhoven, Holland: Naturkundig Laboratorium der N.V. Philips' Gloeilampenfabriken.x
84. VanGeel, W. C., & Schelen, B. J. J. (1957). Some properties of oxide layers produced on aluminium by electrolytic oxidation. *Philips Research Reports, 12*, 240–248.
85. Besser, B. P. (2007). Synopsis of the historical development of Schumann resonances. *Radio Science, 42*(RS2SO2), 1–20.
86. Rummel, T. (1936). *Über Wachstum und Aufbau electrolytisch erzeugeter Aluminiumoxydschichten* (pp. 518–550). Mitteilung aus dem Elektrophysikalischen Laboratorium des Elektrotechnishcen Institiutes, Technische Hochschule, München, 10 Januar 1936.
87. Baumann, W. (1938). *Entstehung und Struktur elektrolytisch erzeugter Aluminiumoxydschichten* (pp. 708–736). Mitteilung aus dem Elektrophysikalischen Laboratorium des Elektrotechnishcen Institiutes, Technische Hochschule, München, 19 Dezember 1938.
88. Edwards, J. D., & Keller, F. (1940). *Formation of anodic coatings on aluminum* (pp. 135–144). Pittsburgh: The Electrochemical Society.
89. Wernick, S. (1948). Theory of anodic oxidation of aluminium, Chapter 5. *Electrolytic polishing and bright plating of metals* (pp. 57–60). London: Alvin Redman Ltd., Whitfield Place.
90. Fraunhofer, J. A. (1976). *Basic metal finishing*. London: Paul Elek (Scientific Books) Ltd.
91. Keller, F., Hunter, M. S., & Robinson, D. L. (1953). Structural features of oxide coatings on aluminum. *Journal of the Electrochemical Society, 100*(9), 411–419.
92. O'Sullivan, & Wood. (1970). The morphology and mechanism of formation of porous anodic films on aluminium. *Proceedings of the Royal Society of London, Series A, Mathematical and Physical Sciences, 317*(1531), 511–543.
93. Hoar, T. P., & Mott, N. F. (1959). A mechanism for the formation of porous anodic oxide films on aluminium. *Journal of Physics and Chemistry of Solids, 9*, 97–99. Pergamon Press.
94. Akahori, H. (1961). Electron microscopic study of growing mechanism of aluminium anodic oxide film. *Journal of Electronmicroscopy, 10*(3), 175–185.
95. Csokan, P. (1980). Nucleation mechanism in oxide formation during anodic oxidation of aluminum. In M. G. Fontana & R. W. Staehle (Eds.), *Advance in corrosion science and technology* (pp. 239–356). New York: Plenum Press.
96. Murphy, J. F., & Michelson, C. E. (1961). *Proceedings of a conference on anodising aluminium at the University of Nottingham*, Convened by the Aluminium Development Association of London, September 12–14, 1961, pp. 83–95.
97. Davies, J. A., Domeij, B., Pringle, J. P. S., & Brown, F. (1965). The migration of metal and oxygen during anodic film formation. *Journal of the Electrochemical Society, 112*(7), 675–680.
98. O'Sullivan, J. P., Hockey, J. A., & Wood, G. C. (1971). Infra-red spectroscopic study of anodic alumina films. *Transaction of the Faraday Society, 67*, 535–541.
99. Brace, A. W. (2000). *The technology of anodizing aluminium* (3rd ed.). Modena, Italy: Interall S.R.L.

100. Thompson, G. E., Furneaux, R. C., Goode, J. S., & Wood, G. C. (1978). Porous anodic film formation on aluminium substrates in phosphoric acid. *Transactions of the Institute of Metal Finishing, 56*(1m), 159–167.
101. Thompson, G. E., & Wood, G. C. (1981). Porous anodic film formation on aluminium. *Nature, 290*, 230–232.
102. Skeldon, P., Shimizu, K., Thompaon, G. E., & Wood, G. C. (1983). Barrier-type anodic films on aluminium in aqueous borate solutions: 2 – Film compositions by Rutherford backscattering spectroscopy and nuclear reaction methods. *Surface and Interface Analysis, 5*(6), 252–263.
103. Thompson, G. E., Xu, Y., Skeldon, P., Shimizu, K., Han, S. H., & Wood, G. C. (1987). Anodic oxidation of aluminium. *Philosophical Magazine, Part B, 55*(6), 651–667.
104. Kniep, R., Lamperter, P., & Steeb, S. (1989). Structure of anodic oxide coatings on aluminium. *Advanced Materials, Communications, 7*, 229–231.
105. Kubicki, J. D., & Apitz, S. E. (1998). Molecular cluster models of aluminum oxide and aluminum hydroxide surfaces. *American Mineralogist, 83*, 1054–1066.
106. Pringle, J. P. S. (1980). The anodic oxidation of superimposed niovium and tantalum layers. *Electrochimica Acta, 25*, 1403–1421.
107. Pringle, J. P. S. (1980). The anodic oxidation of superimposed metallic layers: Theory. *Electrochimica Acta, 25*, 1423–1437.
108. Furneaux, R., Rigby, W. R., & Davidson, A. P. (1989). The formation of controlled-porosity membranes from anodically oxidized aluminium. *Letters to Nature, 337*, 147–149.
109. Masuda, H., & Fukuda, K. (1995). Ordered metal nanohole arrays made by a two-step replication of honeycomb structures of anodic alumina. *Science, 268*, 1466–1468.
110. Wang, H. H., Han, C., Willing, G., & Xiao, Z. (2003). Nanowire and nanotube syntheses through self-assembled nanoporous AAO templates. *Material Research Society Symposium Proceedings, 775*, 4.8.1–4.8.6.
111. Parkhutik, V. (1992). Theoretical modeling of porous oxide growth on aluminium. *Journal of Physics D: Applied Physics, 25*, 1258–1263.
112. Patermarakis, G., & Karayannis, H. S. (1995). The mechanism of growth of porous anodic Al_2O_3 films on aluminium at high film thicknesses. *Electrochimica Acta, 40*(16), 2647–2656.
113. Patermarakis, G. (1996). Transport phenomena inside the pores involved in the kinetics and mechanism of growth of porous anodic Al_2O_3films on aluminium. *Journal of Electroanalytical Chemistry, 404*, 69–76.
114. Patermarakis, G. (1998). Development of a theory for the determination of the compostion of the anodizing solution inside the pores during the growth of porous anodic Al_2O_3 films on aluminium by a transport phenomenon analysis. *Journal of Electroanalytical Chemistry, 447*, 25–41.
115. Patermarakis, G., & Moussoutzanis, K. (2001). Formulation of a criterion predicting the development of uniform and non-uniform abnormal porous anodic alumina coatings and revealing the mechanisms of their appearance and progress. *Corrosion Science, 43*, 1433–1464.
116. Runge, J., & Pomis, A. (2000). Anodic oxide film formation: Relating mechanism to compostion and structure. In *The Proceedings for Aluminium Surface Science and Technology Conference (ASST)*, May 2000, UMIST.
117. Lerner, M. (2003). Is the barrier layer responsible for the cellular structure of aluminum oxide film anodized in sulfuric acid? In *Proceedings of AESF Surfin Conference*, Chicago.
118. Garcia-Vergara, S. J., Iglesias-Rubianes, L., Blanco-Pinzon, C. E., Skeldon, P., Thompson, G. E., & Campestrini, P. (2006). Mecahnical instability and pore generation in anodic alumina. *Proceedings of the Royal Society A, 462*, 2345–2358.
119. Garcia-Vergera, S. J., Skeldon, P., Thompson, G. E., & Habazaki, H. (2006). A flow model of porous anodic film growth on aluminium. *Electrochimica Acta, 52*, 681–687.
120. Hauser, J., & Hebert, K. (May 2009). The role of viscous flow of oxide in the growth of self-ordered porous anodic alumina films. *Nature Materials, 8*, 415–420.

121. Paschanka, M., & Schneider, J. (2011). Origin of self-organization in porous anodic alumina films derived from analogy with Rayleigh-Bénard convection cells. *Journal of Materials Chemistry, 21*, 18761–18767.
122. Paschanka, M., & Schneider, J. (2013). Experimental validation of the novel theory explaining self-organization in porous anodic alumina films. *Physical Chemistry Chemical Physics, 15*, 7070.

Chapter 3
Anodizing as an Industrial Process

3.1 Introduction

All components manufactured from aluminum and its alloys have surfaces with performance requirements that vary based upon the service environment and the application. Whether the requirement is corrosion and/or wear resistance, or simply a beautiful appearance, for many applications, the surface of the aluminum or aluminum alloy component may be anodized to meet the necessary performance requirements. The anodized finish, better identified as the anodic oxide, originates and grows from the surface of the aluminum, producing a uniform, continuous, highly ordered network of individual cells, comprising a layer whose thickness and cell dimensions, and ultimately engineering properties, depend on the electrochemical parameters of the anodizing process.

Many different aluminum and aluminum alloy products: castings; wrought products such as extrusions, rolled products, forgings, and deep drawn products; as well as sintered powder metal products, are designed and specified to have an anodized finish. By capitalizing on the natural phenomenon of passive film formation on aluminum, and by enhancing it electrochemically to build an integral oxide, the anodic oxide finish has become synonymous with surface protection, durability, and decoration of aluminum substrates. All industries that employ the high-strength-to-weight ratio of aluminum and its alloys: aircraft, defense, architecture, automotive, electronics, hydraulics, medical and semiconductor equipment, and other markets not mentioned, look to the anodizing process as an important means to finish a variety of components manufactured in a variety of ways, with the robust and beautiful anodic oxide in order to provide corrosion protection, wear resistance and in many applications, decorative appearance. Integral oxides of substantial thickness have been industrially produced by electrochemical methods for over 100 years [1].

What sets the anodic aluminum oxide (AAO) apart from other metal finishes is that it is not a deposited layer, such as an electrodeposited metal, or an applied layer like a paint or a lacquer. It is also not a chemical conversion coating like a phosphate or

J.M. Runge, *The Metallurgy of Anodizing Aluminum*,
https://doi.org/10.1007/978-3-319-72177-4_3

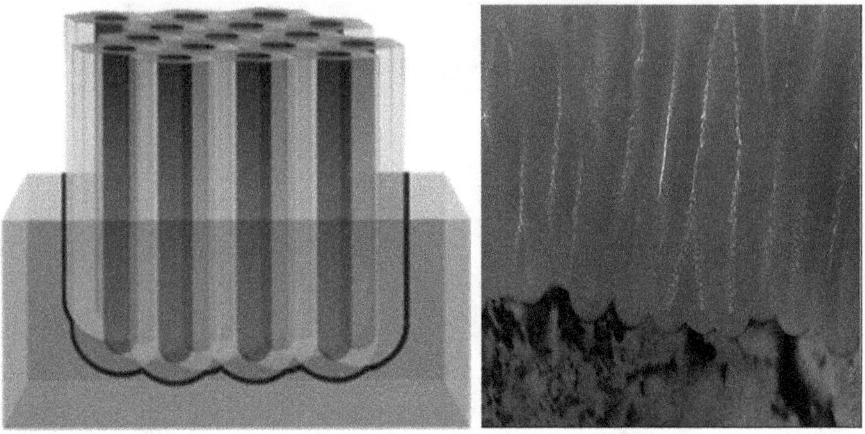

Fig. 3.1 Schematic of the structure of the anodic aluminum oxide, AAO (left), accompanied on the right by a transmission electron microscope image of AAO processed under hard anodizing conditions (low temperature, high current density). The anodic oxide is a self-assembled network of individual columnar cells that have rounded bottoms. Each column of the AAO structure circumscribes its critical feature, the central pore, which is the conduit for ion exchange during oxide growth. In the photomicrograph on the right, the central pore of each columnar cell in the Transmission Electron Microscope (TEM) image appears bright, in contrast with the uniformly gray-appearing columns. The aluminum substrate is at the image bottom, inhomogeneous in appearance, with black speckles and streaks [4]

chromate conversion coating. All conversion coatings are equilibrium processes, with only the chemical potential difference between the electrolyte and substrate driving the chemical reaction that produces the conversion layer; the reaction stops when each free aluminum ion at the surface has been oxidized. To contrast, the anodic oxide is a self-assembled, highly ordered, nanoscale network of oxide cells that has grown from its parent aluminum (alloy) substrate through the electrical stimulation by an external source. Because it is externally driven, anodizing, by definition, is not an equilibrium process: the substrate provides an infinite source of aluminum ions that are "pushed" to the interface by the applied current in order to react with the electrolyte. Nevertheless, by its very nature, AAO has its origins in the natural characteristic for aluminum to form a passive oxide layer, which is an equilibrium process. Aluminum oxide is an effective barrier to the intrusion of external anions, therefore earning the name "barrier layer." Because of this characteristic, the passive layer and the anodic oxide exclusively comprise aluminum and oxygen as aluminum oxide.

Possibly the most significant characteristic of the AAO is the highly ordered structure. AAO comprises many nanoscale columnar cells with rounded bottoms, each circumscribing a central pore. As-grown, the oxide structure is rather precise; the cell walls, bottoms, and pore diameters for oxides grown at specific process parameters are consistent from cell to cell. In fact, the anodizing process can be manipulated or "tuned" to yield structures of specific size, enabling the use of AAO as a template for nanoscale wires or as nanoscale indenters for a variety of applications. In many industrial applications, AAO structural integrity provides a barrier to

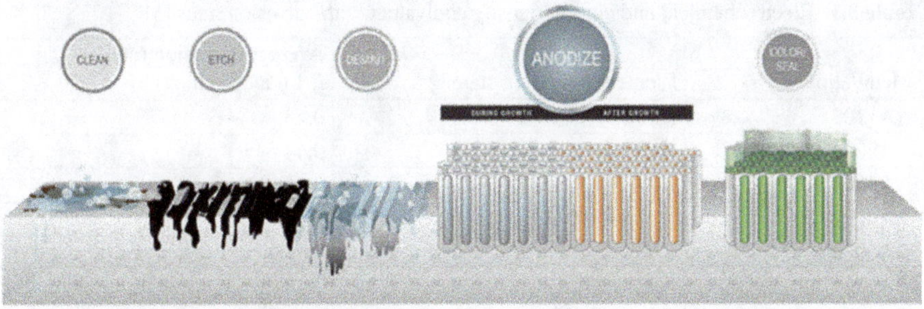

Fig. 3.2 Schematic of the anodizing process. The change in color of the pores from blue to orange represents the change in the character of the oxide during and after growth. During oxide growth, ion exchange along the pore sustains oxide growth; after growth, although ion exchange stops, the pore wall remains ionic, enabling dyeing and sealing, which is represented by the change in color from orange to green

the external environment and corrosion protection over a fixed pH range; the size and residual stress within the AAO structure provide abrasion and wear resistance. The central pore offers reaction sites for bonding with other molecules, providing a surface for dyeing, coloring, and bonding. See Fig. 3.1.

A brief overview and summary of the steps that comprise the industrial process for anodizing aluminum follows and is presented according to the sequence typically followed by most anodizers and follows the basic anodizing course outline presented in the United States annually by the Aluminum Anodizers Council in the Level I class in the School for Anodizers, "Anodizing Essentials" [2]. To familiarize the reader with the structure of various types of anodic aluminum oxide, images highlighting the characteristics of the AAO are presented that have been grown from various electrolytes under various industrial anodizing process conditions. Should the reader require more in-depth process information, the technical details of the anodizing process are well covered in The Technology of Anodizing Aluminium by Arthur W. Brace [3].

3.2 Industrial Anodizing

The industrial anodizing process begins with the preparation of the aluminum substrate. The first step requires the mounting of the components for optimum conductivity and mechanical stability in the electrical circuit. Pretreatment continues by processing the mounted components, or work, through a variety of cleaners, etchants, sometimes brighteners, always with copious rinses, to maximize the amount of free aluminum at the surface for oxidation. Electrochemical oxidation, the actual anodizing, is the step in the process during which the highly ordered oxide finish is grown. Post-treatment consists of rinsing and sometimes dyeing and sealing the anodic oxide structure, finishing the oxide for application. Figure 3.2 provides an overview schematic of the anodizing process.

Table 3.1 Electrochemical and current density equivalent with corrosion rates [5]

Metal/alloy	Element/oxidation state	Average corrosion rate i_a 1 μ a/cm^2 (mpy)
AA1100	Al/3	0.43
AA2024	Al/3, Mg/2 Cu/2	0.44
AA3004	Al/3, Mg/2	0.43
AA5052	Al/3, Mg/2	0.44
AA6070	Al/3, Mg/2	0.43
AA6061	Al/3, Mg/2	0.43
AA7072	Al/3, Zn/2	0.43
AA7075	Al/3, Mg/2, Zn/2, Cu/2	0.44

3.2.1 Alloy Selection

The average corrosion rate for aluminum alloys from each of the main wrought alloy designations is always about the same, meaning the corrosion response for most alloys is consistent. This characteristic is the basis for the success of the anodizing industry, which is based upon the average oxidation performance of aluminum. See Table 3.1. By regarding the lack of variation in corrosion rate for the wide range of the various wrought alloys presented, it is clear that because aluminum is the primary component that dissolves when the anode corrodes all aluminum alloys will anodize. Experience dictates, however, that alloying elements impact *how* the substrate anodizes, with the difficulty increasing with the complexity of the alloy. From this perspective, it is interesting to note that no cast alloys, which are far more complex than the wrought alloys listed, are part of the Table 3.1.

Although the functional characteristics of beautiful appearance, excellent corrosion protection, and rugged mechanical properties in abrasive wear and impact of the AAO are separately or together intended for an engineering application, these characteristics are often considered for the bulk of the component as design requirements, long before the surface reactions that occur during anodizing. This lack of regard sometimes results in alloy selection that is not compatible with growing AAO that has engineering performance or characteristics that match what is desired. Although all aluminum alloys anodize, when designing a component for anodizing, it should be kept in mind that *alloying elements and intermetallic particles don't anodize*. See Fig. 3.3.

3.2.2 Racking or Jigging

Rack design is one of the parts of the anodizing process that is a team effort between the customer, the rack builder, and the anodizer. While the details of rack or jig design are beyond the scope of this book, the following basic considerations are presented.

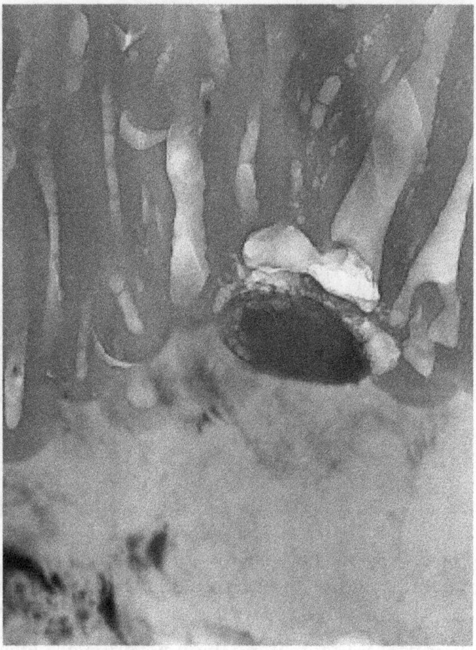

Fig. 3.3 TEM image of a Fe-Cr dispersoid particle in a 6061 aluminum alloy substrate encountering the interface of the anodic oxide during the anodizing reaction. Note the disruption in the normally highly ordered AAO structure by the particle. With continued oxidation, the particle will be incorporated within the structure, but never anodized or chemically bound to it

The most basic function of an anodizing rack or jig is to transport the work to be anodized. However, the demands on racking in any electrochemical circuit create the need for design requirements between the rack, the components, and the electrolyte that deal with three key points: (1) the suitability for the process sequence, including the material that comprises the rack itself; (2) compatibility with pre- and post-treatment solution chemistry; and (3) consideration for the electrical effects that occur in the circuit.

Because anodizing is an electrochemical oxidation process, the mounting, or racking of the components to assure sound electrical contact and anodizing circuit continuity comes before all other concerns about other steps of the process. Racks must be designed to hold the components securely, not only for electrical continuity, but to keep them stable during the transfer between process tanks and while immersed in vigorously agitated solutions. Additionally, racks must be designed to enable adequate rinsing and draining of the entire workload, without retaining pockets of residual solution. Because racking is a mechanical process, the mounting of the components and the means by which the parts are held on the rack, must not only provide sound electrical contact, but must also be done without damaging the components. See Fig. 3.4.

The material of design for the rack must be compatible with the process solutions in the anodizing line. The most typical materials used for rack construction are the higher purity wrought aluminum alloys 6061 or 6063 and the commercially pure grades of titanium. Aluminum has a higher current carrying capacity than titanium, making it ideal for anodizing processes that use higher current density, such as hard anodizing (Type III) processes. Aluminum racks, however, anodize with the workload, limiting

Fig. 3.4 Aluminum alloy parts mounted on a rack, ready to be anodized. (Photo courtesy of Impreglon Surface Engineering, Kaufbeuren, Germany)

their useful life, making them nevertheless ideal for low volume, low production runs. Titanium usually does not react with the anodizing process chemistry, meaning that the applied current density anodizes the components only, not the rack; however, titanium does react with acid etch anodizing pretreatment solutions that contain ammonium bifluoride, or hydrofluoric acid, making it a consideration when selecting the material for rack design; nevertheless, for high volume, long-running production, titanium racks are ideal. Hybrid racks that use both materials; titanium rack with aluminum contacts, for example; or an aluminum rack with titanium contacts can also be designed.

3.2.3 Pretreatment

Even before anodizing, the quality of the substrate surface must be considered. External surface contamination increases the resistance of the anode surface and changes the surface chemical potential. Therefore, the method and means of anode pretreatment prior to anodizing is critical. The substrate must be cleaned and rinsed to rid the surface of any external contamination, such as dirt and grime from manufacturing or storage that may interfere with the formation of a continuous oxide. Rinsing may be the most important step throughout the anodizing process as it removes the residues from cleaning and minimizes contamination from the previous pretreatment steps from being dragged into subsequent substrate pretreatment solutions and the anodizing electrolyte.

The pretreatment that follows cleaning is always performed to minimize alloy effects; that is, to maximize the amount of free aluminum at the surface for oxidation.

The substrate alloy chemistry and microstructure impact the surface resistance because they comprise non-aluminum intermetallic particles, precipitates and phases, and other nonmetallic inclusions and/or surface contamination to the electrochemical circuit that don't anodize. The presence of such particles can change the local chemical potential of the surface drastically from the average values presented in Table 3.1, creating discontinuities in the AAO structure, which increase surface roughness and may even cause burning.

Pretreatment of the surface is also done to enhance finish appearance after anodizing, by cleaning the surface, removing intermetallic particles, and modifying the surface texture. Pretreatment can be done to brighten the surface, by way of chemical polishing or "bright dip" processes, or to deliberately create a matte finish by way of acid etch processes. Even before entering the anodizing process, mechanical treatment to impart a texture for appearance or haptic may be done by way of shot peening, blasting, and/or other forms of surface modification.

Pretreatment processes are mostly done without external electrical polarization or current bias, and therefore, the quality of the substrate microstructure and how it will react with even the simplest cleaning solutions must be regarded as critical to the surface quality of the anode. Without polarization, the slightest driving force at the surface caused by chemical potential differences between the aluminum matrix and an intermetallic particle will produce very high surface corrosion rates, producing pits and developing microstructural features [5].

3.2.4 Rinsing

Because of the myriad of chemical reactions that occur throughout the process, aside from the oxidation step, rinsing is possibly the most important step of the entire anodizing operation. Rinsing stops the chemical action of a treatment step by way of: (1) removing the source of chemical action at the surface; (2) changing the temperature and time of the reaction at the component surface; and (3) changing the pH at the surface by diluting with a clean, neutral solution. Rinsing must also involve draining, in order to prevent reinitiating of chemical attack on the part surface and to minimize the transfer of chemical solution to the next step in the process. The rule of thumb for effective rinsing is that every chemical bath should have 2–3 rinses.

The chemical principle of "like dissolve like" must be employed to optimize the effectiveness and efficiency of a rinse tank. The effectiveness of the rinse may depend upon how rapidly the temperature of the surface is reduced. An immediate cold rinse may not stop the reaction at the surface, because the solution capability at low temperatures is not as high as with warmer rinses. And, with the throughput of each workload, the first rinse tank will take on the pH of the previous chemical treatment tank. However, if one keeps "like dissolves like" in mind, a warm acid rinse for an acid tank followed by a cooler rinse tank with a neutral pH will definitely dilute and remove chemical residue effectively. Exit sprays on the final rinse tank

Fig. 3.5 A rack of hard anodized components exiting a rinse tank after the anodizing process. Supplementary spraying is underway to facilitate rinsing. (Courtesy of Impreglon Surface Engineering, Kaufbeuren, Germany)

offer the chance for a third rinsing cycle in a physical plant that has room for only two rinse tanks. See Fig. 3.5.

Rinse water quality is an important consideration in anodizing, particularly before sealing, when silicates and carbonates can inhibit reactions with the AAO by binding with active ion sites on the porous structure. Therefore, the use of deionized water for the rinse before sealing is considered best practice.

3.2.5 Anodizing

The industrial process of anodizing aluminum can be viewed simplistically as the operation of an electrical circuit, which follows Ohm's Law.

$$V = iR \tag{3.1}$$

In this equation, V is the circuit voltage measured in volts and R is the circuit resistance. In a conventional electrical circuit, the current, i, is typically measured in amps; however, this is because the diameter of a wire is considered negligible. In any electrochemical circuit, the surface area of the entire current carrying workload must be considered in calculating the potential critical for the reaction whether it is for the dissolution of a cathode to produce a deposit on the surface area of an anode, as in

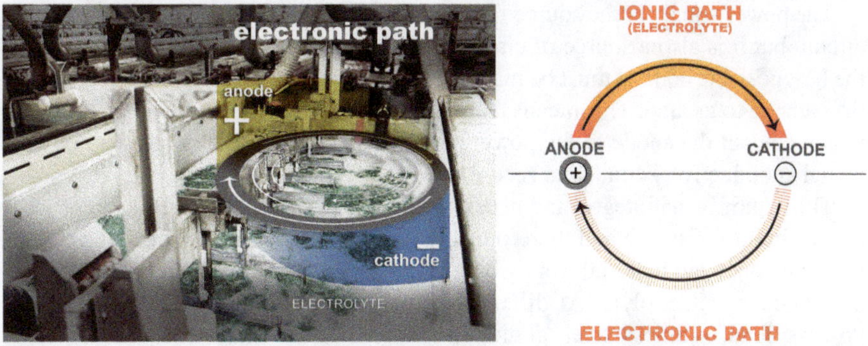

Fig. 3.6 Schematic showing the simple electrical circuit established during anodizing. In anodizing, a power supply is used to set the electronic path (current or voltage bias) in the direction that enables electrotransport of aluminum ions from the anode toward the electrolyte, which provides the ionic path through the electrolyte to the cathode, completing the circuit. The anodic oxide grows on the surface of the anode, when aluminum ions react with oxygen ions from the electrolyte. (Photo courtesy of Impreglon Surface Engineering, Kaufbeuren, Germany)

Fig. 3.7 Photo of an anodizing line. (Photo Courtesy Impreglon Surface Engineering, Kaufbeuren, Germany)

electroplating; or the dissolution of an anode to grow an oxide, as in anodizing. The parameter of current density is particularly important when addressing the anodizing reaction as a corrosion reaction (mass loss due to oxidation) and therefore will be considered in A/unit area.

The elements required for operation are: an electrical power source, an electrolyte (which is usually an acid solution), a cathode (which is usually aluminum or an aluminum alloy, but can be lead, stainless steel, or carbon), and the anode, which is the aluminum substrate from which the anodic oxide will grow. As the oxide grows, it also becomes a part of the anodizing circuit. See Figs. 3.6 and 3.7.

In order to produce the most uniform, continuous and defect-free anodic aluminum oxide, the resistance of the circuit and therefore the resistance of each element of the circuit must be minimized. The resistance of each circuit component is briefly considered in the following paragraphs.

The power supply is the source for the energy necessary to operate the anodizing circuit, but it is also a source of circuit resistance. In order to minimize resistance, the best possible contact must be made from the power supply to the anode and from the cathode to ground. This means that racks or jigs and their respective connections which contact the anode to the power supply and the electrolyte must also be adequately conductive, sound, and secure.

The cathodes, an integral and necessary part of the circuit, should not contribute to the circuit resistance and therefore should be considered from both the perspective of contact and material from which it is made. Cathodes should not dissolve in the anodizing electrolyte. At different times during the history of the anodizing process, lead, stainless steel, aluminum, various aluminum alloys, titanium, and carbon cathodes have been used industrially, with aluminum alloy cathodes being the most commonly used. The best contact possible between the electrolyte and the cathode is a must, as is the efficiency of the cathode to complete the anodizing circuit. Maintenance of the cathodes is therefore an important part of the anodizing.

The electrolyte, by far the biggest conductor in the anodizing circuit, may be comprised of a variety of protonic acids, as a single or mixed acid electrolyte, and may contain additives to chemically aid in reducing electrolyte resistance and/or reducing the interfacial heat of the anodizing reaction. Pure water, having a very low concentration of ions, cannot conduct electricity. Anodizing electrolytes are typically mixtures of strong acids that are soluble in water. Solute anions within the electrolyte carry the current from the anode to the cathode and, depending upon the temperature of the electrolyte, can aid in thermally conducting resistance heat away from the anode surface during oxidation. By increasing acid concentration to increase electrolyte conductivity for a fixed current density, the resistance at the interface of the anode with the electrolyte can be reduced. However, as the acid concentration of the electrolyte increases, the pH of the electrolyte also decreases, increasing the conductivity of the solution causing the resistance of the solution to decrease, and the circuit voltage, or potential, as a response to a specific current density, to decrease.

In anodizing, the electrolyte concentration-potential relationship can lead to problems during industrial processes, as it is commonly believed that higher voltage means faster or more oxide growth, which is not the case. Current density is the parameter that drives the mechanisms that grow the oxide; current produces the necessary electron flow in the substrate that drives ion diffusion and electrotransport of particles and intermetallic compounds to the interface for oxidation to occur, and the voltage is a response to that reaction. Oxidation potential is a characteristic of the anode, which is the aluminum component, and as long as the potential response of the anodizing process exceeds the characteristic potential of the aluminum, oxidation occurs as long as current flows.

The anodic oxide produced by way of a constant voltage process, which allows the current density to vary with the resistance of the finish, is more subject to the microstructure effects of the substrate that change as the interface changes. If voltage is fixed and current is allowed to vary, oxidation kinetics will favor oxide growth in areas that are more aluminum rich, while defects at the interface retard oxide growth, producing

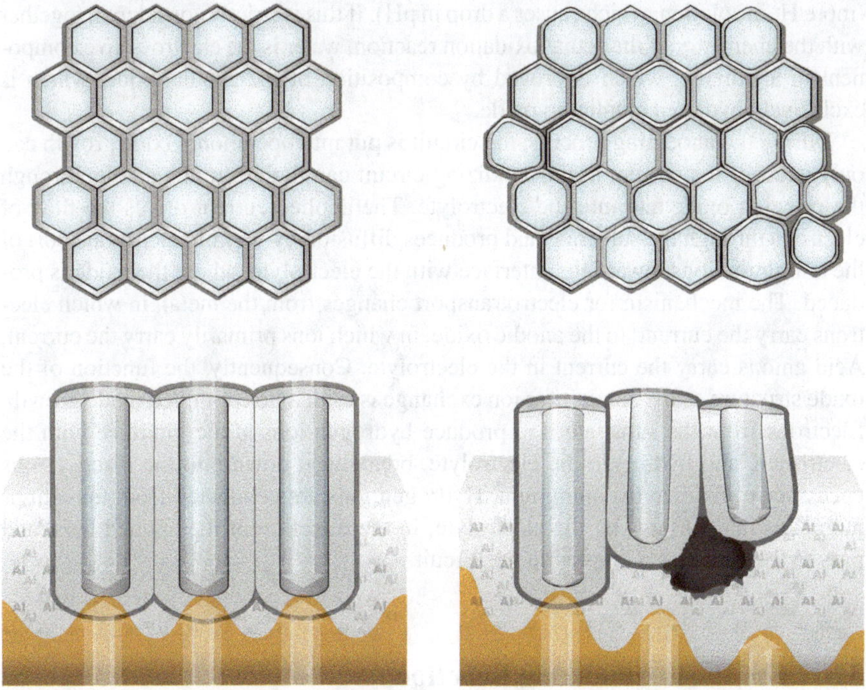

Fig. 3.8 Schematic of the highly ordered aluminum anodic oxide (left) and how the order of the structure can be disrupted by a non-aluminum or nonmetallic inclusion (right). Because these particles do not anodize, their presence changes the local chemical potential for the anodizing reaction. Once they are anodized through or around, order is restored, but the defect is included in the AAO. The details of the reactions that occur while anodizing alloys are presented in Chap. 7

nonuniform oxide growth across the substrate surface, making the oxidation reaction more sensitive to the effects of concentration polarization (see Chaps. 5 and 7). If one anodizes with constant current, even though substrate resistance is still the main limiting factor in oxidation, sustained electron flow over the entire surface overrides most of the effects of microstructural interfacial phenomena, and an average resistance for the entire anode surface can be considered for the oxidation [6]. Current, as the only energy source for oxide growth, is the rate controlling parameter for the oxidation reaction, and the voltage (potential) response can be monitored and noted as a characteristic for finishing a specific alloy. See Fig. 3.8.

The structure of the anodic oxide and how it grows are also affected by the electrolyte counterion (which acid is used to make the electrolyte). Therefore, there are tradeoffs in the chemistry and concentration of the electrolyte that must be understood to fully understand their impact. These will be addressed in Sect. 3.5. Adsorption of the electrolyte counterion on the surfaces of the AAO that are in contact with the electrolyte can alter the AAO structure by inhibiting the ion exchange necessary to sustain the central pore. Regardless of the counterion, it is an electrochemical fact that the oxidizing power of the electrolyte is increased with increased acid concentration (more H^+ in solution, which causes a drop in pH). If this is strictly considered, together with the chemistry of the actual oxidation reaction, water is the electro-active component in anodizing, which is proved by composition of the anodic oxide, which is exclusively hydrated aluminum oxide.

During the anodizing process, the circuit is put into operation. Oxide growth can only occur if the current in the anodizing circuit can flow from the anode, through the growing oxide and into the electrolyte. The applied current drives the flow of electrons through the substrate and produces diffusion by way of electrotransport of the aluminum ions toward the interface with the electrolyte, where the oxide is produced. The mechanism for electrotransport changes from the metal, in which electrons carry the current; to the anodic oxide, in which ions primarily carry the current. Acid anions carry the current in the electrolyte. Consequently, the function of the oxide structure as the conduit for ion exchange is critical to continued oxide growth. Electrons from the substrate also produce hydrogen ions at the interface with the electrolyte, and hydrolyze the electrolyte, breaking it down into the reactive ions necessary to oxidize the aluminum. Hydrogen ions are conducted from the aluminum substrate surface, by the electrolyte, in the direction of the applied bias, and pass to the cathodes, completing the circuit.

3.3 The Basic Anodizing Reaction

When in contact with the applied current and the electrolyte, the source for the growing oxide is the aluminum anode; as such, the oxide consumes the aluminum anode as it grows. Therefore, the interface, from which the oxide grows, continuously changes as it advances into the substrate while new oxide forms, pushing the forming oxide layer outward from the interface. The formation of the oxide at the metal-oxide interface involves the migration of aluminum cations (Al^{3+}) from the metal, across the forming oxide, toward the interface with the electrolyte, and the inward migration of oxygen anions (O^{2-}) from the electrolyte to the metal. Ion exchange is the chief mechanism driven by the electrochemical reaction, which produces the uniquely elegant, self-assembled, highly ordered, columnar nanoscale structure that grows to become the anodic aluminum oxide (AAO) finish.

When aluminum is polarized as the anode, the basic oxidation reaction that occurs at the interface with the electrolyte can be summarized as follows:

$$2Al + 3H_2O \rightarrow Al_2O_3 + 6H^+ + 6e^- \tag{3.2}$$

AAO is not present as corundum, or any crystalline phase of aluminum oxide. It is amorphous, hydrated aluminum oxide present as $Al_2O_3 \cdot H_2O$ and $Al(OH)_3$. This means that as formed, AAO does not have a crystalline structure, but it does have a chemical structure.

For anodizing in a sulfuric acid electrolyte, the reaction proceeds as follows:

Fig. 3.9 Examples of uses of anodized aluminum. Consumer electronics is one of the most common industries that utilize anodized aluminum. The scanning electron microscope image on the right documents the appearance of nickel nanowires that were electroplated in an AAO template [9, 10]

$$4\text{Al}+6\left(\text{H}_2\text{SO}_4\right)\rightarrow 2\left(\text{Al}_2\text{O}_3\right)+6\left(\text{SO}_3^{\;-}\right)+3\left(\text{H}_2\right)\text{g}+6\text{H}^+ +6e^- \tag{3.3}$$

The aqueous sulfuric acid electrolyte is hydrolyzed to enable oxidation of the aluminum, while sulfite ions and hydrogen gas are produced at the interface by way of simultaneous oxidation of the aluminum and electrolysis of the electrolyte. Because the primary reaction produces exclusively aluminum oxide, adsorbate from the electrolyte counterion is incorporated into the portion of the oxide, which is in contact with the electrolyte, and is significant enough to be detected when the anodic oxide is analyzed by way of Energy Dispersive Spectroscopy (EDS). Wernick, Pinner, and Sheasby publish the approximate composition of the total anodic oxide layer as 72% Al_2O_3; 15% H_2O, and 13% SO_3^- [7].

The AAO structure is well documented and is used for a variety of applications, both as a finish integral with the aluminum substrate and as a stand-alone structure. As an integral finish, which has been in use for over 100 years, the anodic oxide provides a dielectric barrier; it also provides wear resistance, corrosion resistance and enables decoration through dyeing and printing. Since the late twentieth century, the unique self-assembled structure of AAO has been used by itself as a dimensionally tunable and inexpensive template for many different nanoscale materials such as nanodots, nanowires, and nanotubes [8]. See Fig. 3.9.

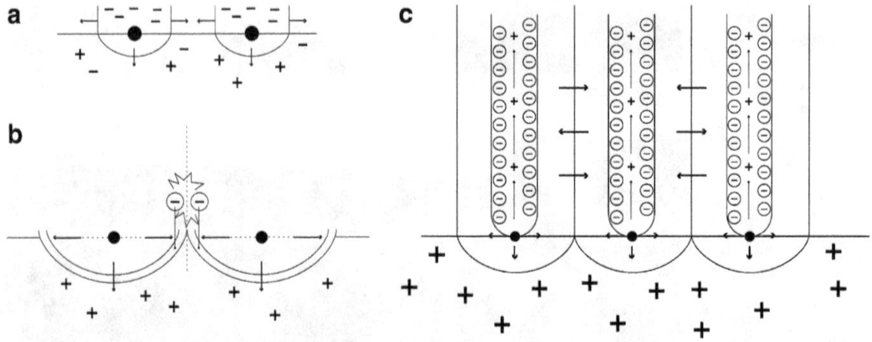

Fig. 3.10 Schematic summarizing AAO nucleation and growth. (**a**) Oxide nuclei, ordered by polarization of the anode substrate, form the base of pore, growing outward while consuming the anode, reconstructing the anode surface with oxide. (**b**) Following total surface reconstruction, with impingement of nearest neighbor oxide elements, or "flakes" repulsive forces between similarly charged oxide "flakes" foster outward growth and initiate pore formation. (**c**) Mass transport and diffusion across column walls form knitlines while repulsive forces within pore maintain the erect nature of the column and dynamic flow of the electrolyte [1]

3.4 Characteristics of the Anodic Aluminum Oxide

In general, the anodic aluminum oxide is a self-assembled, highly ordered network of nanoscale corrosion cells. As formed, anodic oxides formed in sulfuric acid, oxalic acid, and phosphoric acid electrolytes are generally uniform in shape, size, and composition. The cells, regardless of electrolyte, exhibit similar structural characteristics: they are columnar right cylinders with a rounded, closed base and a central pore, comprising only hydrated aluminum oxide (although electrolyte anions adsorb on the structure at the AAO-electrolyte interface). The base and the cell wall dimension are the same. The central pore proceeds from the base of the column to the top of the cell. The cells exhibit distinct boundaries, which are fused together by way of mass transport and diffusion during the oxidation reaction to form a network. The dimensions of each cell are uniform: the wall thickness and pore diameter, and are proportional to the anodizing process parameters of applied current density and potential response, or conversely, the applied "formation" voltage and the current response. See Fig. 3.10.

Because the oxide growth originates at and consumes the aluminum substrate, with the applied current density, the growth front into the substrate is nonequilibrium and dynamic, and the interface between the substrate and the growing oxide is constantly changing. As a consequence, the substrate conditions: surface cleanliness, alloy chemistry, surface texture, and near-surface microstructure profoundly impact nucleation and growth of the anodic oxide. The entire anodic oxide network functions as a barrier layer during formation, to alloying ions at the substrate-oxide interface and to the electrolyte anions at the oxide-electrolyte interface. In engineering application, the AAO provides a barrier to external corrosion and wear.

The anodic oxide grows from the substrate-oxide interface as aluminum ions diffuse to the substrate surface and react with oxygen that has diffused through the structural layers of the finish from the electrolyte. Consequently, AAO exclusively comprises (hydrated) aluminum oxide, $Al(OH)_3$ and AlOOH. Aluminum and oxygen are the only elements on the oxide structure. Alloying elements don't anodize, yet how they are present in the substrate microstructure, as they intersect the substrate-oxide interface during anodizing as well as the overall substrate microstructure, affects how the oxide grows. These aspects of anodic oxide growth and development will be discussed in Chap. 7.

3.4.1 Oxide to Substrate Ratio

When considering anodic oxide growth, it is important to remember that *the oxide growth front proceeds into the substrate.* Therefore, the aluminum substrate is consumed as the oxide grows. The bulk anodic oxide is less dense than the base metal; therefore the total layer thickness is greater than the aluminum consumed, which means the dimension of the substrate (finished component) after anodizing is different (larger) from the dimensions of the original substrate. The amount of change depends on the anodizing process conditions: substrate alloy, temperature, electrolyte type and concentration, and the formation voltage (current density).

In total, the oxide thickness accounts for corrosion of the substrate by the anodizing process; however, there are specific trends for particular anodizing processes that enable general characterization of the characteristics of the anodic oxide. For example, for anodic oxide grown under Type II or technical anodizing conditions, (20–22 °C and a current density of $0.8\ \mathrm{A/dm^2} < i_a < 1.8\ \mathrm{A/dm^2}$), a process typically run in a sulfuric acid electrolyte with wrought substrates with a corresponding alloy loading $\leq 8\%$, approximately 60% of the AAO total thickness is accounted for by substrate consumption; the remaining 40% of the oxide extends beyond the substrate surface. For oxides grown under Type III, or hard anodizing conditions (0–5 °C and a current density of $2.0\ \mathrm{A/dm^2} < i_a < 3.5\ \mathrm{A/dm^2}$), which is a process that typically employs more highly alloyed substrates such as wrought aluminum super alloys and castings with alloy loading exceeding 12%, in a sulfuric acid electrolyte, approximately 50% of the AAO thickness is accounted for by substrate consumption and the balance extends beyond the surface [11].

When considering the effects of anodic oxide growth on the substrate component from the perspective of size, such as final dimensions; abrasion and wear resistance; number of pores per unit grain (oxide density); or even the overall efficiency of the anodizing process, it can be useful to consider the ratio of the anodic oxide volume to the volume of the remaining anode. Other methods, such as coating weight, are also utilized in industry.

Anodizing process efficiency can be calculated by measuring the actual oxide coating weight for a specific anode, and dividing it by the theoretical weight of the oxide as Al_2O_3. Such calculations show the impact of alloy of process efficiency, with more complex alloys, such as 2014 and 2024 exhibiting coating efficiencies of ~80% and alloy 6063 exhibiting a coating efficiency of ~100% [3].

Recently, the use of the Pilling–Bedworth ratio (PBR), or volume expansion factor [12] has been suggested as a means to evaluate pore density. This is especially relevant for the manufacture of highly ordered AAO templates for use in producing nanoscale devices such as wires and dots [13–15]. The PBR was first published in 1923 to quantitatively define the quality of passive oxide layers formed on different metals at high temperatures [16]. The basic assumption for the formation of a passive layer at high temperatures was that oxygen, as O^{2-}, would diffuse through the forming oxide layer to the metal surface. The fundamental premise for the PBR is that the oxide formed is passivating and provides a protecting effect against further surface oxidation for a ratio $1 < \text{PBR} < 2$ (metals such as Al, Ti, and stainless steels); oxides with $\text{PBR} < 1$ are not protective, and oxides with $\text{PBR} > 2$ are brittle and prone to spalling, and therefore provide no protective effect. The calculations made to determine PBR were made on oxides formed in a dry environment with pure metals only, and did not account for the movement of metal ions outward through the oxide, or their role in building the oxide. Alloys other than stainless steel were not measured. It did, however, account for the valence and growth stress that develops in oxide layers and has been used to predict the stress developed at the metal-oxide interface due to volume change as oxide develops.

$$\text{PBR} = \frac{V_{\text{oxide}}}{V_{\text{metal}}} = \frac{M_{\text{oxide}} * \rho_{\text{metal}}}{n * M_{\text{metal}} * \rho_{\text{oxide}}} \tag{3.4}$$

When considering the AAO as a functional oxide for wear applications or as a template for the manufacture of uniform nanoscale wires, tubes, dots, or other types of matrix indentations, the actual volume and density of the oxide must be considered in more detail. The first consideration is the characteristic of the porosity of the AAO itself: the true volume of oxide (V_o), when porous, is less than the apparent volume occupied by the oxide layer (V_{total}). Because the dimensions of the anodic oxide are proportional to the anodizing conditions (pore diameter is proportional to the formation voltage), the dimensions of the individual cells comprising the AAO can be predicted, making V_o a value that can be calculated. Finally, the oxide is not stoichiometric Al_2O_3 and is definitely not dry, but is hydrated: $Al(OH)_3$ and AlOOH. The density of the AAO (3.01 g/cm^3) is definitely different from that of dry, stoichiometric Al_2O_3 (3.97 g/cm^3) [17]. Consequently, when strictly considered, PBR is not appropriate for nonequilibrium reactions such as anodizing, in which there is a dominant effect of ionic activity from the anode because of the applied current density, as well as differences in V_o due to the characteristics of the AAO. However, trends in the values for PBR can give interesting insight as to the functionality of the oxide in wear applications; for example, AAO with a higher PBR for a given alloy and anodizing process conditions indicate higher internal stress and may show higher abrasion resistance than those with a lower PBR.

The PBR for pure aluminum with dry, stoichiometric Al_2O_3 as oxide is 1.29. Brace [3] reports oxide ratios for industrially produced sulfuric acid anodic oxides range, depending upon alloy and anodizing conditions, from 1.30 to 1.65. Xu and Gao [18] clearly determine the impact of substrate alloy on PBR by determining variations in PBR for oxides grown on aluminum-nickel alloys under high temperature conditions. Arurault [12], in comparing anodic oxides grown in boric–sulfuric acid electrolytes at different voltages, showed conclusively that PBR is proportional to the anodizing voltage. A variety of PBR values are also reported from other work for oxides grown in various electrolytes under various anodizing conditions. Assuming all anodes were of the same alloy and in the same microstructural condition, it is clear that the electrolyte has a profound effect on the density of the anodic oxide [12]. Therefore, for the PBR to be relevant for all anodizing applications, it is necessary to standardize PBR values with respect to the anode alloy as well as all anodizing conditions.

3.5 AAO Structural Comparison

In industry, anodic oxides and the anodizing processes have been categorized into types, which exhibit specific properties to suit specific applications. The details of the various processes are well covered in Brace [2]. A brief summary follows to illustrate the relationship of the various AAO structures to their respective processes. The various basic types of anodic oxides set forth by the US military specification MIL-A-8625 [11] (Types I, II, and III) have developed into the basis for many other general international specifications. In the following summary, general industrial references are related to the "Type" designations for clarity and common ground.

3.5.1 AAO Structures Produced in Sulfuric Acid

3.5.1.1 Type II or Technical Anodizing

By far, most anodizing is processed in sulfuric acid electrolytes. The fine, highly ordered, unidirectional, nanoscale porous columnar microstructure of the Type II finish explains why it can be utilized for decorative applications. Type II anodizing is referred to as technical anodizing, "clear coat," decorative anodizing and/or mild anodizing. This is because the process is usually operated at "room temperature," from 20 to 25 °C, and depending on the alloy, other process parameters employ a range of lower sulfuric acid concentrations as compared to other anodizing

Fig. 3.11 Type II anodized and dyed components (Photo Courtesy Impreglon Surface Engineering, Kaufbeuren, Germany)

processes, from 120 to 275 g/L (grams per liter). Lower current densities are used to develop a Type II finish, usually 1.0–1.5 A/dm^2. The anodic oxide yielded by these processes is used primarily for corrosion protection and decoration, and they offer a modicum of abrasion resistance. The anodic oxide processed under these conditions is optically transparent, and the quality of the aluminum microstructure can be developed by the anodizing process, and observed through the oxide after anodizing. Because the oxide is transparent, the structure is dyeable and can be screen-printed to yield a wide range of colors and patterns. Thicknesses of 8–20 μm are usually specified, depending upon the substrate alloy. These finishes are typically sealed to enhance corrosion resistance and to prevent dye from bleeding out of the porous AAO structure. Type II anodizing has a niche in the metal finishing industry for finishes with outstanding clarity and color on aluminum and its alloys. See Fig. 3.11.

The structure of a decorative (Type II) anodic oxide is highly ordered, comprising fine columnar cells with rounded bottoms. The cellular network is oriented by and grows perpendicular to the substrate, and each cell has a central pore that is parallel to the oxide growth direction. Because of the lower density of the total oxide in comparison with the aluminum substrate, for an anodic oxide of total thickness x, approximately 60% is into the metal, and 40% exceeds the former surface of the substrate.

When the oxide encounters a surface defect, such as a tangle of dislocations, a grain boundary, or a component surface feature such as a gentle radius or a corner, deviations in network uniformity are created that are associated with the base metal microstructure. Some of the most common defects in the oxide network are seams that are the result of a change in substrate orientation that appear to "knit" the structure together; other features such as bubbles due to the simultaneous evolution of hydrogen gas from the substrate during anodizing are "frozen" in the oxide struc-

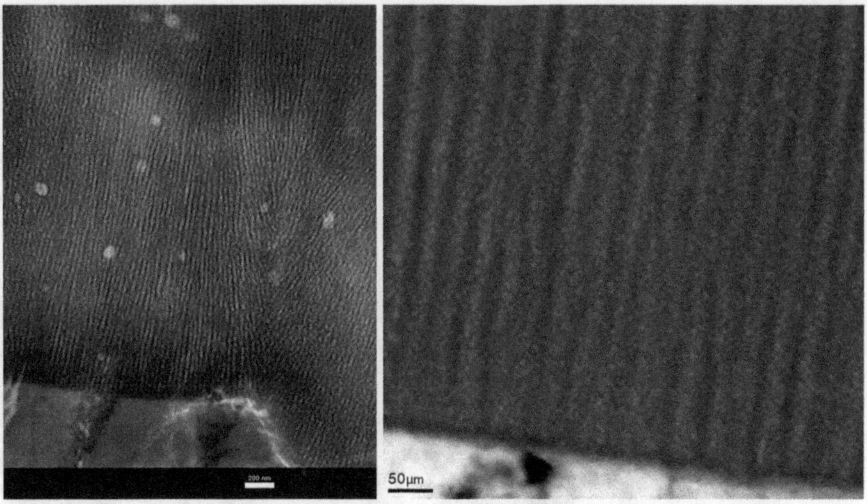

Fig. 3.12 TEM images of a decorative technical anodic oxide (AAO) grown on aluminum alloy 6061 T6. In the image on the left, the columnar structure exhibits changes in the growth pattern, corresponding to the near-surface microstructural features of the substrate aluminum. The column walls and central pores are uniform and fine. The pore diameters measure 10–13 nm and the column walls measure ~6 nm. The pore-to-pore center distances measure approximately 25 nm. Process parameters were sulfuric acid electrolyte = 120 g/L, current density = 1.2 A/dm^2, temperature = 20 °C [19]

ture. Fine bubbles are concentrated within the central pore, and larger bubbles occur randomly, throughout the structure, mostly when complex alloys are anodized, and are a product of the hydrogen evolved when the advancing oxide growth front encounters a defect. These characteristics of anodic oxide growth are found in all types of anodizing processes that produce a cellular structure. See Fig. 3.12.

Low alloy, higher purity aluminum and aluminum alloy substrates tend to be specified for applications in which the clearest AAO is required. These are mostly decorative applications for which the substrate may be processed to yield a specific gloss, specularity, reflectivity, and distinctness of image. The alloys specified are typically wrought alloys and may be precipitation hardenable, such as alloy 6063, or non-heat treatable, such as 5052 or 1050. Most are specified with a T6 temper such that they possess the mechanical properties necessary to perform the engineering function, but such designs are typically appearance driven. The alloys selected exhibit less precipitation in the microstructure as manufactured and consequently have less complex microstructures. When anodized, the oxide yields a high level of clarity and reflectivity. With color from dyeing or from electrolytic coloring, the appearance is brilliant, and because the oxide is integrated with the substrate, it won't chip or peel and has a high level of durability.

A unique application for the Type II anodic oxide is a transition layer for subsequent bonding of an adhesive layer to another, similarly prepared metal substrate or polymer or paper laminate, such as a label or gasket. It is also an effective primer

Fig. 3.13 TEM image of a "flash" anodic oxide on a 6061 T6 aluminum substrate. The AAO was grown under Type II conditions from a phosphoric acid electrolyte. The exposure time was about 45 s. The total oxide thickness is ~40 nm

layer for ink, paint, lacquer, or adhesive. The oxide structure provides a surface with high entropy, highly polar and ionic. A technique called "flash anodizing," based on the process for Type II anodizing, yields thin AAO finishes that measure only several nanometers thick [20]. It is produced at a current density that is rather high in comparison with other Type II processes (about 1.8–2.0 A/dm^2), and the exposure time is less than a minute. The AAO produced by way of flash anodizing develops a uniform and continuous porous oxide from sulfuric, oxalic, or phosphoric acid electrolytes, and is used as a stable and robust interface to enhance bonding. Because it is an integral oxide, it eliminates the concern for filiform corrosion and it doesn't chip or peel like deposited layer. Thin oxides are also the most transparent and are used as the finish for aluminum alloy coil product that will be used as reflectors for overhead lighting and in several automotive, aerospace, and aircraft applications. See Figs. 3.13 and 3.14.

3.5.1.2 Type III or Hard Anodizing

Hard or Type III anodic oxides are thick, hard finishes that are used in applications that require wear resistance. They are typically carried out in sulfuric acid electrolytes, but other electrolytes including phosphoric acid, oxalic acid, and acid mixtures can be used. Electrolyte concentrations up to 350 g/L sulfuric acid are used industrially. The structure of a hard anodic oxide exhibits thicker cell walls and wider pores. Thicknesses of 25–50 μm are typically specified, but finish thicknesses as high as 125 μm are reported when rework has been done to restore component dimensions. To grow such a thick oxide, current densities can be markedly higher than those used

Fig. 3.14 Bright-finished aluminum semi-truck wheels

to grow Type II oxides, about 2.2–3.5 A/dm^2, depending upon how the component is manufactured and the complexity of the alloy. See Figs. 3.15 and 3.16.

The resistance that develops during anodizing is the result of energy input to the process (current density, voltage), creating the characteristic heat of the oxidation reaction. Additional energy develops because of the process components themselves, and resistance changes can help or hinder the oxidation process. For example, as the thickness of the oxide increases, it naturally contributes to the resistance in the anodizing circuit because it has different electrical characteristics than the aluminum substrate. Resistance is also exacerbated by alloy complexity and/or surface contamination, which causes the temperature of the work in the electrolyte to rise that in turn causes the temperature of the electrolyte to rise, regardless of the type of anodizing process. Therefore, electrolyte agitation is an important requirement for all anodizing processes because the thermodynamic heat of oxidation must be kept to a minimum to assure the final properties of the AAO and to help prevent burning.

Higher current density processes tend to produce more resistance heat than lower current density processes, hence the temperature due to resistance heating increases more with hard anodizing than with decorative anodizing. The solubility of the growing oxide increases with the electrolyte temperature, which will change the engineering performance of the oxide in wear applications. Therefore, in order to produce a thicker, harder oxide, the anodizing bath temperature is typically lowered to range from about 0 to 10 °C and agitation by way of the introduction of air or bath recirculation is incorporated to the bath to keep the electrolyte temperature down, keep the work cool, and limit the amount of oxide dissolution.

Fig. 3.15 An assortment of hard anodized components, from both wrought and cast substrates. Note the dark appearance of the finish. This is due to alloying elements becoming trapped within the finish or at the interface between the substrate and the anodic oxide

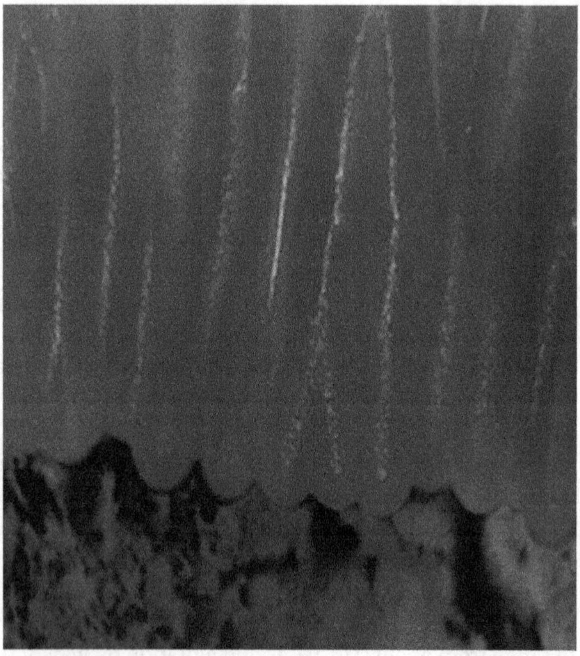

Fig. 3.16 TEM image of a representative structure of Hard (Type III) AAO, grown on aluminum alloy 6061 T6 and anodized in a sulfuric acid electrolyte. Process parameters alone yielded differences in the AAO structure than a Type II anodic oxide. The cell dimensions for the Type III oxide (right) measure 30 nm (pore diameters) and the column walls measure 15 nm. The pore-to-pore center distances measure approximately 90 nm. Electrolyte concentration = 220 g/L, current density = 2.2 A/dm^2, temperature = 0 °C [19]

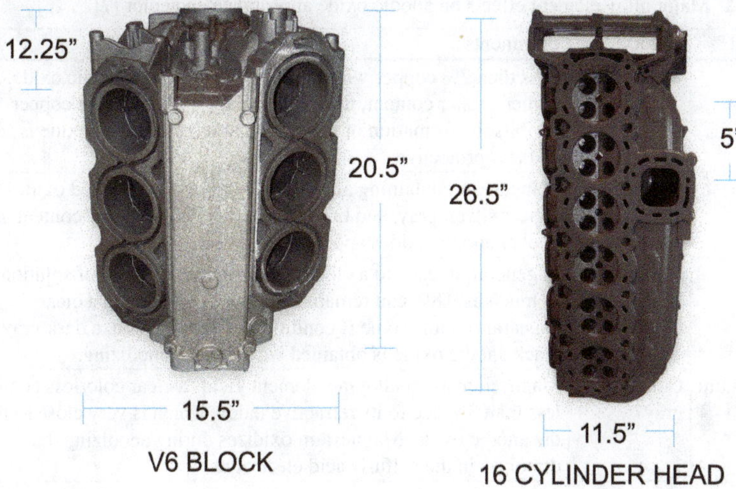

Fig. 3.17 Example of as-cast engine parts, alloy A356 before (left) and after anodizing (right) using hard anodizing or Type III process parameters. The oxide finish thickness measured about 20–25 μm. Note the brown-gray appearance of the anodized finish due to inherent coloring by alloying elements in the cast alloy [21]

Components that are hard anodized are typically used in mechanical applications that require high surface hardness and abrasion resistance; finish clarity is typically not a design requirement for components that are hard anodized. The alloys from which hard anodic oxides are grown are usually more complex (are comprised of more alloying elements) than alloys used for decorative anodizing. These alloys have phases with the elements in solution, and contain non-aluminum particles and intermetallic compounds, which become trapped within the Type III oxide, causing it to take on a color, rather than remain transparent. See Fig. 3.17.

Oxides grown on complex alloys that have a higher alloy content tend to take on color due to the movement of non-aluminum ions along the knitlines between columns and due to adsorption on the surface of the oxide from dissolved metal ions in the electrolyte. Consequently, the anodic oxide takes on color from the oxidation of the various alloying elements, and appears yellow-green to black. Inherent coloring effects are intensified in thicker oxides. Inclusions and intermetallic particles interrupt the ordered structure of a hard anodic (Type III) oxide, which can reduce the corrosion resistance and the wear resistance of the finish. Table 3.2 summarizes the inherent coloring effects that occur with different alloys.

A comparative evaluation of the structures for decorative and hard anodic oxides produced in sulfuric acid electrolytes elucidates distinct relationships between process parameters and oxide characteristics. The structural similarities of the oxides are immediately obvious: the oxide is an assembly of cells that grow in a columnar fashion perpendicular to the aluminum substrate; the cells have curved bottoms, and the oxide thickness at the bottom is the same as the column wall thickness. Each cell has a central pore, which is rather uniform in dimension along the length of the

Table 3.2 Major alloy element effects on anodic oxide appearance and color [7]

Element	Color	Comments
Copper	Black	Less than 2% copper will not impart color to the anodic oxide. At higher copper content, the oxide is dark-appearing and copper inhibits the formation of the columnar structure. The oxide is softer and less protective.
Manganese	Brown	Manganese-containing alloys typically yield discolored oxides that appear silver, gray, and brown up to 1%. Higher alloy content and for thicker anodic oxides appear quite brown.
Silicon	Dark gray, black	In general, it leads to a cloudy appearance when out of solution, but as much as 0.8% can remain dispersed, maintaining a clear appearance under type II conditions. At 5% silicon, a dark gray or black anodic oxide is obtained with type III anodizing.
Magnesium	Colorless, gray	Magnesium as an alloying element yields a clear colorless coating at less than 3% due to its refractive index, which is very close to that of the anodic oxide. Magnesium oxidizes during anodizing, but dissolves in the sulfuric acid electrolyte.
Zinc	Yellowish, Brown	The anodic oxide on zinc-containing alloys up to 5% Zn appears colorless oxide when homogenous, but quite yellow-brown when second phase precipitation is present. Oxide can appear marbled yellow/brown depending on alloy segregation in the surface microstructure.
Iron	Black	As a trace element, iron detracts the specular brightness of the high purity bright anodizing alloys even in very small amounts.

column and from cell to cell in the AAO structure. The central pore extends from the curved bottom of the column to the top of the oxide. Structural deviations can be related to microstructural features in the substrate metal: surface texture, grain boundaries, and tangles of dislocations produce columnar bifurcations, voids, and the detachment of columns from the substrate. See Fig. 3.18.

Process differences yield obvious structural differences. The higher current density, and therefore higher voltage response of the hard anodizing process, produces a coarser oxide structure (thicker cell walls and wider pores) than that produced using the lower current density of the decorative anodizing process. There are fewer oxide cells in AAO produced by hard anodizing process parameters than they are produced by decorative anodizing within a similar field of view, at the same magnification, which implies that there are fewer cells per crystallographic grain of surface with a hard anodic oxide than with a decorative oxide. Table 3.3 presents some basic AAO structural feature measurements for specific anodizing parameters that support the reported visual observations. The observed trends indicate the following process generalizations for anodic oxide growth: Higher current density processes tend to yield thicker column walls and wider pores, regardless of the process temperature.

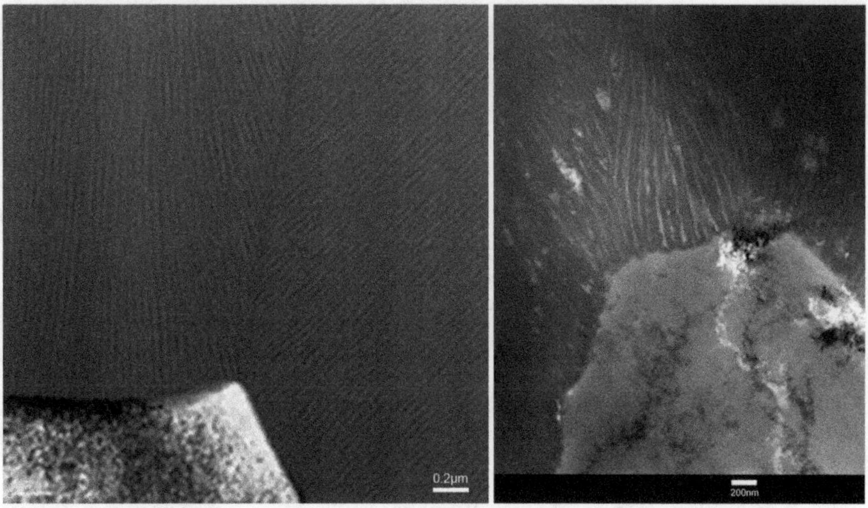

Fig. 3.18 (left) TEM image of a decorative anodic oxide (Type II) as it proceeds along a change in surface orientation. Note the oxide columns have grown perpendicular to the substrate surface, and as the oxide consumed the substrate, a seam was created in the AAO structure that knits the two oxide growth fronts together. (Right) The columnar structure of this Type III oxide exhibits changes in the growth pattern corresponding to the microstructure of the substrate aluminum. Structural deviations can be related to microstructural features in the substrate metal: surface texture, grain boundaries, and tangles of dislocations produce columnar bifurcations, voids, and the detachment of columns from the substrate [19]

Table 3.3 Basic AAO structural feature measurements

Electrolyte acid type and concentration (g/L) and temperature (°C)	Current density (A/dm^2)	Wall thickness (nm)	Pore diameter (nm) and pore spacing
Sulfuric, 180 g/L, 20 °C	0.5	3–4	7 nm, 25 nm
Sulfuric, 120 g/L, 20 °C	1.2	8–10	15 nm, 35 nm
Sulfuric, 180 g/L, 20 °C	1.5	8–10	15 nm, 35 nm
Sulfuric, 220 g/L, 0 °C	2.2	15	30 nm, 60 nm
Oxalic, 10 g/L, 30 °C	2.0	10	40 nm, 100 nm

3.5.2 AAO Structures Produced in Other Electrolytes

According to Brace [3], anodizing in organic acids began during the 1930s in Europe and Japan. Oxalic acid electrolytes are used only to a limited extent industrially in Europe and the United States, due to electrolyte stability issues, but the electrolyte remains popular for anodizing in Japan. Mixed acid electrolytes which incorporate organic acids with sulfuric acid have gained a great deal of popularity for oxide finishes that require enhanced smoothness and hardness, such as automotive and machine parts that are used in wear applications and for semiconductor sputtering chambers and other equipment that requires the AAO have the ability to

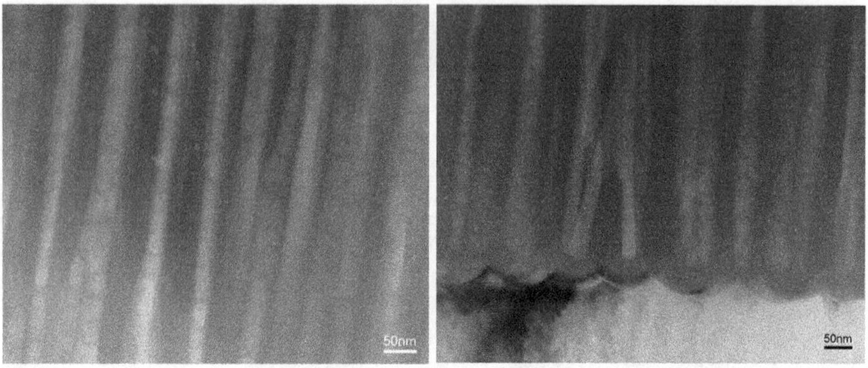

Fig. 3.19 TEM images of a hard (Type III) AAO grown on aluminum alloy 6061 T6 in an oxalic acid electrolyte at the center of the structure (left) and at the interface with the substrate (right). The cellular structure on the right exhibits changes in the growth pattern, developed as columnar branching, corresponding to a tangle of dislocations in the substrate microstructure. The column walls are uniformly coarser than the Type II example in Fig. 3.5, as is the central pore. Process parameters for this particular anodic oxide were as follows: oxalic acid electrolyte concentration = 0.3 M, formation voltage = 40 V, temperature = 0 °C. The pore diameters measure ~40 nm and the cell walls measure ~10 nm, a structure typical for a high voltage (current density) process. The pore-to-pore center distances measure approximately ~100 nm [18]

withstand externally applied voltage without breaking down; in other words, the oxide must have "high dielectric withstand." Trends in AAO structural growth and development from organic acid electrolytes are similar to AAO structures obtained from sulfuric acid electrolytes. See Fig. 3.19.

Phosphoric acid anodizing is used as a pretreatment for structural adhesive bonding in high-humidity environments for the aerospace and aircraft industries. This process was developed by Boeing and is carried out in a 10–12% by weight phosphoric acid electrolyte at 10–15 V. The adhesive bonds achieved with the phosphoric acid AAO have greater durability under adverse conditions than bonds formed with oxides from chromic acid or sulfuric acid electrolytes. Anodic oxides formed in phosphoric acid electrolytes at higher current densities (formation voltages), up to 195 V and lower temperatures (as cold as −5 °C) are utilized for templates for nano-patterning, such as wires, dots, and foils for sensor applications. Because of the scale of the oxide dimensions achieved under hard anodizing conditions, less complex alloys with a high level of metallurgical cleanliness are hard anodized in phosphoric acid electrolytes to yield a relatively coarse anodic oxide structure with regular spacing and pore size. These oxides are used as templates for the manufacture of nanoscale wires, or for other applications that require a regular, highly ordered nanostructure. Table 3.4 presents a summary of various electrolytes, including phosphoric acid, oxalic acid, and a variety of organic acids that can be used to produce template-quality AAO, suitable for uniform nanoscale devices [22]. See Figs. 3.20 and 3.21.

Table 3.4 Major electrolyte formulations used to produce porous oxides on aluminum substrates [22]

Main acid used in electrolyte	Molecular formulas	Concentration (M)	Pore size range (nm)
Acetic	CH_3CO_2H	1.0	Not specified
Citric	$HO_2CCH_2(OH)(CO_2H)$ CH_2CO_2H	0.1–1.0	90–250
Chromic	H_2CrO_4	0.3–0.44	17–100
Glycolic	$CH_2(OH)CO_2H$	1.3	35
Malic	$HO_2CH_2CH(OH)CO_2H$	0.15–0.3	Not specified
Malonic	$CH_2(CO_2H)_2$	0.1–5.0	Not specified
Oxalic	$C_2H_2O_4$	0.2–0.5	20–80
Phosphoric	H_3PO_4	0.04–1.1	30–235
Sulfuric	H_2SO_4	0.18–2.5	12–100
Tartaric	$HO_2CCH(OH)CH(OH)CO_2H$	0.1–3.0	Not specified

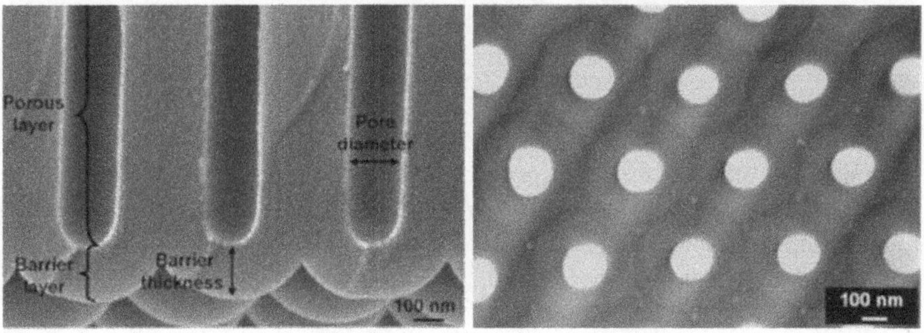

Fig. 3.20 Field emission SEM images of AAO produced in an 8% phosphoric acid electrolyte at −1.5 °C, 185 V for 4 h. The image on the left is a longitudinal cross section of the oxide after the aluminum substrate was etched away. The image on the right is a transverse section through the oxide, showing the distinct honeycomb structure so well recognized as AAO. Note the uniformity of the pore diameter and the dimensions of the "barrier layer" and cell walls [23]

3.5.3 Anodic Oxides without a Cellular Structure

3.5.3.1 Conversion Coatings

The barrier layer phenomenon of the network-forming passive layer is the basis for the reactions that can be employed to enhance the efficacy of the naturally occurring oxide in a variety of applications of the aluminum substrate, including the function as a transition layer for bonding. Because the oxide will not allow the penetration of external anions, the anions can adsorb at active sites on the outer chemical structure of the oxide. "Conversion" of the structure to other compounds by the adsorption of a supplementary anion imparts different or additional properties to the passive layer at the surface. This is an equilibrium process, with no external driving force such as an electrical power supply [24].

Fig. 3.21 Nickel nanowires electroplated in a phosphoric acid anodized AAO template [10]

Formulations for conversion coatings that contain chromates, phosphates, and silicates, to name just a few, are utilized to bolster corrosion resistance and wear resistance, as well as to serve as a primer layer for paints or adhesives. The number of sites on the passive oxide structure, on which supplementary anions can adsorb, limits the reaction and film-building attributes of a conversion coating. The oxidation reaction, driven by the presence of available aluminum, slows down and eventually stops when the aluminum at the interface has been completely reacted. Therefore, conversion coatings are always very thin. As is, the finishes are rarely thicker than the 1–10 nm, which comprise the unmodified passive layer. See Fig. 3.22.

Polymer modifiers can be introduced to the electrolyte that adsorb on the reactive surface of the conversion coating for the purpose of creating a transition layer for subsequent polymer bonding. Such additions to the structure increase the conversion-coating layer thickness by virtue of long hydrocarbon chain length, but are isolated to the surface of the oxide. Nevertheless, conversion coatings are often used as a transition layer between aluminum substrates and subsequent coating and bonding layers such as paints and adhesives, since they are chemically bound to the aluminum surface by way of the naturally occurring passive oxide. Supplementary anions, bound to the surface of the passive layer structure, can also provide a vehicle to form the transition bond, but never intrude to the interface with the aluminum. See Figs. 3.23 and 3.24.

3.5.3.2 Barrier Layer Anodizing

Continuous, non-porous "barrier layer" oxides are the earliest anodic oxides on record as having been researched and for which engineering applications were found. This type of anodic oxide is that which Eugéne Ducretet reported in the late

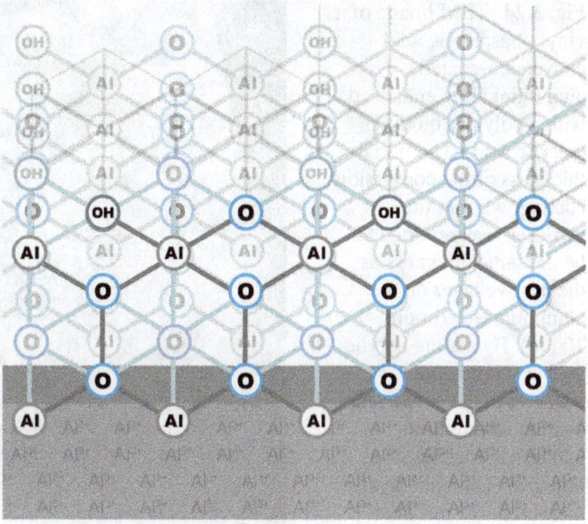

Fig. 3.22 Passive layer chemical structure. The structure is that of hydroxyaluminate, AlO_4^-, the fundamental, nonstoichiometric oxide that comprises the passive layer, and forms initially during anodizing, ripening to develop more stoichiometric AAO close to the aluminum substrate as the oxide grows

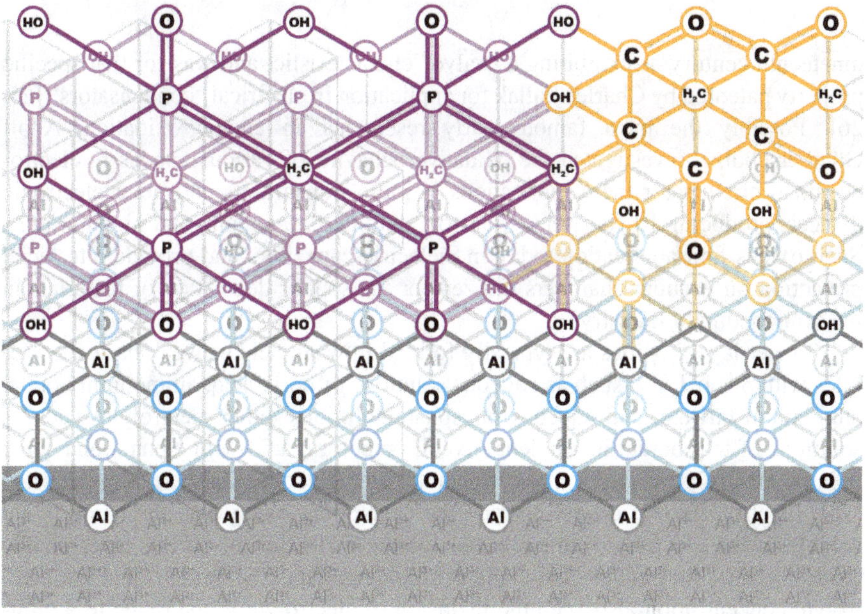

Fig. 3.23 Schematic of passive layer enhanced by the adsorption of vinyl phosphonic acid and acrylic acid, a common conversion layer used as a primer layer for bonding polymers to aluminum [24]

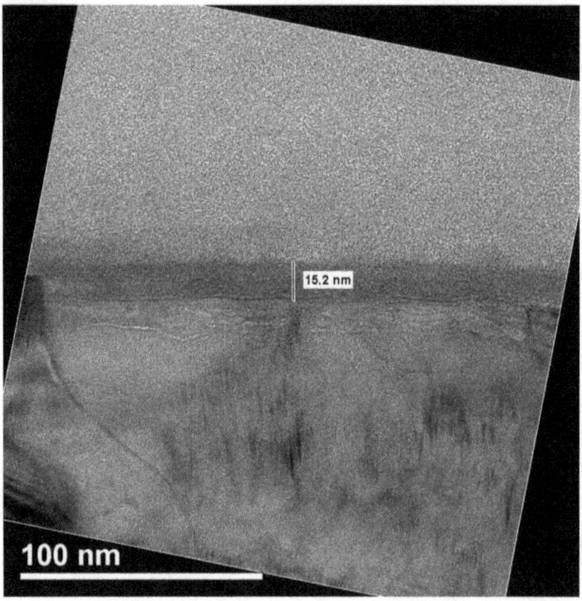

Fig. 3.24 TEM image of a vinyl phosphonic acid conversion-coated aluminum alloy enhanced with a polymer modifier (acrylic acid). The initial thickness of the conversion coating, without the modifier, was reported to be the same thickness as the passive layer on the aluminum (maximum 10 nm). By adding acrylic acid as a long-chain modifier to the conversion-coating formulation, additional thickness was obtained. The final layer measured ~15 nm thick

nineteenth century as exhibiting a "valve" characteristic, and was for that specific property patented by Charles Pollak for application in electrical "condensators" [25, 26]. Possibly the most famous early researcher in anodic oxidation, Adolf Güntherschulze, investigated the oxidative behavior of various anodes in electrochemical circuits for their applications in radio transmitters and other electronic devices, producing barrier-type oxides on aluminum, tantalum, and niobium [27]. Such oxides find application today in dielectric capacitors and provide interfacial protection on vacuum chambers utilized for the sputter deposition of materials in the semiconductor industry.

"Barrier layer" oxides are typically formed in weak acid or weak basic electrolytes with $5 < pH < 7$, such as aqueous solutions of boric acid, ammonium borate, ammonium tartrate, and aqueous phosphate solutions; other electrolytes such as solutions of tetraborate in ethylene glycol, perchloric acid with ethanol and some organic electrolytes; citric, malic, succinic, and glycolic acids are also reported [28]. Electrolytes that do not corrode the developing oxide (i.e., the oxide is not soluble in the electrolyte) are selected for the formation of barrier layer anodic oxides because they enable anodizing at high potentials. The mild pH and low electrolyte concentration also keep the inclusion of electrolyte in the oxide to a minimum. The resultant oxide layer has a compact morphology and is consequently not chemically affected by the electrolyte and the oxide remains insoluble through the anodizing process. The oxides are microcrystalline, exhibiting X-ray diffraction patterns that identify their structures as γ and γ' Al_2O_3. They are characterized by a high dielectric constant and are stable even at high temperatures. The oxide density and thickness are related to the anodizing potential (applied voltage) [29, 30].

As a total layer, compact barrier layer oxides comprise fewer defects than porous anodic oxides and seem not to be affected by the metallurgical quality of the anode. This characteristic supports the theory that the mechanism for the formation of barrier layer oxides may be the result of colloidal precipitation of aluminum hydroxide from the electrolyte, initially forming a gel-like layer by way of peptization of the colloidal hydroxide particles [31–33]. This type of reaction is not supported in a highly acidic electrolyte but could be the result of the effects of anodic oxidation in a complex electrolyte with a neutral pH. Indeed, the formation of negatively charged aluminum hydroxide particles would be pulled back to the parent positively charged anode, ultimately forming a coherent, albeit thin, oxide layer. The amount of energy that develops at the interface between the anode and the oxide due to the applied voltage, with total surface reconstruction of the anode by way of the formation of a coherent insulating layer, is the source for the rapid consolidation of the anodic oxide structure as well as the nature of the development of a crystalline structure [34].

Because all anodic aluminum oxides are ionic solids and exhibit the "valve" characteristics of a diode, all have high ionic conductivity and therefore low electronic conductivity in one direction, from the substrate across the oxide to the electrolyte. Ionic conduction is therefore the main charge transport mechanism at high values of applied voltage. Because the electrochemical circuit for the formation of a barrier-type oxide employs electrolytes of low concentration with corresponding low ionic conductivity, the ionic conduction across the oxide requires a high electric field for the migration of ions, therefore, it is generally accepted that the applied voltage determines the thickness of barrier-type anodic oxide. As the applied voltage exceeds a critical value, there is a transition in ionic mobilities at the interface with the anode and a space charge develops, sometimes resulting in voltage breakthrough, which is rapidly repaired by continued oxide formation, thereby building the oxide thickness, and sometimes creating random pores across the oxide surface. Both Ducretet and Pollak report the voltage discharge characteristic; voltage breakdown became a critical aspect in the investigations performed by Mott [31] in 1904, and Guntherschulze [27] in 1906, who saw the development of "gas" at the anode-oxide interface as the source for porosity in thicker oxides.

In their 2017 paper, "Electronic and structural characterization of barrier-type amorphous aluminium oxide," Evangelisti et al. [35]. provide a comprehensive investigation of barrier-type oxides grown under a variety of conditions. In their investigation of planar and longitudinal cross sections of oxides, as well as electrochemical characterization of oxides formed under different conditions, it was determined that with increasing potential, there was a distinct increase in oxide density. Three distinct anodizing potential domains could be clearly identified in the oxide structures that can be related to the oxide structure and its function as an ionic conductor. At lower anodizing potentials, ionic conduction was facilitated through the low-density oxide structure; however, with increasing oxide density, the ionic conductivity of the oxide was reduced. The dielectric behavior of the oxides formed was likewise studied, correlating the structure properties and lattice and electronic contributions to the dielectric constant, which approached that of bulk crystalline Al_2O_3 for oxides produced at high voltages. Barrier layer oxides can therefore be produced with different structures and dielectric properties by changing the anodizing parameters, much like porous anodic oxides.

3.5.3.3 Plasma Electrolytic Oxidation

A clear development from the technology that produces barrier layer-type oxides, the oldest type of anodic oxides, is the process of Plasma Electrolytic Oxidation (PEO), which from the literature surveyed appears to be about 50 years old [36]. PEO processes have been developed for aluminum, magnesium, titanium, zirconium, and their alloys. The advantage of the PEO process is that it is an electrochemical oxidation process, rather than a high temperature deposition process such as thermal spray, PVD, or CVD processes. Processes that require high ambient temperatures may result in detrimental heat treatment effects that change the mechanical properties of the substrate material. The PEO process is a high voltage arc process that utilizes a fundamental electrolyte of dilute silicate or phosphate-silicate solutions with a neutral pH, which precipitate a coherent, insulating deposit over the anode surface. During this time, an oxide layer begins to form, along with diffusion and mass transport of second phase material to the anode substrate surface. Energy that builds at the oxide-deposit interface is discharged through defects and weak spots in the anodic deposit, as arcs at multiple locations across the surface, producing a dense, thick, continuous oxide layer [37].

As finished, the PEO layer retains some of the initial deposit, so the surface appears rough and irregular. However, the oxide produced by way of PEO is microcrystalline, exhibiting high hardness and, by virtue of its characteristic as an anodic oxide "growth," outstanding adhesion to the substrate, as well as the wear and corrosion resistance. It is a true ceramic, mainly composed of γ-Al_2O_3. Depending upon the electrical parameters of the PEO process, oxides produced by plasma electrolytic processes can also contain α-Al_2O_3, and sometimes, a stable alumina-silica phase called mullite, occurring at the top of the oxide, where it contacts the deposit from the silicate-based electrolyte [38]. Incident hardness measured on PEO layers average 1400 HV. Therefore, anodic aluminum oxide grown by way of PEO can be machined and polished to a high surface finish, enabling a variety of applications that employ its ability to withstand high levels of cyclic fatigue, abrasion, and impact. Deposits such as PVD and electroplated finishes are not as robust in such applications and can blister and peel.

The mechanism for the oxide layer formation by way of PEO processing appears to be the result, at least in part, of thermal oxidation of the substrate. The absence of electrolyte constituents within the oxide layer coupled with the consumption of the substrate by way of anodic oxidation coupled with the diffusion and mass transport of second phase species to the interface with the oxide, without their inclusion in the otherwise dense oxide layer, with only small amount of porosity, supports this conclusion. It is clear that the dielectric strength of the anodic deposit has a direct bearing on the amount of heat produced and sustained at the substrate-oxide interface.

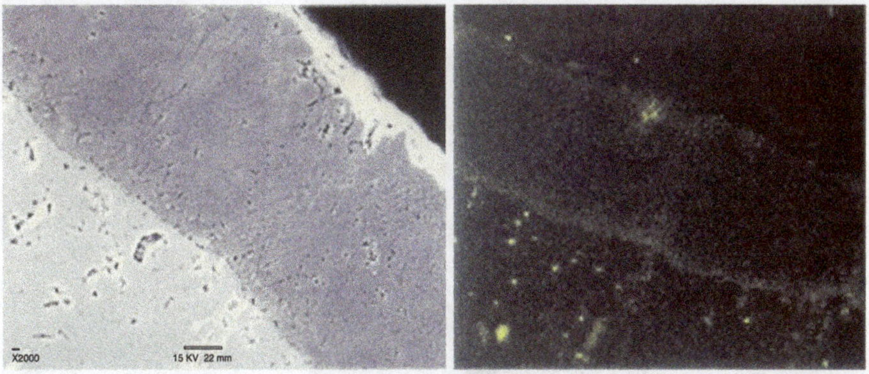

Fig. 3.25 SEM images with a corresponding energy dispersive X-ray analysis (EDS) of a PEO layer on a 6061 T6 aluminum extrusion. Note the density of the oxide layer. The EDS map on the right shows the distribution of silicon through the oxide and entrainment at the anode interface, represented by the concentration of yellow dots at the interface between the aluminum and the oxide layer

The oxide thickness (not including the thickness of the anodic deposit) appears to be reaction time dependent. See Fig. 3.25.

3.5.3.4 Chromic Acid Anodizing

Historically, anodizing processes which employ a chromic acid electrolyte, patented in England by Bengough and Stuart in 1923 [39], are categorized as the Type I process. Type I anodizing was developed to provide corrosion protection and a base for paint for components used in the aircraft and aerospace industries, which by design require reliable performance under high cycle fatigue conditions. Typical anodizing parameters are: 3–5% CrO_3 electrolyte maintained at 40 °C ± 2 °C, the anodizing circuit operation is stepwise, controlled from 10 V to a relatively high voltage of 40–50 V, which is achieved following a gradual voltage increase from a selected starting voltage to the high oxide formation voltage. Other anodizing electrolytes that are used for aerospace applications to avoid the toxicity of the chromate ion employ organic acid electrolytes such as tartaric acid, maleic acid, and/or mixed acid electrolytes such as boric–sulfuric acid or combinations of organic acid with sulfuric acid.

The structure of a chromic acid oxide is thin and compact, about 3–5 μm. The typical columnar cell structure of other anodic oxides produced in sulfuric acid and oxalic acid electrolytes is absent. Instead, AAO grown with the chromic acid electrolyte appears wave-like as opposed to columnar, exhibiting a structure comprising lamellar oxide fragments with interlamellar fine bubbles over a continuous oxide layer at the interface with the substrate, which would typically be identified as a "barrier layer." Similar structures are observed with AAO grown in boric acid electrolytes [40]. See Fig. 3.26.

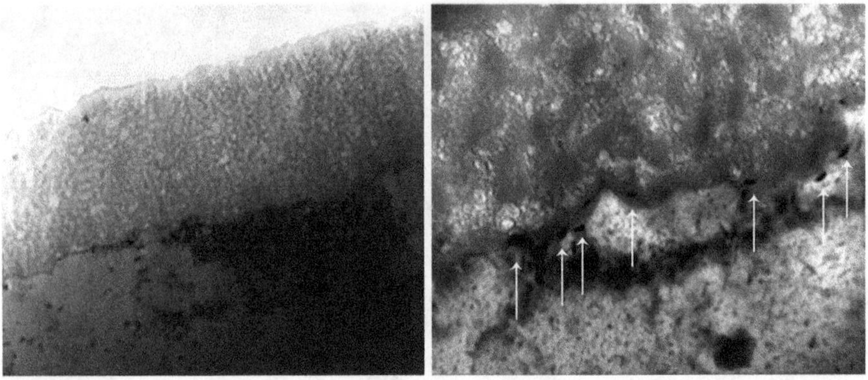

Fig. 3.26 Transmission electron microscope (TEM) photomicrograph of AAO grown in a chromic acid electrolyte (Type I) on a 2024 aluminum alloy substrate. Photo on left is at X15,000; photo on right is at X100,000. Note the absence of a columnar structure; nevertheless, a heavy interfacial oxide is present between the base metal and AAO. Oxide fragments of the same thickness as the interfacial layer are present throughout a less dense foam-like oxide that comprises the rest of the finish. Areas of entrained copper at the aluminum alloy-oxide interface are from the 2024 base metal alloy, and are present as dark particles in the interfacial oxide (see arrows) [41]

The irregular, skewed, compact oxide structure of the Type I finish explains why at even thin finish thicknesses (1–3 μm), the chromic acid finish exhibits good corrosion resistance. Without distinct unidirectional pores, the ingress of the external environment is reduced significantly, making Type I anodic oxides the finish of choice for aerospace/aircraft applications. Further processing the Type I finish with inert seals, or using it as a primer, enhances the corrosion resistance of the finished component even more. Its naturally thin finish thickness is ideal for applications that demand high-tolerance control and fatigue performance. Because the thickness of the AAO is self-limiting when produced in a chromic acid electrolyte, multiple alloys can be anodized simultaneously.

3.6 Post-anodizing Processes: Dyeing and Sealing

During anodizing, the electrolyte contacts the porous structure, while the anion, which is in part oxygen, reacts with aluminum ions diffusing through the finish at the pore surface. Anion species (OH^-, SO_3^-, CrO_3^-, etc.) are adsorbed on the pore walls, facilitating conduction and maintaining the net negative charge of the anodic oxide and the erect nature of each column. These adsorbed anions also provide individual reaction sites for the growing finish, sustaining the ion exchange system known as the "ion pump," which maintains the erect nature of the columns comprising the AAO structure. Even as finished, the porous structure of the AAO remains ionic, which lends itself to processing after anodizing such as dyeing, sealing, or

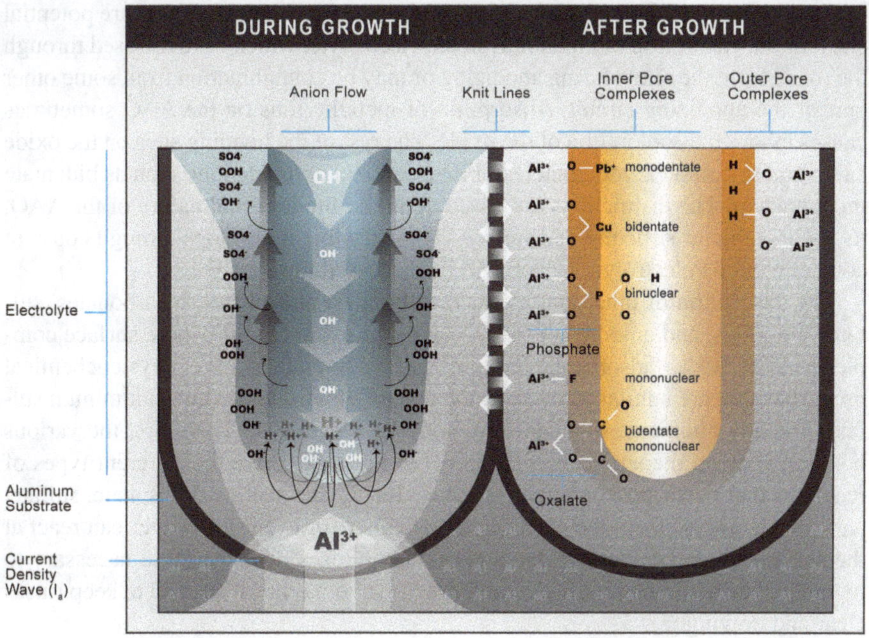

Fig. 3.27 Schematic for the development of the ionic nature of the inside surface of the pore. Following anodizing, the ionic sites remain, providing bonding sites for dyeing and adhesives, as well as a reactive surface for sealing

bonding with supplementary coatings that may be inorganic or organic [42]. See Fig. 3.27.

The ionic nature of the surface of the anodic oxide enables continued reaction with other species in aqueous solution. Some reactions are due to physisorption, and the bond forces are weak, with no chemical bonds forming. These reactions can be reversed with rinsing. Other reactions are due to chemisorption and surface chemical compounds form with the available ligands on the pore surface. The polar surface alone drives hydroxyl-reactions that naturally occur at the pore surface, and are the basis for dyeing and sealing reactions. Spaces in structure accommodate ionic movement, restricting dye reactions, which are on the molecular level, to within the pore or at the finish surface. In other words, AAO does not behave as a sponge. Quite the opposite, based upon the chemical principle of "like dissolves like," hydrous solutions tend to dissolve hydrated structures, if only limited to the surface. This is also the primary reaction for sealing. Temperature, of course, affects the extent of the reductive dissolution reaction, with higher temperatures dilating the structure at a minimum, increasing the surface activity and encouraging more diffusion of the reductant molecules to the anodic oxide surface, promoting both dyeing and sealing reactions.

There are a variety of inner pore surface complexes that are available on the surface of the pore onto which bonding can occur [41]. Some of the complexes are monodentate (A monodentate ligand is a ligand that uses only one pair of electrons to bond to the central metal atom or ion) and some are bidentate. These are potential sites for the absorption of metal ions in the electrolyte, which have diffused through the oxide from the alloy during anodizing or may be contamination from some other part of the anodizing circuit. Adsorption of metallic ions on the AAO sometimes causes inherent discoloration of the oxide. The rest of the bonding sites on the oxide can be more complex: mononuclear, binuclear, and combinations such as bidentate mononuclear. The distinctions are based upon the fundamental nature of the AAO, its semiconductor structure and surface chemistry that it develops through constant contact with the electrolyte, which will be discussed in Chap. 6.

The adsorption of other anions, such as borates, phosphates, bicarbonates, sulfates, fluorides, and anionic colorants, can occur using the inner pore surface complexes of AAO as adsorbents because these compounds have physicochemical properties that are similar to those of the ions of aluminite (hydrous aluminum sulfate) and any aluminate that can form with the electrolyte. Therefore, the various bonding sites on the surfaces of the pores in the AAO drive the different types of reactions that occur post anodizing such as dyeing and sealing. Phosphate, sulfate, and silicate anions from the electrolyte or cleaners or even rinse water can react at the surface blocking sites for dye and seal complexes. It is therefore necessary to maintain the anodizing process to limit drag-out from other steps, and to keep rinses clear and flowing.

3.6.1 Dyeing

Aluminum and aluminum alloy substrates selected for clarity and color tend to exhibit the most homogeneous substrates. Because they have limited compositions, most wrought alloys are limited to less than 8% total alloy additions, and consequently, there is less precipitation. Deformation parameters and aging to a T6 temper typically produce a moderately fine grain size (60–100 μm) that yields a uniform chemical potential across the surface, see Chap. 4. The homogeneity of the substrate, especially as it intersects the surface, produces the most uniform and continuous AAO, with optimal clarity and a fine, dense AAO microstructure. The oxides with the most clarity tend to have wider pores and thinner column walls (ramp from low to relatively higher current densities) and are processed according to Type II anodizing parameters: 18–22 °C, at 1–1.2 A/dm^2. To optimize clarity, the total oxide thickness should be as thin as possible for the design application. It must be kept in mind that thin oxides tend to be iridescent when grown on polished substrates, and must be thick enough to hold enough dye for the color intended. AAO grown under technical anodizing parameters (Type II) presents a highly oriented transparent structure for post-anodizing processes such as dyeing and sealing. See Fig. 3.28.

By increasing the surface area of the pore, by increasing thickness and/or pore diameter, color can be optimized. Dye molecules can be large, and the size of the pore in the AAO can restrict the ingress of large molecules. Dyeing is an equilibrium process, and the movement of dye molecules into the AAO structure is driven

Fig. 3.28 Anodized bicycle headset components. The anodic oxide measured on the clear sample is 2.14 μm, and on the red sample, 11.11 μm. Note the distinctness of image (the reflection of the photographer in the surface of the part) is higher for the part with the thinner oxide. The AAO must be of a thickness to maintain color through dyeing; the thicker oxide yields a beautiful, stable red color [44]

by three concentration gradients between the dye solution and the oxide: (1) the concentration of the dye itself; (2) pH, the dye is a slightly acidic solution and the base of the pore in the AAO is rather basic; (3) temperature, dyeing is a warm process, the AAO is typically rather cool coming out of its final rinse. For all of these reasons, anodizing parameters must be considered that would maximize pore diameter, without developing a thick columnar structure. This would typically mean anodizing at a higher current density relative to normal Type II parameters. Color requires a fine structure with thin walls and a relatively large pore, similar to the transparent finish. Clear finishes are typically thin, about 3–5 μm. A good dark-appearing black-dyed AAO requires greater thickness, experience indicates that a minimum 12 μm thick; best 20–25 μm thick oxide yields the best looking black-dyed component. To achieve a stable color, the oxide requires at least 8 μm [43].

3.6.2 Sealing

Sealing, the final step of the anodizing process, is performed to stabilize the AAO structure by reacting all residual ionic sites. Sealing does much more than "plug the pores" of the AAO; sealing is the actual reductive dissolution of the oxide structure. Hydrothermal and precipitation sealing processes react to dissolve the AAO structure at the surfaces that have been in contact with the electrolyte, including all possible open avenues of ingress from the top surface of the AAO, such as the top surface and the inside surfaces of the pores. This reaction effectively changes the AAO structure at the surface, where it will be in contact with the application environment during service. The seal chemistry, whether hot water, nickel acetate, or nickel fluoride, creates reaction products that remain adsorbed at the oxide surface, or precipitated throughout the seal, increasing the corrosion resistance of the anodic oxide.

Aluminum oxide is amphoteric, that is, it dissolves in both bases and acids. This is clearly illustrated in the Pourbaix diagram for aluminum in which the stability for alumina under equilibrium conditions is shown to be over a pH range from about 4 to about 8 [5]. See Sect. 5.3.1. Conversion of the amorphous AAO surface structure to micro-crystalline pseudo-Boehmite is facilitated by "aging" the structure in a hot water solution (95–100 °C) at pH values between 5 and 7.4 [45]. A fundamental chemical principle: "the addition of a salt of a weak acid to water makes a basic solution," is the premise for the sealing reaction. Sealing baths are typically mildly acidic solutions (pH = 5.5–6.0), but are operated at higher temperatures, over 95 °C. When AAO is reacted with a weak alkaline solution, it forms micro-crystalline pseudo boehmite and hydroxyaluminate ions as a product of the reductive dissolution reaction presented below:

$$Al_2O_3 + H_2O \rightarrow 2AlO(OH) \tag{3.5}$$

Hydroxyaluminate, AlO_4^-, is the fundamental, nonstoichiometric oxide that comprises the passive layer, and forms initially during anodizing, ripening to develop more stoichiometric AAO close to the aluminum substrate as the oxide grows. Sealing is a reverse of this reaction: by continuing to expose the AAO to a mildly basic sealing solution close to the boiling point of water, the reaction proceeds toward dissolution by way of reduction from anodic alumina to pseudo boehmite (reaction 3.1), also forming aluminum hydroxide and hydroxyaluminate ions. Temperature and time enhances the reaction, catalyzing it and promoting dissolution of the AAO structure to the point of collapse at the oxide surface. The randomized and skewed structure together with various hydrated aluminum oxide compounds completely seals the bulk of the AAO from the external environment. See Fig. 3.29.

Because the structure of the AAO is collapsed by the sealing reaction, when the AAO is dyed, sometimes dye can bleed from the AAO during sealing. In the mid 1930s the first patents were filed for precipitation seals, in which additions of a metal salt were made to the sealing bath to prevent the extraction of dye from the oxide structure during sealing. The most common metal salt used for precipitation seals is nickel acetate [3].

The reactions that occur during nickel acetate sealing can be considered to occur as follows [46]. With conventional nickel acetate sealing, the solution is again that of a salt of a weak acid, and the addition of a salt of a weak acid to water makes a basic solution. So, if nickel acetate is considered, Ni^{2+} ions with CH_3COOH will be in solution:

$$Ni^{2+} + CH_3COOH \leftrightarrow Ni^{2+} + CH_3COO^- + H^+ \tag{3.6}$$

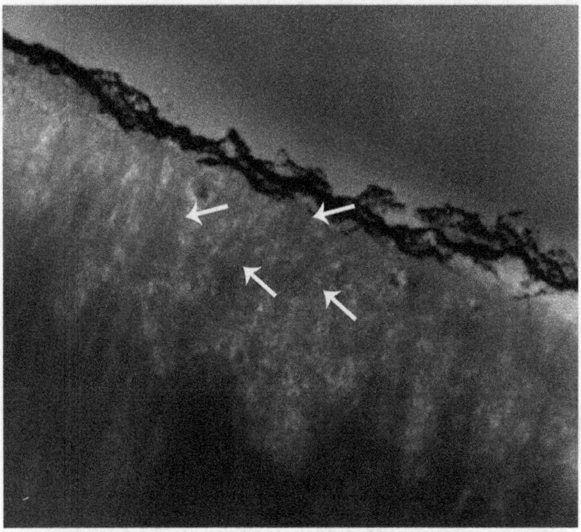

Fig. 3.29 TEM image of a nickel acetate sealed Type III (hard) anodized oxide. The black layer is a layer of sputter-deposited gold, placed to maintain the surface during sample preparation. The blue arrows indicate representative fragments of residual, undissolved AAO columns retained in the seal area, which has been randomized and skewed by the dissolution process of sealing

In solution, the acetate dissociates and the solution creeps basic. This is why additions of acetic acid are made to keep pH within the proper range for operation.

$$Ni^{2+} + CH_3COO^- + H_2O \rightarrow Ni^{2+} + CH_3COOH + OH^- \tag{3.7}$$

When the Ni^{2+} ions are in the presence of OH^- ions, insoluble $Ni(OH)_2$ forms and precipitates out of solution, onto the collapsing structure surfaces, increasing the passivity of the porous structure.

$$Ni^{2+} + OH^- \rightarrow Ni(OH)_2\,(ppt) \tag{3.8}$$

There are a variety of seal formulations that incorporate other metal salts, developed to avoid possible environmental issues with nickel.

Cold seals, based upon nickel fluoride chemistries, can be developed at rather low temperatures (30 °C). Because the temperature of the reaction is low, these seals tend to be shallower than hydrothermal seals. Brace [3] provides a rather comprehensive description of different seal chemistries and processes, as well as the associated reactions and process recommendations.

3.7 Summary and Conclusion

Many different aluminum and aluminum alloy products: castings; wrought products such as extrusions, rolled products, forgings, and deep drawn products; and sintered powder metal products, are designed and specified to have an anodized finish. By capitalizing on the natural phenomenon of passive film formation on aluminum by enhancing it electrochemically to build an integral oxide, the anodic oxide finish has become synonymous with surface protection, durability, and decoration of aluminum substrates. All industries which employ the high-strength-to-weight ratio of aluminum and its alloys, as well as its attractive appearance: aircraft, defense, architecture, automotive, electronics, hydraulics, medical and semiconductor equipment, and other markets, look to the anodizing process as an important means to finish a variety of components manufactured in a variety of ways to provide corrosion protection, wear resistance and in many applications, decorative appearance. Integral oxides of substantial thickness have been industrially produced by electrochemical methods for over 100 years.

Anodizing is an engineering process. By varying the process parameters, the characteristics of the anodic oxide can be "tuned" to meet specific design requirements. All anodizing process steps, outside of actual anodic oxidation, are typically equilibrium processes, with the chemical potential of the substrate driving the surface modification steps during pretreatment, and the dyeing and sealing steps of post-treatment. Only anodic oxidation, because it is driven externally, by an electrical power source, is not considered an equilibrium process. The AAO structural features of pore diameter, pore density (the number of columns per unit grain on the substrate surface), column wall thickness are all governed by the electrical parameters. The aluminum or aluminum alloy substrate, from which the oxide develops and grows, has a direct influence on its structure and quality.

As an industrial process, anodizing has grown with the development of the aluminum industry. As new alloys are developed to meet new and changing engineering demands, so is the anodizing process. One of the main areas for anodizing research is the understanding of the basic science and metallurgy for the mechanism of anodic oxidation. Additionally, new technologies are being developed that use the anodic oxide as a stand-alone product. Lightweighting in all of the transportation industries—planes, trains, and automobiles—as well as rockets and other aerospace applications need aluminum and aluminum alloys, and therefore need anodizing. By studying finishes anodized following standard processes and comparing those finished with different process parameters, a greater understanding of the mechanism of porous oxide formation has been obtained and will continue to grow as long as there is a need for stronger, more corrosion resistant, more wear resistant and more beautiful aluminum surfaces.

References

1. Runge, J., & Pomis, A. (2000, June). Anodic oxide film formation: Relating mechanism to composition and structure. In *Proceedings of the AESF SUR/FIN 2000 Technical Conference, AESF.*
2. Workbook from "Anodizing Essentials", one-day intensive course offered by the Aluminum Anodizers Council (AAC), (2016).
3. Brace, A. W. (2000). *The technology of anodizing aluminium* (3rd ed.). Modena, Italy: Interall Srl.

4. Runge, J. M. (2014). Anodizing for design and function. *Journal of Materials Science and Nanotechnology, 1*(1), S108.
5. Jones, D. A. (1996). *Principles and prevention of corrosion* (2nd ed.). Upper Saddle River, NJ: Prentice Hall.
6. Zhu, H., Dahle, A. K., Zhang, X., & Couper, M. J. (2008). Effect of extrusion microstructure on formation of streaking defects on the surface of anodized aluminum extrusions. In *Proceedings of the 18th Annual Anodizing Conference and Expo of the Aluminum Anodizers Council, San Francisco.*
7. Wernick, S., Pinner, R., & Sheasby, P. G. (1987). *The surface treatment and finishing of aluminum and its alloys, volume 1* (5th ed.). ASM International, Finishing Publications: Middlesex, England.
8. Masuda, H., & Fukuda, K. (1995). Ordered metal nanohole arrays made by two-step replications of honeycomb structures of anodic alumina. *Science, 268*, 9.
9. Apple.com website.
10. Han, C., & Runge, J. (2002). The future of anodizing. In *Proceedings of the Annual Technical Conference and Exposition of the Aluminum Anodizers Council, Oakbrook, Illinois.*
11. Military Specification (MIL)—A—8625F: Anodic Coatings for Aluminum and aluminum Alloys, (1993, Sept 10).
12. Arurault, L. (2008). *Pilling-Bedworth ratio of thick anodic aluminium porous films prepared at high voltages in H_2SO_4 based electrolyte*. Leeds, UK: Maney Publishing, Institute of Metal Finishing.
13. Jessensky, O., Müller, F., & Gösele, U. (1998). Self-organized formation of hexagonal pore arrays in anodic alumina. *Applied Physics Letters, 72*(10), 1173–1175.
14. Thompason, G. E. (1997). Porous anodic alumina: Fabrication, characterization and applications. *Thin Solid Films, 297*(1–2), 192–201.
15. Nielsch, K., Choi, J., Schwirn, D., Wehrspohn, R., & Gösele, U. (2002). Self-ordering regimes of porous alumina: The 10% porosity rule. *Nano Letters, 2*(7), 677–680.
16. Pilling, N. B., & Bedworth, R. E. (1923). The oxidation of metals at high temperatures. *Journal of the Institute of Metals, 29*, 529–591.
17. Gabe, D. R. (2000). Density values for anodic films on aluminium and some observations of pore morphology. *Transactions of the Institute of Metal Finishing, 78*(6), 207–209.
18. Xu, C., & Gao, W. (2000). Pilling-Bedworth ratio for oxidation of alloys. *Materials Research Innovations, 3*, 231–235.
19. Runge, J. (2007). Formation of porous anodic oxide finishes: A new approach and theory. In *Conference Proceedings of Aluminium 2000, Florence.*
20. Levendusky, et al. (2010, June 8). *Corrosion resistant aluminum alloy substrates and methods of producing the same* (US Patent No. 7,732,068).
21. Runge, J., Gilbert, A., Kriesch, G., & Pernick, J. (2006). Hybrid nanostructure of the anodic oxide for polymer bonding: A case study. In *Proceedings of the AAC Annual Technical Conference and Exposition, Toronto, Canada.*
22. Poinern, G., Ali, N., & Fawcett, D. (2011). Progress in nano-engineered anodic aluminum oxide membrane development. *Materials, 4*, 487–526.
23. Le Coz, F., Aururault, L., & Datas, L. (2010). Chemical analysis of a single basic cell of porous anodic aluminum oxide templates. *Materials Characterization, 61*, 283–288.
24. Runge, J. M. (2015). Enhancing anodic aluminum oxide for bonding applications. In *Proceedings of the 24th Annual AAC Conference, San Diego, California.*
25. Ducretet, M. E. (1875). Note sur un Rhéotome Liquide a Direction Constate, Fondé sur une Propriété nouvelle de L'Aluminium. *Journal de Physique Théorique, 4*(1), 84–85.
26. Pollak, C. (1897). German Patent: "Elektrischer Flüssigkeitskondensator mit Aluminiumelektroden", No. 92564, Granted 19 May, 1897.
27. Guntherschulze, A. (1906). Über das Verhalten von Aluminiumanoden. *Annalen der Physik, 21*, 929–954.
28. Sulka, G. D. (2008). Highly ordered anodic porous alumina formation by self-organized anodziing, section 1.2.1 types of anodic oxide film. In A. Eftekhari (Ed.), *Nanostructured materials in electrochemistry* (pp. 7–8). Weinheim: Wiley-VCH Verlag GmbH & Co. KGaA. (Ch. 1).

29. Burgers, W. G. Claasen, A., & Zernike, J. (1932). *Über die chemische Natur der Oxydschichten, welche sich bei anodischer Polarisation auf den Metallen Aluminium, Zirkon, Titan and Tantal bilden* (pp. 593–603). Eindhoven, Holland: Naturkundig Laboratorium der N.V. Philips' Gloeilampenfabriken.
30. VanGeel, W. C., & Schelen, B. J. J. (1957). Some properties of oxide layers produced on aluminium by electrolytic oxidation. *Philips Research Reports, 12*, 240–248.
31. Mott, W. R. (1904). Colloidal precipitation upon aluminium anodes. *Electrochemical Industry, II*(11), 444–447.
32. Murphy, J. F., & Michelson, C. E. (1961). *Proceedings of a Conference on Anodising Aluminium at the University of Nottingham, Convened by the Aluminium Development Association of London, September 12–14, 1961* (pp. 83–95).
33. Paschanka, M., & Schneider, J. (2011). Origin of self-organization in porous anodic alumina films derived from analogy with Rayleigh-Bénard convection cells. *Journal of Materials Chemistry, 21*, 18761–18767.
34. Thompson, G. E., & Wood, G. C. (1983). Anodic films on aluminium, section V. Barrier type anodic films. In J. C. Scully (Ed.), *Corrosion: Aqueous processes and passive films, Treatise on materials science and technology* (Vol. 23, pp. 230–254). London: Academic Press.
35. Evangelisti, F., Stiefel, M., Guseva, O., Patoveh-Nia, R., Hauert, R., Hack, E., et al. (2017). Electronic and structural characterization of barrier-type amorphous aluminium oxide. *Electrochimica Acta, 224*, 503–516.
36. Brown, S. D., Kuna, K. J., & Van, T. B. (1971). Anodic spark deposition from aqueous solution of $NaAlO_2$ and Na_2SiO_3. *Journal of the American Ceramic Society, 54*(8), 384–390.
37. Runge-Marchese, J. M., & Nussbaum, T. (1998). New insights regarding the mechanism of spark anodization processes. In *AESF SURFIN Proceedings* (Vol. 6, pp. 531–540).
38. Dehnavi, V., Liu, X.-Y., Luan, B.-L., Shoesmith, D., & Rohani, S. (2014). Phase transformation in plasma electrolytic oxidation coatings on 6061 aluminum alloy. *Surface & Coatings Technology, 251*, 106–114.
39. Bengough, G., & Stuart, J. (1923, August 2). *Improved process of protecting surfaces of aluminium or aluminium alloys* (British Patent 223,994).
40. Runge, J., & Pomis, A. (2002). Continued development in chrome-free anodic oxide finishes for aluminum: Evaluation of selected mechanical properties. In *Proceedings of the American Electroplaters and Surface Finishers Society Aerospace/Aircraft Forum August 27–29, 2002.*
41. Chesterfield, L., & Runge, J. (2013). Anodizing for design and function. In *Proceedings of the 22nd Annual AAC Conference, Seattle, Washington.*
42. Stumm, W., & Furrer, G. (1989). The dissolution of oxides and aluminum silicates; examples of surface-coordination-controlled kinetics. In W. Stumm (Ed.), *Aquatic surface chemistry; chemical processes at the particle-water interface*. Zurich: Wiley.
43. Chesterfield, L., & Runge, J. (2013). Connecting theory to practice: The science of successfully anodizing aluminum die castings. In *Proceedings of the 19th Annual AAC Conference, Seattle, Washington.*
44. Runge, J. (2012). Trace elements and their impact on surface finishing characteristics of aluminum extrusions. In *Proceedings of ET 2012, Miami, FL.*
45. Vieira Coelho, A.C. (2007). Specific surface area and structures of aluminas from fibrillar pseudo-boehmite. *Revista Materia, 13*(2), 329–341.
46. Hossain T. (2013). Unpublished work.

Chapter 4
Metallurgy Basics for Aluminum Surfaces

4.1 Introduction

When one considers the elegance of the anodizing process, it becomes clear that the essential process components: the aluminum anode, the electrolyte, the cathode material and externally applied current, are where the various process complexities begin. This chapter considers the most essential element of the anodizing process: the aluminum. Because the anodizing reaction is one of substrate (aluminum) consumption with simultaneous oxide growth, the role of the aluminum (alloy) in the reactions that occur at the substrate-electrolyte interface cannot be ignored. The anodic oxide originates at the surface of the substrate aluminum or aluminum alloy; therefore, a basic understanding of aluminum and its alloys is critical to understanding anodizing.

As a metallic element, aluminum has low tensile strength and hardness. In pure metals, the presence of atomic level defects such as vacancies, interstitials, dislocations, and grain boundaries, which are introduced to the metal during solidification and by deformation, alone produces mechanical strength. Increased hardening is achieved when deformation-induced motion, or slip of the crystal, is inhibited by these defects. These various defects develop the microstructure throughout the entire substrate, which includes the surface. In addition to developing the mechanical properties of the aluminum, defects change the thermodynamic and electrochemical properties throughout the microstructure creating variations in chemical potential across the surface that change the response to the anodizing reaction from the atomic to the macroscopic level.

For practical applications, aluminum must be alloyed to overcome mechanical strength limitations of pure aluminum. However, the solid solubility of various metal alloy additions is limited to a few percent, and no element has complete solid solubility with aluminum. As the solid solubility of alloying elements in aluminum is exceeded, mixtures form with varying aluminum content called phases and other

J.M. Runge, *The Metallurgy of Anodizing Aluminum*,
https://doi.org/10.1007/978-3-319-72177-4_4

intermetallic particles called dispersoids and non-aluminum containing precipitates and insoluble additions that fall out of solution, making the microstructure more complex. With deformation, these inhomogeneities impede movement, thereby strengthening and producing hardness throughout the metal. Aluminum alloys have been developed with unsurpassed strength-to-weight ratios and wide ranging applications by way of solid solution and precipitation hardening mechanisms. Microstructural features and defects not only impact the mechanical properties of the metal, just by virtue of their size, population, and distribution, they change the chemical and electrochemical properties of the surface.

Since the late nineteenth century, when engineering applications for aluminum and its alloys were first made a part of industrial designs, alloy development has centered not only around the application, but around how the product will be manufactured. Manufacturing processes such as casting and deformation introduce additional complicating defects to aluminum and aluminum alloy component microstructures. Solidification introduces atomic level defects, and depending upon alloy chemistry and cooling rate, microstructural features (such as phases) form and other intermetallic particles precipitate. Deformation introduces strain energy to the crystal structure, and inasmuch as it produces strength and hardness by producing more defects within the microstructure, the high energy condition brought about by deformation changes the microstructure by altering grain size and changing grain boundary density, developing more metallurgical features, and refining existing ones, which further develop variations in the surface chemical potential. Deformation effects on the microstructure can be relieved or further developed by the heat of the deformation process itself or by additional cold work. Changes to the microstructure induced externally, by deformation and heat treatment, also change the chemical response at the surface.

Clearly, real world substrates are more complex (less ideal) than what is specified in handbooks, which makes the anodizing reaction more complicated than it would be if it were as simple as the chemical reaction for aluminum oxidation as written in the equations presented in Chap. 3. In this chapter, the characteristics of aluminum and its microstructural development will be presented in the context of cast and wrought alloy chemistries; their constitutive phases and particles that comprise each substrate; microstructural feature characteristics and how they are affected by manufacturing. Common alloy and temper designations are presented such that variations in chemistry and heat treatment can be connected to variations in microstructure. This is because the population, distribution, and size of defects and interfaces across the surface, whether they are atomic level, micro- or macroscopic, affect the structure and appearance of the anodic oxide. Such an understanding enables one to develop a virtual picture of the aluminum substrate surface before any actual surface treatment is done, a useful vantage point when it becomes the interface from which the anodic oxide grows, and integral to the design step of alloy selection, when the anodic oxide is the finish of choice.

Table 4.1 Some important properties of the light metals [2]

Property	Aluminum	Magnesium	Titanium	Beryllium
Color	White	Silvery-white	Silvery-white	Grayish
Density (kg/m^3)	2700	1740	4540	1850
Melting point (celsius)	660°	650°	1668°	1283°
Crystal structure	FCC	HCP	HCP, BCC[a]	HCP
Lattice parameter (nm)	0.405	$a = 0.321$ $c = 0.521$	$a = 0.295$ $c = 0.468$	$a = 0.228$ $c = 0.358$
Atomic (weight)	29.98	24.31	47.90	9.01
Tensile (strength) (MPa)	72	78	400	325
Elongation (%)	60	2	27	≤8
Modulus of elasticity (GPa)	70	45	116	303
Electrical conductivity (IACS) (%)	64	38	4	42
Corrosion resistance	Very good	Good	Excellent	Moderate
Magnetic behavior	Nonmagnetic	Nonmagnetic	Nonmagnetic	Nonmagnetic
Specific feature(s)	Nontoxic, nonsparking	Readily oxidizes and burns	Readily oxidizes above 500 °C	Toxic, poor machinability

[a]Above 882 °C with $a = 0.331$ nm

4.2 Aluminum Structure and Properties

Aluminum, magnesium, titanium, and beryllium are classified as *light metals* because, whether used in their pure state or as alloys, their purpose is to reduce the weight of components and structures. The property of *lightness* is relative to the low density of the material, 2.7 g/cm^3 and 1.7 g/cm^3, for aluminum and magnesium, respectively. Upon comparison to the density of other structural metals, 7.9 g/cm^3 for iron and 8.9 g/cm^3 for copper, it is easy to see why the light metals are the preferred materials in the transportation industry, especially in the aerospace and defense industries, where minimizing payload weight can reduce fuel consumption and decrease flight time, which are important economic considerations [1].

Table 4.1 summarizes other important properties of aluminum and the other light metals [2]. When compared to the other light metals, the moderate melting temperature of aluminum enables it to be processed at relatively low temperatures; yet the stability of aluminum over a broad temperature range enables its application in moderately high operating temperature conditions. Aluminum has high electrical and thermal conductivity; in fact, it is the best conductor in weight-to-weight comparison with copper. Aluminum has the amazing capability to efficiently reflect radiant energy, on which the use of aluminum foil is based. Aluminum is easy to

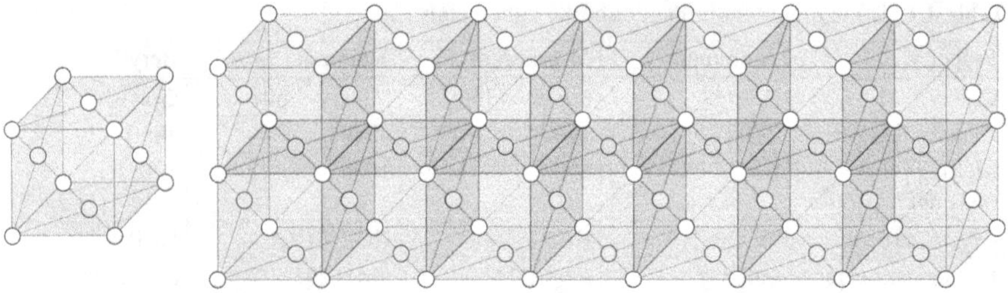

Fig. 4.1 Schematic of the face-centered cubic (FCC) crystal structure of aluminum

machine and the waste from processing is not toxic like that from beryllium or as reactive as magnesium. Because of the ease of extraction and processing, aluminum is not as expensive as titanium or carbon fiber composite materials.

Of the light metals, aluminum is the only one that has a face-centered cubic (FCC) structure, which accounts for many of the unique characteristics of aluminum and aluminum alloys [3]. The aluminum FCC crystal structure is without allotropes, which means it always stays FCC, in spite of remelting, deformation, and heat treatment. FCC structures have long interstitial sites, which accounts for easy alloying, ductility (highest percent elongation for the light metals), ease of deformation along its many slip planes, and resistance to hydrogen embrittlement. Because aluminum has no allotropes, it can be recycled indefinitely without loss of properties. Recycling requires only 5% of the energy and produces only 5% of the CO_2 emissions as compared with primary production of aluminum from bauxite and reduces the waste going to landfill, making aluminum the most cost-effective material to recycle [4]. The ability to recycle indefinitely makes aluminum a truly sustainable engineering material, something to consider when comparing it with other light metals for design applications. See Fig. 4.1.

Pure aluminum derives its limited strength from atomic level defects such as dislocations and vacancies that are introduced to the FCC structure during solidification and deformation. These first-level defects increase the strength of pure aluminum by interrupting the long-range order of a perfect FCC crystal. Atomic level structural and microstructural features, characteristic for all polycrystalline materials, change the chemical potential of the surface and therefore the corrosion resistance to external ions. Grains with varying crystallographic orientation also exhibit different electrochemical potentials, depending on the atomic packing density. In aluminum, the [111] plane is most reactive, because it has the highest packing density, followed by the [110] plane and the [100] planes [5]. For aluminum, the presence of microstructural defects changes etch rates and oxide growth rates that correspond to grain orientation, which are manifested in processes such as cleaning, brightening, etching and anodizing. Point defects such as vacancies and dislocations agglomerate and accumulate at grain boundaries, which change the interfacial energy and confound the electrochemical reaction at the surface, developing the microstructure of even the purest substrates [6]. See Fig. 4.2.

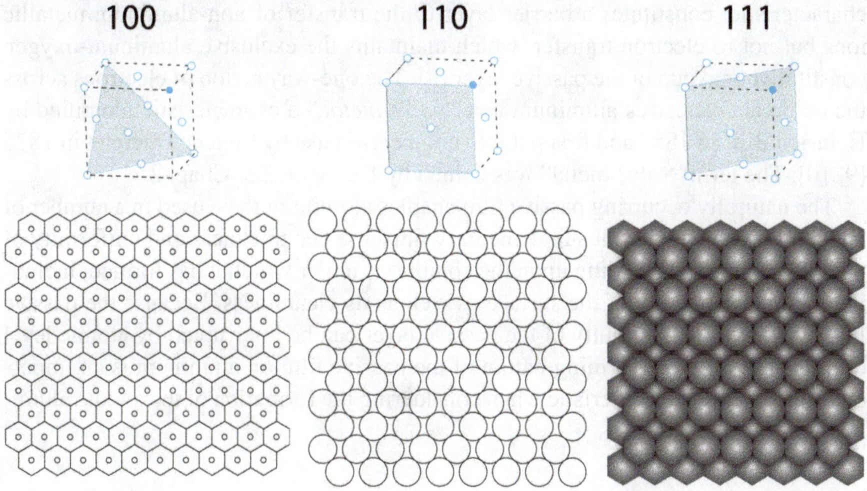

Fig. 4.2 Schematic of atomic packing on different crystallographic planes in the FCC lattice. Because the atomic packing density is highest on the [1] plane (far right), it exhibits the highest chemical sensitivity

4.2.1 The Aluminum Surface

In order to develop a virtual picture of the aluminum microstructure, it must always be considered that the interface created with surface and the corresponding environment (in the case of anodizing, with the electrolyte) is three-dimensional. Two-dimensional representations fall short of reality when considering surface reactions. Surface atoms for all materials are only coordinated with other atoms on one side, and are therefore at a much higher energy state and less firmly bonded than the internal atoms [7]. In addition, atoms in the vicinity of the surface are exposed to a different environment than those in the bulk. The converse is also true: atoms comprising the bulk microstructure have different densities and lower energies than surface atoms. The reactive surface, although potentially expansive, is very thin, on the order of a few nanometers, yet the bulk microstructure is intimately connected to the surface microstructure and therefore to the anodizing reaction. Because the oxide growth interface advances into the bulk during anodizing, the expanse of the reactive surface between the aluminum and the environment continuously changes; therefore, it is not only the surface microstructure that must be considered for anodizing, the near-surface microstructure is equally important.

Aluminum, because of its high surface reactivity, has characteristically high corrosion resistance. Only magnesium has a greater affinity for oxygen than aluminum. Aluminum surface atoms oxidize spontaneously, producing a thin, stable passive film, on the order of 1–10 nm, that prevents further oxidation. When polarized as an anode, electrons from the aluminum easily transfer across the natural oxide, and no ionic tunneling occurs from the outside. The aluminum oxide therefore, as a characteristic, constitutes a barrier layer to the transfer of non-aluminum metallic ions but not to electron transfer, which maintains the exclusive aluminum-oxygen constitutive makeup of the passive layer [8]. The one-way action of electrons across the oxide characterizes aluminum as a "*valve metal*," a characteristic identified by Heinrich Buff in 1857 and first put into engineering use by Eugene Ducretet in 1875 [9, 10]. The term "valve metal" was coined by Ducretet. See Chap. 2.

The naturally occurring passive film enables aluminum to be used in a number of applications as-is, without supplementary finishes. The thickness and uniformity of the film can vary, depending upon the conditions under which it has formed, including the microstructure of the surface as well as its cleanliness. Because the passive layer is very thin, continuity of the passive layer can be interrupted by atomic level defects. The network-forming nature of the passive film on aluminum is an important fundamental characteristic when considering the formation of the anodic aluminum oxide [11].

4.3 Aluminum Alloys

As a pure metal, aluminum has low tensile strength and hardness, so it must be alloyed to overcome mechanical strength limitations. The characteristics of aluminum alloys make them some of the most important for industrial designs. Aluminum alloys of great strength can be produced with an unsurpassed strength-to-weight ratio for designs that might otherwise employ copper, steel, or other light metal alloys. Aluminum alloys have comparatively good ductility and malleability, which gives wrought alloys excellent formability and manufacturability; they have high electrical and thermal conductivity. An unusual characteristic that aluminum alloys possess is excellent mechanical properties at sub-normal temperatures that make them a good choice for aerospace applications. Therefore, aluminum and aluminum alloys have become the material of choice for many applications: aircraft, defense, architecture, automotive, electronics, hydraulics, semiconductor equipment, and outdoor recreation products. These industries all benefit from the high strength-to-weight ratio, fatigue resistance, impact resistance, and corrosion resistance of aluminum and its alloys.

Aluminum readily mixes with other metals in the liquid state because of its FCC structure. However, the solid solubility of various metal alloy additions is limited to a few percent. In fact, no element has complete solid solubility with aluminum. Therefore, alloying elements are added that exceed the solid solubility in aluminum to form mixtures with varying aluminum content called phases; other particulate intermetallic compounds called dispersoids, and non-aluminum containing precipitates form to comprise networks that strengthen and harden the metal. Sometimes insoluble alloy additions such as bismuth and lead, which have no solubility in aluminum at all, are added in order such that they fall out of solution as individual particles, to facilitate mechanical processes such as machining. As metallurgical phenomena, phases and intermetallic particles are present to some extent in all

Table 4.2 Solubility of primary aluminum alloying elements

Element	Maximum solid solubility, wt%	Solubility at room temperature, wt%
Copper	5.65 at 548 °C	0.02
Magnesium	14.9 at 451 °C	2.50
Manganese	1.82 at 657 °C	0.30
Silicon	1.65 at 577 °C	0.10
Zinc	82.8 at 382 °C	2.00

alloys, regardless of the alloy chemistry or the manufacturing process. Table 4.2 presents the solubility for some of the primary elements used for making aluminum alloys at temperatures of maximum solid solubility compared to solubility at room temperature [12].

Because of the passive nature of aluminum, aluminum alloys typically exhibit good corrosion resistance. However, as alloy properties change with the amount and different types of elements added to the aluminum, so does the corrosion resistance. Alloys richer and more diverse in additive elements have more complex microstructures; how and where the particles and phases agglomerate and concentrate during manufacturing impact their size and distribution throughout the aluminum alloy matrix, which in turn impacts the chemical potential throughout the microstructure. The intermetallic compounds which have the greatest impact on alloy performance are those in the greatest proportion, either by size, shape, or population. Variations in chemical potential due to alloy segregation and precipitation are most profound at the surface, where the alloy is in contact with the environment. Intermetallic compounds, richer in alloying elements than the alloy matrix, exhibit differences in chemical potential at adjacent locations across the surface and create significant differences in electrochemical behavior.

Alloys are designed typically for a specific manufacturing process with the thermodynamics of phase and particle formation in mind. Intermetallic phases and particles impact microstructural development, the manufacturing process, as well as alloy performance in the engineering application and service environment. Phase diagrams are useful tools when considering what intermetallic phases and compounds will develop at what temperature. Phase diagrams characterize the metallurgical behavior of metals and alloys in terms of composition and temperature and are helpful when making the connection between the microstructure and the various thermal excursions alloys experience through casting and wrought processing. Understanding thermodynamic connections between alloy chemistry and the manufacturing process are key in understanding the development of the microstructure, as well as mechanical characteristics such as strength and hardness, which are typically characteristics of the bulk component. Microstructural development must also be considered in terms of the surface and how it will impact corrosion resistance and surface treatment. Whether by trial and error or by using computer software, by understanding the thermodynamics of the possible reactions that can occur during alloy development and process development, microstructure can be planned as part of a reliable manufacturing process and total engineering design.

Alloys, cast or wrought, are seldom developed for surface finishing. Variations in composition between the matrix and segregated phases and discrete intermetallic particles change the chemistry of the substrate in general as well as the local chemical potential from point to point across the substrate surface. Different alloys therefore exhibit different responses and sensitivities to corrosion and chemical processing, such as etching, cleaning, even rinsing with water, steps critical to anodizing pretreatment. Since anodizing is an electrochemical process, alloying and microstructure change the response to anodic oxide growth. Each and every microstructural phase, intermetallic compound, grain boundary, and surface defect confound the anodizing reaction to some extent by changing how the oxide nucleates or impeding continuous oxide growth. This is because only aluminum is consumed to yield the anodic oxide; alloying elements, pockets of atomic level defects, porosity and empty space in general, do not anodize. The side reactions that occur at the interface between the aluminum alloy substrate and the anodic oxide are presented in detail in Chap. 7; base metal microstructural considerations for designing with anodized aluminum are presented in Chap. 8.

4.3.1 Microstructural Development in Cast Aluminum Alloys

Two different mechanisms, solid solution hardening and precipitation hardening, increase the tensile strength and hardness of all aluminum alloys. Cast alloys exhibit elements that produce both hardening mechanisms. Solid solution hardening increases strength by introducing lattice strain when non-aluminum atoms occupy sites on the FCC lattice typically occupied by aluminum. Substitutional atoms of different sizes dilate or contract the lattice, producing strain that increases hardness. Further deformation of a structure that is solid solution hardened will produce a dislocation substructure that restricts crystal movement and increases hardness and strength. Precipitation hardening further increases strength by interrupting the long-range order of the FCC lattice by way of the introduction of additional particles that precipitate throughout the structure on the surfaces of other particles and defects during solidification and follow-on heat treatment. Both solid solution hardening and precipitation hardening mechanisms occur to some extent in all alloys, regardless of the casting process; variations in alloy composition typically enhance strengthening characteristics, with the most complex alloys exhibiting the highest strength and hardness. See Fig. 4.3.

As a process, casting can be described as the pouring or injection of molten metal, as a fluid, into a cavity or mold, yielding a uniform part that is, depending upon its composition and purpose, used near net shape as cast components, or further deformed, as wrought components. The properties of the molten fluid, especially molten aluminum, are such that it gets very rapidly covered with an oxide film, which greatly influences casting behavior. Because of the extremely dense and tenacious oxide that forms on aluminum, Schey [14] compares the casting of aluminum as if it were flowing inside a "rather tough bag." Alloying elements that modify

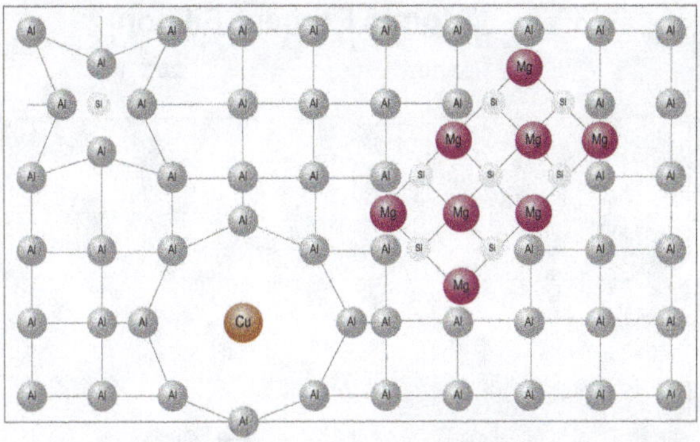

Fig. 4.3 Schematic for solid solution hardening and precipitation hardening. A substitutional atom with a small atomic radius, silicon, produces strain by contracting the lattice; while a substitutional atom with a larger atomic radius than aluminum, such as copper, produces strain by expanding the lattice. In the schematic on the right, a magnesium silicide precipitate has formed interstitially within the aluminum lattice, interfering with the lattice order and producing strain. Most distortion effects caused by atomic level defects within the lattice increase hardness [13]

the oxide greatly affect the casting behavior of aluminum alloys, with silicon being the primary alloy addition for improving flow. All of aluminum die-casting is based upon the addition of silicon to the aluminum melt. See Figs. 4.4 and 4.5.

Solidification occurs from the surface of the mold and proceeds inward, toward the center of the melt. The highest melting temperature alloy composition freezes first, nucleating and growing into the melt as dendritic crystals. Lower melting, more solute-rich phases solidify next, in between the dendrite arms. If solidification is uncontrolled, a microstructural gradient will develop from the mold surface (most aluminum rich) inward to the center (progressively lower melting temperature compositions, with the most alloying elements), producing a condition known as coring. Rapid cooling, regardless of the casting process, produces the most homogeneous microstructures. Atomic level defects are developed during solidification of cast products, with vacancies and dislocations forming during solidification. Rapid solidification “quenches in” these defects, providing surfaces on which intermetallic particles nucleate and grow. See Fig. 4.6.

4.3.1.1 Intermetallic Phase and Particle Formation

The limited solid solubility for most metals in aluminum produces a compositional gradient within the microstructure, which depends upon how the castings solidify. Aluminum-rich phases have a higher melting temperature than other melt constituents, and solidify first as the casting cools. Intermetallic particles that form prior to the solidification of the aluminum dendritic network, especially those that grow freely within the liquid phase, such as primary silicon crystals, form simultaneously with the

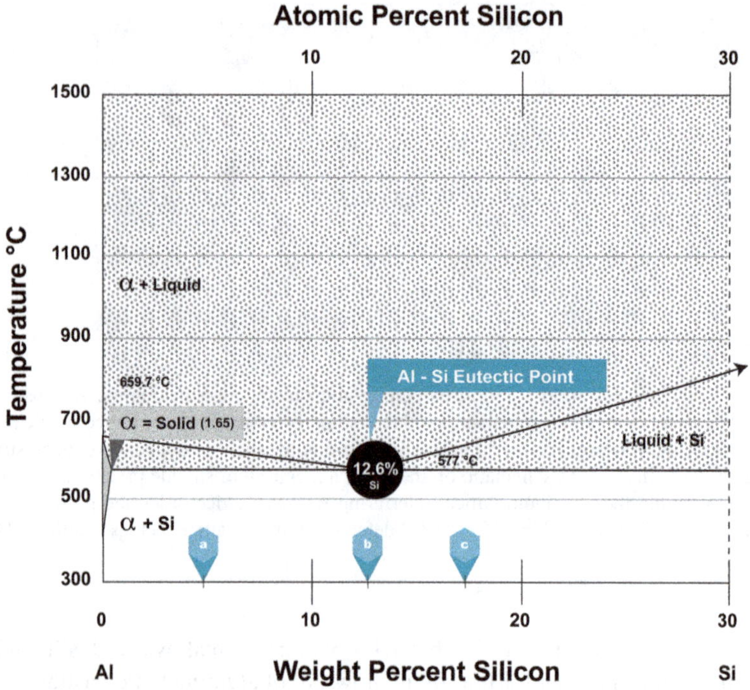

Fig. 4.4 The Aluminum-Silicon binary phase diagram [3]. Arrow indicates the eutectic composition: 12.6%Si. Areas a, b, and c correspond to the microstructures presented in Fig. 4.5

dendritic network and can grow to be relatively large. More solute-rich intermetallic particles consequently become entrained interdendritically, at developing grain boundaries, and form later in the solidification sequence, during or after the period of Al-Si eutectic solidification. These particles are comparatively smaller because they are constrained by the reduced liquid space available to accommodate growth. These effects are profoundly influenced by alloy complexity, the concentration of alloying elements as well as the casting method. Nevertheless, all particles will tend to be larger if the cooling rate is slower. Generally speaking with respect to intermetallic particles in the cast microstructure, the larger the particle, the more detrimental its effects on mechanical properties and subsequent processability of the casting, which includes anodizing.

Three basic types of microstructural particles form during alloy solidification [16]. Primary constituent phases (PCPs) are large when compared to other intermetallic particles, up to 10 microns in size, and are concentrated at the first grain boundaries formed from the melt, between aluminum-rich dendrite arms. As the melt solidifies, particles with lower melting temperature nucleate, precipitate, and grow on the surfaces of PCPs.

Dispersoids are the smallest precipitate particles comprising alloying elements that are highly *insoluble* in aluminum. They tend to form from excess trace elements rejected by the primary constituent phase. Therefore, they begin to form at the billet

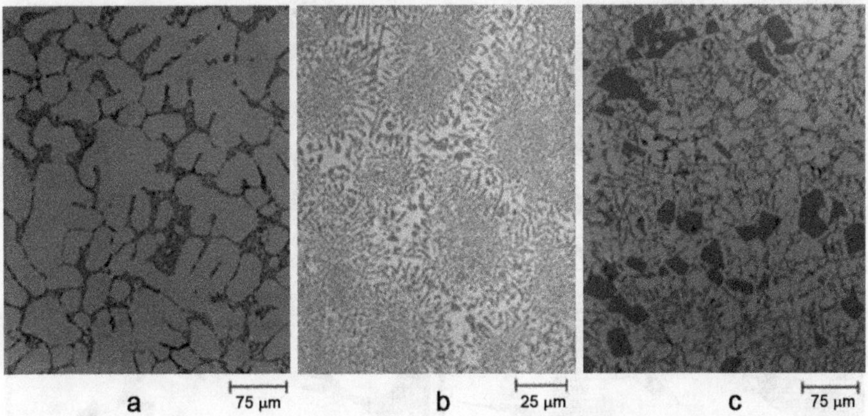

Fig. 4.5 Cast microstructures at three silicon compositions that correspond to the phase diagram presented as Fig. 4.4. (**a**) Alloys with less than 12% Si are referred to as "hypoeutectic." Aluminum-rich dendrites are divided by Al-Si eutectic primary constituent phase (PCP). (**b**) Alloys with close to 12% Si are referred to as "eutectic" (the point at which a direct transformation from liquid to solid phase occurs). The fine-grained structure comprises dark-appearing Al-12.6% Si eutectic PCP divided by the Al-rich matrix forming a characteristic lamellar structure. (**c**) Alloys with greater than 12% Si are "hypereutectic," they exhibit large primary Si crystals that interrupt the microstructural homogeneity. Smaller eutectic needles and Al-rich matrix comprise the rest of the microstructure

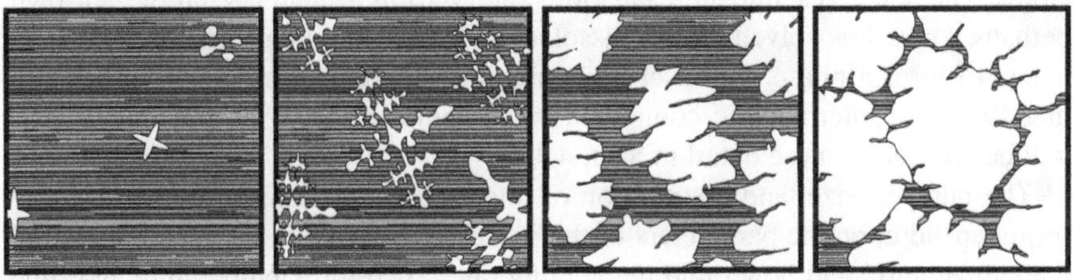

Fig. 4.6 Schematic for solidification from Altenpohl [15]. Nuclei develop in the melt and progressively grow, forming dendrites. The highest melting temperature phases solidify first, and as they grow and develop, the primary constituent phases (PCPs), rich in alloying elements and with lower melting temperature than the parent phase, solidify next and, reside in between the dendrite arms, ultimately forming grain boundaries

level or during preheat, prior to deformation processing, and tend to contain aluminum. They range in size from 0.05 to 0.5 μm. Chromium (Cr), titanium (Ti), iron (Fe), manganese (Mn), and zirconium (Zr) are typical dispersoid formers, with those containing Fe and Mn being the most chemically active in environments where aluminum is used. Dispersoids alone do not increase hardness; the role of dispersoids is that of a recrystallization inhibitor by way of constraining grain growth. Al_3Fe and Al_6Mn are common dispersoid stoichiometries found in many aluminum alloys. See Fig. 4.7.

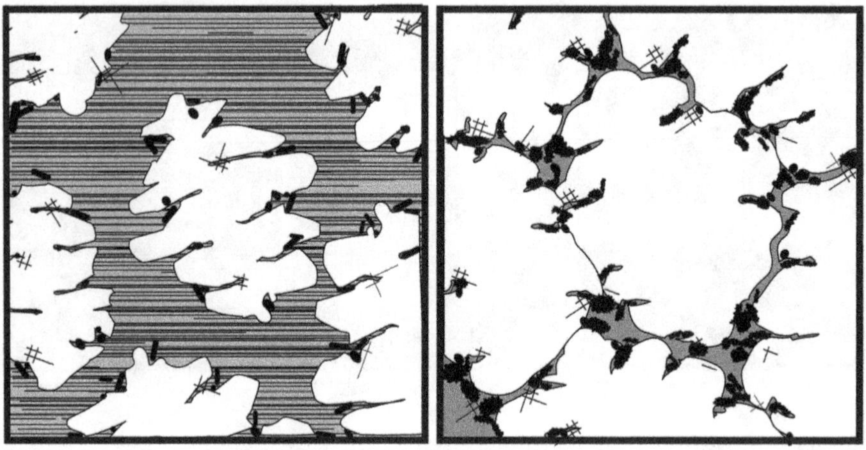

Fig. 4.7 Schematic of cast microstructure development: interdendritic precipitation of primary constituent phases occur between on the aluminum-rich dendrite arms as they solidify. Dispersoids nucleate on the surface of PCPs. They restrict recrystallization, thereby increasing alloy hardness

Other intermetallic particles also precipitate from the melt during solidification. Precipitation can best be described as decomposition of a supersaturated solid solution into its solvent (aluminum) phase and excess rejected solute (alloying element) atoms. The rejected solute atoms agglomerate to form a small crystal of their own, perhaps with a few solvent atoms dissolved in it, or react with some of the solvent atoms to form a phase with a crystal structure different from that of either pure metal (i.e., an intermetallic compound). The agglomerate is the precipitate. This will be covered in more detail in Sect. 4.4.4.

The number, size, and distribution of intermetallic particles and how fast they form depend upon the parameters of the processes external to the metal: how high the temperature, and how rapid the cooling rate. Therefore, temperature and time, above all other parameters, govern intermetallic particle development, which also profoundly affects grain size and shape. Figure 4.8 presents the sequence of intermetallic formation in cast Al-Si-Mg-Cu-Fe alloys over a temperature range from 650 to 450 °C, and over a timespan of 0–400 s. It is interesting to note that in the sequence of intermetallic phase and particle development, the Al-Fe-Si primary constituent phase begins to form as the primary Al dendrites freeze, prior to the solidification of the Al-Si eutectic. The size and distribution of the phases and particles change with the amount of each alloying element and the cooling rate [17].

The energy developed through deformation processing produces enough heat to provide the driving force for microstructural recovery following deformation that with time at sufficient temperature leads to recrystallization and grain growth. Additionally, the energy introduced by way of deformation processing, by way of the speed and amount of reduction that takes place, also impacts precipitation, recrystallization and grain growth. Tempering or aging, which follows solution treatment, does not provide enough of a thermal driving force for grain growth, but effectively develops incoherent precipitates to increase hardness and strength. By

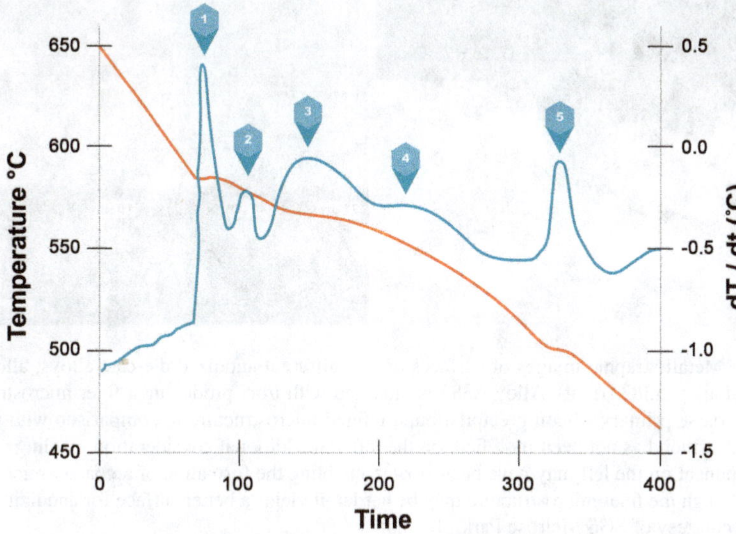

Fig. 4.8 Time-temperature diagram for an Al-Si-Cu-Mg-Fe alloy (similar to cast alloy 383). The labeled peaks correspond to the sequence of reactions that occur to form the following intermetallic phases and particles. (1) Primary aluminum dendrites begin to freeze at ~640 °C. (2) The primary constituent phase, Al_5FeSi forms. (3) Al-Si eutectic freezes. It is interesting to note that the intermetallic phase forms before the eutectic freezes. (4) Complex Mg_2Si Guinier–Preston zones form which correspond later to the formation of precipitates that serve to strengthen during aging. (5) Complex Al_2Cu dispersoid particles form [17]

considering the microstructure in terms of the relative number, size and distribution of atomic level defects, phases, intermetallic compounds and other surface phenomena which comprise aluminum and aluminum alloy substrates, and when and how they occur, a picture develops of how the surface microstructure of a part will appear, how the chemical potential can vary across the surface and how it will react to various surface finishing processes such as anodizing. See Fig. 4.9.

Cast components have limited ductility and can be brittle; therefore, castings are not usually meant for subsequent deformation processing. Other than minimal finishing processes such as surface machining, a casting is typically designed and produced to function near net shape. For this reason, castings can be alloyed beyond what is typical for wrought products; that is, additions of other elements are at a higher amount in cast alloys than the additions for alloys intended for mechanical deformation processes such as extrusion, rolling, or forging. In general, castings contain up to 16% total alloy content, and wrought product contains up to 8%. Wrought aluminum super alloys, typically of the 7XXX series, can be alloyed beyond 14%, and require much attention during deformation processing due to the high alloy content. Alloys containing more than 16% are generally squeeze cast, thixotropic molded or sintered [12]. Increased alloy additions produce correspondingly higher levels of solution phases, intermetallic compounds, and precipitates; as such, cast alloys

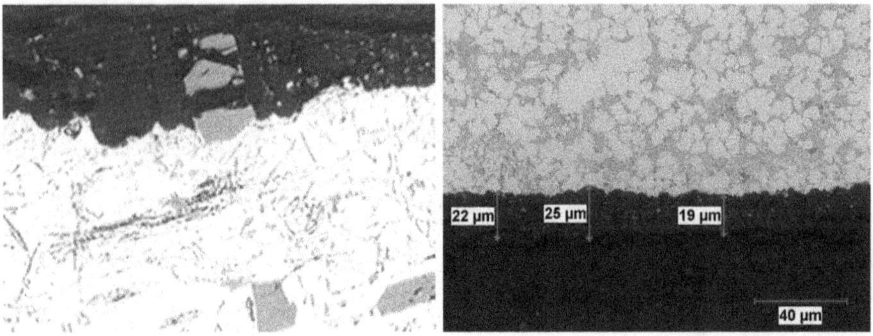

Fig. 4.9 Metallographic images of surfaces of two different anodized die-cast alloys; alloy 383 (left) and alloy A383 (right). Alloy A383 is modified with iron, producing a finer microstructure without coarse primary silicon precipitation, a refined microstructure in comparison with that of alloy 383, which has not been modified, on the left. An additional consideration, cooling rate for the component on the left may have been slower, enabling the formation of a coarser microstructure. Although the finer microstructure may be harder, it yields a better surface for anodizing [18]. (Images courtesy of SGS Melrose Park, Illinois)

exhibit more complex microstructures than their wrought counterparts. Figure 4.10 presents examples of various aluminum alloy castings.

4.3.2 Casting Methods

Casting method typically dictates the cast microstructure by way of product size and cooling method. Sand, permanent and semipermanent mold, die-casting, and other related methods are all utilized today to provide cast aluminum product. The method of choice usually depends upon component size and design, lot size, and alloy requirements. Of these methods, die-casting accounts for almost 70% of the total cast aluminum products available worldwide. Die-casting is a versatile process for producing engineered metal parts by forcing molten metal under high pressure into reusable steel molds. These molds, or dies, can be designed to produce complex shapes with a high degree of accuracy and repeatability. Parts can be sharply defined, with smooth or textured surfaces, and are suitable for a wide variety of attractive and serviceable finishes [19]. Multiple castings can be produced simultaneously, and casting dies are designed to enable rapid solidification and typically yield the smallest components. Because of the rapidity of the die-casting process, microstructures are finer than components cast by other methods. See Fig. 4.11.

Because of the rapidity of the die-casting process, the microstructures produced are finer than components cast by other methods. The size and distribution of the various phases, compounds, and precipitates hinge on: alloy chemistry, which includes the addition of elemental modifiers to the alloy chemistry; the casting method; the method and speed by which the casting is cooled, and whether it is

Fig. 4.10 Examples of cast aluminum parts. The amount of alloy additions for cast product prohibits bulk deformation processing, so castings are designed to function near net shape. Some of the surfaces are machined. Need to get permission from: http://www.mesinc.net/aluminum-die-casting/ Website Image

Fig. 4.11 Example of a hard anodized aluminum die-casting. Note the detail produced in the surface of the casting by the alloy conforming to the surface of the die. Die-casting an attractive alternative to some wrought processes that require finish machining to impart important surface detail. (Images courtesy of SGS Melrose Park, Illinois)

heat-treated (tempered) after it is finished. Precipitation and growth of intermetallic compounds, phase formation, and grain coarsening occur if the casting is slow cooled, or allowed to remain at a high enough temperature after ejection from the die. To keep microstructures fine with complex alloys, the rate of cooling must be as fast as possible, such that the majority of the alloying elements remain in solution. Latent heat retained in castings will cause tempering effects that are manifested in the microstructure by the growth of large primary silicon crystals and coarse precipitates. See Figs. 4.12 and 4.13.

Permanent mold casting involves molds and cores of steel or other metal. Molten aluminum is usually poured into the mold, although a vacuum is sometimes applied. Permanent mold castings can be made stronger than either die or sand castings. Semipermanent mold casting techniques are used when permanent cores would be impossible to remove from the finished part.

The most versatile method for producing aluminum products is sand casting. Lost wax and lost foam processes are examples of sand casting or investment casting methods. The process starts with a pattern that is a replica of the finished casting. Virtually any pattern can be pressed into a fine sand mixture to form the mold into which the aluminum is poured. The pattern is slightly larger than the part to be made, to allow for aluminum shrinkage during solidification and cooling. As compared to die and permanent mold casting, sand casting is slow process but is usually more economical for small quantities, intricate designs or when a very large casting is required. Custom parts, such as dental prostheses and crowns, that are truly one-of-a-kind, are investment cast, a form of sand casting. Because the process is slow, the cooling process for the components is not as rapid as that for die or permanent mold castings. As such, the microstructures for sand castings, in particular large parts, are coarser than they are for castings of similar alloy compositions cast by other methods. See Fig. 4.14.

4.3.3 Cast Alloy Designations

The casting process and the alloy chemistry affect the level of microstructural homogeneity, the defect population, and the variation of chemical potential across a cast component surface. Castings exhibit some of the most complex microstructures because the total alloy element loading is less than or equal to 16% the total alloy content, double the content for most wrought alloys.

When considering how a casting will anodize, it is important to make the connection between the alloy designation, the corresponding composition, and the microstructure near and at the surface. This is because the component surface is the source for the anodic oxide, which continuously changes as the oxide grows. If the casting composition and microstructure is understood, the interface between the component surface and the anodizing electrolyte can be anticipated, such that anodizing process parameters, such as electrolyte concentration, current density and temperature, can be modified to effect optimum anodic oxide growth. The following

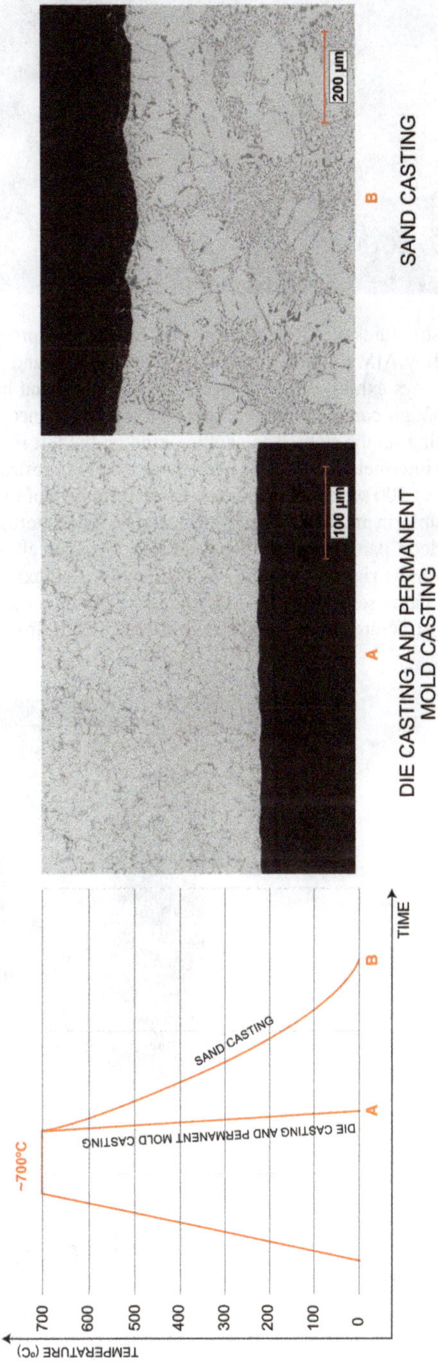

Fig. 4.12 Process-temperature-time diagram comparing casting methods and microstructure [18]

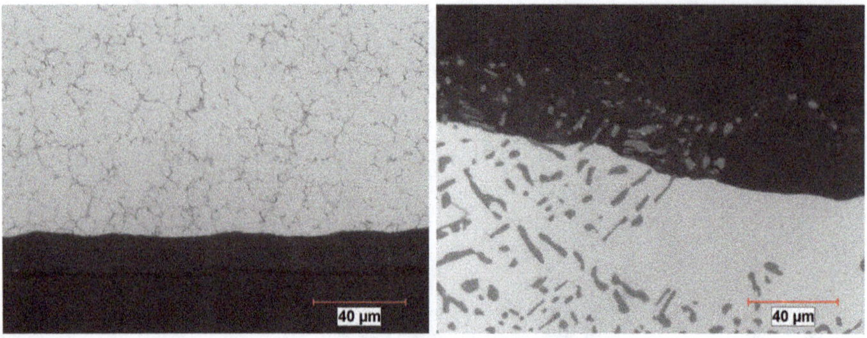

Fig. 4.13 Anodized microstructures of different alloys cast by different processes, as presented in Fig. 4.12. (left) Die-cast alloy AlMg9 (similar to alloy 520), (right) Sand cast alloy AlSi10Mg, similar to alloy 360. The alloys exhibit different levels of coarseness and homogeneity, with the die-casting on the left, although complex, exhibiting a much finer grained, homogeneous micro structure than the sand casting on the right. The anodic oxide on the left is uniformly thick oxide without inclusions from the intermetallic particles in the base metal microstructure. This is because intermetallic particles in most 500 series castings, comprised primarily of magnesium silicide and magnesium oxide, are soluble in most anodizing electrolytes. The microstructure on the right exhibits coarse primary silicon particles and silicon needles. Because alloying elements do not anodize, the anodic oxide on thc right is nonuniform in thickness and exhibits crack equivalents throughout the oxide from paths set up by the mass transport of silicon particles from the base metal into the anodic oxide structure [18]. (Images courtesy of SGS Melrose Park, Illinois)

Fig. 4.14 Photo of a cast alloy A356 engine block [20]. The component was cast by way of the lost foam process

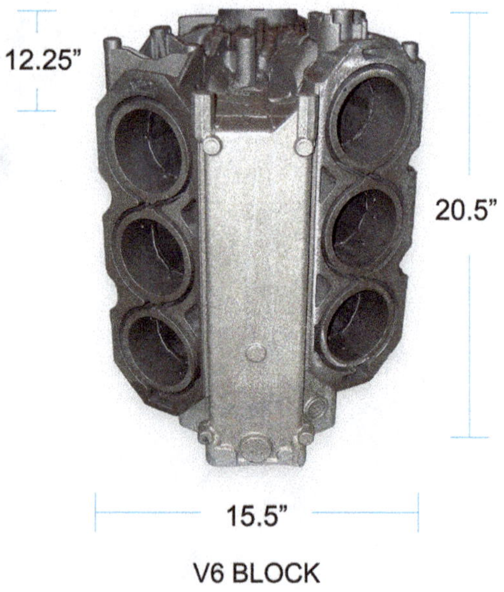

Table 4.3 Cast aluminum alloy designation system and available cast product forms

Alloy designation	Main alloying element(s)	Available product forms
200	Copper	Ingot, Sand, and Permanent Mold Castings
300	Silicon, with added copper and/or magnesium	Ingot, Sand, Permanent Mold and Die-Castings
400	Silicon	Ingot, Sand, Permanent Mold and Die-Castings
500	Magnesium	Ingot, Sand, Permanent Mold and Die-Castings
700	Zinc	Ingot, Sand, Permanent Mold and Few Die-Castings
800	Tin	Ingot, Sand, and Permanent Mold Castings

Data taken from ASM Specialty Handbook, Aluminum and Aluminum Alloys, ASM International, Materials Park, OH, 1993 [3]

summary of cast alloy designations, compositions, and alloying elements is included for the understanding of three things: (1) how they affect the casting process, (2) how they affect the mechanical properties of the finished casting, and (3) how they affect the surface, and therefore the anodizing process.

The American National Standard Alloy and Temper Designation Systems for Aluminum (ANSI H35.1/H35.1(M)-2013), and International Alloy Designation System (IADS) comprise a three-digit basic designation, according to primary alloy additions with a fourth digit as a decimal point to identify aluminum and aluminum alloys in the form of castings and foundry ingot. The first digit indicates the alloy group, the second and third digits indicate a particular alloy chemistry [21, 22]. Cast aluminum alloys are designated by a 3-digit code that corresponds to primary alloying additions that produce the commonly desired properties: aluminum (Al) 99.0% and greater (1xx.x-series); copper (Cu) (2xx.x-series); silicon (Si) with added copper (Cu) and/or magnesium (Mg) (3xx.x-series); silicon (Si) (4xx.x-series); magnesium (Mg) (5xx.x-series); zinc (Zn) (7xx.xseries); tin (Sn) (8xx.x-series). For cast alloys, the unused series is 6xx.x; the other element series is 9xx.x. Table 4.3 presents the US cast aluminum alloy designation system and the available product forms.

300 Series: Silicon, with added Copper and/or Magnesium: The 300 series castings are the most commonly produced and the highest-volume-usage alloys. Because they comprise multiple primary alloying elements, in addition to aluminum, the 300 series alloys typically exhibit the most complex microstructures and therefore present some of the most complex surfaces to the anodizing electrolyte.

300 series alloys contain, in addition to aluminum: silicon, magnesium, and/or copper. In general, they fall into three types: Al-Si-Mg, Al-Si-Cu, or Al-Si-Cu-Mg. Silicon contents, depending on the alloy and the cast process, may range from 5 to 16%. Copper contents range from 0% to a maximum of 4.5%. Magnesium content ranges from 0.3 to 0.6% for high strength alloys to 1.0% for piston alloys 332 and 336.

The addition of 1–3% zinc in 300 series die-castings (alloy 380) results in the highest strength cast aluminum based alloys commercially available [23].

Copper, as an alloy addition to aluminum castings, decreases the surface tension of the molten pour, thereby increasing the fluidity of the melt, which aids in producing a casting that is free of hot shortness and porosity. Within the microstructure, copper refines the grain size; that is, the presence of copper in the microstructure prevents excessive grain growth during cooling and heat treatment. Both copper and magnesium increase strength and hardness in the as-cast condition through increased solid solution hardening. Much greater increases in hardness are brought about by artificial aging (precipitation hardening). Precipitate structures are based on Mg_2Si, and/or combinations of dispersoids and phases Al_2Cu and Al_2CuMg. In alloys that contain zinc, Al-Zn-Mg phases form as well as $Mg(Zn, Cu, Al)_2$. These phases are distributed throughout the cast microstructure, which means they also intersect the surface.

Copper, magnesium, and/or zinc atoms are able to reside on the aluminum lattice in alloy solution. They are also cathodic with respect to the standard oxidation potential for aluminum. This means that when present at the interface with the growing anodic oxide, they do not anodize and can inhibit the anodizing reaction. In contrast, dispersoids and precipitates such as Mg_2Si are not coherent with the aluminum lattice, so even though they do not anodize, depending upon their size, population and distribution, they don't confound the reaction to the same extent as an alloying element in solution; instead, they disrupt the structural order of the AAO, which can be catastrophic to the function of the oxide. This is because the anodizing reaction proceeds around incoherent intermetallic compounds, and they are taken into the anodic aluminum oxide (AAO) structure [24].

400 Series: Silicon: The 400 series alloys are based on the binary aluminum-silicon phase diagram and contain from 5 to 12% Si. Alloys with these compositions find many applications where combinations of moderate strength, high ductility, and impact resistance are required. The Al-Si system is a simple eutectic with limited terminal solubility and is the basis for the 400 series alloys. See Fig. 4.4. The limited solid solubility of silicon in aluminum produces Al-Si compositions that will exhibit approximately 1% Si in solid solution as a continuous phase, and the rest will fall out as essentially pure silicon particles that, depending upon cooling rate, can range in size from small particles to large needles and plates.

Silicon is the alloying element that makes the commercial viability of the high volume aluminum casting industry possible. This important benefit derives from the effects of silicon and aluminum molten mixtures which exhibit increased fluidity, reduced cracking, and improved feeding to minimize shrinkage porosity. Alloys with the eutectic composition (Al-12%Si) exhibit highest fluidity during casting. Silicon content between 4% to the eutectic level of 12% reduces scrap losses, permits production of much more intricate designs with greater variations in section thickness and yields castings with higher surface and internal quality (forms sound outer surface layer); because of this, 400 series castings can be used in applications where pressure tightness is a requirement.

500 Series: Magnesium: The aluminum-magnesium alloys are single-phase binary alloys with moderate-to-high strength and toughness properties. Because magnesium has an even higher affinity for oxygen than aluminum, if the melt is not sufficiently de-gassed prior to pouring, the presence of magnesium as an alloy addition inhibits the flow of the molten pour during casting. To ensure best corrosion resistance, a low impurity content in the initial pour, both solid and gas, is required. This is sometimes difficult to achieve because magnesium is the only element that has a higher affinity for oxygen than aluminum. This requirement mandates that 500 series alloys be prepared from high quality metals and handled with great care in the foundry, giving rise to questions about whether the 500 series castings can be effectively recycled. Minor amounts of silicon are added to these alloys to precipitate Mg_2Si intermetallic compounds for increased mechanical strength.

High corrosion resistance, especially to seawater and marine atmospheres, is the primary advantage of castings made of Al-Mg alloys. These alloys are also the most readily anodized and yield a favorable appearance after anodizing. The 500 series alloys are suitable for welded assemblies and are the best suited when an aluminum casting is needed for decorative or architectural applications. Aluminum-magnesium alloys also exhibit good machinability.

700 Series: Zinc: Aluminum-zinc alloys were among the first to be commercially die cast at the beginning of the twentieth century. The alloy ratios were zinc rich, and the castings were characterized by their tendency for hot cracking (cracking in the mold while cooling). Since this time, aluminum-rich alloys that contain additional elements to zinc offer the highest combination of strength and fatigue resistance, based upon intermetallic compounds that form with magnesium, copper, chromium, and manganese. As a solid solution strengthening addition, zinc resides on the aluminum lattice similar to copper and magnesium. According to Mondolfo in alloys with less than 3% zinc, all of the zinc remains in solid solution [25]. Zinc can also be added to 300 series cast product.

4.3.4 Common Trace Alloy Additions for Aluminum Castings

Other alloying elements, besides copper, silicon, magnesium, and combinations thereof, are added to castings to produce second phase constituents and intermetallic particles that modify the aluminum-silicon structure and increase strength and hardness. Depending upon their aluminum content, these phases and particles anodize at different rates than the aluminum matrix or do not anodize at all, confounding the anodizing reaction. A brief summary of common alloy additions and their role in the cast process and microstructure follows [21].

Iron: Iron is always present in aluminum as an impurity from initial smelting by way of the Bayer Process of converting bauxite ore to alumina in preparation for electrolytic reduction by the Hall–Héroult process. Primary aluminum typically contains between 0.02 and 0.15% (by weight) iron with an average of 0.07–0.10% (by weight). The solubility of iron in the solid state is very low in aluminum (~0.05%), therefore, most of the iron present in aluminum over this amount appears as part of a primary constituent phase or as part of intermetallic compounds in combination with aluminum and other elements.

Small amounts of iron effectively modify the cast microstructure. The primary constituent phase comprising aluminum-silicon-iron strengthens and increases the wear resistance of cast product. Because iron reacts with the aluminum and silicon to make a second phase early in the solidification process, microstructures tend to be finer and less complex when alloys have reduced iron content because it gets "used up" before it can make other intermetallic compounds that may otherwise produce a more complex microstructure. As iron content increases in cast aluminum alloys, the ductility decreases. Modified 300 series castings (A360, A380, and A383) contain less iron, which increase their ductility and make them easier to anodize.

Manganese: Manganese always follows iron as a trace element. The presence of manganese in the melt reduces the formation of large Al-Fe-Si primary constituent phase (PCP), which on its own can be quite brittle. By inhibiting the growth of large PCPs during solidification, the formation of large Al-Si eutectic grains is also reduced, which thereby increases ductility. Manganese binds with Fe to stabilize the $FeAl_6$ phase in heat treatable alloys, which modifies the microstructure by keeping the intermetallic phase from becoming needle-like or plate-like. By keeping brittle phases small, the addition of Mn as a modifier acts to reduce die wear and makes the manganese-containing alloys easier to anodize; it also increases ductility and toughness (impact strength). During casting, manganese additions increase fluidity of the molten alloy and decrease shrinkage with solidification, which produce a more continuous surface microstructure that facilitates surface finishing of all types.

Chromium and Titanium: Both may be added in small amounts to inhibit nucleation of new grains as well as grain growth. Together with magnesium, chromium and titanium combine with aluminum to produce precipitates, as dispersoids, that enhance mechanical properties. Such additions keep grain size small which enhances the surface characteristics for anodizing. Dispersoids form early in the casting process, just after the primary constituent phase, and are usually insoluble through solution treatment. Even though they are technically "precipitates," dispersoids should not be confused with the precipitates that form in low temperature aging processes.

Table 4.4 presents the compositions of some commonly specified aluminum die-cast alloys with corresponding, most similar, European and Japanese designations.

4.3.4.1 Structural Modifiers for Cast Microstructures

Calcium, Sodium, Strontium, and Antimony: These elements are added in small amounts to modify the aluminum-silicon microstructure when the silicon content is less than 12% (hypoeutectic). This modification produces a more fibrous and less needle-like Al-Si structure, which is more fluid during casting, and finer grained when it intersects the surface for anodizing. Alloys 360, 380, their modified versions (A and B), all 500 series alloys, and some 400 series alloys are hypoeutectic; refer to Table 4.4 for specific die-cast alloy chemistries.

Table 4.4 Compositions of some commonly specified aluminum die-cast alloys

US designation	European (E) and/or Japanese (J) designation	Si	Fe	Cu	Mn	Mg	Cr	Ni	Zn	Sn	Ti	Other
359	AlSi9 (E)	8.5–9.5	0.2 max	0.2 max	0.1 max	0.5–0.7	n/a	n/a	0.1 max	n/a	0.2 max	0.15
360		9.0–10.0	2.0	0.6	0.35	0.7	n/a	0.5	0.5	0.15	n/a	0.25
A360	AlSi10Mg(Fe) (E), ADC3 (J)	9.0–10.0	1.3	0.6	0.35	0.4–0.6	n/a	0.5	0.5	0.15	n/a	0.25
380	AlSi8Cu3 (E), ADC10Z (J)	7.5–9.5	2.0	3.0–4.0	0.5	0.1	n/a	0.5	3.0	0.35	n/a	0.5
A380	AlSi9Cu3(Fe)(Zn) (E)	7.5–9.5	1.3	3.0–4.0	0.5	0.1	n/a	0.5	3.0	0.35	n/a	0.5
	ADC10 (J)	7.5–9.5	1.3	3.0–4.0	0.5	0.1	n/a	0.5	1.0	0.35	n/a	0.5
383	AlSi9 Cu3(Fe)(Zn) (E), ADC12Z (J)	9.5–11.5	1.3	2.0–3.0	0.5	0.1	n/a	0.3	3.0	0.15	n/a	0.5
390		16.0–18.0	1.3	4.0–5.0	0.1	0.45–0.65	n/a	n/a	0.1	n/a	0.2	0.2
B390	ADC14(J)	16.0–18.0	1.3	4.0–5.0	0.5	0.45–0.65	n/a	0.1	1.5	n/a	0.2	0.2
413	AlSi12(Fe) (E), ADC1 (J)	11.0–13.0	2.0	1.0	0.35	0.1	n/a	0.5	0.5	0.15	n/a	0.25
A413		11.0–13.0	1.3	1.0	0.35	0.1	n/a	0.5	0.5	0.15	n/a	0.25
C443		4.5–6.0	2.0	0.6	0.35	0.1	n/a	0.5	0.5	0.15	n/a	0.25
516	ADC6 (J)	0.3–1.5	0.35–1.0	0.3	0.15–0.4	2.5–4.5	n/a	0.25–0.4	0.2	0.1	0.1–0.2	0.1
520	AlMg9	0.25	0.3	0.25	0.15	9.5–10.6	n/a	n/a	0.15	n/a	n/a	0.15
712	5.8Zn0.6 Mg0.5Cr0.2Ti	0.3	0.5	0.25	0.1	0.5–0.65	0.4–0.6	n/a	5.0–6.5	n/a	0.15–0.25	0.02

Data taken from ASM Specialty Handbook, *Aluminum and Aluminum Alloys*, ASM International, Materials Park, OH, 1993 [3]

Phosphorus: Phosphorus is added to modify the structure of hypereutectic (Si content is greater than 12%) Al-Si alloys. Alloy 383, 390, and alloy 413 are examples of hypereutectic compositions that could benefit from phosphorus modification. Modification helps to eliminate large, coarse primary silicon crystals that are harmful in the casting and machining of such compositions. The elimination of coarse primary silicon crystals also enhances the integrity of the AAO as larger silicon crystals, when included in the finish, reduce the cohesive strength of the finish and therefore its wear and impact resistance.

Whenever a cast microstructure is modified by way of the intentional addition of a modifier, it is reflected in the alloy designation with a suffix that follows and is separated from the foundry designation by a hyphen (for example, A356.1-S). The suffix is a single letter, corresponding to the chemical symbol for each alloy modifier: N for Sodium (Na); S for Strontium (Sr); C for Calcium (Ca); P for Phosphorus (P). In cases where more than one modifier is intentionally added, the suffix letter shall identify only the modifying element of greatest concentration. The modifying element's concentration is not counted as part of the Others, Each or Others, Total, part of the alloy chemistry [22].

4.4 Microstructural Development in Wrought Aluminum Alloys

Wrought alloys are designed to be deformed and therefore must possess a minimum ductility commensurate with the manufacturing process under consideration for engineering application [14]. Common bulk deformation processes are: forging, extrusion, rolling, and drawing. All deformation processes involve the pushing, pulling, hammering, or rolling of material over or through a tool or die to a desired fixed shape, thickness, or cross section. Whether a hot or cold process, the material responds to deformation by flowing in the direction of the applied force, or by cracking. When the bulk of any component is deformed and the material flows, the surface of the component is affected, whether or not the material comes in direct contact with the tool.

The starting point for deformation processes to produce a manufactured, wrought component is typically a cast billet or cut-to-size log or slab which is deformed to desired tolerances or for further deformation and finishing. Consequently, the microstructure of the billet directly impacts the quality of the wrought product. The speed and friction between a tool or a die and the component profoundly affects the surface of the component, introducing strain energy that provides the driving force to mechanically and thermally alter the microstructure. Microstructural development, together with the corresponding mechanical and chemical properties, continues throughout the manufacturing process. As the metal flows in the deformation direction, microstructural changes occur with the applied stress and through excursions over elevated temperatures that orient the grain structure, enable recovery and recrystallization, causing both grain growth and precipitation of intermetallic particles. See Fig. 4.15.

To facilitate deformation, wrought alloys are designed with less alloying elements than cast alloys in order to take advantage of the many slip systems of the FCC aluminum primary crystal. Common wrought aluminum alloys are alloyed to a maximum of 8% total additions [12]; new wrought super alloys are being constantly developed that exhibit enhanced strength and fatigue resistance, and these alloys contain up to 14% total alloy additions. Through the selection and amount of different alloy additions, and by virtue of how the alloy additions come together to form the characteristic intermetallic phases, particles, and compounds, results in two basic classes of wrought alloys: non-heat treatable (NHT) and precipitation hardenable (PH).

Non-heat treatable wrought aluminum alloys are based on the characteristic development of hardness and strength by the simultaneous and mutually interactive influences of elements in solid solution and the dislocation structures introduced by deformation, which inhibit dynamic recovery by strain hardening, without heat treatment. The chief industries that use these alloys are the major markets that use flat-rolled products: transportation, packaging, beverage cans, building and construction. All foil and coil product use the non-heat treatable alloys, in applications where reflectivity and bright finishes are desired [26].

Precipitation hardenable wrought aluminum alloys are based on the characteristic development of hardness and strength by the formation of precipitates through a low temperature heat treatment process after deformation. This process is called age hardening or tempering. Precipitation can best be described as decomposition of a supersaturated solid solution into its parent phase and excess rejected solute atoms [27]. The rejected solute atoms agglomerate to form a small crystal of their own, perhaps with a few solvent atoms dissolved in it, or react with some of the solvent atoms to form a phase with a crystal structure different from that of either pure metal (i.e., an intermetallic compound). The agglomerate is the precipitate. Precipitates enhance the hardness of the matrix by increasing local stress within the microstructure and by restricting dislocation movement across the crystal. The chief industries that use these alloys are the major markets that use extruded products: transportation, automation, building and construction. Many architectural products are typically precipitation hardenable alloys, specifically the 6XXX series, which rely on the formation of Mg_2Si precipitates for hardness and strength.

The microstructure, mechanical properties, and chemical potential of both wrought alloy classes are subject to change with thermal excursions during deformation or during heat treatment to provide stress relief through annealing. However, only the precipitation hardenable alloys are deliberately formulated to develop precipitates during post-deformation heat treatment called tempering or age hardening. It is important to keep in mind that only annealing, as a heat treatment or through the temperatures reached through deformation, will change the grain structure through the processes of recovery, recrystallization, or grain growth. The temperature at which tempering, or aging, although sufficient to develop precipitation, is carried out at too low temperature to change the grain structure or relieve stress.

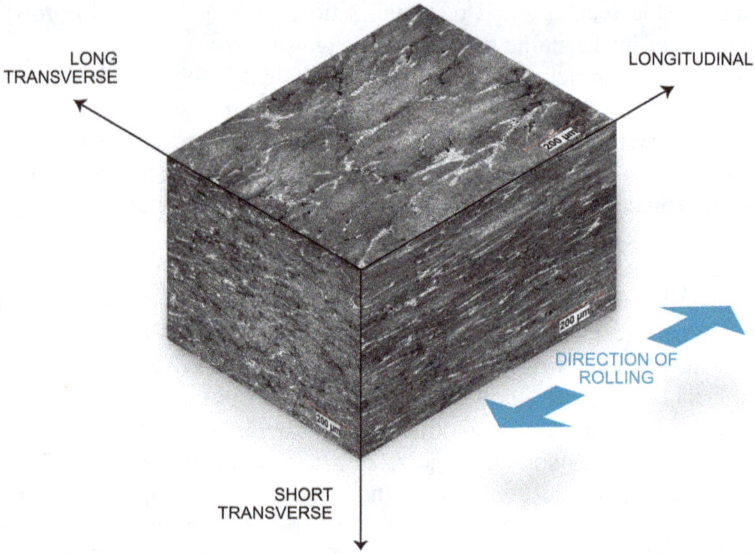

Fig. 4.15 Microstructure development of a 6XXX alloy through deformation by way of roll forging [3]

The ANSI and IADS assign each wrought alloy a four-digit number [21, 22]. The first digit is assigned on the basis of the major alloying element(s). For most wrought alloy chemistries, the second digit indicates a modification with 0 indicating no modification; but for the 1XXX series alone, the second digit indicates minimum aluminum purity [27]. The last two digits serve to identify the specific alloy. Similar to the cast aluminum alloys, the rest of the wrought aluminum alloys are grouped according to the major alloying elements. Table 4.5 presents a list of the wrought (4-digit) aluminum alloy designations, their major alloy constituent, and how the major alloying element is contained in the microstructure, in solution or as a precipitate. For wrought alloys, the unused series is 9xxx.

4.4.1 Microstructural Development in Non-heat Treatable Alloys

Four different alloy designations of non-heat treatable (NHT) wrought alloys have been developed around the characteristic of strain hardening: the 1XXX, 3XXX, 5XXX, and some 8XXX series alloys. Deformation-induced work hardening, recovery, and recrystallization, which govern mechanical properties and surface characteristics of NHT alloys, are influenced by the elements in solid solution (alloy chemistry), as well as the phases formed during casting by way of the formation of primary constituents phases (PCPs) and billet handling, prior to deformation.

Table 4.5 Wrought aluminum alloy designation system

Alloy designation	Major alloying element	Characteristic alloy solution
1XXX	Aluminum (Al) 99.0% and greater	Highly formable, Not heat treatable, often alloyed with small amounts of iron and silicon to make primary constituent phase AlFeSi
2XXX	Copper	Some Cu in solution with aluminum, heat treatable due to other alloy additions, which make the alloys precipitation hardenable, makes $AlCu_2$ and Al_2CuMg precipitates. High strength alloys
3XXX	Manganese	Mn in solution with aluminum, cold work hardenable, corrosion resistant, limited ductility. Non-heat treatable alloys
5XXX	Magnesium	In solution with aluminum, cold work hardenable, corrosion resistant, reflective, and most complex of non-heat treatable alloys
6XXX	Magnesium, silicon	Precipitation hardenable, heat treatable, most commonly used wrought alloy group, strength and formability
7XXX	Zinc	Precipitation hardenable, but some in elements in solution. Most complex wrought alloys. Heat treatable, high strength, and fatigue resistance
8XXX	Other elements, for example, iron, nickel, and tin	The characteristics of this group of alloys vary with the major alloying elements. High iron and silicon alloys, some lithium-containing alloys

Data taken from ASM Specialty Handbook, *Aluminum and Aluminum Alloys*, ASM International, Materials Park, OH, 1993 [3]

Dispersoids begin to precipitate on the surfaces of PCPs during solidification; the amount is determined by the cooling rate (slower cooling rate—more precipitation). If the billet is preheated to homogenize prior to deformation, dispersoids form as the alloy cools prior to and during deformation, with their formation limited by the rate of cooling (fewer dispersoids precipitate with more rapid cooling).

Aside from the 5XXX series alloys, the alloy content for most NHT alloys is low, lower than the 8% maximum alloy loading for wrought aluminum alloys. Therefore, the corresponding microstructures tend to be rather homogeneous, with most alloying elements in solid solution and small primary constituent phases and dispersoids distributed throughout. Because of the general homogeneity of the microstructures, these alloys typically anodize well.

The homogeneity of the NHT alloys is reflected in the broad regions of single-phase solid solubility in the phase diagram. At 2.5%, magnesium has the highest solid solubility of all primary alloying elements at room temperature. With most 5XXX series alloys containing less than 7% magnesium, it is easy to see that a single-phase solid solution is stable over a rather broad magnesium content and over a moderate temperature range; however, the solvus is steep, especially for low alloy compositions, meaning precipitation occurs rapidly at cold working temperatures. Nevertheless, 5XXX series alloys, and indeed, all NHT alloys tend to exhibit rather homogeneous microstructures. See Fig. 4.16.

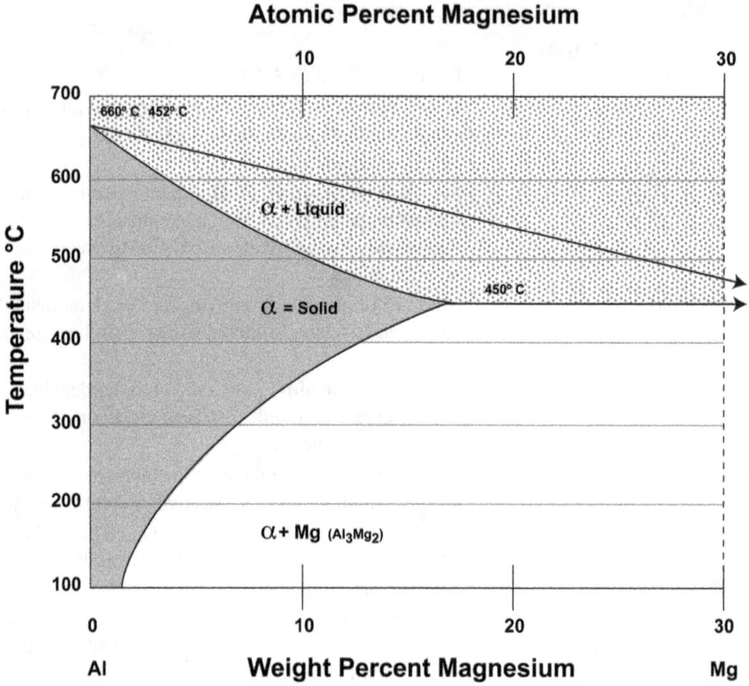

Fig. 4.16 Aluminum-Magnesium phase diagram. Note the broad range of solid solubility of magnesium in aluminum, up to ~15% at 450 °C. The solvus is steep, meaning precipitation will occur rapidly at cold working temperatures

The phenomena of recovery and recrystallization during subsequent processing of the non-heat treatable alloys play an important role in the strength levels achieved, the accompanying ductility, the type of solution that is formed, and the degree of order achieved in the solution [26]. Important obstacles to dislocation movement during deformation are: solute atoms, other dislocations, particles of a second phase, grain boundaries and sub-grain boundaries. The magnitude and intrinsic severity of the elastic interaction between solute atoms and dislocations depend on the change in solute concentration across a deformed region containing the stress field of the dislocation. With increased concentration, dislocation-solute atom elastic interaction effects increase. As deformation occurs, these features continually intersect and move along the surface, changing the surface microstructure as well as the local chemical potential from point to point across the surface. See Fig. 4.17.

The NHT alloys can exhibit stability and strength similar to precipitation hardenable aluminum alloys, but they are more malleable and formable. The major alloying elements for the NHT alloys are: silicon, iron, manganese, and magnesium. They are added singularly, as a primary alloying element, or in various combinations to form primary constituent phases and as minor additions, to form dispersoids. Alloy additions, present as solute atoms on the primary aluminum lattice,

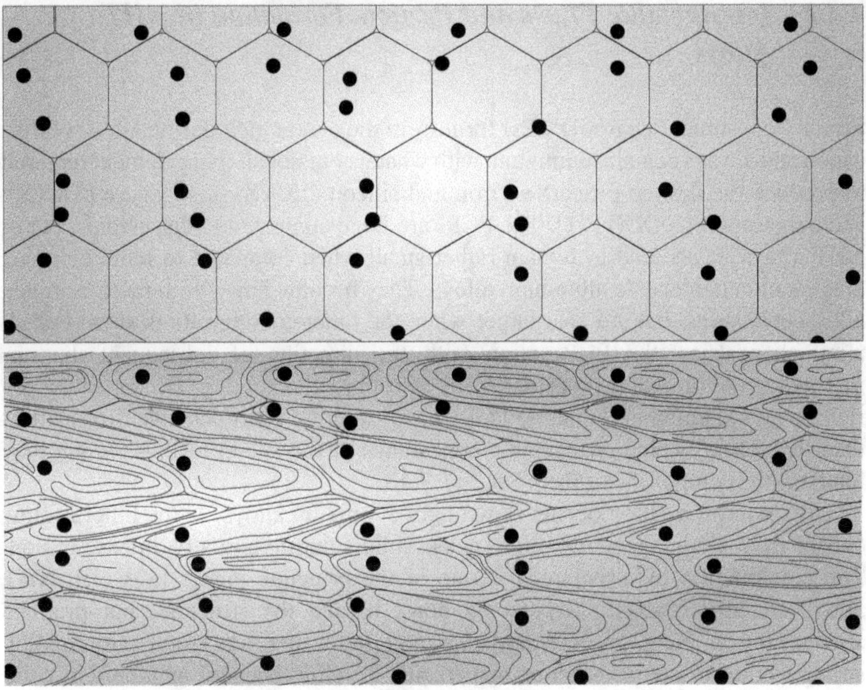

Fig. 4.17 Schematic of a microstructure with long-range order showing dislocation-solute atom elastic interaction with deformation. The top schematic represents a microstructure with precipitation with long-range order, prior to deformation. The bottom schematic represents what happens to the microstructure as deformation occurs: dislocations intersect solute particles as they move along the surface, strengthening the structure as a fibrous microstructure is produced. Disruption in the microstructural homogeneity caused by cold work increases hardness in NHT alloys but also produces change in the chemical potential throughout the microstructure, especially at the surface

together with the second phase particles that form within the microstructure, provide strength by hindering the progress of dislocations through the lattice structure such that a higher applied stress is required to continue deformation. Strength is also increased by other atomic level defects produced by thermal treatment, or which form due to recovery from earlier deformation processing, such as dislocations, vacancies, and grain boundaries, which pre-exist deformation, also act as nucleation sites for dispersoids during cooling and provide obstacles to dislocation movement. The microstructural features that develop at the surface, because of their inherent chemical potential differences, are often replicated in the anodic oxide produced during anodizing [24].

4.4.2 Intermetallic Phase and Particle Formation in NHT Alloys

Primary constituent phases (PCPs) form from the parent melt during alloy solidification and always contain aluminum, with whatever major alloying element required to produce the desired properties: iron and silicon (1XXX), manganese (3XXX), and magnesium (5XXX) [21]. The PCPs are large compared to dispersoids, but in NHT alloys, PCPs tend to remain rather small when compared to more complex precipitation hardenable aluminum alloys. They become larger with more complex alloy chemistries, but are also larger when the billet-cooling rate is slow. For all alloy chemistries, the PCPs align with the direction of subsequent thermo-mechanical deformation processes such as rolling and forging. Because they are rich in non-aluminum elements, PCPs have different electrochemical properties than the matrix, which changes the appearance of the anodized finish when they intersect the component surface. See Fig. 4.18.

Dispersoid formation occurs in wrought aluminum alloys, as the billet is slow cooled from a homogeneous single-phase solid solution into a two-phase region. The precipitation of dispersoids is one of the principle means by which NHT alloys are strengthened; dispersoids don't harden the structure, but produce strength by pinning grain boundaries and inhibiting recrystallization. Alloy strengthening occurs as soon as dispersoids develop, because as atoms in solid solution rearrange themselves into solute-rich clusters, called Guinier–Preston Zones (GPZs), strain is generated due to the difference in equilibrium lattice spacing, and/or because the atomic diameter of the solute-rich regions and the FCC parent matrix differ. GPZs are not ordered, and the difference in internal lattice energy (as stacking faults) and the rest of the matrix cause strengthening on all lattices, regardless of the amount of order. GPZs mature into dispersoids as slow cooling during preheat, prior to cold working. However, if deformation cuts through forming GPZs, no strengthening occurs because they present an insufficient barrier to movement in the crystal lattice [28].

4.4.3 Designations for NHT Alloys

Four different alloy designations of non-heat treatable (NHT) wrought alloys have been developed around the characteristic of solid solution strengthening and cold work hardening: the 1XXX, 3XXX, 5XXX, and some of the 8XXX series alloys [21, 22]. The 1XXX alloys are alloyed with small amounts of silicon and iron, with a 99% aluminum minimum. The primary alloy addition for the 3XXX series is manganese; and for the 5XXX series, magnesium. The 8XXX series are alloyed with other elements that are not covered by the other series. Microstructure control of the non-heat treatable (NHT) wrought alloys is achieved through selection of appropriate alloy chemistry, solidification and thermal and deformation processing.

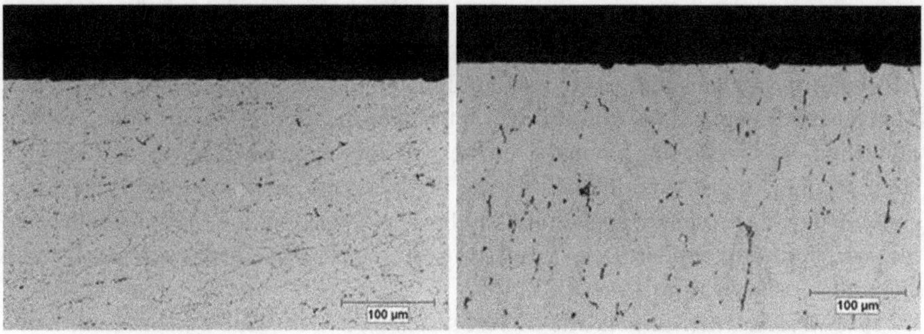

Fig. 4.18 Microstructures of roll-forged 5052 alloy samples that show the effects of different process parameters. Alloy 5052 has approximately 2.4% magnesium as the primary alloying element. Both samples are anodized. The sample on the left was deformed and more rapidly cooled during forging. The microstructure on the right is coarser, with aligned primary phase and precipitation contributing to the pitting condition at the surface. At magnesium concentrations greater than 3.5%, or when the forging is not properly cooled during deformation processing, Mg_2Al_3 or Mg_5Al_8 precipitates heterogeneously form and favor grain boundaries, reducing corrosion resistance, producing a sensitized condition [29]. Note, in the image on the right, due to the orientation of the sample microstructure with respect to the surface from which the oxide grew, the distinct hemispherical pits at the surface, which correspond to sensitized grain boundaries (end grain attack). Also note the microstructure exhibits precipitate-free zones surrounding the grain boundaries. This indicates that the sample was possibly slow cooled from the solution treatment temperature. Courtesy of Anderson Shumaker

4.4.3.1 Alloy Designations

1000 Series; High Purity Aluminum Alloys: The 1000 series alloys have the fewest solute or precipitation alloying elements. These alloys have low strength because of their high purity and therefore few barriers to dislocation mobility. They are ductile and easy to deform. Iron and silicon are the primary alloy constituents with less than 1% added to increase strength with small, stable AlFeSi phases forming. $AlFe_3$ dispersoids also form in 1XXX series alloys, which help to pin grain boundaries and thereby restrict recrystallization. The 1XXX series alloys find application in foils, capacitors, and lighting applications. These alloys are extremely homogeneous, have superior corrosion resistance, and anodize well.

Because of their low alloy content, NHT alloys of the 1XXX series, tend to exhibit long-range order (LRO), in which dislocations tend to agglomerate, rather than advance through the lattice. This phenomenon limits the strengthening in these alloys because there is not enough disorder in the structure to enable dislocation pinning. With LRO, there is little tendency for groupings greater than two (2) dislocations. The distribution of dislocation pairs is homogeneous on the primary slip planes having the greatest critical resolved shear stress in a solution with LRO, so strengthening effects are at a minimum [28].

3000 Series; Manganese: Manganese is the primary alloying element in the 3000 series alloys. Like all of the 1XXX and 5XXX series alloys, the 3XXX series alloys derive their strength from strain hardening (cold work). One to two percent manganese resides on the FCC lattice in solid solution, producing a medium strength alloy that is ductile, with good formability. Iron and silicon are added to form dispersoids $Al_{12}(Fe,Mn)_3Si$, which are critical in controlling grain size and mechanical properties. Manganese dispersoids ($AlMn_3$) also form to stabilize grain size by restricting recrystallization, thus increasing strength. Magnesium additions enhance the solid solution strengthening, and iron additions increase hardness and polishing effects. Beverage cans are the most common application for the 3XXX series alloys (3104), but they are also used for home appliances and cookware (3003). Hard anodized cookware is typically manufactured from alloy 3003.

5000 Series; Magnesium: The 5XXX series alloys, of which magnesium is the primary alloying element, are the strongest of the non-heat treatable alloys and are a good example of solid solution strengthening. Magnesium has high solubility in the aluminum matrix, 1.9% at 1000 °C, in liquid solution, and 15.35% at 450 °C, in solid solution. This is higher than silicon, iron, or manganese. The magnitude and intrinsic amount of the elastic interaction between solute atoms (magnesium) and dislocations (deformation) depend on the change in concentration across a region containing the resultant stress field from deformation. This increases with the concentration of the solute—so the more complex the alloy, and the higher the solubility of the solute on the lattice, the greatest amount of strengthening because there is more dislocation-solute atom interaction.

At magnesium concentrations greater than 3.5%, Mg_2Al_3 or Mg_5Al_8 particles precipitate heterogeneously and favor grain boundaries, reducing corrosion resistance, producing a sensitized condition. At room temperature, the precipitate rate is low, and sensitization practically doesn't happen; however, at temperatures greater than 80 °C, grain boundary precipitation happens quickly. Similarly, microstructural bands that appear as feathery marks can disrupt the surface of 5XXX series alloys following stretching after rolling. This condition is called Lüdering and is due to the interaction of dislocations and Mg atoms during plastic deformation. Anodizing highlights this condition [28].

8000 Series; Other High Solute Alloys: The 8XXX series alloys contain higher amounts of iron, up to 1.5% with silicon (8079); others are alloyed with Fe and Mn. Higher strength foils derive their strength and ductility from fine intermetallic particle size and higher volume fraction. In as-cast billet with 0.5%Mn, rods of (Fe,Mn) Al_6 are formed which break down with cold work to a very fine dispersion of primary constituent particles. In combination with annealing, these particles control the microstructure during recovery and recrystallization. The presence of bands of primary constituent phases throughout the microstructure and intersecting the surface will produce dark-appearing streaks in the anodized finish, especially with increased alloy content.

Table 4.6 presents the compositions of some commonly specified non-heat treatable wrought aluminum alloys and their most similar European and Japanese designations [29].

Table 4.6 Commonly specified non-heat treatable wrought aluminum alloys [30]

US designation	ISO and/or Japanese designation (JIS)	Si	Fe	Cu	Mn	Mg	Cr	Zn	V	Ti	Ga	Other each	Other total
1050	Al99,5	≤0.25	≤0.40	≤0.05	≤0.05	≤0.05		≤0.05	≤0.05	≤0.03		0.03	
1085		≤0.10	≤0.12	≤0.03	≤0.02	≤0.02		≤0.03	≤0.05	≤0.02	≤0.03	0.01	
1100	A1100 (JIS)	Si + Fe = 0.95		0.05–0.20	≤0.50			≤0.10				0.05	0.15
3003	AlMn1Cu; A3003 (JIS)	≤0.60	≤0.70	0.05–0.20	1.0–1.5			≤0.10		+Zr 0.10		0.05	0.15
3104		0.60	0.80	0.05–0.25	0.8–1.4	0.8–1.3	…	0.25	0.05	0.10	0.05	0.05	0.15
3105	AlMn0,5Mg0,5	≤0.60	≤0.70	≤0.30	0.3–0.8	0.2–0.8	≤0.20	≤0.40		≤0.10		0.05	0.15
5019	AlMg5	≤0.40	≤0.50	≤0.10	0.1–0.6	4.5–5.6	≤0.20	≤0.20		≤0.20		0.05	0.15
5052	AlMg2,5	≤0.25	≤0.40	≤0.10	≤0.10	2.2–2.8	0.15–0.35	≤0.10				0.05	0.15
5083		0.40	0.40	0.10	0.40–1.0	4.0–4.9	0.05–0.25	0.25		0.15		0.05	0.15
5456		0.25	0.40	0.10	0.50–1.0	4.7–5.5	0.05–0.20	0.25				0.05	0.15
5657		0.08	≤0.10	≤0.10	≤0.03	0.6–1.0		0.05			0.03	≤0.02	≤0.50
8011		0.5–0.9	0.6–1.0	≤0.10	≤0.20	≤0.05	≤0.05	≤0.10		0.08		0.05	0.15
8079		0.05–0.3	0.7–1.3	≤0.05				≤0.10				0.05	0.15

4.4.3.2 Temper Designations

Following the alloy chemistry designation, wrought NHT alloys are further identified by H designations to indicate the level of cold work (CW, or strain hardening) and possible thermal treatment. The H designation is followed by two or more numbers, which indicate the level of strain hardening quantified by a desired minimum ultimate tensile strength or degree of hardness. The level of strength achieved is compared with the standard limits published for the various alloys by the Aluminum Association. Table 4.7 presents the H Temper Strain Hardening Code based upon the American National Standard Alloy and Temper Designation Systems for Aluminum, ANSI H35.1/H35.1 (M)-2017 approved in May of 2017 [22].

The strain hardening code for NHT alloys can be interpreted as follows. The first number after the H indicates the specific combination of basic operations on the material as: H1, strain hardened only; H2, strain hardened over the desired amount and partially annealed to the desired strength level; H3 material is strain hardened and stabilized with low temperature thermal treatment or stabilized by way of the heat of deformation; H4 is strain hardened and then lacquered or painted. The second number following the designation H1, H2, H3, and H4 indicates how much the alloy was strain hardened (percent of reduction by cold work) as indicated by the minimum value of ultimate tensile strength. Numeral 8 is assigned to the hardest tempers normally produced. For example, H18 is defined as having 75% cold work (full hard), with H12 (quarter hard), H14 (half hard), and H16 (three-quarters hard) proportionately reduced. In practice, H18 is sometimes defined as having 80% CW which makes the math work out more easily. Material harder than that with a H18 temper is actually quite popular. Beverage can body stock is commonly specified as 3104-H19, while the ends and tabs are 5182-H49. These harder tempers have CW in excess of 80%, but technically must exceed the strength of the corresponding HX8 temper by at least 2 ksi. Any subsequent numbers indicate what special practices may have been performed following strain hardening and are variations from the first two numbers.

Examples of how temper designations for NHT alloys are defined are presented as follows:

- 3003 H12—Strain hardened approximately 25%, no other heat treatment
- 3005 H26—Strain hardened and partially annealed to approximately 75% effective strain hardening
- 5052 H32—Strain hardened and stabilized to approximately 25% effective strain hardening
- 5052 H42—Strain hardened and given a special treatment (coated by a process requiring heat that will affect the properties of the base material) to approximately 25% effective strain hardening

The NHT alloys may be the only class where surface finish through anodizing is a design consideration. Alloy chemistries, for example, 1085, 3002, and 5657, are formulated to have a low volume fraction of constituent phases that reduce surface pitting and enhance brightness. Furthermore, anodizing quality (AQ) sheet (5005 or its

Table 4.7 H temper strain hardening code [22]

H temper designation	Definition
H1	Strain hardened only, no supplementary thermal treatment. The second digit indicates the degree of strain hardening
H2	Strain hardened followed by a partial high temperature recrystallization anneal to take the mechanical properties back to a stable level less than that achieved by cold work. When this temper is used, the alloy has been deliberately cold worked to a level higher than the desired amount and then annealed to achieve a specific level of strength. The second digit indicates the degree of strain hardening remaining after the product has been partially annealed
H3	Strain hardened followed by a thermal stabilization treatment. This means that the alloy is heated to a moderate temperature and held for a period of time to stabilize and avoid time-dependent age softening (recovery) at room temperature, especially prevalent in the 5XXX series alloys. The heat of follow-on deformation processing can also achieve this. The second digit indicates the degree of strain hardening remaining after stabilization treatment
H4	Strain hardened followed by thermal treatment due to the heat from curing a lacquered or painted coating. The heat of the treatment must be sufficient to alter the properties of the alloy by reducing the hardening introduced by cold work and stabilizing the microstructure. The second digit indicates the degree of strain hardening remaining after the product has been subjected to the heat of cure. Corresponding H2X or H3X mechanical property limits apply
HX2	¼ hard (strain hardened approximately 25%)
HX4	½ hard (strain hardened approximately 50%)
HX6	¾ hard (strain hardened approximately 75%)
HX8	More than ¾ hard (hardest temper normally produced)

derivatives) is not only formulated but also processed to yield nondirectional, nonstreaky, matte finishes by way of careful control of recrystallization and grain size of the microstructure.

4.4.4 Microstructural Development in Precipitation Hardenable Alloys

Four different alloy designations of precipitation hardenable (PH) wrought alloys have been developed around the characteristic of age hardening, the 2XXX, 4XXX, 6XXX, and 7XXX series alloys [21, 22]. Deformation-induced work hardening, recovery, and recrystallization which develop the primary microstructure and surface characteristics (formation of primary constituents phases and dispersoids) are enhanced by the formation of precipitates after deformation, during tempering. The alloy content for the precipitation hardenable alloys ranges from low to the 8% alloy loading maximum for common wrought aluminum alloys and up to 14% for super alloys. Therefore, the corresponding microstructures become more complex, depending upon alloy content and manufacturing process, but are more homogeneous than cast alloys. Depending upon the alloy chemistry, some of the primary alloying elements will be in solid solution with the aluminum, while primary constituent phases, dispersoids, and precipitates form throughout the microstructure. These alloys exhibit variations in their resistance to anodic oxide growth, with the less complex alloy compositions exhibiting the lowest barrier to anodizing, and the more complex alloys anodizing with more difficulty.

Copper, magnesium, silicon, and zinc are the primary alloying elements for the PH alloys. As with NHT alloys, constituent phases form from the parent melt at forming grain boundaries (between dendrites of aluminum-rich matrix) during alloy solidification (cast processes) and always contain the parent and the major alloying elements. Therefore primary constituent phases are comprised from, aluminum with the primary alloying element required to produce the desired properties: copper (2XXX), silicon (4XXX), magnesium (plus silicon, 6XXX), and/or zinc (7XXX). The primary constituent phase is coherent with the aluminum matrix, which means that they exhibit the minimum possible surface energy. They are large compared to dispersoids and precipitates; they become aligned with the direction of subsequent thermo-mechanical deformation processes such as rolling, forging, or extrusion which means these phases are often grouped in bands. Because they are rich in non-aluminum elements, the bands have different electrochemical properties than the matrix. As an electrochemical process, anodizing develops these bands differently from the matrix and they appear as dark streaks and can increase surface roughness, objectionable defects developed by the anodizing process that are completely related to the metallurgical condition of the aluminum base material. See Fig. 4.19.

Similar to the NHT alloys, dispersoids in PH alloys nucleate on the surfaces of the primary constituent phases and form at relatively high temperatures, for example, as an alloy is cooled slowly from a single-phase solid solution into a two-phase region. Dispersoid formation is an important consideration during alloy design because of their characteristic as a recrystallization inhibitor; therefore their presence enhances alloy strength by constraining grain growth. They are the smallest particles that form within the microstructure and comprise trace-alloying elements with other solute atoms and sometimes aluminum. Dispersoids nucleate in large numbers about the primary constituent phases, and are very stable and highly insoluble in aluminum. Because of this feature, the interfaces between dispersoids and the matrix or dispersoids and primary constituent phases are electrochemically sensitive. See Fig. 4.20.

4.4.4.1 Precipitation in Precipitation Hardenable Alloys

Precipitation during aging is the phenomenon that characterizes heat treatable alloys and separates them from the non-heat treatable alloys. Precipitation occurs at the lowest temperatures, and particles form at temperatures below the recrystallization and stress relief temperature. The phenomenon occurs heterogeneously, and nucleation is at preferential sites throughout the microstructure. PH alloys can be

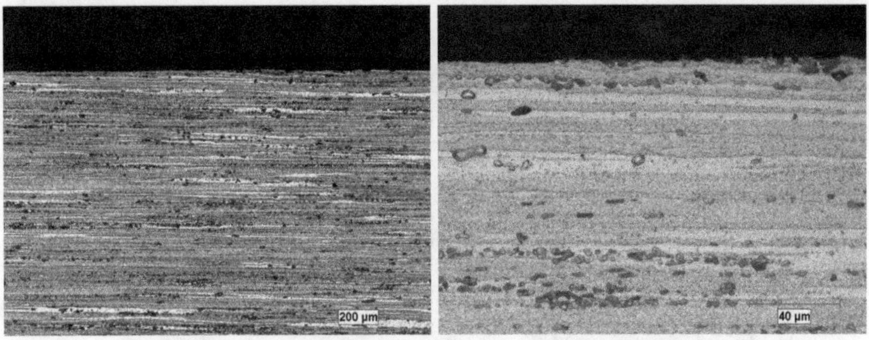

Fig. 4.19 Anodized 6082 T6 extruded microstructure. The fibrous microstructure is characteristic for more complex alloys with higher mechanical strength. The microstructure contains a high population of "stringers," or primary constituent phases, broken up and aligned parallel to the direction of deformation. Upon intersecting the surface during anodizing, the stringers were included in the AAO, producing an increased surface roughness. Courtesy of Impreglon Surface Engineering

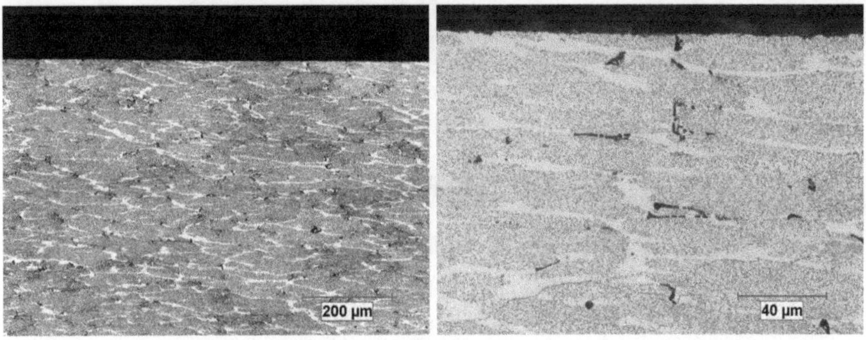

Fig. 4.20 As wrought product, processed at high temperatures, forgings can exhibit all of the microstructural particles within the microstructure. This 6061 T6 forging exhibits transgranular precipitation of dispersoids and Mg_2Si precipitates, with large, dark-appearing primary constituent phases residing at the grain boundaries. Courtesy of Accuride

characterized by phase diagrams such as the one presented as Fig. 4.21, the pseudo-binary phase diagram for Al-Mg_2Si, which is useful for the 6XXX series alloys [30].

The precipitation sequence comprises the nucleation, growth, and development of an equilibrium precipitate. Since precipitation is from a supersaturated solution, the driving force for the formation of the equilibrium precipitate is large and is dependent upon the amount of undercooling achieved through quenching (the larger the amount of undercooling, the larger the driving force). Nucleation is heterogeneous within the supersaturated solution because of the presence of quenched-in defects such as vacancies, dislocations, and grain boundaries. The driving force for the heterogeneous nucleation of precipitates is due to the change in volume free energy due to strain on the lattice created as precipitation occurs and the particles

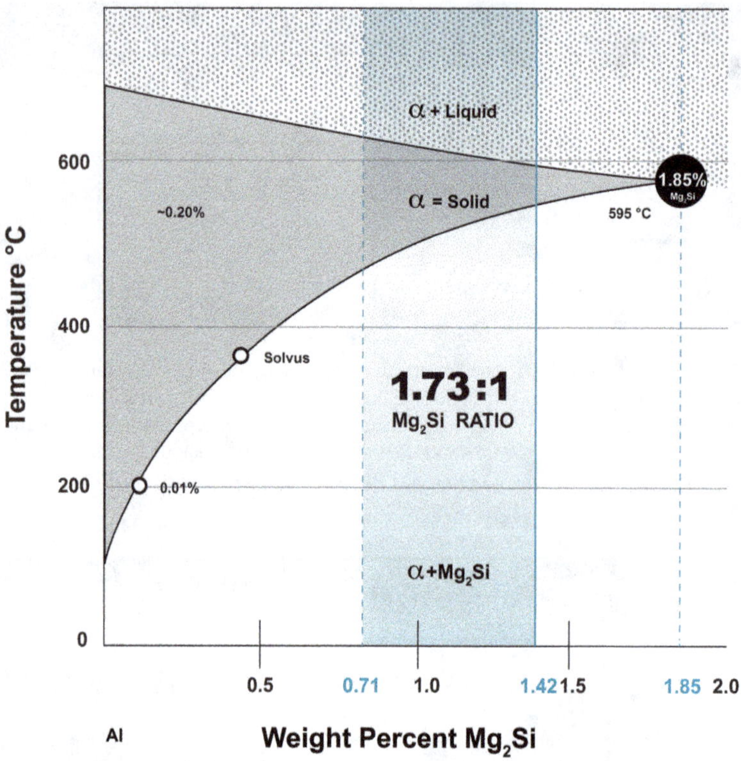

Fig. 4.21 Pseudo-binary Al-Mg_2Si phase diagram for an Mg:Si weight ratio of 1.73:1. The displacement of the average composition of the solid under nonequilibrium conditions corresponding to industrial solidification practices is shown [31]

grow, distorting the lattice order. Alloy chemistry and the amount of strain energy due to deformation also affects the driving force for precipitate nucleation. In the case of 6000 series alloys, the magnesium to silicon ratio must be sufficient to precipitate an Mg_2Si particle. The driving force induced by cold work reduces the energy required for precipitation by producing dislocations and vacancies, which offer surfaces on which precipitates nucleate [27].

The precipitation hardening process can be generalized as follows and is presented in Fig. 4.22. The process is described as follows: solution treatment occurs at a relatively high temperature within the single-phase region to dissolve the alloying elements into the matrix.

1. Rapid cooling or quenching, usually to room temperature, to obtain a supersaturated solid solution of these elements in the aluminum matrix.
2. Controlled decomposition of the supersaturated solid solution to form a finely dispersed precipitate, by aging for appropriate times at one and sometimes two intermediate temperatures.

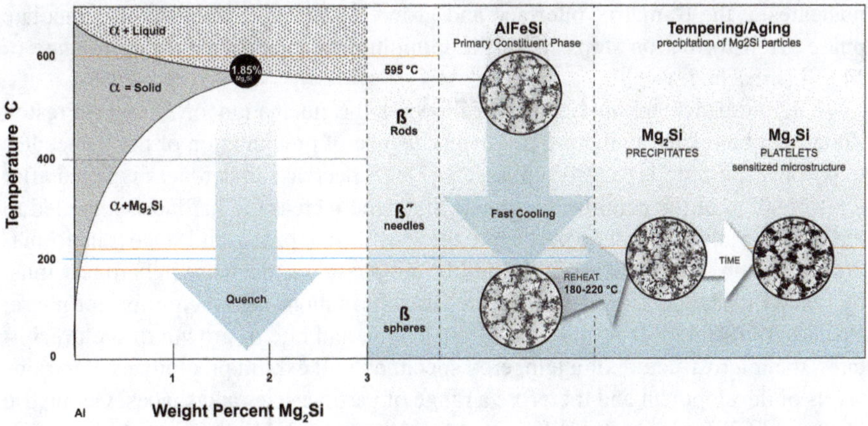

Fig. 4.22 Pseudo Al-Mg_2Si phase diagram showing the development of precipitates with rapid cooling and aging [32]

Precipitates develop with a transition through intermediate phases originating with the nucleation of Guinier–Preston zones (GPZs) [27]. GPZs are ordered, solute-rich clusters of atoms that retain the structure of the matrix and are coherent with it. Because of the coherent nature, they have a lower interfacial energy than the equilibrium phase. Mondolfo [25] indicated for the 6XXX series alloys that the GPZs are spherical, which also reduces the lattice misfit strain, so with GPZs, there is no appreciable increase in alloy hardness. The reduced strain reduces the barrier to nucleation for the GPZs, which enables the low temperature of an aging or tempering process.

GPZs form prior to intermediate phases which simultaneously: (1) have higher interfacial energy *and* (2) require additional energy for solute interdiffusion. The latter requirement is more prevalent in the 6XXX series alloys than in any other aluminum alloy as they require the activation energy to form the equilibrium intermetallic phase Mg_2Si.

Following the formation of the spherical GPZs, the intermediate phases precipitate. For the 6XXX series alloys, the precipitation sequence leading to the equilibrium phase, Al-Mg_2Si, is preceded by the formation of intermediate phases β'' and β'. In describing the precipitation sequence, Mondolfo [25] suggests that the GPZs elongate in the [100] direction. The intermediate phase, β'', has a monoclinic structure and a needle-like shape. The structure contains a large number of vacancies and exerts a compressive force on the matrix. The corresponding stress developed through this force increases the hardness of the alloy.

The β'' needles grow to become the next intermediate phase, β', which have a rod-like shape. β' is semi-coherent with the matrix and has a cubic or hexagonal lattice. The departure from coherency increases the compressive force on the matrix, further increasing the hardness of the alloy and changing the local chemical potential from point to point across the microstructure. Peak hardness is reached just before the Mg_2Si platelets (equilibrium phase) form. The equilibrium phase nucleates at the β′-matrix interface and grows at the expense of the intermediate phase. Transformation stops when the minimum free energy equilibrium state of Mg_2Si is reached.

A definite increase in hardness following the nucleation of Guinier–Preston Zones can be related to the progressive sequence of precipitation of the β″ needles to β′ rods, and β′ rods to Mg_2Si platelets [27]. A decrease in hardness is noted after the formation of the equilibrium phase. Since the increase in hardness is related to the growth of the precipitate phase into the equilibrium phase, and since transformation ceases upon formation of the equilibrium phase, the decrease in hardness must be related to the continued growth, not transformation, of the equilibrium phase. Because of differences in the time of nucleation and rate of growth of the precipitates, the microstructure of a tempered specimen will exhibit precipitates at various levels of development and therefore a range of particle sizes and shapes. Geometric clusters of GPZs also have a different chemical potential than the surrounding grain, and when they intersect the surface, clusters of GPZs can be developed through chemical processing. See Fig. 4.23.

Although completion of the precipitation process to form an equilibrium particle achieves the minimum free energy of the transformation, instability within the matrix persists because the total interfacial free energy is not at a minimum; in other words, precipitation is only uniform after solution treatment and rapid quenching, followed by a controlled temper or aging process. That doesn't usually occur with deformation processing performed on a day-to-day basis in industry. In extrusion, for example, the workpiece is not usually re-solution treated prior to tempering. Therefore, the high density of smaller equilibrium precipitates from the initial transformation will continue to coarsen into a lower density of larger equilibrium precipitates to minimize the total interfacial area. The precipitate coarsening produces the decrease in hardness. Dislocation pinning effectiveness (the resistance of the crystal to deformation because of the presence of dislocations within the crystal structure) decreases with the decrease in precipitate population. For the precipitation hardenable aluminum alloys, precipitate coarsening will decrease the corrosion resistance and therefore increase the alloy's susceptibility to chemical attack because of increased capillarity at the precipitate-matrix interface. Other material properties such as hardness, ultimate tensile strength, and resistivity also change as precipitates ripen [25]. See Fig. 4.24.

Precipitates within a tempered specimen are at the same temperature and have identical structures and compositions, yet, because they continually develop throughout the manufacturing process, precipitates at different stages of development and consequently different size and shape have different chemical potentials and differing amounts of strain imposed on the lattice. The higher chemical potential in the smaller precipitates causes diffusion in the direction of the largest precipitates, away from the smaller precipitates. Therefore, the smaller precipitates shrink and disappear while the larger precipitates grow or "*ripen*." Because the reaction is diffusion controlled, the rate of coarsening will increase with increasing temperature. Excessive coarsening occurs during heat treatment when the material is

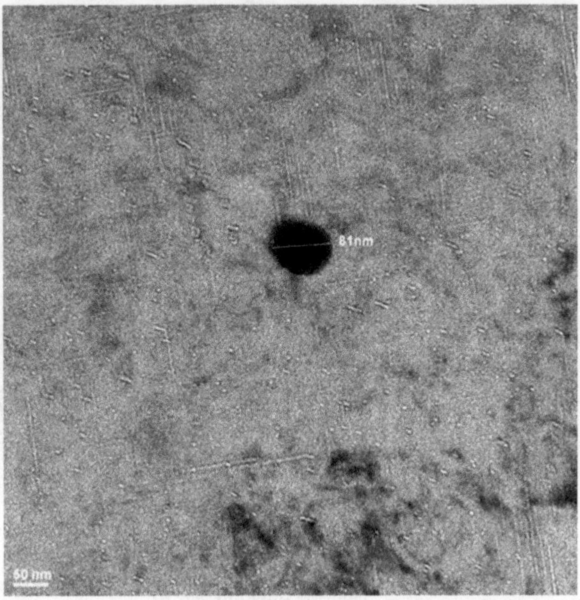

Fig. 4.23 TEM image of clusters of precipitates at various stages of formation within a 6063 T6 alloy. Hashtag-shaped particles are Guinier–Preston zones, the precursors to actual Mg_2Si particles

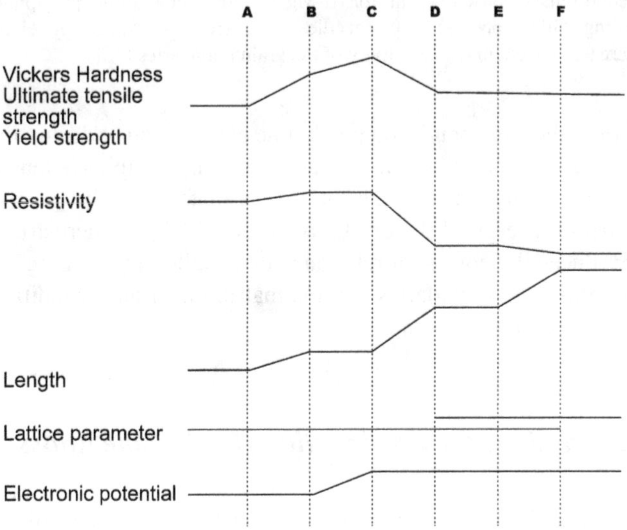

Fig. 4.24 Schematic of property changes during aging of aluminum-magnesium silicide alloys according to Mondolfo. The schematic is for a constant aging temperature over an increasing amount of time, from time A to time F. Beginning at time A, precipitates nucleate and grow to a peak hardness at time C, that plummets from time C to time D. From time D to time F, the microstructure is sensitized. Note the lattice parameter stays the same throughout aging (grain size doesn't change); also note that the electrolytic potential markedly climbs from time B to time C, when the precipitates coarsen and move to the grain boundaries, producing a radical change in chemical resistance. Note how resistivity plunges from time C to time D, showing the increase in chemical sensitivity (conductivity increase between the grain boundary and the grain) [25]

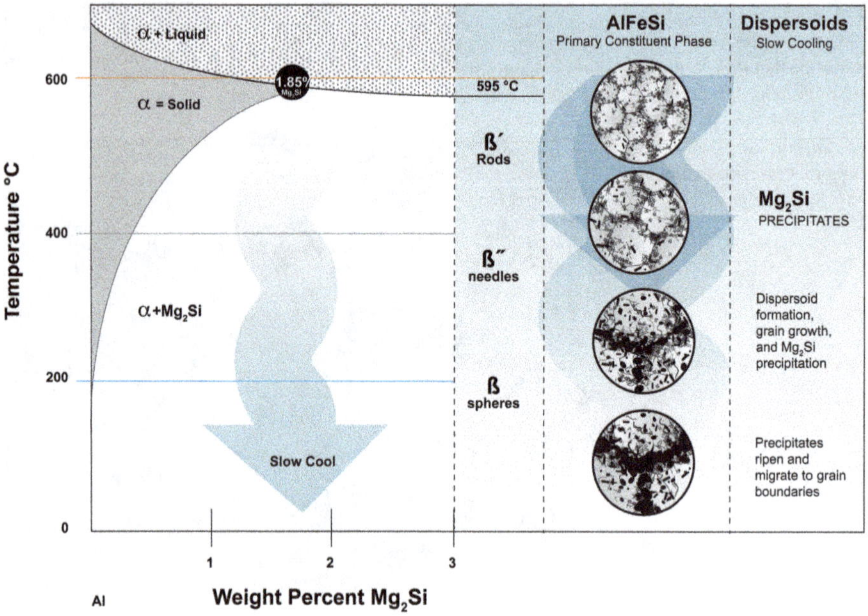

Fig. 4.25 Pseudo-binary phase diagram for Al-Mg_2Si alloys showing the precipitation sequence with slow cooling: GPZ (spheres) → β″ (needles) → β′ (rods) → Mg_2Si (platelets). Precipitate coarsening increases the chemical sensitivity of the grain boundaries [32]

allowed to "*over-age*" or spends too much time at temperature during the precipitation hardening process. In addition to coarsening, the precipitates tend to move by way of solid-state transport, through the matrix material to the grain boundaries, where the intrinsic energy of the crystal is lower [33]. This microstructural condition further reduces the corrosion resistance of the alloy by creating tiny galvanic cells between the grain boundaries and the matrix material, a condition known as "*sensitization*." See Figs. 4.25 and 4.26.

4.4.5 Designations for Precipitation Hardenable Alloys

The ANSI and IADS designations for the precipitation hardenable alloys are as follows: the primary alloying element for the 2XXX series alloys is copper; the 4XXX series, silicon; the 6XXX series, magnesium and silicon; and the 7XXX series, zinc [21, 22]. These alloys represent the most complex compositions of the wrought aluminum alloys. Each class of alloys behaves differently, with composition and structure dictating the working characteristics and subsequent properties. The precipitation hardenable alloys exhibit the effects of cold work through deformation processing but exhibit the unique characteristic of developing precipitates through subsequent heat treatment, which impart additional strength and hardness.

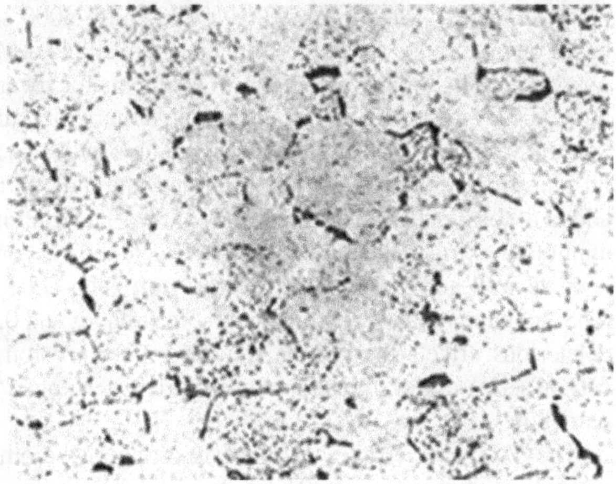

Fig. 4.26 Optical photomicrograph of a sensitized 6061-T6 microstructure (×400, etchant: Tucker's reagent) [1]. The microstructure is typical for an overaged condition, and corresponds to a structure that would have developed between time C and D in Fig. 4.24

4.4.5.1 Alloy Designations

2000 Series; Copper: Copper is the primary alloying element in the 2000 series alloys. The alloys are precipitation hardenable. Developed for strength and hardness, the presence of copper in the microstructure refines the grain size by preventing excessive grain growth during cooling and heat treatment, which produces strength. $AlCu_2$ and Al_2CuMg are precipitates which form during age hardening (tempering). The 2000 series alloys always contain some iron, which forms the AlFeSi primary phase. At additions of greater than 0.5%, iron reduces the tensile strength by forming Cu_2FeAl_7, which decreases the amount of copper available for precipitation hardening. The presence of copper in solid solution reduces the corrosion resistance of the 2XXX series alloys and inhibits the anodizing reaction. The aluminum-copper alloys are some of the oldest aluminum alloys. They were developed for and continued to be used for military applications such as armor, but also find use in the aerospace and aircraft industries.

4000 Series; Silicon: Silicon is the primary alloying element for the 4000 series alloys and is added to almost all wrought aluminum alloys to enhance strength. Magnesium, nickel, phosphorus, and copper are minor alloying additions. Magnesium is present from 1 to 2% to enable the formation of Mg_2Si precipitates. Silicon is present in solid solution up to 1% of the total alloy loading, forms intermetallic precipitates with copper, and falls out of solution as primary silicon crystals. Phosphorus may be present to prevent them from becoming plate-like. The eutectic composition for aluminum with silicon is 12.3% by weight, and with this composition, the fluidity of the alloy is optimized at the lowest possible temperature. Alloy 4004 contains up to 10% Si to exhibit hypoeutectic silicon and flow to enable its use as a filler metal. The 4000 series alloys are characterized by low ductility and low formability. The 4000 series alloys have limited use; it is used as a cladding alloy on brazing sheet and as a filler metal for welding.

6000 Series; Magnesium-Silicon: The entire 6000 series of wrought alloys are based on precipitate structures of magnesium silicide (Mg_2Si). Review of the binary phase diagrams for the major alloy constituents for specific precipitation hardenable alloy 6061; Al-Mg, Al-Si and Mg-Si, the ternary diagrams for Al-Mg-Si, as well as literature dealing with the 6XXX series alloys indicates the major solute intermetallic phase that forms is Mg_2Si. Balanced amounts of magnesium and silicon are added to form quasi-binary Al-Mg_2Si alloys. These alloys comprise the majority of aluminum extrusions and exhibit good weldability, corrosion resistance, typically anodize well and are immune to stress corrosion cracking. Alloy 6061 (Al–1Mg–0.6Si) contains 0.25% copper to improve strength and 0.2% chromium to offset any deleterious effects the copper may have on corrosion resistance. Like all of the 6XXX series alloys, the strength of 6061 is developed through precipitation (age) hardening.

The 6XXX series alloys are not as strong as the other precipitation hardenable wrought aluminum alloys (2XXX and 7XXX) but they have good formability and weldability. With low melting temperature alloy additions such as lead and bismuth, the 6XXX alloys are easy to machine. The 6XXX alloys are the "work horse" alloys for extrusion and find applications in automotive, architecture, construction, offshore structures, and heating and ventilation.

7000 Series; Zinc: In general, wrought aluminum alloys that contain additional elements to zinc (the 7XXX series) offer the highest combination of strength and fatigue resistance, based upon intermetallic compounds that form with magnesium, copper, chromium, and manganese. Because of its low melting temperature and vapor pressure, zinc is never added alone to wrought aluminum alloys; copper and magnesium are often added to form Al-Zn-Mg phases as well as $Mg(Zn,Cu,Al)_2$, and/or combinations of the phases Al_2Cu and Al_2CuMg. These intermetallic compounds are distributed throughout the microstructure, which means they also intersect the surface, changing the surface chemical potential which reduces overall corrosion resistance and complexes the anodizing reaction.

Copper, magnesium, and/or zinc atoms reside on the aluminum lattice in alloy solution. In fact, as a solid solution strengthening addition, zinc resides on the aluminum lattice similar to copper and magnesium. According to Mondolfo et al., in alloys with less than 3% zinc, all remains in solid solution. This means that when present at the surface in solid solution with the parent aluminum, copper, magnesium, and zinc atoms will inhibit the anodizing reaction.

Table 4.8 presents the compositions of some commonly specified precipitation hardenable wrought aluminum alloys and their most similar European and Japanese designations.

4.4.5.2 Temper Designations

Following the alloy designation, the precipitation hardenable wrought alloys are further identified by a T designation to indicate the temper, a thermal treatment with or without supplementary strain hardening, to produce a stable temper. One or more

Table 4.8 Common precipitation hardenable wrought aluminum alloys [30]

US designation	ISO and/or Japanese designation (JIS)	Si	Fe	Cu	Mn	Mg	Cr	Zn	Bi/Pb	Ti/Zr	Ga	Other each	Other total
2011	AlCu6BiPb	≤0.40	≤0.70	5.0–6.0				0.30	0.2–0.6	≤0.03		0.05	0.15
2017	USA	0.2–0.8	≤0.70	3.5–4.5	0.4–1.0	0.4–0.8	0.1	0.25		0.15		0.05	0.15
2024	AlCu4Mg1	0.5	0.5	3.8–4.9	0.3–0.9	1.2–1.8	0.1	0.25				0.05	0.15
2195	USA	0.12	0.15	3.7–4.3	0.25	0.25–0.8	…	0.25	**Ag** 0.25–0.6	0.10/0.08–0.16	**Li** 0.8–1.2	0.05	0.15
2219	USA	0.20	0.30	5.8–6.8	0.20–0.40	0.02	…	0.10	**V** 0.05–0.15	0.02–0.10/0.10–0.25		0.05	0.15
6061	AlMn0,5 Mg0,5	0.4–0.8	0.7	0.15–0.40	0.15	0.8–1.2	0.4–0.35	≤0.25		≤0.15		0.05	0.15
6063	AlMg5	0.2–0.6	0.35	0.10	0.10	0.45–0.9	0.1	0.1				0.05	0.15
6463	USA	0.2–0.6	0.15	0.20	0.05	0.45–0.9		0.05				0.05	0.15
6082	AlMg2,5	0.7–1.3	0.50	0.10	0.4–1.0	0.6–1.2	0.25	0.20				≤0.02	≤0.50
7003	A7003 (JIS)	0.3	0.35	0.20	0.30	0.5–1.0	0.2	5.0–6.5	**Zr** 0.05–0.25	0.2		0.05	0.15
7050	USA												
7075	AlZn5,5MgCu	0.4	0.5	1.2–2.0	0.3	2.1–2.9	0.18–0.28	5.1–6.1		0.2		0.05	0.15
7129	USA	0.15	0.3	0.5–0.9	0.1	1.3–2.0	0.1	4.2–5.2	**V** 0.05	0.05	0.03	0.05	0.15
7136	USA	0.12	0.15	1.9–2.5	0.05	1.8–2.5	0.05	8.4–9.4		0.1/0.10–0.20		0.05	0.15

Table 4.9 T temper precipitation hardening code [22]

T temper designation	Definition
T1	Cooled from an elevated temperature shaping process and naturally aged to a stable condition
T2	Cooled from an elevated temperature shaping process, cold worked, and naturally aged to a stable condition
T3	Solution heat-treated, cold worked, and naturally aged to a stable condition
T4	Solution heat-treated and naturally aged to a stable condition
T5	Quenched from an elevated temperature shaping process and artificially aged
T6	Solution heat-treated, then artificially aged
T7	Solution heat-treated, then overaged/stabilized
T8	Solution treated, cold worked, then artificially aged
T9	Solution treated, artificially aged then cold worked
T10	Quenched from an elevated temperature shaping process, cold worked, then artificially aged

digits always follow the T designation. For all T tempers, solution heat treatment is achieved by heating cast or wrought products to a suitable temperature, holding for enough time to allow constituents to enter into solid solution. The product will be rapidly cooled (quenched) to develop a supersaturated solution to take advantage of subsequent precipitation and the associated hardening that usually takes place by artificially aging, which involves reheating to a moderate temperature for a specific period of time, during which precipitates form within the microstructure without further annealing effects. Table 4.9 presents the basic T tempers with corresponding definitions.

Additional digits that follow the primary temper designations T1 through T10 further define the T tempers. This is done to identify a variation in treatment that significantly alters the product characteristics through stress relief by way of stretching, compressing or a combination of these mechanical treatments. The properties of the product are changed beyond what would be obtained with the basic single-digit temper treatment. The code for stress relief by way of stretching, compression, or a combination of these treatments applies to precipitation hardenable wrought product, processed by different deformation methods, after solution heat treatment or after quenching from an elevated temperature deformation process in order to produce a stable condition or "permanent set." The products receive no further straightening after the additional temper treatment, except for the TX511 product, which may receive minor straightening after stretching to comply with standard tolerances. Product that is stress relieved by a combination of stretching and compressing, the TX54 temper, is typically restruck cold in the die after stretching. Table 4.10 presents the code and definitions for the additional digits for T tempers.

Table 4.10 Code and definition for additional digits on temper designations for precipitation hardenable alloys [22]

Temper	Product type	Definition (amount permanent set)
Stress relieved by stretching		
TX51	Plate	1½–3%
	Rolled or cold-finished rod and bar	1–3%
	Die or ring forgings and rolled rings	1–5%
TX510	Extruded rod, bar, profiles (shapes), tube	1–3%
	Drawn tube	½–3%
TX511	Extruded rod, bar, profiles (shapes), tube	1–3%
	Drawn tube	½–3%
Stress relieved by compression TX52	All wrought products	1–5%
Stress relieved by a combination of stretching and compression TX54	Die Forgings	Applies to all die forgings that are restruck cold in the finishing die after stretching. NO permanent set specified

4.5 Trace Alloy Additions in All Wrought Aluminum Alloys

While main alloying additions develop primary mechanical properties which designate the various aluminum alloy classifications, trace elements are what bring the alloy composition together, refining the characteristics of tensile strength, ductility, corrosion resistance, fatigue resistance, and even surface quality to the finished product. These small amounts make a big impact on alloy performance; shifts in trace element ratios can drastically change the alloy microstructure. The summary presented is according to Davis [21].

The mechanism by which trace elements function is to combine with major alloying elements to form intermetallic compounds as precipitates, constituent phases and particles or dispersoids. These various microstructural phenomena act to: pin grain boundaries and inhibit recrystallization, thereby imparting strength; pin dislocations and impede lattice movement during deformation, thereby imparting rigidity (reducing ductility); increase passivity, thereby increasing corrosion resistance.

Iron: The most common trace element added to aluminum alloys is iron. The solubility of iron in the solid state is very low in aluminum (~0.05%), therefore, most of the iron present in aluminum over this amount appears as an intermetallic second phase in combination with aluminum and other elements. Intermetallic compounds comprising aluminum-silicon-iron strengthen and increase the wear resistance of cast product and comprise the primary constituent phase of most wrought alloys. As with castings, because iron reacts with the aluminum and silicon to make

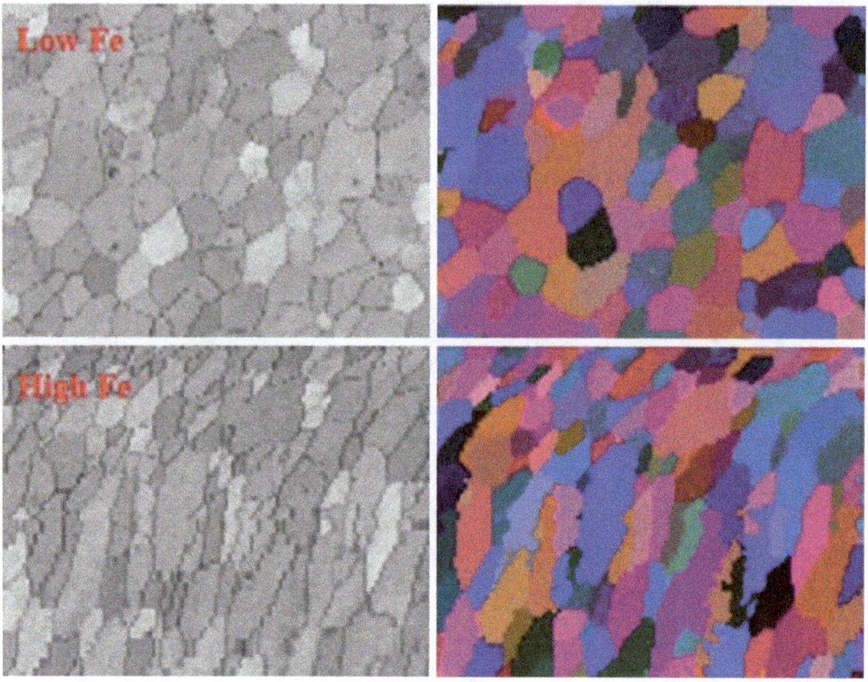

Fig. 4.27 Electron Beam Surface Diffraction of a 6061 T6 extrusion with varying levels of iron shows the effects of higher iron on the microstructure. Iron effectively pins the grain boundaries, producing a microstructure that is resistant to recrystallization. Note how the grains of the High Fe sample exhibit retained directionality while the Low Fe sample exhibits a more equiaxed structure due to recrystallization [34]

a second phase, the structures tend to be finer than alloys with reduced iron content. In 3000 series alloys, iron modifies the microstructure by keeping the intermetallic phase AlFeSi from becoming needle-like or plate-like, which reduces die wear and makes the alloys easier to anodize. $AlFe_3$ is well-documented particle that forms in aluminum alloy microstructures that adds strength by inhibiting recrystallization and imparting creep resistance. See Fig. 4.27.

Manganese: Manganese follows iron as the next most common trace element found in aluminum alloys. As a trace element, manganese binds with Fe to stabilize the $FeAl_6$ phase in heat treatable alloys, forming AlMnFe. The presence of $AlMn_6$ is an intermetallic compound that produces increased tensile strength and increases the recrystallization temperature. Iron and silicon are added to form dispersoids $Al_{12}(Fe,Mn)_3Si$, which are critical in controlling grain size and mechanical properties. Manganese dispersoids ($AlMn_3$) also form to stabilize grain size by restricting recrystallization, increasing strength. At concentrations higher than 1%, aluminum-manganese intermetallic particles will reduce alloy ductility.

Chromium: Chromium has a slow diffusion rate and forms fine dispersoids in wrought products. Dispersoids inhibit nucleation and grain growth. Chromium is used to control grain structure by inhibiting grain growth in aluminum-magnesium alloys, and to prevent recrystallization in aluminum-magnesium-silicon or aluminum-magnesium-zinc alloys during hot working or heat treatment. Chromium and titanium are added in small amounts to form dispersoids which inhibit the nucleation of new grains and inhibit grain growth. Together with magnesium, chromium and titanium produce precipitates that enhance mechanical properties by keeping grains small and equiaxed through recrystallization. Such additions can enhance the surface characteristics for anodizing.

Lithium: Lithium is the lightest alloying element added to make certain 2000 series aluminum alloys. Lithium decreases density, ductility, and toughness but produces the lightest weight high strength aluminum alloys. As an alloy addition, 1–3% lithium is typically added to produce low-density aluminum alloys. The amount of lithium added in order to increase "lightness" is often counter-alloyed with copper in order to produce more ductility. Particles of Al_2LiCu (T_1 precipitates) and AlLiZr dispersoids, in zirconium-containing alloys, are documented as changing the fracture characteristics of lithium-containing aluminum alloys. Lithium-containing alloys are part of both the 2000 series and 8000 series alloys.

Calcium: Calcium is added to precipitation hardenable alloys to form $CaSi_2$ dispersoids. Calcium silicide formation inhibits age hardening by reducing the formation of Mg_2Si precipitates. Their presence increases conductivity slightly, increases strength but decreases ductility.

Nickel: Nickel is never added alone to aluminum. Al-Li-Ni alloys are formulated to operate at high temperatures.

Vanadium, Zirconium, and Titanium: Increases recrystallization temperature.

Other Insoluble Trace Elements: Lead, bismuth, tin, and cadmium are trace elements that are typically added to 2000 and 6000 series alloys, specifically alloys 2011 and 6262. Because they are insoluble in the aluminum matrix and have low melting points, the addition of these elements in small amounts improves metal removal by producing small chips rather than long curls of metal waste during machining (lathe, mill, hone, etc.). The disadvantage of adding lead and cadmium is that they are toxic, and their incorporation as an alloying element restricts the application and disposal (recyclability) of the component.

4.5.1 Energy Effects on Microstructure

4.5.1.1 Thermal Effects

Microstructural characteristics provide insight to the process history of the component, and thermal history has a decisive effect on the microstructure of cast and wrought products. Microstructure varies depending upon alloy, process temperature, and forming technique; as alloy complexity increases, the number of intermetallic phases and particles increases. Nevertheless, as dwell time increases through deformation processing, the workpiece is exposed to temperatures that effect

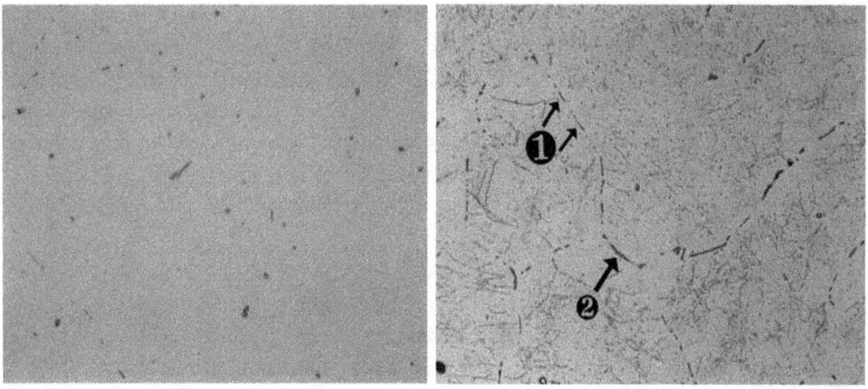

Fig. 4.28 Documentary photomicrographs of alloy 6061 billet microstructures after homogenization. The structure on the left is well homogenized with globular AlFeSi PCP well broken up and distributed. The structure on the right is not well homogenized, most likely slow cooled following homogenization. Arrow 1 indicates coarse needles of Al-Fe-Si primary constituent phase at the cast grain boundaries of the billet and Arrow 2 indicates coarse Mg_2Si. Small precipitates comprising dispersoids and soluble Mg_2Si particles are present throughout the microstructure [13]. Courtesy of AAC

structural changes that wouldn't happen if the process were faster. More time at high temperature enables more time for structural change: recrystallization of deformed grains occurs, the effects of strain energy are relieved, more precipitates form, and grains grow. These trends are consistent for all types of alloys, both non-heat treatable and precipitation hardenable.

Wrought products always start as billet castings; therefore, the condition of the billet is critical to developing a quality wrought component, as well as to the life of the deformation tool. Billets are typically DC cast, quenched, and then sent on for further processing by way of cutting into logs for extrusion or rolling into slabs, sheet, or foil [12]. See Fig. 4.28.

Grains originate and grow from the cast microstructure, with the primary constituent phase typically defining the initial grain boundaries. Deformation, regardless of the process used, further develops the grain structure of the aluminum: orienting it in the process direction, causing grain growth at the surface where the workpiece contacts the deformation tool, which leads to the recrystallization of grains, and nucleation of new grains. A thermal gradient develops from the core of the workpiece to the surface, where it contacts the tool. Thermal effects are accentuated at the aluminum-tool interface, causing the development of a microstructural gradient from the center of the workpiece to the surface of the tool or die. This gradient can be visualized microstructurally in the formation of a coarse peripheral zone, in which the heat of deformation at the tool-workpiece interface is manifested in larger grains than at the core of the workpiece. This phenomenon is testament to the thermal effects of deformation on microstructural development. Depending on the application of the workpiece, the coarse peripheral zone

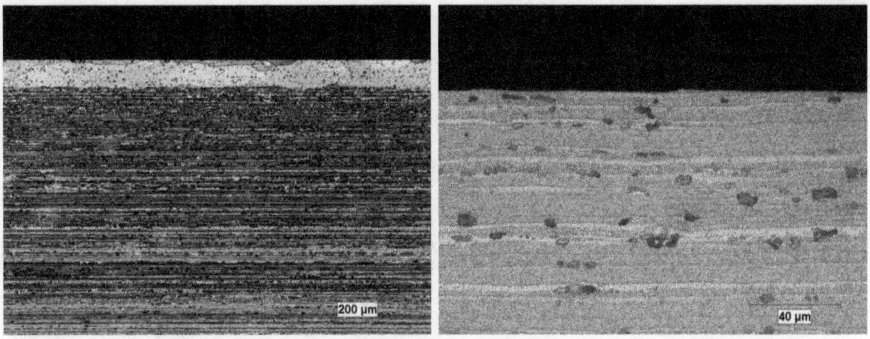

Fig. 4.29 7XXX series extruded aluminum super alloy as-solution treated and tempered, showing the coarse peripheral zone in contrast with the fibrous, dynamically recrystallized microstructure retained from deformation. The coarse grain size adjacent to the surface develops due to the heat of deformation at the extrusion die-extrudate (workpiece) interface. The elongated grains conform to the deformation direction. A mixture of broken primary constituent phase, which appear black, and precipitates, which appear as light gray specks, decorate the microstructure primarily at the grain boundaries. Courtesy of Universal Alloy Corporation

may be removed by machining in order to utilize the more uniform, underlying microstructure. See Fig. 4.29.

4.5.1.2 Deformation Effects

Regardless of whether the manufacturing process is hot or cold, during deformation processing, the material will seek a flow pattern that results in minimum energy expenditure. The stress developed within the material due to this flow is called the flow stress. For a component to have the most uniform-appearing surface, the flow stress must be maintained and monitored for the temperature, strains and strain rates for the specific process, including the alloy and the tool. Each aspect will affect the surface microstructure from the size and distribution of phases, precipitates and intermetallic compounds, grain size to surface defects induced by friction and yielding.

The amount of total deformation is not usually reflected in the microstructure of a wrought product because even as deformation introduces strain energy to the crystal structure, and inasmuch as it produces strength and hardness by producing more defects within the microstructure, the high energy condition brought about by deformation relieves the strain of cold work in the microstructure by orienting the grain structure, altering grain size, and changing grain boundary density. This is because the temperatures of the process overlap with the temperatures required to anneal the microstructure, which relieve stress and cause grain growth. By adjusting the rate of the process, the time through this critical temperature range can be extended or limited, and the effects, observed in the microstructure, can be controlled.

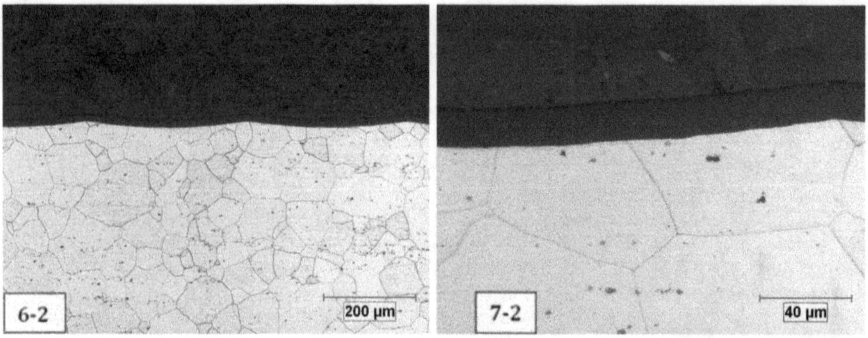

Fig. 4.30 Metallographic images of extruded alloy 7129 showing an equiaxed microstructure with normal levels of precipitation and residual directionality from deformation processing. The anodic oxide is uniform and continuous, displaying no visual defects. Courtesy of MD Building Products

Deformation effects on the microstructure can be relieved or further developed by thermal treatment or by additional cold work.

Therefore the kinetics, or speed, of deformation also known as strain rate—a very important source of thermodynamic energy—impacts flow stress and therefore dramatically impacts the microstructure. The speed and reduction parameters set by the deformation process produce energy that have distinct effects on the microstructure, similar to heat treatment. A rapid strain rate will affect recrystallization sooner, producing a finer grain size at the surface and preventing grain growth from occurring on the deformation tool. This is an advantage when a homogeneous microstructure is desired. In combination with a homogenized billet, a rapid process through the solvus temperature by way of a strain rate with an optimized extrusion ratio will produce a microstructure that is as homogeneous as possible, with a fine grain size and little grain boundary precipitation of intermetallic compounds and precipitates, the *best* microstructure for anodizing. See Fig. 4.30.

Mechanical properties such as yield strength and ductility can be related to microstructure: higher strength alloys exhibit more fibrous microstructures and retained directionality from deformation processing, while ductile, low alloy components tend to exhibit recovery from deformation processing, displaying equiaxed microstructures. High strength alloys such as the higher alloyed precipitation hardenable alloys of 6061, 6082, and 7075 are typically processed to yield a fine-grained, fibrous-appearing microstructure. Lower alloy, higher purity aluminum substrates such as 6063, 5052, and 1050 are typically processed to yield a fully recrystallized, equiaxed structure, with a moderate grain size (60–100 μ).

Dynamic heat treatment effects that occur during deformation are visible in the grain structure as annealing effects enable the microstructure to recover and recrystallize, causing grain growth and precipitation, while follow-on aging affects the size, shape, and distribution of precipitates. Reducing the strain rate for complex alloy chemistries enables recrystallization during deformation, producing larger equiaxed grains in the microstructure.

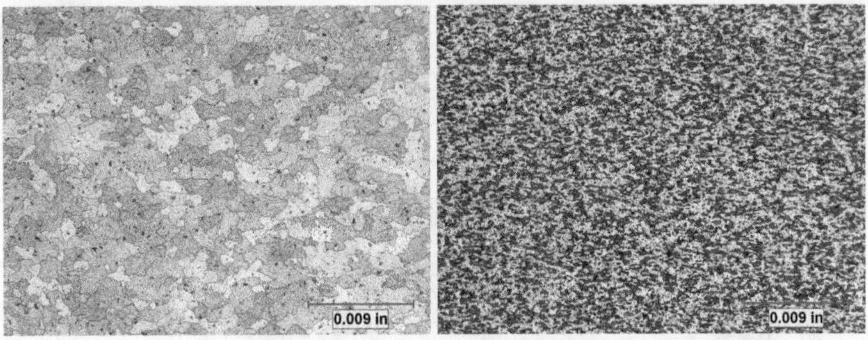

Fig. 4.31 Microstructures of 7075 T6 extrusions manufactured for the same component. Chemical analysis determined both extrusions conformed to the requirements of alloy 7075. The microstructure on the left yielded a reflective anodic oxide. The microstructure on the left had a chemistry with the high vanadium (0.0069% as compared with 0.0038%), which raises the recrystallization temperature, limiting recovery during the heat of processing, follow-on heat treatment, or even mechanical stress relief by way of stretching or straightening after extrusion. The microstructure on the right therefore exhibits retained deformation and ultra-fine grains due to dynamic recrystallization; it also exhibited a muddy, dull appearance after anodizing [35]. Courtesy of Cane Creek Cyclery

Trace element additions can vary widely within a particular alloy, yet still be within the chemical requirements for a specific alloy. While trace elements most often are added to preserve alloy strength by limiting recrystallization, variations can produce drastic differences in microstructure for similar components, even if they are manufactured from the same alloy. These differences are often not noticed until anodizing develops the microstructure. See Fig. 4.31.

4.5.2 *Metallurgical Processes in the Production of Wrought Aluminum Products*

From the microstructure of the billet to the finished product, following deformation processing, there are a number of intermediate steps to which the aluminum (alloy) workpiece is subjected that also contribute to the final appearance of the microstructure. While the steps may vary between manufacturing processes, in general, they are all reflected in the summary below, and yield appearance conventions that can be connected to a specific point in the thermal history of the component. This connection between the appearance of the surface microstructure, and the appearance of the anodic oxide can be helpful in specifying a specific type of manufacturing process for a component, or help in trouble shooting, when the appearance of the component is not what is expected or desired.

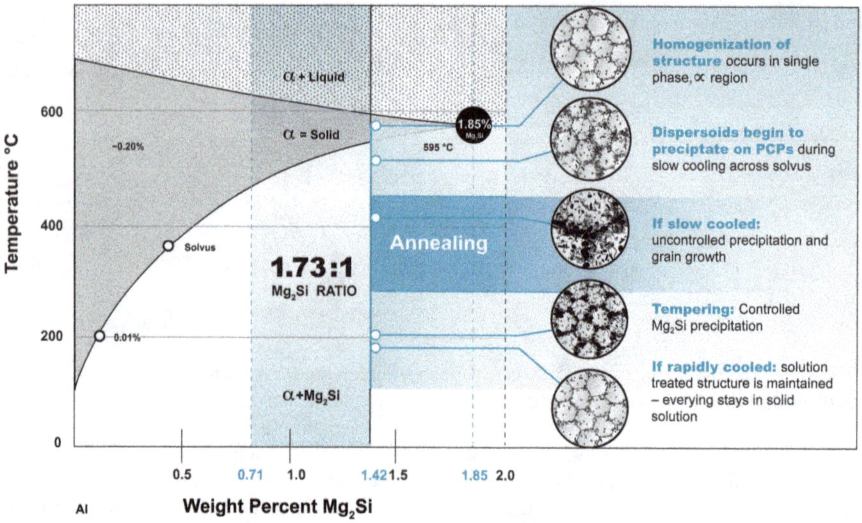

Fig. 4.32 Pseudo Al-Mg_2Si phase diagram with important thermodynamic events of precipitation, recrystallization, and grain growth. If one considers these events in the context of the various temperature excursions that occur during manufacturing, such as when a casting solidifies and cools, or during a deformation process such as extrusion or forging, in terms of time and kinetics, it is easy to develop a picture of the microstructure at each step [31]

The following process summary is based upon a similar summary, "Metallurgical Processes in the Production of Aluminum Semi-finished Products," published in Altenpohl [15], and includes definitions and comments summarized in the ANSI specification [22]. While the conventions were originally presented to help the reader become oriented with respect to microstructural changes due to specific process steps, the specific purpose in this section is to help the reader develop a virtual picture of the component surface, with the intention of finishing it by way of anodic oxidation as described in Chap. 3 of this book. Superimposing the corresponding specific thermodynamic events, specifically, those of solution treatment, precipitation, and grain growth, at the appropriate temperatures with the corresponding phase diagram, enables one to picture how the microstructure will develop over the course of the manufacturing process [31]. See Fig. 4.32.

Preheating Prior to Deformation: Depending upon the temperature selected, the preheat may bring the workpiece to the proper temperature for better formability, can homogenize the structure by bringing alloying elements in solution within the cast structure, and remove high stresses from the cast billet that can lead to cracking during deformation. This is typically carried out well above the solvus temperature for the specific alloy (minimum of 10–20 °C higher). However, the lower the preheat temperature, the more precipitation occurs. Slow cooling below the solvus during processing can produce precipitate-free zones around grain boundaries, which greatly alter the appearance of the surface microstructure.

Annealing: Annealing is the only follow-on heat treatment for which the main purpose is to transform the grain structure. Recrystallization and stress relief (recovery) occur within the annealing temperature range and additional precipitation of intermetallic particles may take place. With a slow applied strain rate, following recrystallization, grain growth also occurs. A temperature between 300 and 450 °C is usually selected.

Solution Heat Treating and Quenching (post deformation, prior to tempering): The solution treatment process for cast or wrought products is a solid-state process that requires heating the alloy to a suitably high temperature, within the single phase region of the phase diagram, and holding at that temperature long enough to allow constituents to enter into solid solution. Quenching is a rapid cooling process that immediately follows solution treatment. Quench rate is critical to keep alloying elements in solution and to prevent premature precipitation and grain growth; the quench media must be cold (as cold as liquid nitrogen). Components must stay cold, and not be allowed to reheat, or tempering will proceed uncontrolled. Depending upon the alloy composition, the temperature of the solution heat treatment may be as high as the homogenization treatment prior to deformation processing, or even higher, well above the solvus temperature. Care must be exercised to avoid incipient melting of intermetallic compounds that may have formed. This is true especially for aluminum super alloys, with compositions exceeding the typical 8% alloy loading maximum.

Tempering: Low temperature process, 150–180 °C. Designed to increase the hardness and strength of the material through precipitation. No other grain growth or annealing occurs. As a follow-on treatment to solution treatment, tempering allows for controlled development of the microstructure and mechanical properties and enables the development of new and smaller precipitates.

4.6 Summary and Conclusion

By considering in general the various aluminum alloy chemistries and how their respective microstructures develop through casting, deformation, and heat treatment, a picture develops of the component surface that varies with the component metallurgy and the various processes involved in its manufacturing. The metallurgy of aluminum governs the mechanical properties and the chemical resistance of the component. Microstructure, which is developed by the alloy chemistry, fabrication method (whether the component is cast or wrought), and the condition: as-fabricated, heat-treated, solution treated, or cold worked, determines how the surface microstructure develops and therefore the surface chemical potential of the component. The surface microstructure dictates how well the component anodizes.

Unalloyed, the aluminum FCC crystal has planes with different atomic packing densities, which alone present differences in chemical potential. A hierarchy of defects forms through casting and deformation of aluminum, which develop a microstructure that will ultimately comprise a component designed to fulfill or support a specific application. Basic atomic level defects, vacancies and dislocations, form in pure aluminum during casting and increase with deformation. These first-level defects interrupt the long-range order of the crystal and produce a level of strength that exceeds that of a perfect crystal. Atomic level defects migrate and cluster around grain boundaries, an important microstructural feature comprising the polycrystalline substrate. Even at this most basic level, the substrate as an anode exhibits a surface with variations in local chemical potential, which changes the response to the anodizing reaction.

Alloying provides strength and hardness to aluminum but adds another level of complexity to the defects that develop in a substrate. Each element added to the alloy mixture will either substitute for an aluminum atom on the FCC lattice or, if the amount exceeds the solubility in aluminum, will segregate to form intermetallic phases or incoherent particles. The atomic level defects, which provide strength to the pure aluminum lattice, also provide surfaces for nucleation on which alloy mixtures of phases and particles can form. The rate at which nucleation occurs depends upon the amount of strain energy and the crystallographic plane on which nucleation occurs.

Individual intermetallic particles change the local chemical potential from point to point across the surface; depending upon the size, amount, and location, networks of particles can change the chemical potential across the surface of an entire substrate. Changes in chemical potential alter how the oxidation reaction proceeds at the anode surface, as well as the appearance and performance of the anodic oxide for two reasons: 1) changes in chemical potential will alter the surface pitting potential and 2) by the very nature of the anodic oxide, alloying elements don't anodize.

Manufacturing processes, both cast and wrought, introduce defects to the microstructure that serve for nucleation of particles, but also new grains, as there is energy sufficient to drive recovery, recrystallization, and grain growth within the microstructure. Time and temperature are the drivers for nucleation, precipitation, and grain structure changes; they happen haphazardly through an uncontrolled manufacturing process, or orderly, through controlled cast, deformation, and heat treatment processes of solution treatment, annealing, and tempering. The rate of cooling (quenching) from high temperatures impacts the level of microstructural homogeneity; faster cooling retains the solution treated condition; to contrast, precipitation occurs during slower cooling. In all cases, the microstructures are characteristic of the processes that produce them, and careful investigation can reveal the source for deficiencies in performance, whether they are mechanical, corrosion, or cosmetic failures.

Through the consideration of microstructural development in the context of cast and wrought alloy chemistries; their constitutive phases and particles that comprise each substrate; feature characteristics; and how they are affected by manufacturing, a picture of the substrate surface microstructure and therefore a concept of how well an anodic oxide will grow from it can be developed. Recognition of the condition of the microstructure in connection with the process can aid in understanding the appearance of the surface microstructure and consequently, its appearance after anodizing. The reactions that occur at the interface with the aluminum (alloy) surface during anodizing that develop the substrate surface microstructure, including actual case studies, will be discussed and presented in detail in the following chapters.

References

1. Runge, J. M. (2013). Base metal microstructural considerations for anodizing wrought aluminum alloys. *The Proceedings of Aluminium 2000*, Milan.
2. Sharma, C. P. (2004). *Engineering materials, properties and applications of metals and alloys*. New Delhi: Prentice-Hall of India.
3. Davis, J. R. (Ed.). (1993). *ASM specialty handbook, aluminum and aluminum alloys*. Materials Park, OH: ASM International.
4. Website of the Aluminum Association: Recycling. Retrieved from aluminum.org.
5. Parsons, R. (1987). Chapter 2: The electric double layer at the solid-solution interface. In W. Stumm (Ed.), *Aquatic surface chemistry*. New York: Wiley.
6. Murr, L. (1975). *Interfacial phenomena in metals and alloys*. Reading, MA: Addison-Wesley.
7. Van Vlack, L. H. (1980). *Elements of materials science and engineering*. Reading, MA: Addison-Wesley.
8. Sato, N. (2011). Chapter 1: Basics of corrosion chemistry. In S. K. Sharma (Ed.), *Green corrosion chemistry and engineering: Opportunities and challenges*. New York: Wiley.
9. Buff, H. (1857). Ueber das electrische Verhalten des Aluminiums. *Annalen der Chemie und Pharmacie, 102*(3), 265–284.
10. Ducretet, E. (1875). Note sur un rhéotome liquide a direction constante, fondé sur une propriété nouvelle de l'aluminium. *Électrodes D'Aluminium, 4*(1), 84–85.
11. MacDonald, D. A new phase space interpretation of the formation of anodic oxide films.
12. Polmear, I. (2006). *Light alloys, from traditional alloys to nanocrystals*. Oxford, UK: Elsevier.
13. Ellard, B. (2010). Metallurgy basics for aluminum surfaces. *AAC Anodizing Essentials Workshop*.
14. Schey, J. A. (1977). *Introduction to manufacturing processes*. New York: McGraw-Hill.
15. Altenpohl, D. (1998). *Aluminum: Technology, application, and environment—A profile of a modern metal* (6th ed.). Washington, DC: The Aluminum Association.
16. Birbilis, N., & Buchheit, R. (2005). Electrochemical characteristics of intermetallic phases in aluminum alloys and experimental survey and discussion. *Journal of the Electrochemical Society, 152*(4), B140–B151.
17. Taylor, J. A. (2012). Iron-containing intermetallic phases in Al-Si based casting alloys. *Procedia Materials Science, 1*, 19–33.
18. Runge, J. Understanding the interface: Meeting the challenges of anodizing cast substrates. In: *Proceedings of the 2016 AAC Annual Technical Conference and Exposition*, Montreal, Canada.
19. Chesterfield, L., & Runge, J. (2010). Connecting theory to practice, the science of successfully anodizing aluminum die castings. In: *Proceedings of the 19th Annual Technical Conference and Exposition of the Aluminum Anodizers Council*, Montreal.
20. Runge, J., Gilbert, A., Kriesch, G., & Pernick, J. (2006). Hybrid nanostructure of the anodic oxide for polymer bonding: A case study. In: *Proceedings oft he AAC Annual Conference and Exposition*, Toronto.
21. Davis, J. R. (Ed.). (2001). *Alloying, understanding the basics*. Materials Park, OH: ASM International.
22. American national standard alloy and temper designation systems for aluminum. *ANSI H35.1/H35.1 (M)-2017*, Secretariat, The Aluminum Association, Inc., American National Standards Institute, May 12, 2017, pp. 7–8.
23. Runge, J. M., & Hossain, T. (2015). Interfacial phenomena in 7000 series alloys and their impact on the anodic oxide. In: *Proceedings of Aluminium Two Thousand World Congress and International Conference on Extrusion and Benchmark ICEB 2015*. Elsevier Science Direct.
24. Runge, J. M. (2008). Interfacial phenomena and anodizing: Ramifications and process solutions. In: *Proceedings of the 17th Annual AAC Conference*, San Francisco, CA.
25. Mondolfo, L. F. (1976). *Aluminum alloys, structure & properties*. London: Butterworth & Co.

26. Sanders, R. E., Hollinshead, P. A., & Simielli, E. A. (2004). Industrial development of non-heat treatable aluminum alloys. *Materials Forum* (vol. 28).
27. Porter, D. A., & Easterling, K. E. (2004). *Phase transformations in metals and alloys*. Boca Raton, FL: CRC Press.
28. Sanders, R. E., Jr., Baumann, S. F., & Stumpf, H. C. (2000). *Aluminum alloys: Their physical and mechanical properties* (Vol. 3). TransTech Publications.
29. Searles, J., Buchheit, R., & Mills, M. (2000). *The electrochemical characteristics of Al_3Mg_2 and stress corrosion cracking of AA5083 (Al–4.5Mg–1.0Mn)*. The Ohio State University.
30. Kaufman, J. G. (2000). *Introduction to aluminum alloys and tempers*. Materials Park, OH: ASM International.
31. Asensio-Lozano, J., Suárez-Peña, B., & VanderVoort, G. (2014). Effect of processing steps on the mechanical properties and surface appearance of 6063 aluminium extruded products. *Materials*, pp. 4224–4242.
32. Runge, J. Base metal microstructural considerations for anodizing aluminum extrusions. In: *Proceedings of Eleventh International Aluminum Extrusion Technology Seminar, (ET 2016)*, Chicago, IL, May 2016. Extrusion Technology for Aluminum Profiles Foundation.
33. Lupis, C. H. P. (1983). *Chemical thermodynamics of materials*. New York: Elsevier.
34. Zhu, H., Dahle, A. K., Zhang, X., & Couper, M. J. (2008). Effect of extrusion microstructure on formation of streaking defects on the surface of anodized aluminum extrusions. In: *Proceedings of the 18th Annual Anodizing Conference and Expo of the Aluminum Anodizers Council*, San Francisco.
35. Runge, J. (2012). Trace elements and their impact on surface finishing characteristics of aluminum extrusions. In: *Proceedings of ET 2012*, Miami.

Chapter 5
Anodizing as a Corrosion Process

5.1 Introduction

The relationship between corrosion science and aluminum anodizing is obvious: the formation of an oxide layer represents a form of corrosion because the metal surface is consumed in its formation. In order to develop a model for the process of anodizing aluminum based upon corrosion science, a clear link needs to be established between the electrochemical reactions and the thermodynamics at the interface between the aluminum substrate and the electrolyte. Most anodizing research and handbook values that are relied upon to make this connection have been made under standard equilibrium conditions at standard temperature and pressure (STP), 25 °C (298 K) at 1 atm, and on finished anodic oxide structures. When considering the AAO as a finish or functional oxide, equilibrium conditions can be accepted. However, additional conditions must be taken into consideration as the anodic oxide grows and develops, which is a dynamic nonequilibrium process.

In this chapter, the various parts of the corrosion model are developed from first principles for unalloyed, isotropic aluminum, beginning with equilibrium processes, which show the oxidation potential for aluminum is definitely linked to the anodic dissolution of the substrate, producing an oxide on the substrate surface at the interface with the electrolyte. The formation of the passive oxide layer is presented as the basis for the formation of a coherent protective oxide under atmospheric conditions and explained using the thermodynamic concept of work function, which takes into account the electrochemical driving forces of the environment external to the anode and the chemical potential of the aluminum surface. Model development continues with nonequilibrium processes at STP, which accounts for the conditions that enable the formation of stable anodic aluminum oxide at a pH of less than 2. The effects of the anodizing process parameters on oxide formation: polarization, the imposition of an electrical bias, electrolyte concentration, temperature and current density are discussed with respect to reaction kinetics and resistance effects utilizing an

J.M. Runge, *The Metallurgy of Anodizing Aluminum*,
https://doi.org/10.1007/978-3-319-72177-4_5

expanded form of Ohm's Law and the concept of work function, which uses the parameters of the total anodizing circuit, known as the Tafel equation.

This chapter generally follows the treatment of corrosion as it relates to electrochemical circuits by Jones [1], who organized general corrosion principles according to Stern and Geary [2] and later by Fontana and Greene [3], developing a sound pedagogical foundation from which clear connections between science and practice can be drawn. The connections drawn between corrosion and the anodizing process throughout the chapter are the author's, and are therefore clearly based, offering many explanations for how a stable anodic aluminum oxide structure nucleates, grows, and develops. An account of the unique structure and properties of the anodic oxide is provided in Chap. 6.

5.2 Equilibrium Corrosion Processes: Aluminum Oxidation

Almost all metallic corrosion processes involve transfer of electronic charge in aqueous solutions. So it is with anodizing aluminum in aqueous electrolytes using any combination of process parameters. In industry, an aqueous electrolyte of a strong protonic acid is typically utilized. However, the electrolyte anion is not consumed by the anodizing reaction; in fact, electrolytes can be reused continuously if the dissolved aluminum is removed as it builds up in solution over time. Furthermore, analysis of AAO shows electrolyte is incorporated as a contaminant within the oxide structure only where it is in contact with the electrolyte. For example, XPS analysis of AAO grown in sulfuric acid electrolytes determined no sulfate ions at the interface with the aluminum substrate [4]. Therefore, the basic electro-active component of the aluminum oxidation reaction in the treatment of anodic oxidation as a corrosion process is water. Nevertheless, the electrolyte concentration and the type of anion(s) in its formulation have an impact on the structure and properties of the anodic oxide and in how the anodizing circuit is completed; the role of the electrolyte must be considered separately from the basic anodizing reaction.

The most fundamental anodizing reaction for aluminum oxidation is presented as Eq. 5.1.

$$2\text{Al} + 3\text{H}_2\text{O} \rightarrow \text{Al}_2\text{O}_3 + 3\text{H}_2 \tag{5.1}$$

In ionic form, the reaction is:

$$2\text{Al} + 6\text{H}^+ + 3\text{O}^{2-} \rightarrow 2\text{Al}^{3+} + 3\text{O}^{2-} + 3\text{H}_2 \tag{5.1a}$$

Eliminating O^{2-} from both sides of the equation gives:

$$2\text{Al} + 6\text{H}^+ \rightarrow 2\text{Al}^{3+} + 3\text{H}_2 \tag{5.2}$$

As written, the oxidation reaction for aluminum could occur in any protonic acid, and for this reason, water can confidently be considered the electro-active component in the corrosion reaction. Reaction 5.2 can be separated into *half-cell reactions* as follows:

$$\mathrm{Al} \rightarrow \mathrm{Al}^{3+} + 3e^{-} \quad \text{anodic reaction} \tag{5.3}$$

$$6\mathrm{H}^{+} + 6e^{-} \rightarrow 3\mathrm{H}_2 \quad \text{cathodic reaction} \tag{5.4}$$

Reaction 5.3, defined as the anodic reaction, is an oxidation reaction in which the aluminum valence increases from 0 to +3, liberating electrons, e^-; while reaction 5.4, defined as the cathodic reaction, is a reduction reaction in which the oxidation state of hydrogen decreases to zero, consuming electrons.

The separation of Eq. 5.2 to give Eqs. 5.3 and 5.4 with the exchange of electrons, e^-, cannot be considered an electrochemical reaction unless there is some evidence that electrons are actually involved. In actual practice, when an excess of electrons is supplied to the aluminum substrate (making the aluminum negative, a cathode), it is always observed that the rate of corrosion, expressed by the anodic reaction 5.3, is decreased while the rate of hydrogen evolution is increased, represented by the cathodic reaction 5.4. To contrast, all aqueous corrosion reactions involve an anodic reaction such as 5.3, and application of a positive potential (as in anodizing) creates an electron deficiency in the metal, which accelerates anodic dissolution. Therefore, the role of electrons in corrosion reactions is proven, and corrosion reactions, including anodizing, can be classified electrochemical in nature.

5.2.1 Thermodynamics of Equilibrium Oxidation Processes

Having proven that aluminum oxidation is an electrochemical process, which involves the dissolution of the aluminum anode, Eq. 5.3, and the development of hydrogen gas, Eq. 5.4, the energy produced by each half-cell reaction, which is connected to the electrochemical potential of the substrate, can be accounted for by the fundamental thermodynamic relationship for Gibbs free energy:

$$\Delta G = -nFE \tag{5.5}$$

where n is the number of electrons (concentration equivalent) exchanged in the reaction and F is Faraday's constant, 96,500 coulombs per equivalent. The *physical significance* of such a relationship is that with thermodynamic accounting of the energy for an electrochemical reaction, we have the fundamental relationship in which a charge, nF, taken reversibly at equilibrium through a potential, E, corresponds to a change in the condition of the substrate. This is significant for anodizing, during which the aluminum metal dissolves to its ionic state, with a corresponding change in energy, ΔG. To complete the reaction, we have the energy

Table 5.1 Standard electromotive force potentials (reduction potentials) for aluminum and common aluminum alloying elements

	Reaction	Standard potential, e^o (volts vs. SHE)
Noble	$O_2 + 4H^+ + 4e^- = 2H_2O$ (pH = 0)	+1.229
	$O_2 + 2H_2O + 4e^- = 4OH^-$	+0.820
	$Ag^+ + e^- = Ag$	+0.799
	$Fe^{3+} + e^- = Fe^{2+}$	+0.771
	$Cu^{2+} + 2e^- = Cu$	+0.342
	$2H^+ + 2e^- = H_2$	0.000
	$Pb^{2+} + 2e^- = Pb$	−0.126
	$Fe^{2+} + 2e^- = Fe$	−0.447
	$Cr^{3+} + 3e^- = Cr$	−0.744
	$Zn^{2+} + 2e^- = Zn$	−0.828
	$AlO_2^- + 4H^+ + 3e^- = Al + 2H_2O$	−1.262 [3]
	$Al_2O_3 + 6H^+ + 6e^- = 2Al + 3H_2O$	−1.550 [3]
	$Al^{3+} + 3e^- = Al$	−1.662
Active	$Mg^{2+} + 2e^- = Mg$	−2.372

of reaction from the hydrolysis of water, forming hydroxide ions such that they can react with aluminum ions to form anodic alumina (AAO) and water. In all cases, the energy required for the reaction is a sum of the energy of the reaction external to the substrate, within the electrolyte, and the energy of the reaction at the substrate surface.

$$2H_2O + 2e^- \rightarrow H_2 + 2OH^- \tag{5.6}$$

$$Al^{3+} + 3OH^- \rightarrow AlOOH + H_2O \tag{5.7}$$

Each half-cell reaction has a characteristic energy. The energy associated with common half-cell reactions, such as 5.3 and 5.4, is tabulated in a listing of standard half-cell electrode potentials that comprises the electromotive force (emf) Series. A partial list of half-cell reactions characteristic for aluminum and many of elements utilized as constituents for aluminum alloys is presented as Table 5.1. The elements at the top of the list are more "noble" than those near the bottom, which are more "active." The reactions are presented as reduction, rather than oxidation, reactions and the more positive values are for more noble elements and the more negative are for active elements. When one regards the activity of aluminum in the context of Table 5.1, it is clear that aluminum oxidizes readily, and it therefore forms a characteristic passive oxide under equilibrium conditions and STP. It is also noteworthy that of all common alloying elements for aluminum, which are part of the table below, magnesium is the only element more active than aluminum. The effects of the various characteristic oxidation potentials impact oxide formation when anodizing aluminum alloys, and will be discussed in following chapters.

5.2.2 Aluminum Oxidation Under Atmospheric Conditions: The Passive Layer

In general, whenever a metal surface undergoes chemisorption, in other words, reacts with the environment to form a product, corrosion occurs. When aluminum is exposed to the atmosphere, it corrodes preferentially as oxide nuclei form at sites on the surface where the energy required for adsorption is lowest. They grow and develop as long as there is available free aluminum to oxidize, and the thermodynamic work function of the metal is such that the oxidation reaction is favored [5]. The process continues until a coherent oxide layer or passive film is formed. The formation of a passive oxide on aluminum is a characteristic of the element and prevents further attack of the aluminum by other ionic species at STP by excluding them from the oxide structure.

Work function for the formation of the passive oxide layer, E_w, can be defined as the difference in energy between the electrostatic potential outside the forming oxide layer, −eV, and the electrochemical potential of the anode, μ [5].

$$E_w = -\mathrm{eV} - \mu \tag{5.8}$$

The activity, −eV, of the external electrostatic potential (the electrolyte) will vary with the type of atmosphere and amount of contamination; however, since the actual electro-active component of oxidation at the aluminum surface is water, the anodizing reaction product will always be aluminum oxide. For an equilibrium reaction, the rate at the interface is governed by the amount of available aluminum ions; as long as E_w is not zero, the presence of Al^{3+} ions is assured and there is a potential difference between the activity of the electrolyte and the average value for the standard chemical potential for aluminum from Table 5.1, $\mu = -1.66$ V. Under these conditions, an oxidation reaction occurs that is equilibrium in nature. Regardless of the total substrate microstructure that governs the average oxidation reaction rate for the bulk of the anode, β, the chemical potential, μ, varies locally across the surface, caused by variations in surface energy due to grains with differing crystallographic orientations and atomic level point defects such as dislocations, steps, and kinks. These slight deviations in μ cause changes in the local oxidation kinetics and therefore, local variations in passive layer thickness. See Fig. 5.1.

Understanding work function as a thermodynamic relationship is important to understanding why corrosion can occur in nature by way of chemical potential as the driving force, without an external electrical stimulus; and how by enhancing the driving force in the presence of an electrical stimulus, corrosion can be accelerated. In atmospheric corrosion, as long as there is a driving force, oxidation (corrosion) proceeds on the substrate surface and oxide growth occurs as the individual nuclei consume the aluminum and other oxide nuclei, reconstructing the surface as an oxide monolayer.

In atmospheric oxidation, surface reconstruction proceeds extremely rapidly on clean aluminum as the formation of a stable native oxide or passive film proceeds, developing a thickness of several molecular layers. The natural oxide layer, formed under atmospheric influence, always contains a certain amount of aluminum oxyhydroxide

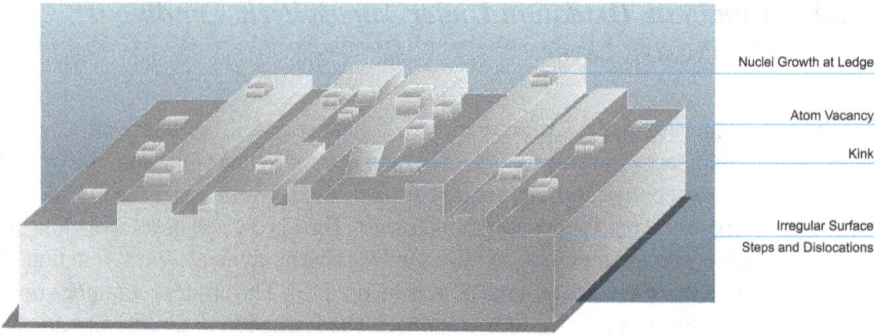

Fig. 5.1 Schematic of atomic level defects according to Murr [5]. Under equilibrium conditions, anodic charge is naturally concentrated at surface asperities, accelerating corrosion at these preferred sites across the substrate

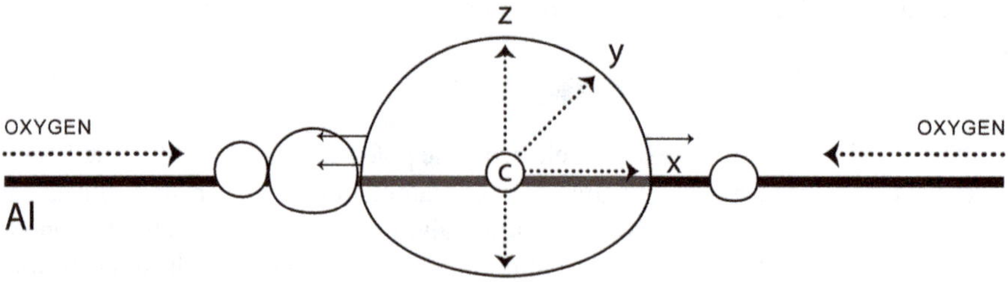

Fig. 5.2 Individual oxide particles or nodes, represented by the hemisphere originating at point c at the center of the schematic, initially form by way of diffusion and reaction of atmospheric oxygen with the base metal. The nodes grow laterally (in the *x* direction), outward from the substrate (*y*, *z* directions), and into the aluminum (negative z direction) by way of the potential difference between the oxide and the base metal, and by way of the potential difference between large nodes and small nodes. Lateral growth dominates the oxidation reaction until the aluminum surface is totally consumed, restructured with oxide

AlO(OH) and water bonded physically or by chemisorption. The initial reaction is preferential, at energetically favorable reaction sites, with chemisorption first occurring at certain surface structures, crystallographic orientations, and other defects. For the oxide layer to grow, diffusion of either the oxygen from the atmosphere or the metal through the oxide layer, or both, is required. The nuclei mature and increase in size (ripen) to form stable oxide nodes that grow due to the chemical potential difference between the oxide and the base metal; the larger nodes grow laterally and perpendicularly at the expense of the smaller nodes to form a layer. The resultant layer is very thin, and hence sensitive to the environment in which it forms [6]. See Figs. 5.2 and 5.3.

Under atmospheric conditions, the oxidation reaction ceases as the monolayer exhibits continuity over the aluminum surface, which means the electrostatic potential outside of the forming oxide layer, −eV, is equivalent to the chemical potential of the surface, μ. In other words, the oxidation reaction ends when the work function for the aluminum, E_w, has reached equilibrium (equals zero). Once a stable passive oxide forms, it may be transformed by hydration to hydroxide or hydrate compounds in

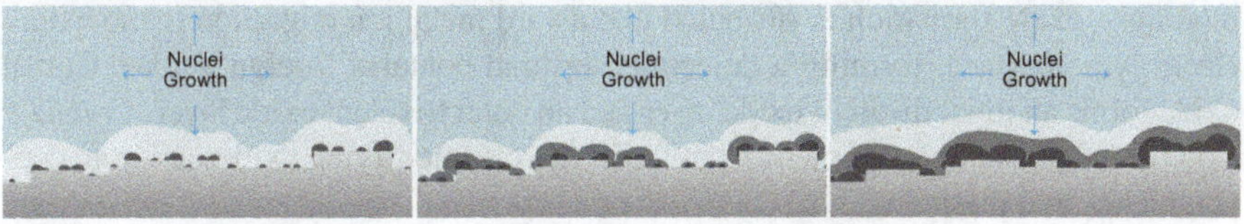

Fig. 5.3 Schematic for preferential nucleation and progressive formation of the passive oxide layer on aluminum. The initial reaction is preferential, at energetically favorable reaction sites, with chemisorption first occurring at certain surface structures such as sharp corners, ledges, or burrs. Along with energetically favorable surface defects, crystallographic orientations most densely packed with aluminum atoms are also favored for preferential oxidation

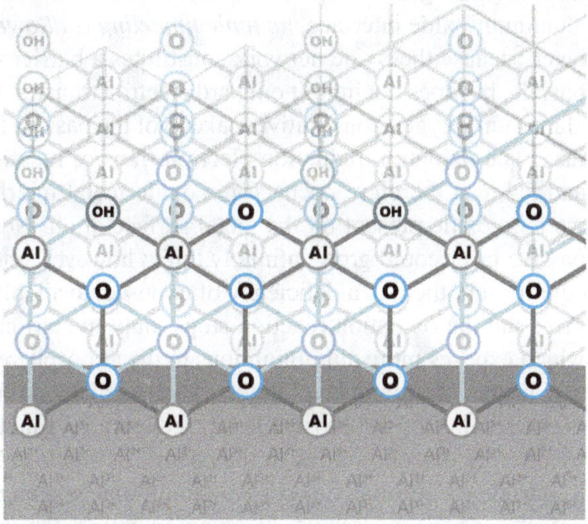

Fig. 5.4 Schematic for the structure of the passive oxide layer on aluminum as AlO_{4-} tetrahedral structures forming a continuous chainlike coupling over the surface

chainlike coupling, forming a reactive structure between the atmosphere and the base metal. However, because the passive film is extremely thin, on the order of 1–10 nm, electrons easily transfer across the film, irrespective of whether the film is an insulator or a semiconductor. The passive layer provides corrosion resistance to the underlying aluminum, until it is scratched or otherwise disturbed; however, if the atmospheric conditions do not change, oxidation begins again, a characteristic referred to as "self-healing." See Fig. 5.4.

Sato [7] classifies passive oxide films into two categories: the network *former* and the network *modifier*. Aluminum falls into the category of a network former, meaning it forms a layer comprising a single oxide type (Al_2O_3). According to Sato, the anodic formation of network-forming passive oxides occurs due to the inward migration of oxide ions to the metal-oxide interface, producing a compact layer that contains no foreign anions other than oxide ions. Other examples of metals that form network-forming oxides are titanium, zirconium, silicon, and tantalum. To contrast, the growth of passive layers on metals that are classified as network modifiers, which form a multilayered oxide film of progressive stoichiometry, such as cobalt. For network modifiers, oxide formation is attributed to outward metal ion migration to the oxide-electrolyte interface, forming a defective film that contains foreign anions. Cobalt oxide forms an inner divalent oxide layer and an outer trivalent oxide layer ($Co/CoO/Co_2O_3$). Other examples of metals that form network-modifying oxides are iron, copper, nickel, and platinum [7].

The concept of the formation of a progressive networking-forming oxide layer is important to understanding some of the key characteristics and properties of the anodic aluminum oxide and how the anodic oxide nucleates and grows. The structure of the passive film is characterized as amorphous. Nevertheless, as the network of preferential oxide nuclei particles forms, a layer develops that itself has a chemical structure. This structure provides the path for ionic conduction and continued oxide growth. As the oxide structure develops, forming a layer at the aluminum-oxide interface, *no ionic tunneling is allowed to occur* across the passive layer because the oxide network constitutes a barrier layer to ion transfer from the outside, but does not inhibit outward electron transfer, which maintains an exclusive aluminum-oxygen constitutive makeup of the passive layer at the metal-oxide interface. The oxidation-reduction-electron transfer reactions, represented by Eqs. 5.3 and 5.4, can therefore occur at the oxide-metal interface continuously, as if there were no oxide present on the metal surface at all. Normally, this would imply the passive layer could grow infinitely thick; however, once the work function reaches equilibrium, there is a deficiency of exposed, available aluminum ions at the interface, and the oxidation reaction stops, only to start again if the layer is damaged, which exposes more aluminum ions to the atmosphere for the oxidation reaction.

Clusters of atomic level defects such as dislocations or vacancies can interrupt the continuity of the passive layer, reducing its function as a barrier to corrosion by offering discontinuities in the chemical structure that allow for foreign anions to intrude the normally exclusive aluminum-oxygen passive layer structure. Such defects can be overcome by supplementing the passive layer structure with protective anions that adsorb on the surface of the reactive passive layer and bridge point defects. It is on this fundamental structural characteristic of the passive oxide layer on aluminum that chemical conversion coatings have been developed to enhance the passive layer, which are commonly used in industry (see Chap. 3).

5.3 Thermodynamics of Nonequilibrium Corrosion Processes

Thermodynamics can be used to determine the energy changes in terms of the electrochemical potential when the oxidation reaction is at equilibrium. However, most real world situations in which alloys, rather than pure metals, manufactured and processed in a variety of ways for a variety of applications present non-ideal substrates to the environment, which means most corrosion reactions deviate from the standard state. The standard state requires all reactants and products to be at the same concentration (unit activity). This never happens in industrial anodizing; therefore, the next step in understanding the electrochemical cell during aluminum anodizing is to address the reactions that occur in a nonstandard state, in which the reactants and therefore the products vary in concentration. Simultaneous and competing reactions occur which change the energy of the reaction. Energy changes provide the driving or confounding force that determines whether or not there will be a chemical reaction. Thermodynamics therefore can also be used to predict the effects of deviation from the standard state and show how conditions in an electrochemical cell can be adjusted to regulate corrosion, which is the growth of the anodic oxide.

To link thermodynamics to the anodizing process, the oxidation of the aluminum substrate must be regarded as nonstandard, because the aluminum substrate, even unalloyed, is imperfect and poses many deviations from the standard state. Anodizing aluminum alloys complicates the aluminum oxidation reaction further as alloying elements other than magnesium will not oxidize under typical anodizing process parameters, due to the fact that their oxidation potential is higher than that for aluminum. However, other corrosion reactions with alloying elements compete with the oxidation of aluminum and occur simultaneously over the anode surface, presenting deviations in interfacial resistance to the anodizing reaction. The basic chemical reactants and products for such reactions deviate from unit activity (exhibit different concentrations), and therefore, deviate from the standard state.

In order to consider these reactions in general, for an aqueous reaction of reactant A to yield product B (5.9), the Nernst Equation 5.10 can be used to calculate half-cell potentials that deviate from standard state.

$$a\mathrm{A} + m\mathrm{H}^{+} + ne^{-} = b\mathrm{B} + d\mathrm{H_2O} \tag{5.9}$$

$$E = E^{\circ} + \frac{2.3RT}{nF}\log\frac{(\mathrm{A})^{a}(\mathrm{H}^{+})^{m}}{(\mathrm{B})^{b}(\mathrm{H_2O})^{d}} \tag{5.10}$$

The Nernst Equation expands on the concept of work function by relating the potential due to free energy changes in the standard reaction potential, E°, to the nonstandard, E, and shows that the half-cell electrode potential becomes more positive (noble) as the activities (concentrations) of oxidized species (A) and (H^+) increase, which coincides with what occurs in anodizing practice. It is this relationship that enables the construction of potential-pH diagrams also known as Pourbaix diagrams. A complete derivation of the Nernst equation is available in most corrosion text books and should be reviewed if the reader is interested [1].

To simplify Eq. 5.9 such that electrolyte pH becomes a working variable in calculating the potential for the corrosion reaction, a fixed value for $2.3RT/F$ is calculated as 0.059 V at 25 °C (standard temperature), when the appropriate values for the other constants are substituted. The activity of water in aqueous solutions = 1, and to relate the activities of the reactant A and the product B to the activity of the oxidizer (H^+), it is important to remember that $\mathrm{pH} = -\log(\mathrm{H}^+)$.

$$E = E^{\circ} + \frac{0.059}{n}\log\frac{(\mathrm{A})^{a}}{(\mathrm{B})^{b}} - \frac{m}{n}0.059\ \mathrm{pH} \tag{5.11}$$

Significant trends in the importance of H^+ as an oxidizer, which is the same as electrolyte concentration, are elucidated with the Nernst equation in this form. The activities (concentrations) for the reactant, A, and the product, B, can be equal, or they can be absent from any half-cell reaction. If the oxidizer alone is considered,

$$2\mathrm{H}^{+} + 2e^{-} = \mathrm{H}_2 \tag{5.4}$$

The Nernst equation becomes, assuming that the activity of H_2 is in its standard state (=1 atm),

$$E_{\mathrm{H}^{+}/\mathrm{H}_2} = E^{\mathrm{o}}{}_{\mathrm{H}^{+}/\mathrm{H}_2} - 0.059\ \mathrm{pH} \tag{5.12}$$

This basic treatment shows that the activity of the oxidizer, H^+, increases as the acid concentration of the electrolyte increases; consequently, pH decreases, and the potential, $E_{\mathrm{H}^{+}/\mathrm{H}_2}$, increases. Therefore, potential is commonly considered to be a measure of the oxidizing power of the solution. This is true in anodizing practice, when considering the concentration of acid electrolytes, as the activity of any dissolved oxidizer increases (as concentration increases), more positive potentials are measured experimentally, which can hinder or help when anodizing complex aluminum alloys.

5.3.1 *The Potential/pH (Pourbaix) Diagram for Aluminum Under Atmospheric Conditions*

Passive oxide layer stability is governed by the atmospheric conditions that comprise the environment (the electrolyte) in which it forms, as well as the surface microstructure of the aluminum. The potential-pH diagram (Pourbaix diagram), calculated by using the Nernst Equation, can be considered a thermodynamic map showing conditions of solution oxidizing power (potential) and acid or alkaline pH for the various possible phases that are stable in an aqueous electrochemical system, such as anodizing. All possible equilibria involving the thermodynamic system are considered in constructing a diagram. A Pourbaix diagram does not take reaction kinetics into account; however, it does show reactions and reaction products that will be present when equilibrium is reached, providing all appropriate reactions have been included. Boundary lines on the diagram dividing areas of stability for different phases can be calculated and plotted, mapping out the conditions under which a metal: (1) will corrode, (2) will be immune to corrosion, or (3) will exhibit passive behavior. Diagrams are typically calculated under the conditions of standard temperature 25 °C and pressure 1 atm.

It is interesting to analyze the stability of aluminum oxide as a passive layer, and as an anodic oxide, under standard and nonstandard conditions. There are three general equations by which aluminum may react anodically in the presence of water.

By way of oxidation to aqueous cations:

$$\mathrm{Al} = \mathrm{Al}^{3+} + 3e^{-} \quad e^{\circ} = -1.66\ \mathrm{V} \tag{5.3}$$

To aluminum hydroxide or oxide:

$$2\mathrm{Al} + 3\mathrm{H_2O} = \mathrm{Al_2O_3} + 6\mathrm{H}^{+} + 6e^{-} \quad e^{\circ} = -1.55\ \mathrm{V} \tag{5.13}$$

And to aqueous anions:

$$\mathrm{Al} + 2\mathrm{H_2O} = \mathrm{AlO_2}^{-} + 4\mathrm{H}^{+} + 3e^{-} \quad e^{\circ} = -1.26\ \mathrm{V} \tag{5.14}$$

The basic Pourbaix diagram for aluminum, under standard conditions, utilizes these three equilibrium reactions to show that Al^{3+} is stable at a low pH; and that the aluminate ion, AlO_2^-, is stable at a high pH, which means that aluminum will dissolve in water, in other words, corrode to form these species, over specific potential ranges.

The diagram is calculated using a basis of an activity of 1 for the concentration of Al^{3+}, at the standard oxidation potential of $e_{\mathrm{Al/Al^{3+}}} = -1.66\ \mathrm{V}$ at all pH. Under typical atmospheric (equilibrium) conditions, which form a passive aluminum oxide film, this condition represents an unrealistically high concentration of Al^{3+} in solution causing corrosion; therefore, a lower value for aluminum ion concentration 10^{-6} is taken, giving an even lower value for the standard oxidation potential $e_{\mathrm{Al/Al^{3+}}} = -1.78\ \mathrm{V}$ at all pH. This result shows that the tendency for the aluminum substrate to form a passive film under atmospheric conditions is even greater than it is to anodize. This is because anodic polarization produced by the electrical bias during anodizing enhances the tendency for anodic dissolution, producing more Al^{3+} than what is produced under atmospheric conditions. Both conditions are presented on the diagram, showing equilibrium for reaction (3) at unit activity and at atmospheric activity = 10^{-6}. See Fig. 5.5.

Under equilibrium conditions, the Pourbaix diagram shows that Al_2O_3 is stable over an intermediate pH range of ~4 to ~8. This immediately begs the question of how a stable oxide can be formed in electrolytes with pH lower than 4. This demonstrates clearly the nonstandard nature of the anodizing process. It is important to realize that the diagram constructed under standard atmospheric conditions presents the range of potential and pH for stable aluminum oxide already formed, and gives insight as to the environmental conditions under which it will remain stable. Therefore, the usefulness of the diagram is not only for passive layer oxides, it is useful as a map that shows the general stability of functional anodic oxides when placed in their work environments.

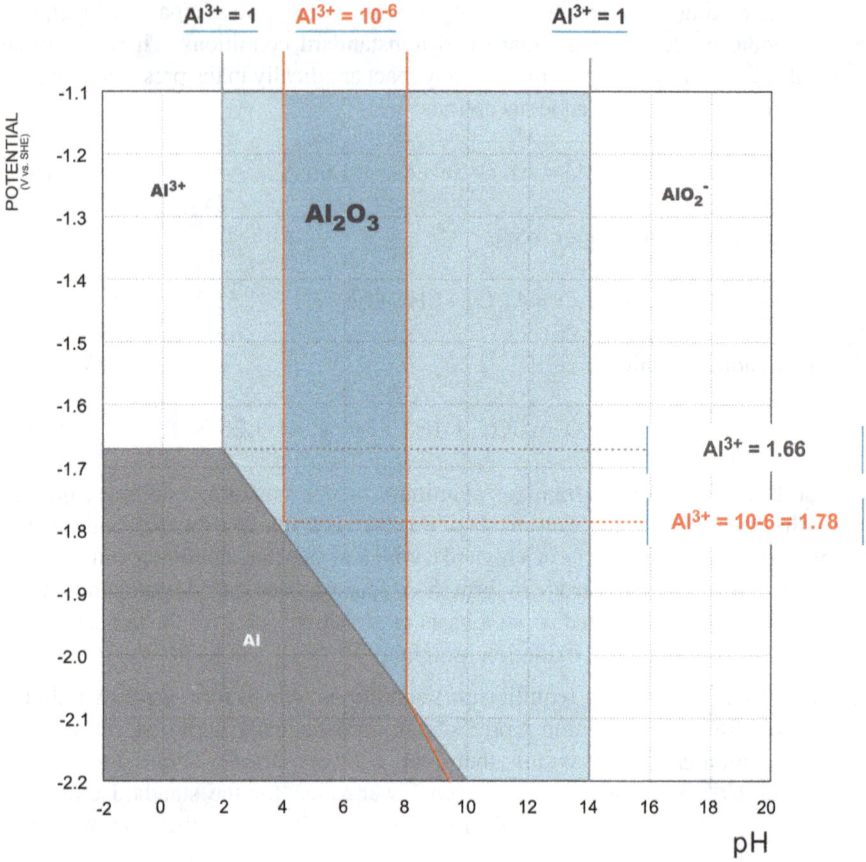

Fig. 5.5 The potential/pH (Pourbaix) diagram for aluminum. The stability of aluminum oxide is shown to range between ~4 < pH < ~8

5.3.2 *The Potential-pH Diagram for Aluminum Under Dynamic Conditions*

Without the imposition of an electrical bias, corrosion is controlled by the diffusion of dissolved oxygen from the atmosphere to the aluminum surface, where it will react very rapidly with available Al^{3+} ions. With the imposition of a current bias oriented such that the aluminum is the anode, the charge at the surface becomes ordered (polarized) and the flow of aluminum ions to the surface is dramatically increased. Corrosion then becomes controlled by the growth of the anodic oxide. Therefore, the success of the anodizing process depends upon the initial formation of a stable oxide under dynamic conditions.

The goal of the anodizing process is to produce a stable oxide finish suitable for the rigors of application, yet the potential and the pH of the anodizing process are such that, according to the Pourbaix diagram constructed under standard temperature and pressure, presented in Fig. 5.5, a stable oxide cannot be grown in industrial electrolytes that have a pH much less than 4 (a typical anodizing electrolyte will have a pH $\ll -2$!). It in fact shows that aluminum oxide is stable over a pH range of ~4 to ~8, which contradicts what is accomplished daily in industry, and also clearly shows that anodizing is a nonstandard, nonequilibrium process. Because it is obvious that one can produce a stable oxide in an industrial electrolyte, the question of how stable anodic oxide can be developed in highly acidic electrolytes immediately comes to mind and the validity for using a potential-pH diagram in anodizing is also brought into question.

In an electrolyte with a pH that is $\ll 2$, under equilibrium conditions and without polarization, the slightest driving force produces aluminum corrosion at very high rates, especially at an interface with surface defects and alloy additions. The interfacial resistance of the oxidation reaction increases the heat of reaction substantially. As the temperature due to the heat of reaction increases, the amount of base metal dissolution increases spontaneously, producing corrosion at preferential sites across the anode surface. The corresponding cathodic reaction produces hydrogen ions, which serves to decrease the pH at the interface even more, enabling the reaction to continue. No oxide can be sustained under such conditions; uncontrolled, such reactions can produce damaging effects. In industrial practice, these conditions are typical for pretreatment prior to anodizing, in order to remove any native oxide or reaction products that may inhibit the formation of a uniform, continuous anodic oxide, or to chemically brighten the surface. Other pretreatment methods that use alkaline formulations, or more mildly acidic solutions, are also used in the anodizing industry; as equilibrium processes, if not carefully controlled, can produce damage to the anode surface prior to anodizing.

In industrial anodizing, the pH of the anodizing electrolyte is much less than 2, which means consideration must be given to other anodizing process parameters in order to understand how to produce a stable oxide without damaging the anode, under nonequilibrium, dynamic conditions. Foremost, the imposition of the applied current density must be considered because it is the unique parameter by which the oxidation reaction becomes nonstandard, as well as the driver for the potential response *E*. The other parameters of the Nernst equation are also important to consider: temperature, as the heat of oxidation during anodizing, as well as polarization, *N*; each parameter impacts the stability boundaries for anodic aluminum oxide (AAO) formed as the reaction proceeds.

By considering how the anodizing process produces interfacial conditions that deviate from the standard conditions used to construct the Pourbaix diagram for aluminum, another potential-pH diagram can be constructed that expands the pH range over which stable oxide can form. In order to construct a Pourbaix diagram that is more germane to anodizing, it is important to consider that in creating an electrical circuit for anodizing aluminum, the reaction will be driven by an amount of Al^{3+} ions that exceeds what is typical for atmospheric corrosion, therefore, it is reasonable to consider the standard oxidation potential of -1.66 V at all values of pH for the corrosion reaction. This removes the stability line for Al_2O_3 at an Al^{3+} concentration of 10^{-6}, which is calculated to represent atmospheric corrosion, and

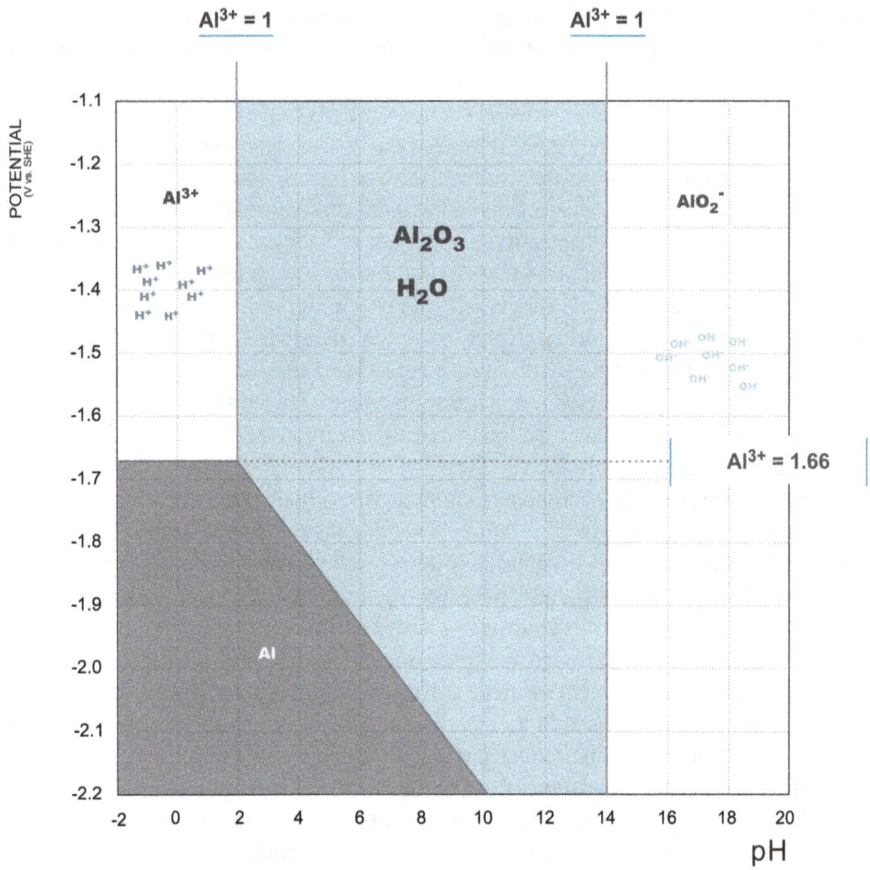

Fig. 5.6 Modified Pourbaix diagram, stability boundaries are placed to reflect the oxidation potential as −1.66 V for all pH. This reflects better conditions for anodizing, in which an infinite amount of aluminum ions is provided to the anode surface by the applied current density. Under nonequilibrium conditions, that raise the pH range for the stability for aluminum oxide is increased to ~2 < pH < ~14

reduces the pH for stable oxide formation from a pH = 4 to a pH = 2; and raises the pH for stable oxide formation from a pH = ~8 to a pH = ~14. See Fig. 5.6.

5.3.3 Formation of Stable Aluminum Oxide Under Dynamic Conditions: Polarization Effects

By considering the anodizing process parameters with the Nernst equation, the Pourbaix diagram can be adjusted to expand the region over which aluminum oxide is stable, which is helpful when considering the effects of the parameters for chemical operations that follow anodizing. Nevertheless, with the adjustments to the Pourbaix diagram made possible by the applied current density, which changes the oxidation reaction from an equilibrium reaction to a nonequilibrium reaction, even though the stability range is expanded, it is still not clear how stable anodic aluminum oxides can be formed under anodizing conditions. With the introduction of an applied current density to the anode, which in turn has a complementary potential response, perhaps the most important effect occurs regarding the moderation of the dynamic nature of the anodizing reaction and the formation of stable anodic oxides: anodic polarization of the substrate surface.

Polarization provides an ordered charge to the anode surface that overrides many of the interfacial effects presented by the anode surface. The applied current density pushes aluminum ions to the surface, producing ordered nodes of Al^{3+} ion concentration across the polarized substrate surface. The electrons produced from the anode as the aluminum is dissolved, produce electrolysis at the interface with the substrate; hydrogen and hydroxide ions are segregated at the surface, and a complementary pH wave forms in the electrolyte above the surface. Because of these conditions, the pH over the polarized surface varies locally, according to the charge density set by polarization. At areas of aluminum ion concentration on the anode surface, due to surface polarization and electrolysis of the aqueous electrolyte, the adjacent solution pH is increased due to the attraction of OH^- ions, which raise the pH local to the nodes of aluminum ions to accommodate the formation of stable AAO. Areas of lower pH correspond to the areas between the areas of polarized concentration of aluminum ions, where hydrogen ions are formed by way of electrolysis. Polarization of the surface therefore provides the conditions for forming an ordered, stable oxide under nonequilibrium conditions. Under these conditions, it is assumed that all aqueous electrolytes will be electrolyzed, and that all OH^- will be consumed in the oxidation reaction to form aluminum oxide, and that H^+ will travel to the cathodes to complete the electrochemical circuit. Any aqueous species that may remain in the oxide will have been re-adsorbed on the oxide surface at the oxide-electrolyte interface. These effects will be discussed in more detail in Chap. 6 (see Fig. 5.7).

5.4 Aluminum Oxidation Under Dynamic Conditions: Anodizing

The fundamental link between the passive layer on aluminum and the anodic aluminum oxide is the formation and growth of an oxide by way of network formation. The oxide network formed by way of passivation has an exclusive aluminum-oxygen structure that is a barrier to the movement or tunneling of other ions from the outside, inward toward the surface, which makes the passive layer a functional barrier to corrosion by other anions. However, because the passive layer is grown by virtue of the work function between the environment and the aluminum substrate under equilibrium conditions, it remains thin and subject to mechanical damage.

The oxide network formed by way of anodizing exhibits the same exclusive aluminum-oxygen composition as the passive layer. The obvious difference between

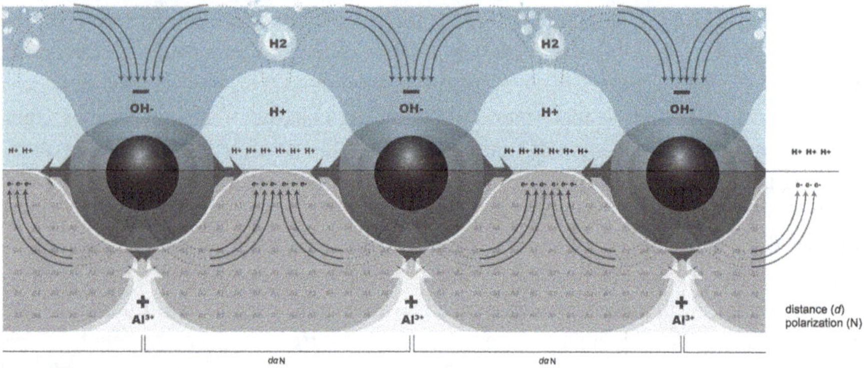

Fig. 5.7 Schematic showing the effects of polarization on the substrate and initial oxidation. Electrolysis of the aqueous electrolyte at the interface with the polarized, corroding anode produces separation of the hydrogen (H^+) and hydroxide ions (OH^-), producing alternating charge concentration adjacent to the surface. The pH varies locally according to the concentration of ionic charge, with concentrations of hydroxide ions producing a higher pH adjacent to concentrations of aluminum ions, enabling the formation of stable oxide in an acid environment that would normally not support such a reaction

the ambient corrosion that forms the passive layer and the process to form the anodic aluminum oxide is the change in work function by the imposition of an electrical bias from an external power source, making the formation of an anodic oxide *a controllable, engineering process*. This distinct difference enables control of the current flow within the substrate and polarizes the substrate surface, therefore increasing the corrosion rate by stimulating the movement of Al^{3+} ions to the substrate surface. Anodic polarization orders the charge at the substrate surface enabling the development of a self-assembled, ordered oxide at the interface with the electrolyte.

The interface created between the substrate surface and the electrolyte comprises the most important combination of variables that govern the structure and properties of the anodic oxide finish. Oxidation is controlled by (1) thermodynamics and (2) diffusion. The thermodynamic conditions which comprise the interface: the electrolyte and the surface, must be considered when modeling oxidation. Diffusion occurs on both sides of the interface. In the electrolyte, diffusion occurs as the components of electrolyzed water (H^+ and OH^-) react with electrons pushed from the anode to form hydrogen gas, the cathodic reaction, Eq. 5.4; and with aluminum ions to form oxide, the anodic reaction, see Eq. 5.3. The applied current density stimulates, in addition to electron flow, the diffusion of metal ions in the anode, as well as electrotransport of intermetallic particles. Unmixing of substitutional alloys occurs that can confound the oxidation reaction by changing the kinetics at the substrate surface. This is particularly important when considering oxidation reactions that occur when anodizing alloys, and the resultant structure of the anodic oxide.

Until this point, when considering anodizing as a corrosion process, the imposition of an electrical bias has sufficed in making the anodizing reaction nonstandard, and explains how, by way of surface polarization, when operating in the regime of the pH values for known industrial electrolytes, a stable oxide can be formed. The electrochemical and thermodynamics of the corrosion reaction to produce an anodic oxide will now be considered in terms of kinetics, allowing for the development of important engineering considerations for the anodizing process: polarization and reaction rate, which dominate AAO structural development, thereby governing the "type" of oxide that grows, developing particular structure and properties for a variety of applications. Alloying elements have a distinct impact on the anodizing reaction and should have a corresponding impact on making alloy selection an important consideration for engineering designs that incorporate anodized aluminum components.

The next section develops the corrosion model for anodic aluminum oxidation by considering an expanded version of the thermodynamic work function called the Tafel equation. By making the connections between the anodizing process parameters and the oxidation of the surface, anodizing becomes an engineering process that can be tuned to yield specific structural characteristics and properties. Details of how the structure and characteristics of the anodic oxide develop as it nucleates and grows are presented in Chap. 6; alloy effects and interfacial phenomena will be discussed in Chap. 7.

5.4.1 The Tafel Equation as a Means to Understand the Anodizing Reaction

The bridge between corrosion science and metal finishing is the Tafel equation, which is based upon the Nernst equation and in the thermodynamic concept of work function. The Tafel equation takes into consideration the polarization (potential) of the total electrochemical reaction by breaking down the corrosion reaction into the terms of work function: the polarization (potential) at the substrate surface, which will be referred to as surface activation polarization and the polarization (potential) external to the substrate, which will be referred to as concentration polarization. Each part of the Tafel equation contains the components of the interface in terms of the anodizing process parameters: total polarization as the result of the imposition of an electrical bias, temperature, and current density are included within the equation. The characteristics of the anode and the electrolyte are addressed within the experimental quantities of exchange current density and limiting current density. By understanding how each parameter functions with respect to reaction kinetics and resistance effects, the entire anodizing process can be better understood, from surface preparation to basic oxide nucleation and growth, enabling effective troubleshooting when an established process exhibits deviations from typical results [8].

Using the Tafel equation for modeling anodic oxidation of aluminum differs from using it for most other corrosion reactions that involve anodic dissolution because with anodizing, the oxidation reaction involves intentional anodic dissolution by an impressed current; in other words, anodizing is driven by the electrical bias provided by the power supply. The characteristics of all circuit components are affected by the electrical bias; the electrolyte, the substrate, and the interfaces they form, which all constantly change with the growth of the stable anodic oxide. The total anodic polarization, N_{total}, which is the sum of the surface activation polarization, η_{surface}, and the concentration polarization, η_{conc}, represents the oxidation reaction:

$$N_{\text{total}} = \eta_{\text{surface}} + \eta_{\text{conc}} \tag{5.15}$$

By considering Ohm's Law and the thermodynamic work function for electrochemical reactions (the Nernst Equation), the above equation can be expanded to the complete Tafel equation:

$$N = \beta_{\text{a}} \log\left(\frac{i_{\text{a}}}{i_0}\right) + \frac{2.3^{*} RT}{nF} \log\left(1 - \frac{i_{\text{a}}}{i_{\text{L}}}\right) \tag{5.16}$$

A more intuitive simplification of the Tafel equation is that it is an expanded form of Ohm's First Law, with each product representing a portion of the electrochemical reaction:

$$V = iR \tag{5.17}$$

By keeping this simplified view of an electrical circuit for the electrochemical oxidation reaction, first in terms of the surface portion of the circuit, and then in terms of the electrolyte, a distinct picture of the anodizing reaction develops in terms of the process parameters. The interface that forms between the electrolyte and the substrate constantly changes as the oxide grows, ultimately forming two interfaces that are addressed by the parts of the equation—one at the substrate and the oxide (surface activation polarization) and one at the oxide and the electrolyte (concentration polarization). When consequently viewed in this way, the Ohmic resistance is incorporated into both sides of the equation due to surface microstructure characteristics such as segregated phases or intermetallic particles; surface contamination, such as process residues; and all anodizing process parameters that contribute to the heat of oxidation, T, including the thickness of the oxide itself. This concept is valuable for troubleshooting unusual events in anodizing, such as defects in oxide growth that arise due to changes at the interfaces.

The Tafel equation was used in the past by Deal and Grove to explain the oxidation of silicon [9]. Using the Tafel equation to understand the kinetics of the oxidation reaction with respect to the total electrochemical reaction can be considered similar for both aluminum and silicon [8]. In the following sections, the parameters of the Tafel equation are defined in terms of the anodizing process for aluminum.

5.4.1.1 Surface Activation Polarization

$N_{surface}$: *The surface polarization of the anode.* By way of producing an ordered charge at the substrate surface, $N_{surface}$ sets the spacing of the initial oxide nuclei, which grow to form the ordered columnar structure that comprises anodic oxide network. The distance between the oxide nuclei is proportional to the polarization, or overpotential, which is the value by which the standard oxidation potential of the metal, in this case the aluminum, at −1.66 V, is exceeded by the potential response of the circuit due to the applied current density, i_a. Growth of the anodic oxide is driven by the applied current density, i_a. The electrochemical characteristic of average corrosion rate is taken into consideration with the Tafel Constant, β_a, and the microstructure and surface characteristics of the aluminum alloy are taken into consideration by the exchange current density, i_o. This is represented by the first part of the Tafel equation

$$N_{surface} = \beta_a \log\left(\frac{i_a}{i_o}\right) \quad (5.16a)$$

i_a: *Applied Current Density.* Current density, amps per unit area, rather than current, is proportional to corrosion rate and most useful when considering that a metal finishing process can mean anodizing many small components, such as jewelry, or anodizing something as large as an airplane wing frame. The concept of current density is easy to imagine when considering pushing the same current through different sized surface areas—the same current concentrated into a smaller surface area would result in a faster corrosion rate. In other words, the rate of oxidation is inversely proportional to area for the same applied current. The applied current density, i_a, drives electron flow and diffusion of metal ions to the substrate-AAO interface, establishing the oxide growth rate. This can be an applied value, if the circuit is galvanostatic, or a resultant value, if the circuit is potentiostatic.

The applied current density externally drives both the surface activation polarization and the concentration polarization parts of the Tafel equation. Without i_a, the finish would not grow beyond a monolayer because the applied bias drives the metal ions from the substrate through the forming ionic solid, which becomes the anodic oxide; i_a is typically sufficient to drive the oxidation reaction uniformly and continuously at the anode surface.

β_a: *Tafel Constant.* Under equilibrium conditions and at STP, β_a is a constant for a given metal and environment and is temperature dependent, and is typically around 0.1 V for $T = 25$ °C. Since an electrical bias is employed to drive anodizing, it is a nonstandard process, so the slope β_a becomes more important. With changes in temperature, β_a will increase, and so will the corrosion (oxidation) rate. Under nonequilibrium conditions, β_a can be regarded as an average slope, or resistance of the aluminum substrate based upon the alloy composition and metallurgical condition, the size of the anode, and the temperature. Different alloys will have a higher or lower average value for β_a depending upon the level of alloy loading. Under equilibrium conditions, β_a for aluminum alloys ranges from 0.43 to 0.44 V.

i_0: *Exchange Current Density.* Under equilibrium conditions, exchange current density, i_0, is the current produced at the interface between the electrode and the corrosion product by way of the difference in chemical potential. Oxidation reaction kinetics are governed by i_0; therefore, they are highly sensitive to the properties of the surface. The chemical potential differences between surface features across the corroding surface also develop exchange current density. For example, alloy compositional differences, such as microstructural phases and particles as well as concentrations of atomic level defects such as vacancies and dislocations, can lead to complex local differences in corrosion rate. Therefore, exchange current density, i_0, is affected foremost by the nature of the surface on what it occurs.

Under nonequilibrium conditions, surface polarization effects are dominated by the applied current density, and the effects of exchange current density are relegated to discrete locations across the substrate surface where microstructural defects are concentrated. Even though anodizing is a nonequilibrium process, polarization is well ordered over the substrate surface; nevertheless, the concept of exchange current density is particularly important because the growth front for the anodic oxide proceeds *into the substrate.* This means that the interface at which the oxidation reaction occurs is constantly changing.

Consequently, surface activation polarization effects are reflected in the ever-changing exchange current density, i_0, which fluctuates with the changing chemical potential at the interface between the aluminum and the growing oxide. Changes in i_0 occur as anodizing proceeds, and how the anodic oxide grows may vary significantly from point to point across the interface for many different reasons that involve the alloy chemistry and the condition of the component microstructure. Resistance heat, T, which is part of the concentration polarization portion of the Tafel equation, also develops with surface activation polarization, by way of interfacial resistance effects that develop i_0, and can substantially affect the structure of the developing anodic oxide.

Different alloys will have local areas that exhibit higher or lower average i_0 depending upon the condition of the substrate surface. Alloying additions, as well as variations in the condition of the microstructure induced by the manufacturing method and heat treatment affect the local activation energy required for the corrosion reaction to continue, and are out of the control of the anodizer. For example, microstructural defects, caused by inadequate quenching following casting, excessive cold work, precipitation, and/or manufacturing process failures all impact i_0, changing the rate of the oxidation process local to the specific defect. Indeed, when the effects of the exchange current density are considered in terms of inhibiting the oxidation process, i_0 also impacts the limiting current density, i_L, which is in the second half of the Tafel equation, by creating concentration polarization effects. The condition of the substrate surface is critical to the oxidation reaction because in all types of aluminum anodizing, regardless of alloy or electrolyte, the thermodynamics of the process seek to oxidize free aluminum, and it is the thermodynamics of the surface that sets i_0.

Most alloying elements and nonmetallic inclusions do not oxidize under aluminum anodizing conditions, and variations in composition, microstructure, and defect population all change i_0, and confound the consumption of the anode and impact uniform oxide growth. It is important to realize that compositional and microstructural variations can occur within the same alloy specification. This is why components manufactured from the same alloy, through the same anodizing process, on the same rack, sometimes at the same time, exhibit different finish appearances, and sometimes different finish performance. How interfacial phenomena affect the growth of the anodic oxide will be addressed in detail in Chap. 7.

$\left(\frac{i_a}{i_o}\right)$: *the ratio of applied current density to exchange current density.* Consideration of the current density parameters, i_a, and the exchange current density, i_o, individually, in terms of oxide growth, elucidates their role as a ratio. In general, the ratio effects govern the kinetics of oxide growth.

If the ratio of applied current density to exchange current density is considered alone, it can be concluded that as long as the exchange current density is different from the applied current density, oxidation will proceed. The moment they are equivalent, the anodizing reaction will stop, because the surface polarization term, $N_{surface} = 0$ (see Eq. 5.16a). If there is no i_o, in the case where the substrate is masked to prevent localized oxidation, the ratio will be undefined, no surface polarization effects result and no oxidation takes place at all because concentration polarization effects dominate (the electrochemical circuit is open). On an isotropic substrate, i_o will be constant because the chemical potential difference between the aluminum and the anodic oxide alone drives the corrosion reaction and i_a will function as the actual current density for the process; resistance effects will be at a minimum.

To keep the oxidation reaction going, i_o must differ from i_a. i_o varies continuously during the anodizing process due to the way the anodic oxide grows: consuming the anode and proceeding into the anode surface. Additionally, to sustain oxidation, the value for i_o must be very small, $0 < i_o \ll i_a$. The oxidation reaction is always affected by the ever-changing characteristics of the anode surface, especially at locations where microstructural features and defects occur. Electron flow supplied by the applied current density, i_a, stimulates ion flow to the interface necessary for oxide growth and is not dependent on the anode surface. It is a quantity fixed by the operator setting the power supply. Yet, variations in surface polarization that deviate from the total polarization necessary to anodize, caused by variations in surface microstructure, affect i_o and therefore the kinetics of the corrosion that follows due to local chemical potential differences introduced by alloying elements in solution with the aluminum matrix, intermetallic particles, precipitates, and nonmetallic inclusions. Similarly, surface topographical variations such as contamination, sharp edges, machined burrs, and texture variations alter i_o and the total oxidation potential.

5.4.1.2 Concentration Polarization

The interface created between the substrate surface and the electrolyte is the most important combination of variables that govern the structure and properties of the finished anodic oxide. In order to understand the total effects of the dynamic nature of the anodizing reaction, it is necessary to consider both sides of the interface as ever-changing solutions. The changes in the substrate surface as the oxide grows are well covered by the concept of surface activation polarization in the previous section. However, the growing oxide and the electrolyte, which are also important parts of the total electrical circuit comprising the anodizing process, must also be considered. It is important to consider the following questions: how does the electrolyte change at the interface as the oxide nucleates and grows? What impact does the movement of electrons and the corresponding transport of ions and intermetallic particles through the forming oxide have on the interface? How does the electrolyte change as the surface changes? Can the process be modified by way of changes in the electrolyte to limit the effects of the exchange current density? How does this change the oxidation reaction? The elements necessary to answer these questions are set forth in the second part of the Tafel equation which represents the concentration polarization of the oxidation reaction.

Concentration polarization for anodic oxidation during corrosion reactions under atmospheric conditions in order to form the passive layer is usually ignored because an unlimited supply of metal atoms is available at the interface. Diffusion of oxygen to the anode surface where the metal atoms are abundant controls the corrosion reaction. However, in anodizing, especially in the case of complex alloy substrates, concentration polarization of the electrolyte cannot be ignored because the application of i_a produces high corrosion rates, and conditions may occur at the interface between the oxide and electrolyte that limit the transport of soluble products away from the interface. Once transport from the surface is *limited*, Concentration Polarization Effects (CPE) develop. This transport value is known as the limiting current density, and it is expressed by i_L.

Hindrance of the corrosion circuit produces concentration polarization; in fact, anything that blocks the flow of hydrogen ions in the oxidation reaction away from the oxide surface causes concentration polarization and changes the oxidation kinetics. One example for how concentration polarization effects develop is the absence of exchange current density, i_o, caused by masking, which blocks the oxidation of aluminum at the surface. Masking can be deliberate, by coating a surface with a paint or epoxy prior to anodizing; or can be accidental, when a surface is contaminated by residual grease or oil. Another example of how concentration polarization effects can develop during anodizing is by way of the adsorption of metal ions from the electrolyte onto the surface of the forming oxide, which prevent the ion exchange necessary for continued oxide growth. A third means by which concentration polarization effect develops is actually how the condition was given its name: surface asperities such as sharp corners in the surface morphology, present by design or as burrs, laps or topographic roughness introduced by blasting or other texturing methods, are sites for charge concentration that inhibit oxide formation. For uniform and continuous oxide growth, conditions that contribute to concentration polarization effects must be minimized to grow anodic oxides of functional thicknesses.

During anodizing, the applied current density, i_a, continues to drive the oxidation reaction beyond the thickness of the passive layer by providing a constant supply of aluminum ions to the interface with the growing oxide, thereby producing i_o. The oxide structure provides the path for ionic conduction and continued oxide growth.

With continued growth beyond a critical thickness that occurs with total reconstruction of the surface by oxide, the structure of the oxide alone accommodates the transport of soluble products away from the interface and proceeds to form a network of columns, each comprising the most important characteristic of the anodic oxide: the central pore; not by dissolution of the oxide, but through the formation of individual ion exchange cells, each corresponding to one of the major oxide nucleation points set by surface polarization, enabling the formation of anodic oxides of substantial thickness. Therefore, the corrosion rate is only reduced by anodic oxide formation, not blocked.

The decrease in corrosion rate caused by blocking the flow of soluble products away from the interface, including hydrogen, is *Concentration Polarization*, N_{conc}, and is expressed as a function of current density by

$$N_{\text{conc}} = \frac{2.3RT}{nF}\log\left(1-\frac{i_{\text{a}}}{i_{\text{L}}}\right) \tag{5.16b}$$

In the concentration polarization portion of the Tafel equation, the anodizing process parameters of temperature, electrolyte concentration, anode-cathode distance, and agitation are considered.

coefficient $= \frac{2.3RT}{nF}$: In Sect. 5.2, it was shown that the electro-active component of the oxidation reaction for aluminum could occur in any protonic acid, and for this reason, the cathodic reduction to form hydrogen gas will be considered the reaction occurring to deplete the electrolyte of the hydrogen ions adjacent to the anode. Therefore, n = number of equivalents per ion = 1, F = Faraday's constant = 96,500 coulombs/mole, and T = temperature, as the heat of oxidation.

T: *temperature*. Temperature in the Tafel equation represents the heat of the oxidation reaction. This is not related to the temperature of the electrolyte but the interfacial temperature that forms due to the heat of reaction. As the temperature increases, the corrosion rate increases. As discussed in Sect. 5.4.1, the heat of reaction will govern the local pH during oxidation and can dictate the stability of the oxide itself. In order to exercise a level of control over the effects of increased temperature due to the heat of reaction, the concentration and temperature of the electrolyte can be adjusted to help regulate the interfacial temperature. By increasing electrolyte concentration, the conductivity of the solution is increased, drawing down the heat of resistance by carrying soluble products from the interface. Reducing the temperature of the electrolyte aids in reducing the overall heat of the process. These process adjustments, together with agitation of the electrolyte, help to maintain the dominant effects of polarization and to prevent burning at contact points with racks, or at concentrations of surface defects in the anode microstructure.

i_{a} = *Applied Current Density*. i_{a} is set by the process operator. When considering concentration polarization effects, how the electrolyte changes as the oxidation reaction proceeds across the interface, from the substrate to the electrolyte, in the near-surface region, is different from how the electrolyte in total reacts to the applied current density, i_{a}. With i_{a}, the anodic reaction pushes electrons into the electrolyte, producing an electron deficiency in the anode, resulting in surface activation polarization, which oxidizes aluminum metal to aluminum ions and electrons; the electrons, in turn, are consumed by hydrogen ions in the electrolyte and form hydrogen gas, which escapes into the electrolyte (reaction 5.4). As i_a increases, the rate of reaction increases; any hindrances that inhibit circuit completion affect the limiting current density, i_L, creating concentration polarization effects. Therefore, i_a is related to i_L.

i_L: *Limiting Current Density*. i_L is the measure of a maximum reaction rate that cannot be exceeded because of a limited diffusion rate of H^+ in solution; in other words, the dissolution of Al to Al^{3+} at the anode is somehow blocked. The reduction of hydrogen ions to hydrogen gas at the cathode (reaction 5.4) completes an important connection in the corrosion circuit. If the cathodic reduction of hydrogen ions in the electrolyte and/or their flow to the cathode is hindered, then electrical continuity across the interface is hindered; the electrical efficiency of the circuit is reduced and resistance to the flow of i_a increases. At this point, a limiting current density, i_L, approaches zero, and the oxide growth rate will be reduced. In worst cases, when $i_L = 0$, the circuit can be completely shut down.

5.4.2 *Concentration Polarization Effects*

Concentration Polarization Effects (CPE) increase the heat at the reaction interface and can result in localized voltage breakdown, producing significant defects in the oxide layer. On the other hand, even though the development of i_L can represent problems within the anodizing circuit, it can also have an important engineering application. For example, i_L is deliberately created in anodizing processes that employ masking to inhibit the oxidation reaction in areas where the mask has been applied.

According to Jones, the limiting current density, i_L, can be calculated from the parameters of the corrosion reaction.

$$i_L = \frac{D_z n F C_B}{\delta} \tag{5.18}$$

where n and F are defined above for the coefficient, and D_z is the diffusivity of the reacting species, which is hydrogen when anodizing isotropic aluminum. C_B is the concentration of the electrolyte and δ is the virtual distance between the anode and the cathode.

The sources that limit or block the cathodic reduction of hydrogen at the interface between the substrate and the electrolyte need to be recognized and understood in order to make the proper adjustments. Since the oxidation reaction comprises both the anodic dissolution of aluminum and the cathodic reduction of hydrogen, if one or the other is blocked, the limiting current density, i_L, will react in kind, by approaching zero, and concentration polarization, N_{conc}, will become undefined, blocking the oxidation reaction. Therefore, hindrances from both sides of the oxide must be addressed, at the interface with the substrate, and at the interface with the electrolyte. It is important to remember that the growth of the

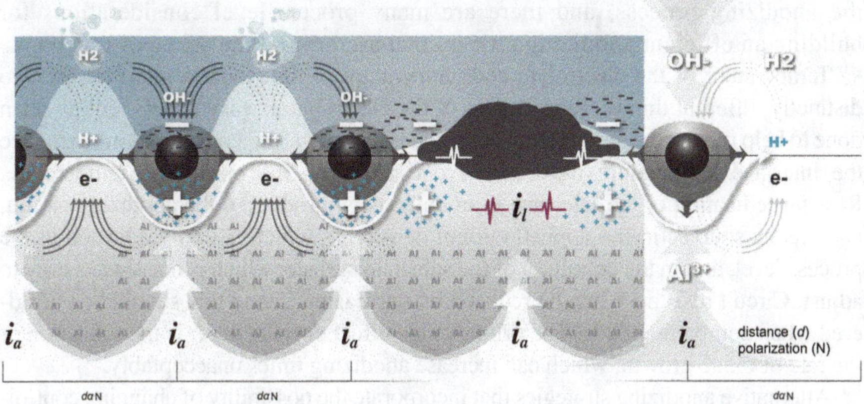

Fig. 5.8 Schematic of how polarization of the surface breaks down due to contamination on the surface creating concentration polarization effects. Interruption of continuity in the electrochemical circuit, causing a blockage of H^+ ion flow, causes i_L to go to zero, and the oxidation reaction is locally blocked

oxide itself will always pose some hindrance to the interfacial ion exchange between the substrate and the electrolyte, especially toward the end of the process when the oxide is quite thick. CPE can be unavoidable, yet process parameters such as applied current density, i_a, electrolyte temperature, T, electrolyte concentration, C_B, and the distance between the anode and the cathode together with electrolyte agitation which control δ can be controlled to minimize them.

According to Eq. 5.18, adjustments can be made to an anodizing circuit that exhibits CPE to aid in preventing i_L from getting too small. For example, if the electrolyte concentration, C_B, is increased (more H^+), pH will decrease and i_L will be increased; if the temperature is increased, D_z will increase, consequently increasing i_L; δ, the theoretical distance between the anode and the cathode, can be decreased by increasing solution agitation in order to increase i_L. The physical distance between the anode and the cathode can be changed, too. The limiting current density should never approach or drop to zero, nor should it ever be equal to or exceed the applied current density, i_a. If the process parameters are such that i_L is kept within this range of current density values, the anodizing circuit in total will not exhibit Concentration Polarization Effects (CPE). See Fig. 5.8.

The sources for CPE can be related to the mechanics of the anodizing process, which can be as simple as a bad electrical contact; with the fixes, most of the time, being just as simple: adjusting the attachment of the component to the rack. The design of the anodizing tank and the rack that takes into consideration the distance between the anode and the cathode as well as the component design to limit the amount of shielding by the electric field in the tank will help to limit the effects of CPE. Inadequate temperature and/or agitation in the electrolyte, which allow for the reduction of the limiting current density, are other process-related sources to consider. Effective tank design is the key to the efficacy of the anodizing process, and there are many process-level considerations for building an efficient anodizing process that are beyond the scope of this book.

Temperature of the electrolyte and temperature of the reaction interface are two distinctly different things. Temperature of the electrolyte is easy to set—and is often done to help reduce the temperature at the interface. However, considerations to reduce the interface temperature must go beyond adjusting the electrolyte temperature. Resistance heating of the interface is an obvious by-product of the anodizing circuit, and has its sources in the applied current density, i_a, which is easily adjusted at the process level, and in the condition of the substrate surface, which is not always easy to adjust. Circuit resistance can be reduced by decreasing i_a, and this should be considered when combatting CPE. The primary side effect of reducing i_a is the reduction of the rate of oxide growth, which can increase anodizing times unacceptably.

Alternative anodizing strategies that incorporate the possibility of changing controllable process parameters may indicate the need for a gradual increase of i_a, or current density ramp, to the anodizing formation voltage. Ramping allows for the dissipation of resistance heat with oxide growth into the substrate. A current density pulse incorporating periodic reductions in i_a over the anodizing time can dissipate stored charge in the anodic oxide, which causes heat, without reducing the time to a required finish thickness, and produce the conditions for stable oxide growth from complex surfaces. Consumption of the surface by the oxide under controlled rate by a ramp can often overcome the CPE of the surface conditions, that may involve residual surface contamination that was not removed during pretreatment but that have nevertheless blocked the anodic oxidation of aluminum (production of Al^{3+}), and produce a sound oxide layer as the reaction proceeds into the near-surface microstructure of the anode.

The limiting current density, i_L, and exchange current density, i_o, are related by the effects of alloy complexity. Highly alloyed aluminum components, such as castings and high alloy wrought product, will always present a more complex microstructure at and near the surface, producing i_o. However, when non-aluminum elements in the form of intermetallic particles block the anodic portion of the oxidation reaction by inhibiting the dissolution of aluminum, i_L is reduced, producing transient CPE. Once the size of the particle is exceeded by the consumption of the adjacent substrate, the anodizing reaction will continue to proceed under the particle. For elements in solution, electrotransport within the anode stimulated by i_a results in a large flux of electrons on solute atoms [10], meaning alloying elements will make diffusive jumps interstitially, on the aluminum crystal lattice, to the interface with the oxide where they will not oxidize, but become entrained, blocking the local movement of aluminum ions toward the reaction interface. In this respect, the surface activation portion of the reaction interface can create a circuit that exhibits CPE, and adjustments to the process need to be made to cope with these challenges. These effects will be covered in detail in Chap. 7.

5.5 Relating the Tafel Equation to Anodic Oxide Growth and AAO Structure Conventions

The characteristics of the anode as well as the electrolyte and the process parameters for the anodizing process are reflected in the Tafel equation, enabling the development of a picture of a functioning electrochemical circuit, which functions according to Ohm's Law. The Tafel equation also considers the solutions on both sides of the interface: the solid solution of the anode, with the alloy constituents and microstructure in mind; and the electrolyte, with the rest of the parameters of the anodizing circuit in mind. The next summary paragraphs are written to aid in the development of the picture.

By imposing an external electrical bias to the anode, surface polarization, N_{surface}, overrides the resistance of most surface microstructure phenomena, producing regular spacing of the oxide nuclei, which develop into the highly ordered AAO. See Fig. 5.9.

Polarization, N, can also be described as the potential change, $E - e$, between the applied potential for the anodizing circuit, E (or the potential response if the applied current density i_a is fixed), and the half-cell potential for aluminum dissolution, e, which has been presented throughout the chapter as Eq. 5.3. Total anodic polarization can therefore be referred to as overpotential, the amount by which the applied potential must exceed for the anodizing reaction to proceed. The value for N must exceed the standard oxidation potential of aluminum, −1.66 V.

N governs the regular spacing of the individual oxide cells that comprise the AAO network. If N is large, the repulsive forces for each polarized location across the surface will be large, and the oxide nuclei will be spaced further apart. Consequently, if N is large, there will be fewer oxide cells that develop per unit grain of substrate surface. The converse is also true. With a deliberate increase in work function, through the imposition of an external electrical bias, i_a, the oxide nuclei become a network of hydrated, charged flakes that originate at their polarized centers; growing into the substrate and away from their polarized centers, consuming the aluminum until they impinge on one another, establishing the hexagonal basis structure that is thermodynamically of lowest energy. At this point, the anode surface is entirely reconstructed by oxide.

The significance of the spacing, set by N_{surface}, develops as the oxide grows: the spacing of polarized oxide centers and the rate of oxide growth, which is set by the applied current density, i_a, determine wall thickness of each discrete AAO cell. As the corrosion reaction continues, and the oxide flakes impinge on one another, an orga-

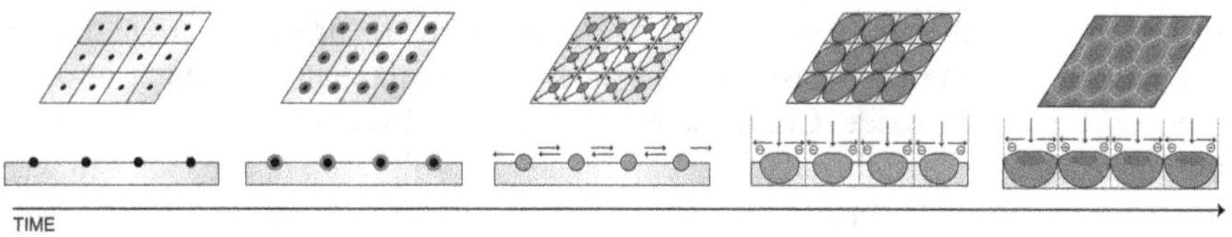

Fig. 5.9 Two-dimensional schematic of oxide nucleation and initial surface reconstruction [11]

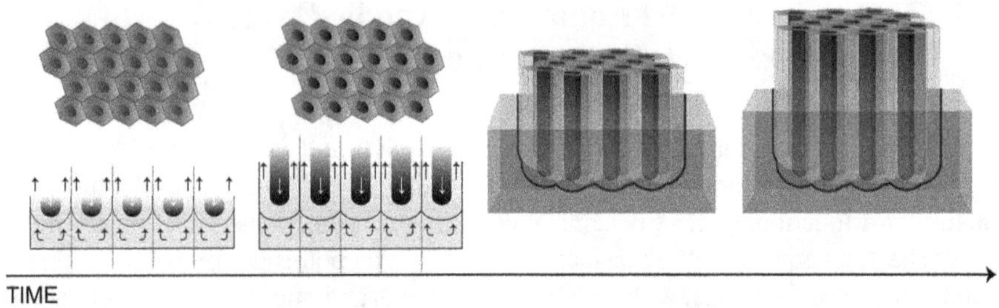

Fig. 5.10 Two-dimensional schematic continued from this figure, showing AAO growth following surface reconstruction. The schematic is both planar and cross-sectional. Individual corrosion nuclei continue to grow with the imposition of an external electrical bias, impinging on one another, growing upward while consuming the aluminum substrate. The central pore is sustained by the polarization of the individual corrosion cells that comprise the oxide network [11]

nized growth front is established downward into the aluminum. Electrostriction occurs, which couples the strain between the opposing edges of each of the impinging flakes that builds with the continued oxide growth, with the polarization forces that occur on the surface of each oxide flake. As growth continues, the shape of the flakes changes as polarization forces repel each other on what becomes the inside surface of individual cup-like cells, comprising a unique, self-assembled network of nanoscale corrosion cells. Each cell of the anodic oxide comprises the initial nucleation point at its rounded base, meaning that pore center-to-pore center distances are set by $N_{surface}$. The nucleation points grow by way of i_a to reconstruct the anode surface; however, the quality of the interface has a profound influence on how the oxidation reaction proceeds, meaning that the kinetics of the oxidation reaction that are set by $\left(\frac{i_a}{i_o}\right)$ will set the thickness of the oxide base and column walls. The oxide flakes become the column walls that circumscribe a central pore (the inside surface of the cup), whose formation and diameter are governed by the repulsive forces of the total polarization effects, N_{total}, which originated at the surface and sustained the finish growth. The scalloped appearance of the AAO network is the natural result of how the finish grows. It is important to understand that the newest oxide is that grown from the interface and the oldest oxide is that in contact with the electrolyte. See Fig. 5.10.

How individual oxide cell features form following a given initial surface polarization, N, is summarized as follows: applied current density, i_a, (amps per unit surface area of substrate) sets the growth rate for the anodic oxide. If current density is high, the growth rate is fast, and the time to oxide flake impingement is short, producing a thinner oxide. Consequently, the AAO cell walls will be thinner when the current density is higher. A lower current density will result in a thicker cell wall for the same given surface polarization, because there is more time before each oxide flake impinges on its nearest neighbors that are growing at the same rate. From the moment the oxide flakes impinge, the cell walls cease to coarsen (cease to become thicker) because substrate consumption by the oxide is translated to down-

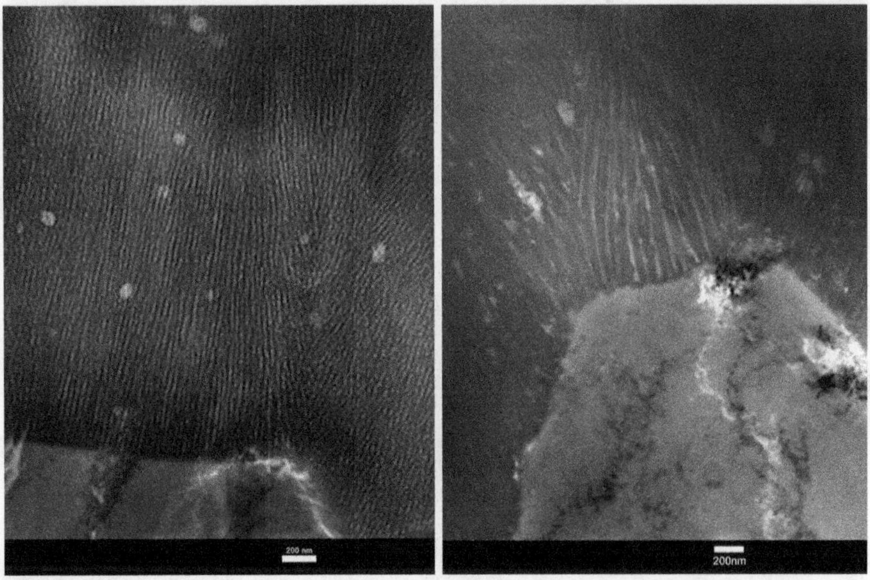

Fig. 5.11 By modifying process parameters of temperature and current density, obvious differences in the AAO structure are clearly developed. The structure on the left was processed at 20 °C, at 1.2 A/dm^2 and with a 120 g/L sulfuric acid electrolyte. The structure on the right was processed at 0 °C, at 2.2 A/dm^2 and with a 220 g/L sulfuric acid electrolyte. The substrate material for both samples was aluminum alloy 6061 T6 (AlMgSi1). Note the wider pore and thicker cell walls for the structure on the right, processed with higher values for both current density and electrolyte concentration

ward (substrate consumption) rather than outward growth. At this point, the characteristic central pore begins to form. A higher current density for a given surface resistance will result in a higher potential response; once the flakes impinge and the growth front proceeds downward, the repulsive forces created by the resultant polarization effects within the central pore will be greater, sustaining a wider pore and a corresponding higher residual stress within the AAO structure. The converse follows as well, with AAO grown at a lower current density (for the same substrate β_a and electrolyte concentration as the high current density example) will have a narrower central pore, and the AAO structure will have a lower residual stress. See Fig. 5.11.

Exchange current density, i_o, which is a function of the anode composition and microstructure, also plays an important part in the development of surface activation polarization, and must be understood as the substrate alloy complexity increases. The level and concentration of alloy additions, constituent segregation, and other microstructural defects compound the effects of the substrate on exchange current density, and therefore impact the structure and quality of the oxide structure. Of the process parameters, the exchange current density governed by the surface microstructure of the substrate is by far the most complicated, as it changes as the oxide growth front advances into the aluminum alloy microstructure while the substrate is consumed. By taking the complex picture of the various possible types of substrate surfaces, based upon alloy composition and manufacturing process (as presented in Chap. 6), to the reactions that occur at the ever-changing interface with the electrolyte (Chap. 7), an understanding of the relationships between the AAO structure and the microstructure of the aluminum substrate from which it is grown can be further developed.

Concentration polarization effects, which can be expressed in terms of the current density of the anodizing process as limiting current density, i_L, is one of the foundations from which anodizing strategies are formulated. i_L represents the amount of resistance to continued oxidation that develops at the anode-oxide interface or at the oxide-electrolyte interface. i_L can never be equal to i_a, but should never be too small, and must never be allowed to reach zero. When $i_L = 0$, the egress of hydrogen ions has been blocked, preventing completion of the anodizing circuit. If the reaction is blocked deliberately, by way of a mask, or inadvertently, by way of contamination, i_L will drop to zero and the anodizing reaction will stop local to the blockage. As the process continues, oxide growth alone will hinder continued operation of the anodizing circuit, which either mandates that the reaction be terminated to prevent damage to the anodic oxide, or that adjustments be made to the process that enable it to continue, to grow even thicker oxide.

Process adjustments to temperature, electrolyte concentration, distance between the anode and the cathode, and electrolyte agitation therefore play an important role in avoiding electrochemical circuit resistance effects to oxide growth and development. Supplying the interface with enough hydrogen ions by way of adjusting the electrolyte concentration, to enable the cathodic part of the aluminum oxidation reaction, minimizes interfacial resistance. A higher acid concentration will provide more hydrogen ions to the interface, reduce the pH of the electrolyte, and boost the conductivity of the electrolyte to aid in the extraction of interfacial heat from the reaction interface. Process temperature can also be set to minimize the heat of reaction at the interface and is aided by movement of the electrolyte through agitation. As with all strategic process adjustments, changes must be done sensibly and with caution.

It is important to differentiate N_{total}, as the sustaining source for the pore diameter from $N_{surface}$, as the source for the spacing of the AAO cells. N_{total} is the sum of the surface polarization, $N_{surface}$, and the concentration polarization, N_{conc}. All of the anodizing parameters that impact oxide growth by way of impacting the exchange current density or by producing concentration polarization effects will change the polarization response as the oxide grows, and N_{total} can reflect the changes in shifts in the pore diameter. A typical DC anodizing process will yield an oxide with a uniform pore diameter because N_{total} will reflect the average conditions set forth by the process parameters. To contrast, deliberate changes in i_a during anodizing will be reflected in the structural characteristics of the oxide, because the N_{total} will change as the growth process for the oxide changes. In other words, a higher value for N_{total} will produce a wide pore and a lower value of N_{total} will produce a narrow pore. Cycling over a current density range will produce a pore that ranges in diameter for a single cell; local areas of high resistance will produce a wider pore for a given current density in the area adjacent to the defect(s) that increased the heat of reaction. See Fig. 5.12.

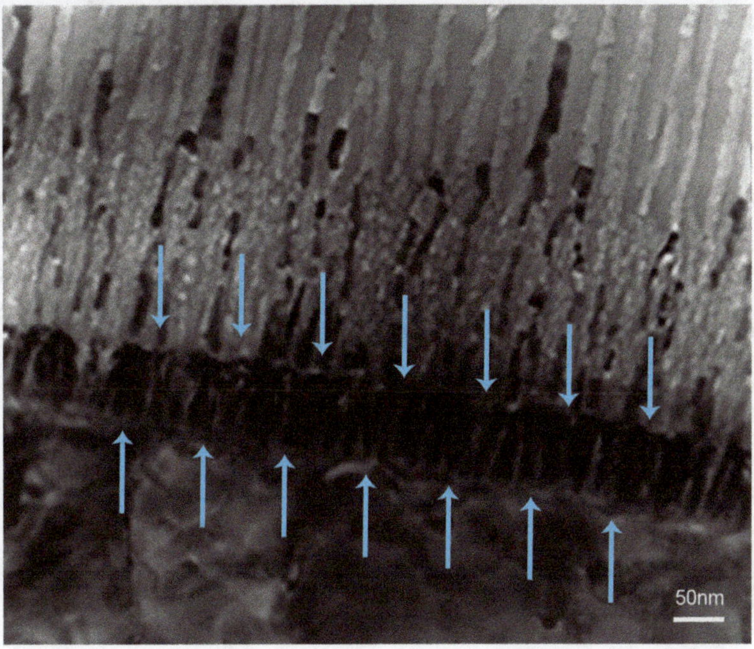

Fig. 5.12 Deliberate changes to the electrical parameters during anodizing change the structure of the anodic oxide. In this case, interrupting the process deliberately after "Oxide 1," which occupies that top half of the photomicrograph. Breaking and reestablishing contact and continuing anodizing enabled the growth of a second, distinct oxide layer, "Oxide 2" beneath Oxide 1, bracketed by blue arrows. Such processes change the optical quality of the total oxide finish, especially for electrolytic coloring purposes

Whether anodizing a complex die-cast alloy for automotive applications or a high purity foil for AAO patterning, the treatment of anodic oxidation as a corrosion process is consequent in making anodizing an engineering process, one that enables tuning the structure such that it yields distinct characteristics to fulfill design and application requirements. The Tafel equation allows predictions for how the anodic oxide will grow and develop, as well as explanations for observed anodizing phenomena. Understanding $N_{surface,}$ surface activation polarization enables a clear understanding of the impact the anode has on the process and helps to answer questions when parts from the same lot anodize differently. Understanding N_{conc}, concentration polarization enables understanding of how the other process parameters impact the anodic oxide. The total circuit polarization, N_{total}, which is the sum of activation polarization and concentration polarization, dictates the dimensions of the fundamental columns, which comprise the AAO and can be used to explain the various anodizing processes and anodic oxides that form.

References

1. Jones, D. A. (1996). *Principles and prevention of corrosion* (2nd ed.). Upper Saddle River, NJ: Prentice Hall.
2. Stern, M., & Geary, A. L. (1957). Electrochemical Polarization: I. A theoretical analysis of the shape of polarization curves. *Journal of the Electrochemical Society, 104*(1), 56–63.
3. Fontana, M. G., & Greene, N. D. (1986). *Corrosion engineering*. New York: McGraw Hill.
4. Runge, J. M., & Pomis, A. J. (2000). Anodic oxide film formation relating mechanism to composition and structure. In *Proceedings ASST*, Manchester, England.
5. Murr, L. (1975). *Interfacial phenomena in metals and alloys*. Reading, MA: Addison-Wesley.
6. Csokan, P. (1980). Nucleation mechanism in oxide formation during anodic oxidation of aluminum. In M. G. Fontana (Ed.), *Advances in corrosion science and technology*. New York: Plenum Press.
7. Sato, N. (2012). Basics of corrosion chemistry. In S. K. Sharma (Ed.), *Green corrosion chemistry and engineering: opportunities and challenges* (1st ed.). Weinheim: Wiley-VCH.
8. Runge, J. M. (2014). Anodizing for design and function. *Journal of Materials Science and Nanotechnology, 1*, 108.
9. Deal, B. E., & Grove, A. S. (1965). General relationship for the thermal oxidation of silicon. *Journal of Applied Physics, 36*, 3770–3778.
10. Shewmon, P. (1989). *Diffusion in solids*. Salem, MA: TMS Publication.
11. Runge, J. M. (2008). Interfacial phenomena and anodizing: ramifications and process solutions. In *Proceedings of the 17th Annual AAC Conference*, San Francisco, CA.

Chapter 6
Anodic Aluminum Oxide Growth and Structure

6.1 Introduction

When AAO is admired, possibly the most amazing characteristics are which gives it a distinct honeycomb appearance: the structural order and the most prominent feature, the pore at the center of each columnar cell that comprises it. Another important feature which, because of its subtlety, is often overlooked, is the intercellular "knitline" that circumscribes each cellular element comprising the AAO. Each cellular feature can be related to engineering properties of the AAO: dimensional stability of the porous network, dyeability, corrosion resistance, and abrasion resistance, which emphasizes the criticality for understanding the mechanism of formation and growth. In this theoretical treatment, the approach to the anodic oxide structure is that AAO is a self-assembled, highly ordered, nanoscale network of individual cells. Therefore, the individual cells must be regarded from the point of origin to understand how and why (1) the oxide is ordered and (2) the oxide network continues to grow after the substrate surface is reconstructed by oxide.

Building on the theory for anodic oxide nucleation and growth presented in Chap. 5, this chapter presents a mechanistic approach to the formation and growth of the anodic oxide, from surface polarization and surface reconstruction to total porous anodic oxide development and growth. To contrast, while Chap. 5 discusses why the anodic oxide develops and grows, this chapter discusses how it grows. Consideration of the conditions surrounding the anodizing process, specifically, the effects of polarization and the applied current bias, enables clear explanations for: why the AAO structure is ordered; how the significant features, the central pore and knitlines of the anodic oxide, develop as it grows; and how these features continue to function in the oxide network to support continued growth. A model for anodic oxide formation is presented which ties initial surface oxidation together with porous oxide growth, resulting in a highly ordered, self-assembled network of individual oxide cells. See Fig. 6.1.

J.M. Runge, *The Metallurgy of Anodizing Aluminum*,
https://doi.org/10.1007/978-3-319-72177-4_6

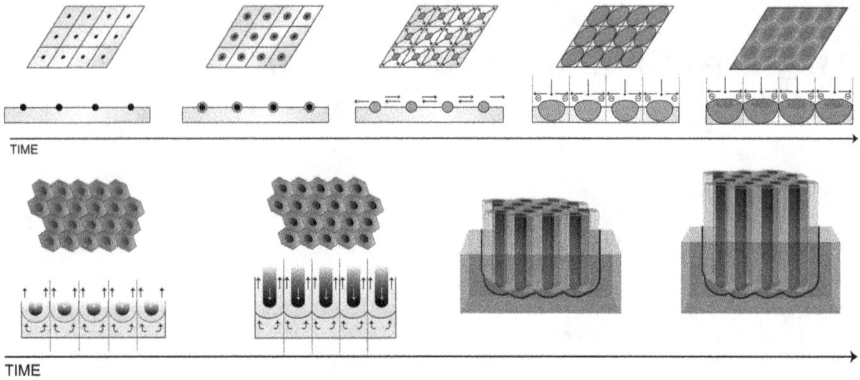

Fig. 6.1 Two-dimensional schematic of AAO nucleation and growth. The schematic is both planar and cross-sectional, showing that as the oxide nuclei grow and develop, the oxide consumes the aluminum substrate both perpendicularly, into the bulk of the substrate, and laterally (as shown by the arrows). When the surface is completely reconstructed, the oxide has assembled itself into a layer of individual oxide elements that have assumed a hexagonal shape as the equi-spaced elements impinge on one another. With continued oxide growth, the structure develops into a network of individual cells, each with a central pore [1]

6.2 Anodizing as a Manufacturing Process

Often the best way to understand the origin of something is to consider, in depth, its manufacturing process. What is involved in *making* anodic aluminum oxide? In nature, when aluminum is exposed to the atmosphere, a passive oxide layer forms on the surface that is only a few nanometers thick. Oxidation proceeds while the aluminum anode reacts with the atmosphere, which is functioning as the electrolyte. Oxide growth requires diffusion of oxygen ions from the environment external to the aluminum substrate (the electrolyte) and counterdiffusion of surface metal ions from the aluminum to the interface with the electrolyte through the forming oxide layer. Without an applied electrical bias, the oxide layer grows in excess of a monolayer to an equilibrium thickness of a passive oxide layer, and stops with dissolution at the interface with the atmosphere and continued growth from the metal interface. At this time, the net reaction rate for passive layer formation is zero; aluminum is oxidized at the same rate as the oxide forms, which identifies passive layer formation as an equilibrium process.

Limited by the equilibrium nature of the process of native oxide formation, the nucleation of the oxide is entirely preferential; in other words, governed by the surface microstructure. The passive oxide nucleates preferentially at substrate defects and other features on the surface that are thermodynamically favored, such as surface asperities, and interfaces between inert particles and the more active aluminum. Therefore, microstructural defects profoundly affect the structural integrity of the native oxide. The native passive oxide layer can be discontinuous and nonuniform, depending upon the substrate alloy composition and surface finish.

No porous oxide of the scale or order of anodic aluminum oxide (AAO) naturally forms without an external electrical bias. For the most conventional case of DC anodizing, the circuit elements of power, substrate (aluminum anode), a cathode, and electrolyte are necessary to manufacture AAO. A fifth component of the anodizing circuit is introduced by the development and growth of the anodic oxide. The reactions that occur: (1) at the interface with the oxide and the aluminum (alloy) substrate, (2) at the interface with the oxide and the electrolyte, and (3) through the oxide as it grows, are all critical to understanding anodizing as a manufacturing process. All of the reaction parameters can be described in terms of the Tafel equation, which is presented in detail in Chap. 5.

Because the electrical bias is a necessary component of the manufacturing process to form a porous oxide, the applied current density, i_a, that comes from the power supply, is the unique component to *manufacturing* AAO, and an integral engineering component of the anodizing process. With constant i_a, a steady-state process, rather than equilibrium process, develops in which an infinite amount of aluminum ions is produced by way of oxidation, while an infinite amount of hydroxide and hydrogen ions is produced by way of electrolysis of the electrolyte. Both electrons and ions carry the charge for the oxidation reaction through the anode, which polarizes the surface. Polarization develops an ordered charge density on the anode surface that overrides the energy of surface phenomena, which typically determine the points at which the oxide nucleates. Polarization not only sets nucleation, it produces regular spacing of the oxide nuclei that develop into the highly ordered AAO. Therefore, surface polarization must be considered as it relates to the organization of charge on the surface, and how it functions with the electrolyte.

As the AAO grows and develops, it becomes a functional part of its own manufacturing process. The interfaces that bracket the oxide, and each individual cell in the AAO network, become critical boundaries to consider as they produce unique impact on continuous oxide growth. Ions alone carry current through the forming oxide. An important engineering consideration for manufacturing AAO in terms of alloy selection, manufacturing process and metallurgical quality of the anode, is the exchange current density, i_0, which develops at the interface with the oxide and is a function of the surface quality of the substrate. Since i_0 is a function of the anode surface, it governs the kinetics of the oxidation reaction at the interface and therefore the structural quality of the developing oxide. Lastly, the limiting current density, i_L, must be considered as it is integral to completing the anodizing circuit. i_L is a function of the reactions that occur at the oxide-electrolyte interface, governing circuit efficiency and critical to completing the anodizing circuit at the cathode, and therefore drives the decision as to the electrolyte type and concentration.

The next essential component of the anodizing process is the electrolyte. Its function is twofold: (1) it is the source for aqueous anions, as hydroxide ions, with which aluminum ions react to form the anodic oxide and (2) it provides electrical continuity, acting as a giant wire to conduct the hydrogen ions, produced by way of electrolysis at the anode surface from the anode to the cathode, completing the anodizing circuit. Ion exchange between the substrate and the electrolyte through the oxide is the primary mechanism by which oxidation occurs due to electrolysis via i_a. While i_a provides an infinite supply of aluminum ions to the interface with the forming oxide, when electrolyzed, the electrolyte provides an infinite supply of hydroxide ions. The composition and concentration of the electrolyte deal with the conductivity of the electrolyte and the structure of the forming oxide; therefore, the entire efficiency of the anodizing circuit, governed by i_L, depends on the electrolyte.

Consideration of the anode is always necessary, because in industry, this is the component to be anodized. It is therefore the most essential component for the anodizing process. In this chapter, the substrate under consideration is isotropic aluminum, which is face-centered cubic (FCC), unalloyed, and without allotropes. Although they will be referred to from time to time in the following discussion, no alloying elements, defects, or deformation interfere with anodic oxide growth when anodizing isotropic aluminum. Without impediment, anodizing should theoretically proceed as long as there is aluminum to oxidize. However, as the anodic oxide grows and develops, reactions occur at the interface with the metal, through the forming oxide, and at the interface with the electrolyte. Each of these reactions is important to address and understand, such that manufacturing oxides on real world substrates, that are far more complex, can be better understood.

As this chapter proceeds, each aspect critical to manufacturing the anodic oxide will be explored in order to explain the mechanisms that act and react to reliably produce the structure and yield the properties that are depended upon when specifying AAO as the finish of choice for any engineering application.

6.3 Theoretical Description of Anodic Aluminum Oxide Formation

Building upon the foundation of anodizing as a corrosion process, the following sections detail the mechanisms that act to make the anodic oxide. The fundamental corrosion, thermodynamics, diffusion, and interfacial phenomena, specifically related to surface reconstruction, can be credited to Jones [2], Lupis [3], Shewmon [4], and Murr [5], respectively. All connections to the anodizing process are the author's.

6.3.1 Anodic Oxide Nucleation

6.3.1.1 Surface Polarization and Oxide Nucleation

As a metallic substrate for anodizing, aluminum has "free charges"—electrons in the structure that are able to move around. Some electrons are not attached to any particular atom, and some comprise tightly bound electron clouds that function as interstitial bonding arms on the FCC structure. In this respect, a metal can behave like both a conductor and a semiconductor. In aluminum, the bonding arms are comparatively long, which enables easy alloying and ductility (see Chap. 4).

When aluminum is placed in an electric field, the electron cloud that comprises each bonding arm on the FCC lattice is distorted instantaneously so that each behaves like a charged dipole, orienting the dipoles so that the bulk of the substrate functions to maintain a neutral charge, while the surface exhibits a net positive or net negative charge. In anodizing, the distortion and orientation of the charges in the aluminum, which produces charge neutrality in the bulk, simultaneously polarizes the surface, producing an ordered charge density over the surface comprising many discrete, evenly spaced positive charges [6]. See Fig. 6.2.

With the applied current density, i_a (see Chap. 5), electrons initially carry all of the electric current in the aluminum substrate. If we consider first the situation of an anodic oxide growing on an ideal aluminum substrate: all cation interstitials on the anode lattice will be of the same size, and carry the same charge, because they will all be aluminum. The ratio of electrons to atomic (ionic) currents is high. Atomic transport, which in the case of an isotropic aluminum substrate, comprises only aluminum ions, results from the impact of the flux of electrons on interstitial atoms on the FCC lattice. These interstitial atoms make large diffusive jumps. The moving electrons hit the interstitial atoms, and, transferring their momentum, bias the diffusive jumps *in the direction of* i_a. The continued biased movement of electrons also inhibits jumps from occurring in the opposite direction; consequently, the substrate surface is flooded with aluminum ions and is positively charged as an anode. These effects are called *electrotransport* [4].

Polarization, as a response to i_a, sets the charge density at the surface during anodizing, and as free aluminum ions in the metal are being driven to the interface by i_a, Al^{3+} nodes form at the surface that are evenly spaced, and symmetric with respect to one another. Simultaneously, electrons are ejected from the surface into the electrolyte, causing electrolysis. Most of the electrons are consumed by hydrogen ions as electrolysis occurs in the electrolyte, and travel to the cathode where they evolve as hydrogen gas, but the rest are pulled back toward the positive charge distribution at the surface. This is because not all of the electrons have enough energy to immediately escape the charge density at the surface, due to the presence of so many metal ions (primarily Al^{3+}), which draws them back toward the surface. Electrons therefore end up oscillating over the positive charge distribution, producing a charge density wave at the substrate-oxide interface, complementary to the Al^{3+} nodes on the surface of the substrate. See Fig. 6.3.

The resultant condition is called anodic polarization, which represents one of the driving forces for corrosion and in the specific case of anodizing, for oxidation, by the anodic reaction:

$$Al \rightarrow Al^{3+} + 3e^{-} \tag{6.3}$$

The oxidation reaction occurs at the substrate surface as i_a pushes electrons that drag aluminum ions from the anode bulk, occurring primarily at locations where the reaction is organized by the resultant charge density wave, set by polarization of the

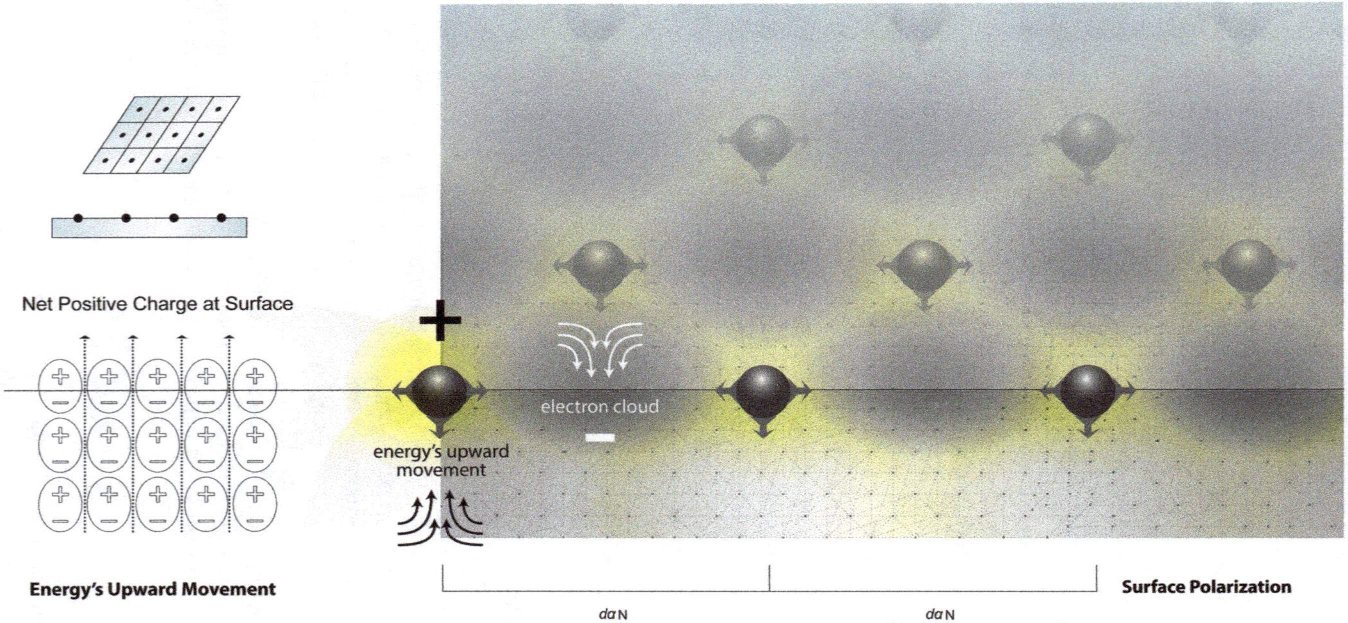

Fig. 6.2 When the substrate is placed in an electric field, E, the charges on the FCC atomic lattice become oriented such that the bulk of the substrate maintains charge neutrality; however, when anodically polarized, the surface has a net positive charge [6]

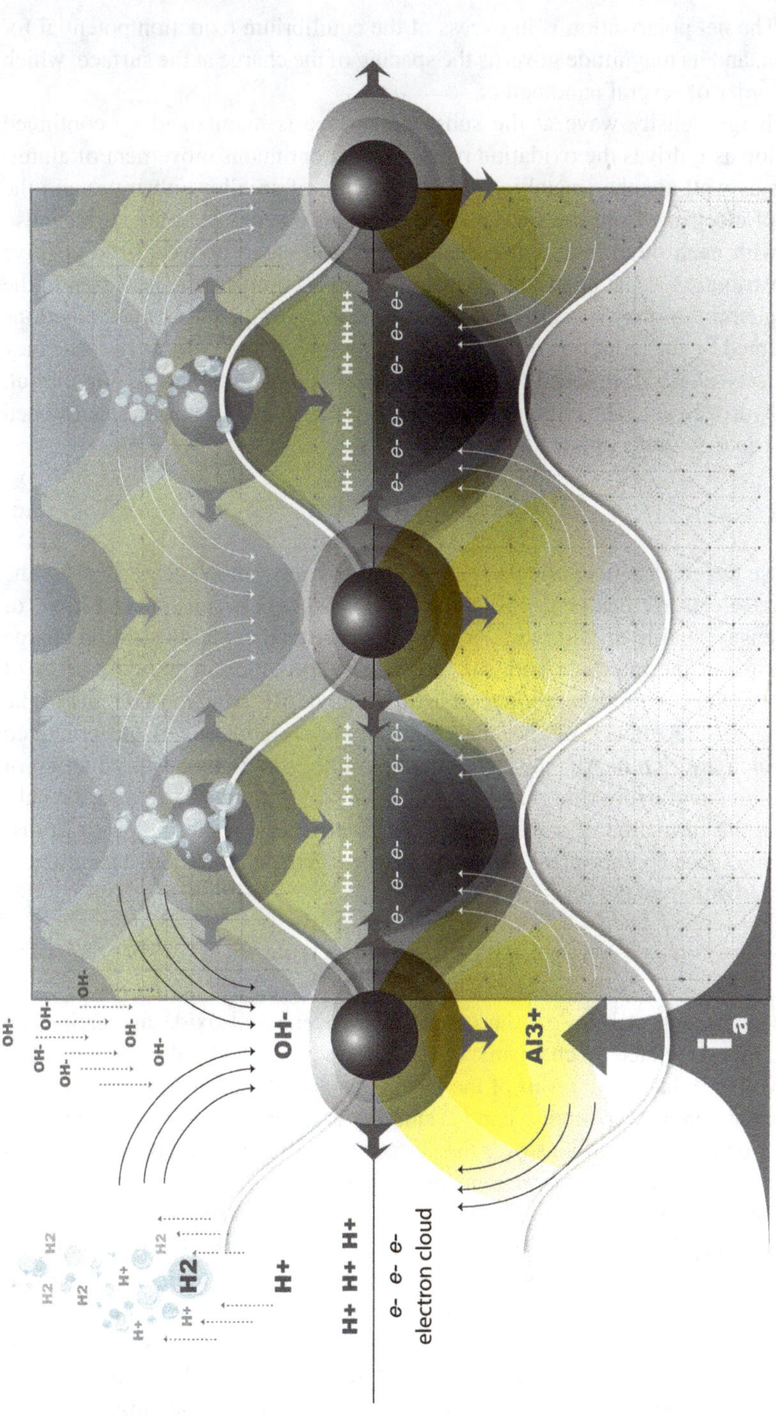

Fig. 6.3 Schematic of a current density surface wave, which is a result of surface polarization. The current density wave at the substrate surface is the result of electrons, ejected from the surface, electrolyzing the aqueous electrolyte, segregating constitutive charged particles and interacting with the positive surface charge that exists due to anodic polarization

surface. The net polarization is in excess of the equilibrium oxidation potential for aluminum, and its magnitude governs the spacing of the charge at the surface, which is on the order of several nanometers.

The charge density wave at the substrate surface is maintained by continued polarization as i_a drives the oxidation reaction: the continuous movement of aluminum ions as well as other metallic ions (in the case of an alloy substrate) and the ejection of electrons from the surface. The electrons ejected from the surface also interact with each other, maintaining a free electron cloud above of the positive charge distribution at the polarized surface. Electrons complete the next step in the anodizing circuit as they react to decompose (electrolyze) the electrolyte: electrons are consumed by the reduction of water to form hydrogen ions, which, because they are positively charged, proceed to the cathodes to complete the anodizing circuit, forming hydrogen gas. The remaining negatively charged hydroxide ions are trapped at the interface with the anode.

$$2H_2O + 2e^- \rightarrow H_2 + 2OH^- \tag{5.6}$$

Because the electric field effects are concentrated at polarized centers across the surface, the electron cloud tapers off at interstitial spaces between concentrations of charge density, and hydroxide ions become concentrated at saddles in the charge density wave at the interface with the anode. The formation of periodic areas of hydroxide ion concentration produces a locally higher pH, relative to the pH of the electrolyte, in the saddles of the charge density wave adjacent to the polarized nodes, which are rich in Al^{3+} ions. It is here, with the simultaneous production of hydrogen ions and hydroxide ions in the electrolyte at the interface with the substrate, that the formation of stable anodic aluminum oxide nuclei proceeds at discrete, evenly spaced points on the substrate surface where Al^{3+} ions have accumulated due to polarization by the reaction:

$$Al^{3+} + 3OH^- \rightarrow AlOOH + H_2O \tag{5.7}$$

The spacing and stability of the centers of concentrated oxide nucleation are governed by two related mechanisms: (1) spacing is governed by the surface polarization (voltage) that is the result of the charge density at the surface applied by i_a, set by the external power supply; and (2) stability is assured by way of the periodic accumulation of aluminum ions over the surface that are in alignment with concentrations of OH^- ions in the charge density wave, producing energetically favorable reaction sites for the formation of stable oxide nuclei, rather than at other surface defects that might be preferred if the substrate wasn't polarized. Evenly spaced agglomerates of oxide molecules are formed at the sites of aluminum ion accumulation, which immediately develop into evenly spaced oxide nuclei. This reaction governs the oxygen pressure at the interface with the substrate by the reduction of water at the surface. Most important, the charge density wave produces the variation in pH local to the surface, necessary to produce and sustain the anodizing conditions that support the formation of stable oxide under process conditions that are not in line with the stability boundaries for aluminum oxide set forth by the Pourbaix diagram (see Chap. 5).

Polarization is the result of i_a; therefore, the size and number of polarized centers across a substrate surface are governed by its magnitude. Since polarization affects the surface at the atomic level, several uniformly spaced concentrations of oxide nuclei centers form per grain aluminum. For example, oxides formed at lower current density processes, such as decorative anodic oxides, have correspondingly lower surface potentials (reduced polarization forces), and exhibit more centers of oxide nuclei per unit grain. This is because with a lower i_a, there is less repulsion of accumulated like charges across the surface and fewer aluminum ions available to react at each node; therefore, more, smaller nuclei form per crystallographic grain on the surface of the anode at lower current densities. Conversely, oxides formed at higher current densities (higher voltage), such as hard anodic oxides, exhibit fewer nuclei per unit grain aluminum because polarization forces set by the higher current density set them further apart. There are fewer, larger nuclei per unit grain at the anode surface with a higher i_a, with more repulsion of accumulated like charges and more aluminum ions available to react. Summarizing: the spacing of oxide nuclei is closer together at lower charge (current) densities, i_a, as in decorative or Type II anodizing (average pore center-to-pore center spacing = 25 nm) and further apart at higher current densities, i_a, as in hard or Type III anodizing (average pore center-to-pore center spacing = 80–100 nm). See Fig. 6.4.

6.3.2 *Surface Reconstruction*

The oxidation reaction, although concentrated at the polarized centers on the substrate surface, is not entirely limited to them. Additional oxide also nucleates in progressively reduced amounts about the primary oxidation sites on the polarized aluminum surface according to the amplitude of charge density wave at the anode-electrolyte interface and the distance between the polarized centers. As the oxidation reaction continues on the substrate surface at the interface with the electrolyte, the larger oxide centers at each polarized node grow at the expense of the smaller ones, causing oxide growth and development to proceed outward, from each polarized center, producing stable oxide elements of constantly increasing size. As the developing oxide consumes the aluminum substrate, the individual oxide elements grow in width and depth, taking on a flake-like morphology. For this reason, as the discussion continues, the oxide elements will be referred to as flakes.

While considering oxide growth on an isotropic aluminum surface, it is important to remember that the driving force for individual oxide flake growth and continuous aluminum consumption for flakes of identical composition originates with (1) surface tension and (2) chemical potential, both of which are higher in larger flakes than in smaller flakes. Additionally, not all of the flakes that develop as the result of oxidation are stoichiometric oxide, which also produces chemical potential differences that contribute to the driving force for oxide growth. Consequently,

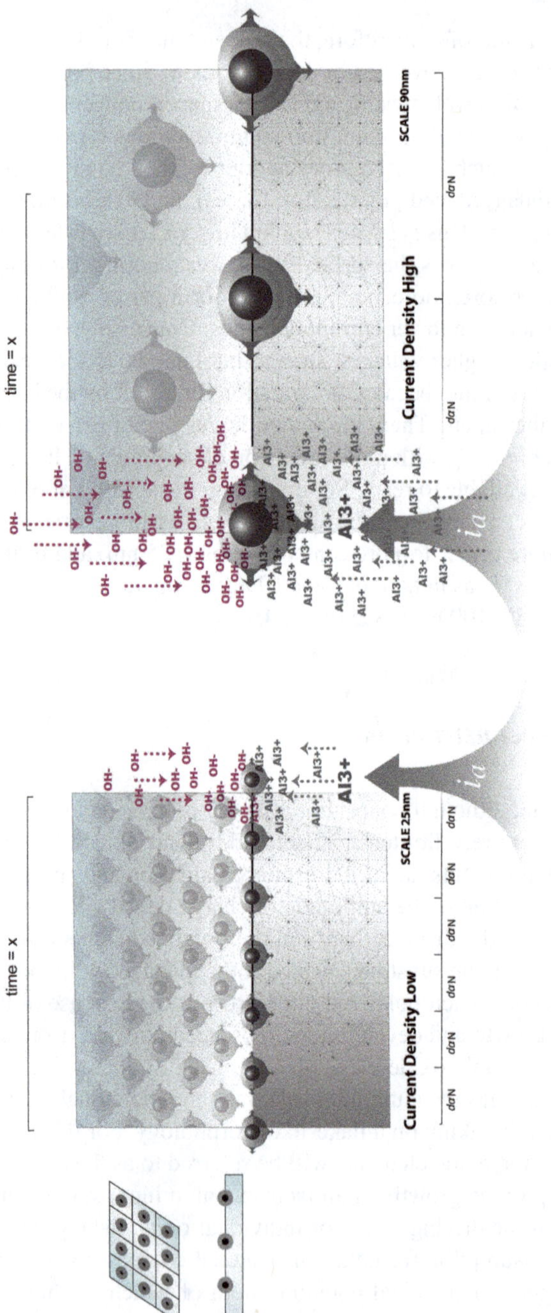

Fig. 6.4 Differences between polarization at high and low current densities. At low current density (left) there is a lower net charge on the surface producing a higher density of oxide nuclei per unit area, as compared with the high current density schematic (right), in which the higher net charge enables fewer oxide nuclei per unit area. The scale of each schematic reflects actual measurements from Type II (low current density) and Type III (high current density) anodic oxides

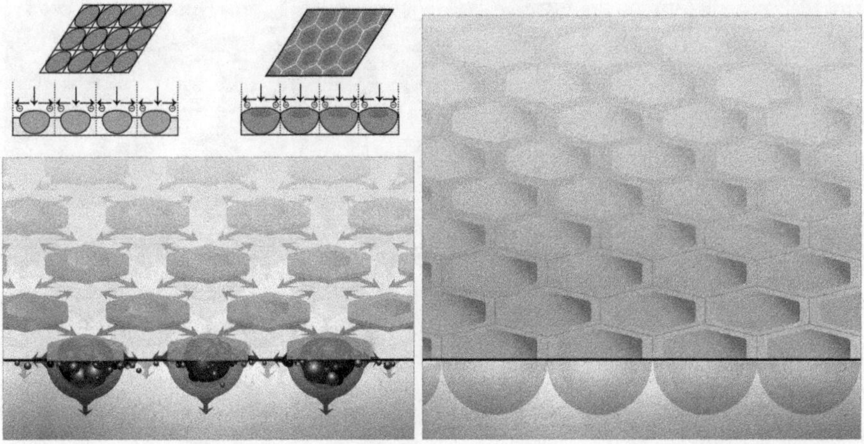

Fig. 6.5 Schematic showing progressive surface reconstruction by oxide nuclei, forming flakes that grow and develop from centers of aluminum ion concentration set by polarization of the anode. Minor nuclei that form about each polarized center provide the chemical potential that drives lateral growth of the oxide, with the larger oxide flakes growing at the expense of the smaller surrounding nuclei. Constraint between the oxide flakes produces small pores at the center of each flake as growth continues downward, into the substrate. Polarization forces sustain the pore diameter

smaller, less stoichiometric oxide nuclei of different sizes cannot be in equilibrium with neighboring larger, more stoichiometric flakes. The absence of equilibrium adds to the driving force for oxide growth as larger nuclei consume smaller nuclei, until a primary oxide flake of critical size and stable stoichiometry develops. The substrate surface is therefore steadily reconstructed by oxide from uniformly spaced polarized points of origin.

With continued application of the current density, i_a, oxidation continues, and the primary oxide nodes continue to grow: laterally and parallel to one another; perpendicularly, consuming the aluminum anode; as well as symmetrically from each polarized center on the substrate surface, consuming other small oxide nuclei which have developed on the surface, consuming any remaining unreacted aluminum substrate. Larger flakes consume smaller flakes until the surface is completely reconstructed by oxide. In this way, the surface is reconstructed by "waves" of oxide growth. As the minor waves of oxide growth are consumed by the larger, primary wave, originating from the centers of polarization across the surface, they develop a contiguous oxide layer comprising ordered, equiaxed oxide flakes. Under most conditions, the initial reconstructed surface oxide layer resembles a floor covered with hexagonal tiles. Constraint between oxide flakes results in the formation of growth stress, causing the development of minor pores at their centers. With continued oxide growth, the anode surface, reconstructed by discrete oxide flakes, becomes the network of porous oxide columns so well recognized as the anodic aluminum oxide. See Fig. 6.5.

If one considers the surface polarization in terms of how the charge density is organized across the surface, it is easy to visualize how the concentration of

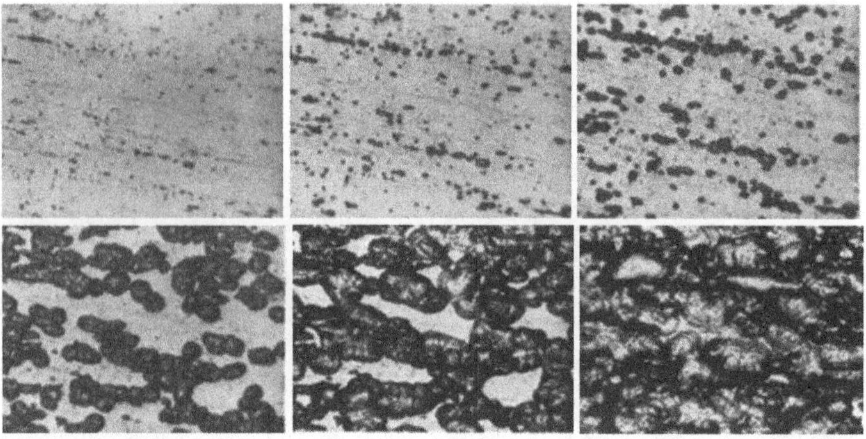

Fig. 6.6 Progressive oxide development over a period of ~10 s in a 5% sulfuric acid electrolyte with 36–38 V at ±1 °C. In dilute electrolytes at low current density, nucleation is preferential rather than precisely ordered by polarization. In the first photomicrograph (upper left-hand corner), preferential nuclei, corresponding to the rolling direction of the substrate, are observed in the first half-second of exposure after switching on the current, and grow outward, becoming circular oxide "flakes" which ultimately reconstruct the anode surface [7]

aluminum ions accumulates at polarized nodes, and then tapers interstitially, between symmetric nodes. The charge density wave corresponds to and complements polarization as electrolysis continues; hydroxide ions are concentrated with aluminum ions at each polarized node, and move with the oxide growth front as it consumes the substrate. This condition provides the concentration gradient for continuous lateral growth and consumption of the aluminum by the developing oxide layer. Growth occurs outward from each polarized oxidation center, reconstructing the surface by forming progressively larger circular oxide flakes that come together to form a hexagonal network. In his seminal work, performing stop-action-microscopy of anodic oxide growth, Csokan [7] documented lateral growth, describing it as occurring in continuous waves of secondary oxide growth. See Fig. 6.6.

6.3.2.1 Thermodynamics of Surface Reconstruction

The thermodynamic mechanism by which growth and development of the nucleus to a particle of critical size is well known is Ostwald ripening. Ostwald ripening is represented by Thomson's equation, which, for the case of a polarized surface, explains the driving force for the growth of particles comprising α within or on the surface of a volume of material V_i^α, in terms of the surface or interfacial tension, σ_{growth}, related to the growth of a particle with radius, r, and chemical potential, μ_i. The following relationship describes the formation of the primary oxide particle of radius r_{ox} on a substrate of volume V_i^α [3].

For two particles of identical oxide, α, one with radius = r' and the other with radius = r'':

$$\left(\mu_i^\alpha\right)_{r'} - \left(\mu_i^\alpha\right)_{r''} = V_i^\alpha 2\sigma_{\text{growth}}\left(\frac{1}{r'} - \frac{1}{r''}\right) \tag{6.1}$$

The relationship between chemical potential of the forming oxide, $\mu_{i,}$ and how it relates to the origin of anodic oxide growth from an *initial point of concentrated oxide nucleation* is significant because the condition and composition of the surface governs the chemical potential across the anode surface. Therefore, the substrate alloy chemistry and microstructure impact how the anodic oxide grows.

On isotropic aluminum, however, the sources of chemical potential are: (1) the difference between oxide particles of different radii and (2) the difference between the aluminum and the oxide. Therefore, with constant value for applied current density, i_a, and steady-state oxidation, the different sizes of oxide nuclei and more developed flakes produce lateral growth that proceeds in waves, with the larger flakes progressively consuming the smaller, adjacent nuclei until dominant flakes develop, each proceeding from points of symmetry set by the polarized nodes established by i_a. When substrate surface coverage is complete by a network of dominant oxide particles, ordered by their polarized centers, and there are no leftover, smaller oxide nuclei left to be consumed, and all $r'' = \infty$, and r becomes the radius of each dominant oxide flake comprising the network. Because in anodizing r can only be comprised of *oxide*:

$$\left(\mu_i^\alpha\right)_{r'} - \left(\mu_i^\alpha\right)_{r'=\infty} = V_i^\alpha 2\frac{\sigma_{\text{growth}}}{r_{\text{ox}}} \tag{6.2}$$

To better describe the steady-state relationship between chemical potential and the growth of the oxide layer from first principles, it is important to relate the chemical potential to the applied current density, i_a. As the dominant oxide flakes consume all surrounding minor nuclei and proceed to the point of impingement, i_a continually stimulates metallic ion diffusion in the direction of the current bias, enhancing chemical potential effects across the substrate surface due to the transfer of aluminum ions from the metal to the structure of the forming oxide at the interface, which produces the exchange current density, i_o. Chemical potential is therefore proportional to the exchange current density with respect to the surface area over which i_o develops. Because both chemical potential and exchange current density are properties of the substrate:

$$\Delta\mu_i \alpha i_o \tag{6.3}$$

Therefore, for an oxide element that originates from a polarized portion of the anode volume of V_i^α, and for an applied current density, i_a, that produces a chemical potential response μ_i at the interface with that portion of the substrate, a corresponding exchange current density i_o is produced that can be related to the growth stress produced in a single element of the oxide during surface reconstruction.

$$\frac{i_a}{i_o} = V_i^{\alpha} 2 \frac{\sigma_{growth}}{r_{ox}} \tag{6.4}$$

Solving for σ_{growth}:

$$\sigma_{growth} = \frac{i_a r_{ox}}{2 i_o V_i^{\alpha}} \tag{6.5}$$

This relationship shows the thermodynamic growth stress in a single oxide element with radius r_{ox} comprising the reconstructed surface, and represents the growth strain per unit surface area of the anode that arises in the oxide preceding electrostriction.

V_i^{α} can be rewritten to reflect the product of the surface area and the thickness of the aluminum substrate, h.

$$V_i^{\alpha} = \text{SurfaceArea}(h) \tag{6.6}$$

The thermodynamic growth stress, σ_{growth}, produced at the interface with the aluminum and at the interfaces between each oxide element, increases in each oxide element as oxidation reconstructs the surface and as substrate volume changes. The volume of the substrate, V_i^{α}, is reduced by the growth of the total oxide, because as the surface area of the substrate remains the same, the thickness of free aluminum (alloy) substrate, h, is reduced as it is consumed by oxide. Equation 6.5 can be rewritten to account for oxide growth, accounting for the decrease in substrate volume as a source for increased growth stress. The quantity is negative because the volume of the growing oxide reduces the volume of the substrate.

$$\sigma_{growth} = -\frac{i_a r_{ox}}{2 i_o} \int_{h}^{r_{ox}} \frac{1}{\text{SurfaceArea}(h)} dh \tag{6.7}$$

As shown in Eq. 6.7, as oxide volume increases to the point of surface reconstruction, the volume of aluminum comprising the substrate is reduced, and σ_{growth} increases in the growing oxide. In total, when considering the changes that occur during anodizing, for a given anode volume, the ratio of oxide volume to the volume of the substrate changes, as the thickness of the oxide grows. The surface area of the substrate nevertheless remains constant, making the change in oxide volume directly related to the increase in oxide thickness. With total surface reconstruction on an isotropic substrate, steady-state oxidation of the aluminum produces a network layer of uniform thickness; that is, each oxide flake comprising the network is of uniform size and thickness. This means that at the point in time when surface recon-

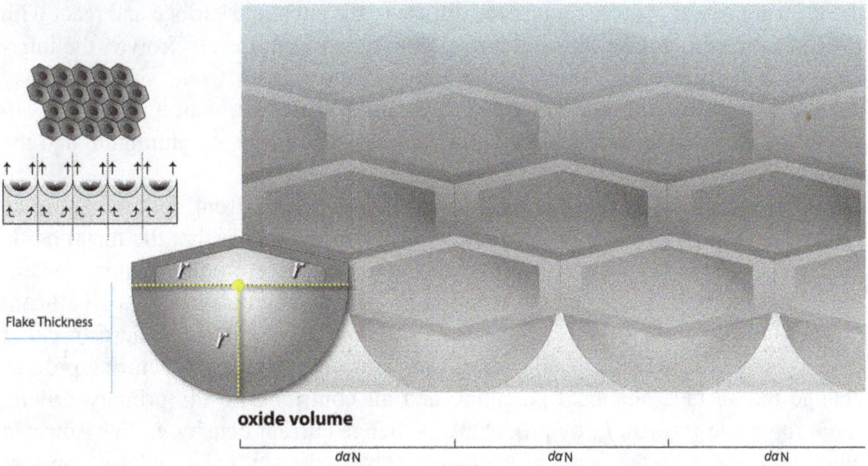

Fig. 6.7 Diagram of reconstructed surface showing r_{ox} as it relates to the total volume of the oxide at surface reconstruction

struction is complete, on isotropic aluminum, σ_{growth} is uniform through each oxide element and through the thickness of the oxide network. The growth stress due to network formation places the substrate surface in compression.

With total surface reconstruction, the oxide network has consumed the entire anode surface and each flake comprising the network constrains its nearest neighbor, producing a hexagonal surface structure. However, because the oxidation reaction is initiated at points of symmetry set by polarization, more aluminum is naturally consumed at these centers due to the increased amount of aluminum ions, to a depth proportional to the applied current density, yielding the hemispherical base for each oxide element. Therefore, r_{ox} is not only the radius and side of each hexagon comprising the surface, but is also the radius of the hemisphere at the base of each oxide element, which remains constant, representing the transient thickness of the oxide layer at the point of total surface reconstruction. See Fig. 6.7.

6.3.2.2 Chemical Potential Effects on Surface Reconstruction

In the early stages of surface reconstruction, aluminum ions move onto the structure of the forming oxide while electrons and hydrogen ions escape between the developing oxide flakes. Although polarization sets the origins for the location of the dominant oxidation reaction, it is the chemical potential, as exchange current density i_0, that drives the early part of the reaction. Chemical potential not only includes the driving force between aluminum and aluminum oxide, but between aluminum oxide nuclei of different size and stability. With total surface reconstruction, the centers of the charge density wave, set by polarization, remain at the interface with the growing oxide. As aluminum ions diffuse to the substrate surface and react with hydroxide ions that have diffused through the finish from the electrolyte, the interface with the substrate, which is the actual oxide growth front, simultaneously moves to consume the aluminum. The substrate, therefore, continues to be consumed by way of the chemical potential difference between the aluminum and the forming oxide.

When one considers that the total oxide network growth front, following surface reconstruction, proceeds into the aluminum, it becomes clear that the metal-oxide interface of an alloy substrate continuously changes, therefore the chemical potential at the interface varies as the reaction proceeds, which can have a significant effect on the structure of the anodic oxide. All atomic level defects, microstructural defects, and external interfacial phenomena induced by the manufacturing process change the surface chemical potential, and all contribute to the primary driving force for oxide growth, i_a, by producing exchange current density, i_0. For isotropic aluminum, the simplest case of exchange current density exists with a constant value for i_0 across the surface of the substrate because the sole chemical potential effect is equivalent to the difference in chemical potential between the metal and the forming oxide.

On more complex substrates, chemical potential effects can produce resistance to the oxidation reaction. Deviations in surface chemical potential are caused by phases and particles of different composition and different oxidation potentials and typically act to block the flow of Al^{3+} ions, causing i_0 to approach zero. Consequently, substrate microstructure defects can be the biggest impediment to *ordered* oxide growth and structural change in the oxide may occur due to local variations in chemical potential. Nevertheless, polarization effects tend to dominate the surface oxidation reaction, producing an overall average exchange current density that maintains structural order, otherwise the formation of an ordered oxide by way of anodizing, particularly with complex aluminum alloys, would be impossible.

As discussions regarding anodic polarization and oxide nucleation and growth proceed in this chapter, all considerations will be for an ideal aluminum surface, therefore the assumption is made that the interface will behave isotropically. Nevertheless, it is appropriate to remark that the surface of a real world substrate will always behave anisotropically, as the microstructural condition at the surface of an alloy is substantially different from the surface microstructure of pure aluminum with no defects. From this point of view, for any aluminum substrate, regardless of alloy composition, the concept of surface tension alone (the difference in chemical potential between the aluminum and oxide) as a mechanical driving force for oxide growth becomes difficult to accept because chemical potential effects will dominate the effects of differing particle size. Interfacial aspects of AAO nucleation and growth with respect to alloy composition and microstructure will be discussed in Chap. 7.

6.4 Anodic Oxide Growth

6.4.1 Growth Stress and Volume

The thickness of the covering oxide layer at the point in time when the surface is totally reconstructed by oxide depends on the distance between the polarized nodes and the steady state between the processes of aluminum corrosion (dissolution) and oxide formation, which is controlled by the applied current density. Steady state is different from equilibrium; since the external conditions driving the corrosion reaction exceed equilibrium with the continuous flooding of the surface with aluminum ions, the continuous growth of the anodic aluminum oxide under constant current conditions (anodizing) is a steady-state process. Steady-state oxide growth enables the anode surface to become an ordered network of hydrated, charged oxide flakes such that when they impinge on one another, they establish the hexagonal basis structure that is thermodynamically of lowest energy. See Figs. 6.8 and 6.9.

Because oxide nuclei size and spacing are charge (current) density dependent, the initial oxide layer thickness also depends upon the current density. The kinetics of the oxidation reaction are governed by the amount of charge polarizing the surface developed by i_a, and the exchange current density developed at the interface with the growing oxide. On isotropic aluminum, the polarization on the surface, $N_{Surface}$, is produced by the ratio of applied current density, i_a, to the exchange current density, i_o, which sets the primary nucleation centers produced by the chemical potential difference between the aluminum substrate and the primary oxide nodes comprising the AAO. If the oxide nuclei centers are far apart and the current density is low, particle growth rate is slower; consequently, there is more time needed for nuclei to grow outward from their centers and consume the aluminum, producing thicker oxide flakes during surface reconstruction. For a situation in which the nuclei centers are initially the same distance apart as in the previous example, but the current density is high, particle growth is faster and there is less time required for the nuclei to grow outward and impinge on its nearest neighbors, a corresponding thinner flake will be produced at the point of surface reconstruction. See Fig. 6.10.

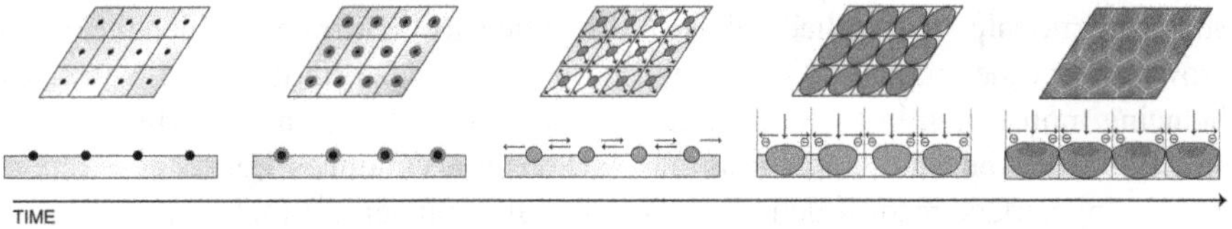

Fig. 6.8 2-D schematic for surface reconstruction of aluminum showing the uniform distribution of oxide elements which eventually comprise the anodic oxide

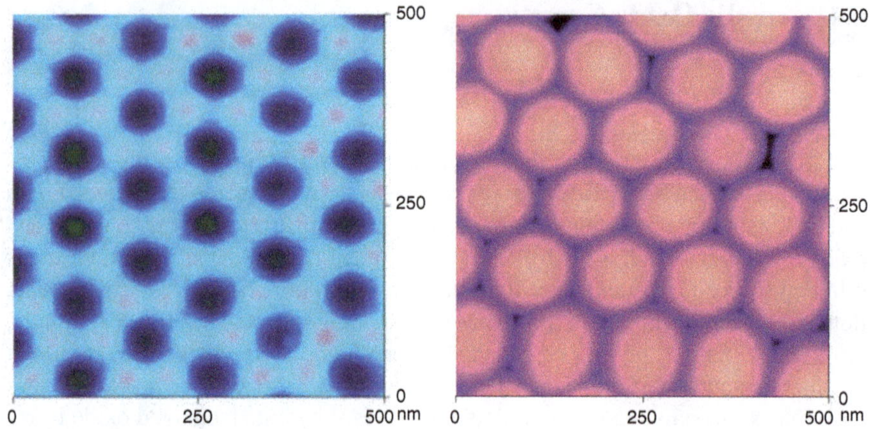

Fig. 6.9 (left) Atomic Force Microscopic (AFM) image of the surface of a commercially pure aluminum anode surface (Alloy 1100) after removal of the anodic oxide with H_2CrO_4 (1.8%wt) + H_3PO_4 (6%wt), at 60 °C for 12 h (etchant dissolves the AAO but does not attack the aluminum substrate). (right) AFM image of the underside of an anodic oxide prepared by using saturated $HgCl_2$ or 0.1 M $CuCl_2$ (etchant removes the aluminum substrate, but doesn't attack the AAO). The images clearly show the uniformity (order and depth) of the oxide. Polarization established the regular spacing between each column. A constant current density provided the vehicle for uniform oxide growth. The image of the underside of the AAO (right) documents the rounded base of each cell and the distinct hexagonal pattern produced by the array of columns. The different hues in the image represent the height progression along the contour of the base of the array. Note that the bases are rounded and the knitlines circumscribe each column [8]

The growth stress produced in the oxide during anodizing profoundly impacts its engineering properties. In nature, the thermodynamic growth stress, σ_{growth}, associated with the formation of an equilibrium reaction layer, such as an oxide, is due to the concentration gradient created by surface chemical potential differences alone. Therefore, when oxidation stops with total surface reconstruction, forming a passive layer, the stress level in the oxide is low, which yields a thin layer with very low abrasion resistance.

The anodizing reaction yields anodic aluminum oxide with consistent material properties. The dimensions of each cell comprising the growing oxide are uniform and proportional to the applied voltage (current density). The rate at which the oxide grows affects the amount of stress induced in the oxide, with higher growth rates inducing more strain in each oxide cell, particularly at the point of surface reconstruction when the flakes impinge as the oxide reaction continues. Çapraz et al. report that during surface reconstruction, at the point of impingement (before pore formation) compressive stress accumulated to several GPa within a 3–5 nm thick layer at the anode-oxide interface, while the stress in the oxide interior remained low [9].

Faster growth rates produce more constraint, and corresponding higher internal strain. Therefore, the amount of constraint between neighboring cells affects residual stress in the total oxide, which can be related to the abrasion resistance. These

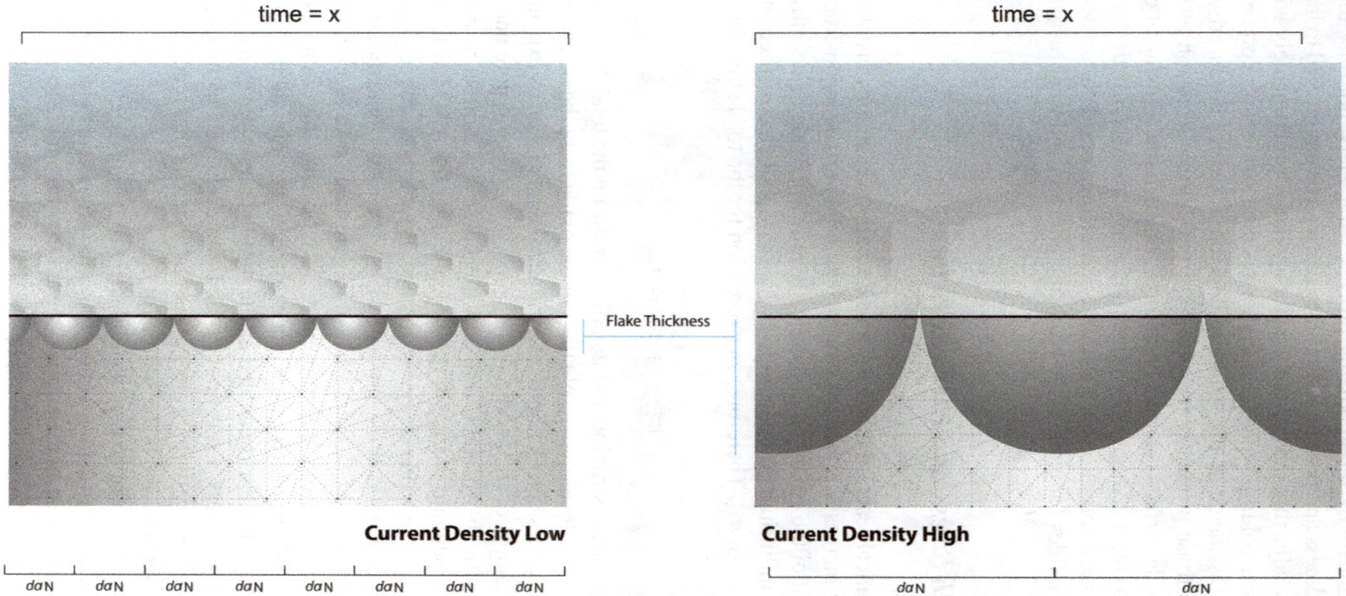

Fig. 6.10 Schematic representing the moment in oxide development when surface is reconstructed prior to electrostriction. At lower current densities the oxide is thinner, but there are more oxide cells that comprise the total oxide layer

effects are proportional to the size of the oxide columnar structure, meaning that oxides can be "tuned" by the anodizing process parameters, not only to exhibit a specific wall thickness and pore diameter; they can also be engineered to have higher residual strain by adjusting the current density upward. Proportionately wider pores result, with thinner column walls, and higher residual strain with correspondingly higher abrasion resistance of the structure develop in oxides produced with these parameters, than those produced at a lower current density. Abrasion resistance of the aluminum oxide surface can thus be tailored, through the adjustment of anodizing parameters, to meet a variety of engineering applications. Growth stress as well as the volume of oxide produced is a critical engineering consideration when manufacturing anodic oxides for application requiring abrasion and wear resistance.

6.4.2 Growth Stress and Electrostriction

In an electrochemical reaction driven by an external power supply, instability of the total oxide layer increases rather than stops with total surface reconstruction [10]. The impingement of the polarized, dynamically growing oxide flakes introduces an additional, nonequilibrium component to σ_{growth} by way of electrostriction, creating electrostriction stress, σ_{ES}. Therefore, an expression for the total stress in the anodic oxide can be written.

$$\sigma_{growth} + \sigma_{ES} = \sigma_{Total} \tag{6.8}$$

For explaining electrostriction and its effect as σ_{ES} on the total AAO structure, it is best to consider the development of stress in the oxide from the perspective of the single cell. This enables correlation to experimental data as well as explanations for how structural features of the anodic oxide, such as the central pore and knitlines, develop. σ_{ES} is distinctly different from σ_{growth}, because it is a direct response to the magnitude of i_a, producing change in the structure after the AAO has reconstructed the anode surface, as the oxide elements impinge. Prior to surface reconstruction, as long as the thermodynamic stress produced by i_a drives the oxidation reaction by way of the development of i_o, there can be no deformation or reorientation of the total AAO structure on isotropic aluminum. In fact, because there is no electrostriction stress in the oxide prior to surface reconstruction, the formation of the rounded bases of each oxide element is enabled as the AAO naturally takes on the shape of least thermodynamic stress. Constraint of the individual oxide elements, or flakes, produces the hexagonal pattern of the self-assembled cells that reconstruct the aluminum substrate surface, and is the source for σ_{ES}.

Electrostriction is a property of all dielectric materials, and is caused by displacement of ions in the structure when exposed to an external electric field [11]. It is critical to remember that as the applied current flows from the anode to the oxide, that there is a shift in how the current is carried. In the anode, electrons carry the current, and in the oxide, ions carry the current. Therefore, in the case of forming AAO, which has the properties of an ionic solid, positive ions (Al^{3+}) are displaced in the direction of the applied electric field, while negative ions (OH^-) are displaced in the opposite direction. Furthermore, in AAO, displacement originates from a point of symmetry: the polarized center set at the beginning of the anodizing process. Ion displacement accumulates throughout the "bulk" of each discrete oxide flake, developing σ_{ES}, planar, with respect to the anode surface, and orthogonally, through the thickness of the oxide. The result is an overall electrostrictive strain (elongation) in the direction of the electric field, set by the bias of the applied current density, i_a. Mechanical strain and polarization forces counteract one another and proceed along the interface with the aluminum substrate as each oxide flake comprising the network develops and grows, and at the point of impingement between nearest neighbor flakes, in addition to σ_{growth}, σ_{ES} is produced.

Strain vs. electric field plots for anodic alumina exhibit a quadratic relationship, which assures the oxide is a linear dielectric. Therefore, for a single oxide cell at the metal-oxide interface, the electrostrictive stress/strain, σ_{ES}, is compressive and proportional to the square of the applied electric field or induced polarization from the point of symmetry, which is the polarized center set at the start of the anodizing process [11–15].

$$\sigma_{ES} = -\left(\frac{\nu}{1-\nu}\right)\left(\frac{\varepsilon_0 \varepsilon}{2}\right)E^2 \tag{6.9}$$

In this expression, ν is the Poisson coefficient of the oxide flake, ε is the relative dielectric constant of the anodic oxide, and ε_o is the dielectric permittivity of free space. The entire expression is negative, because the oxide places the surface of the substrate in compression. Because the coefficients are all constants and characteristics of the AAO, the product is represented in the next equation as Q. The electric field, E, is a second-order variable, the result of the applied current density, i_a, and can be directly related to the total polarization response of the anode, N_{total}, expressed by the Tafel equation (Eq. 5.17). Because each cell in the oxide originates at a center of symmetry due to the polarization of the surface, the electromechanical stress/strain expression can be simplified for a single cell in the anodic oxide [11].

$$\sigma_{ES} = -QN_{Total}^2 \tag{6.10}$$

Therefore, as AAO continues to grow beyond surface reconstruction, when we consider the total stress that forms in the oxide as it accumulates in the oxide with continued i_a, we begin with the following expression.

$$\sigma_{Total} = \frac{i_a r_{ox}}{2 i_o V_i^{\alpha}} - QN_{Total}^2 \tag{6.11}$$

As AAO continues to grow beyond surface reconstruction, σ_{growth} is dominated by σ_{ES}, and the effects of σ_{growth} are diminished. The polarization, N_{Total}, is the total polarization accounted for by the Tafel equation. It is appropriate to use this expression to account for σ_{Total} because: (1) the interfacial polarization, $N_{surface}$, includes the effects of interfacial features that develop i_o, which affect σ_{ES} and (2) the so-called concentration polarization, N_{conc}, includes the effects of the alignment of dipoles along the direction of i_a, which creates the electric field through the thickness of the oxide, and the development of the limiting current density, i_L, at the interface with the electrolyte. The second contribution develops the so-called *dielectrostriction* through the thickness of the oxide, which exceeds the thermodynamic growth stress as well as σ_{ES} at the interface, which is why continued oxide growth produces distortion in the oxide in the direction of the applied current density (electric field) [12, 13].

6.5 Structural Development of the Anodic Oxide Network

6.5.1 Formation of the Central Pore

Continued application of the current density, i_a, produces continued oxide growth, causing neighboring oxide flakes to actually push against one another, continually producing more mechanical strain in the individual oxide flakes. The uniformity of the compressive stress across the surface prevents the individual oxide flakes that comprise the growing network from sliding over one another or collapsing on top of one another.

The tensile component of σ_{ES}, related to the dielectric nature of the forming oxide, also referred to as dielectrostriction, develops through the thickness of the growing oxide, orthogonal to the substrate. The contribution of additional tensile stress to the total stress in the oxide induces uniform deformation in each oxide flake at the point of impingement with the nearest neighbor flakes, which proceeds as growth continues into the aluminum substrate. Electromechanical strain develops at each interface of each element comprising the growing oxide. Deformation occurs in each oxide element comprising the AAO network and causes the edges of the flakes flow in the direction of the applied current, upward, away from the substrate. As the oxide growth front proceeds into the aluminum substrate, the edges of each impinging flake become seams as "dimples" begin to form at the center of each flake in the developing oxide network.

Polarization forces, acting on the inside of dimple at the center of each individual flake, push against the mechanical forces in the oxide; as they repel one another from opposing oxide surfaces, *a cup-like cell forms*. As mechanical forces build in each oxide flake or element, electrostriction effects increase. The inside surface of each cup remains erect, sustained by the polarization voltage that established its initial center. Polarization forces act to keep each cup in compression with respect to one another, forming seams between each forming cell. The surface of the oxide is also in compression with respect to the plane of the substrate. The alignment of the dipoles in the ionic solid and ionic conduction through the thickness of the oxide, due to i_a and the continued oxidation reaction, act to maintain dielectrostriction.

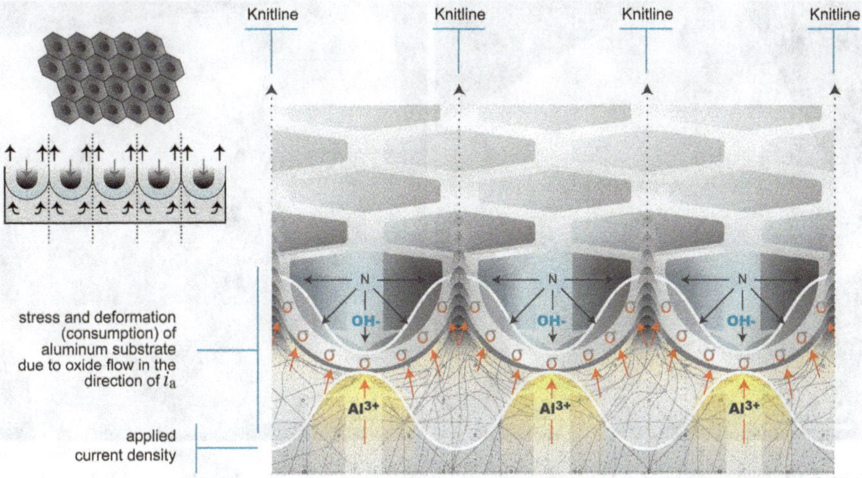

Fig. 6.11 Schematic of the transition from surface reconstruction to pore formation via electrostriction at impingement. Two-dimensional schematic continued from Fig. 6.5, showing AAO growth following surface reconstruction. The schematic is both planar and cross-sectional. Individual corrosion nuclei continue to grow with the imposition of an external electrical bias, impinging on one another, creating growth stress that enables the formation of individual oxide cells that grow upward by way of the interaction of electrostriction and polarization forces, while simultaneously consuming the aluminum substrate. The central pore is sustained by the polarization, N, of the individual corrosion "flakes." This schematic is more detailed showing force arrows that form the "virtual die" through which the AAO flows to form the cup-like cell [16]

The wall of each cup-like cell grows to form the individual columnar cells that comprise the oxide network. The effects of electrostriction are permanent and uniform on the shape of each cell in the oxide network. The polarization effects on the inside of the cup surface, reflecting the resultant corrosion potential (voltage), dictate the pore diameter. The thickness of the base and wall of each cell in the network is proportional to the voltage. Experimental data from constant current anodizing studies show that σ_{ES} increases linearly with the voltage response [12]. The consistency and reliability with which anodic alumina can be manufactured shows that the relationship between electrostriction and polarization renders a tunable structure of precise dimensions. See Figs. 6.11 and 6.12.

Much experimental work has been done that demonstrates movement or flow of the anodic oxide as the source for order and pore formation, and shows that the stress developed in the oxide, by way of electrostriction, is an important contributing factor to the structural development of AAO. Studies utilizing various tracers incorporated into the anodic oxide from the aluminum substrate confirm oxide growth by way of flow from the pore bases to the cell walls. Tracers are redistributed during the oxidation reaction without major loss to the electrolyte. Pore formation initiates along with oxide flow, with an increase in voltage associated with the

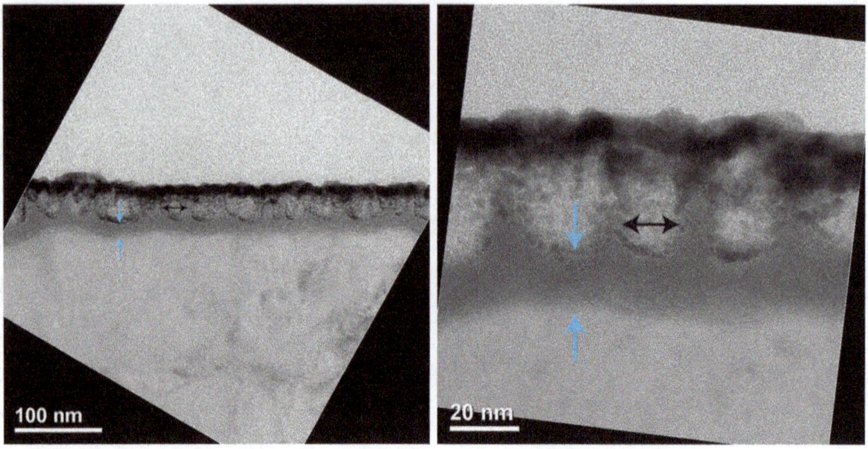

Fig. 6.12 TEM image of AAO in the early stages of outward growth (phosphoric acid electrolyte). The oxidation reaction was stopped after about 30 s. The transition from surface reconstruction to the formation of cup-like cells had just begun. Blue arrows bracket the base of the columnar structure; black arrow indicates the diameter of a developing pore [17]. Courtesy of Accuride

development of the "major pores." Many recent studies also regard the role of stress in the anodic oxide as critical to the development of self-ordering and pore formation. Innovative methods for measuring stress through the oxide demonstrate the change in stress in the oxide as it grows and develops [9, 18–22].

6.5.1.1 Function of the Central Pore

As the columnar oxide network is established by way of electrostriction, the surface of the central pore becomes the most important interface for the oxidation reaction, as it enables current-carrying ions to leave the forming oxide and proceed to the cathode. As the oxide grows, the pore length increases, too, and increases the surface area over which ion exchange can occur. Ion exchange is the basic process by which the oxide functions as part of the anodizing circuit. Consequently, the function of the central pore in each oxide cell, as the anodizing circuit element for ion exchange, is critical to continued oxide growth.

Continued electrolysis of water molecules at the base of the pore in each oxide cell provides the necessary hydroxide ions that are pulled through the ionic structure to the anode surface where they react to continually form oxide, together with hydrogen ions, which flow to the cathodes to complete the circuit. The source for electrolysis is the electrons that are continually being pushed outward to the metal-oxide interface by the applied current density. Therefore, electrons always carry some electricity through the forming oxide in spite of its ionic nature. Consequently, OH^- ions are concentrated at the base of the pore, and pH is locally higher than it is

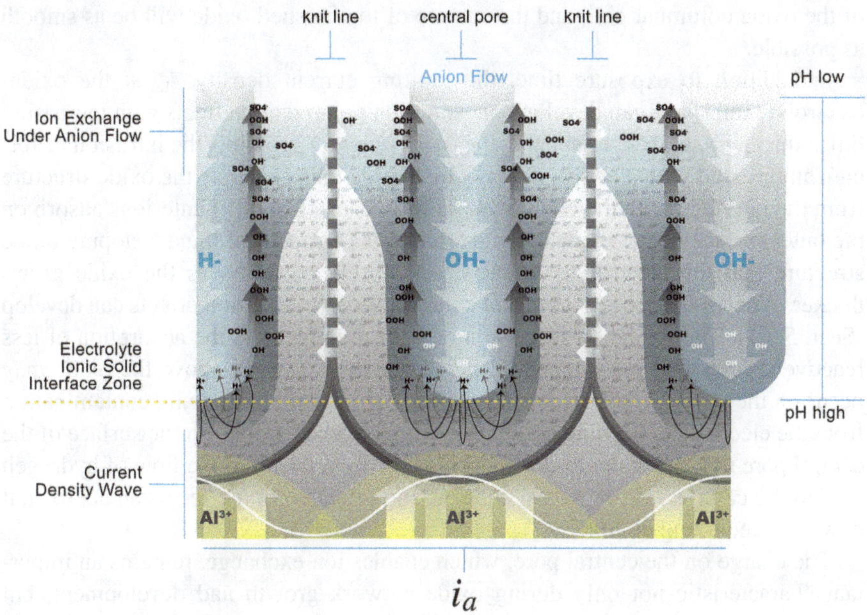

Fig. 6.13 Schematic showing the formation and maintenance of the central pore with ion exchange. Note the formation of a pH gradient *due to continued electrolysis of the electrolyte at the aluminum-AAO interface*. The concentration of OH^- ions at the base of the pore produces a higher pH adjacent to the oxide, enabling continued formation of stable oxide. A lower pH is present at the top of the forming oxide, due to the highly acidic electrolyte

at the oxide surface adjacent to the electrolyte. This creates a pH gradient from the top of the oxide to the bottom of the central pore that facilitates an "ion pump" in and out of each oxide cell. The push–pull pump of ion exchange across the forming oxide produces convection within the pore of each oxide cell [23, 24]. See Fig. 6.13.

The increase in oxide surface area developed by the central pore is directly related to: (1) the polarization effects that are a response to i_a; (2) the corresponding resistance at the interface with the substrate that develops as a result of the exchange current density, i_o, and (3) the height of the cellular network (total oxide thickness). Adsorption of electrolyte counterions on the outer surface of the central pore creates concentration polarization effects, and the limiting current density, i_L, may inhibit completion of the circuit, causing variations in cell height that produce increased surface roughness of the finished oxide. If the effects of i_o and i_L are minimal, as they would be with isotropic aluminum, and the conductivity of the electrolyte is at a maximum, resistance effects and the heat of oxidation will be minimized, and the oxidation reaction will proceed as if there is no oxide growing at all. Under these conditions, the oxidation reaction can proceed, theoretically, until all of the aluminum is consumed, and only the exposure time need be regarded to arrive at an oxide of a specific thickness. The pore diameter will be uniform from the top to the bottom of the oxide columnar cell, and the surface of the finished oxide will be as smooth as possible.

In addition to exposure time, the limiting current density, i_L, at the oxide-electrolyte interface, can develop concentration polarization effects with increasing finish thickness. Characteristically, the AAO structure prohibits the intrusion of foreign anions, and with electrolysis, only hydroxide ions penetrate the oxide structure from the interface with the electrolyte. However, electrolyte counterions adsorb on the outer surface of the oxide structure, along the surfaces of the developing oxide structure that the electrolyte contacts. Adsorption increases as the oxide grows thicker. When this occurs, significant concentration polarization effects can develop (Sect. 5.3.1). The oxidation reaction no longer dominates as the adsorption of less reactive species that are trapped in the charge density wave above the oxide may occur on the pore surface, such as dissolved cations and other ionic contamination from the electrolyte, forming complexes with adsorbed anions on the surface of the central pore that block continued ion exchange by preventing the flow of hydrogen ions to the cathodes. In this way, concentration polarization effects can act to shut down the anodizing circuit.

The charge on the central pore, which enables ion exchange, remains an important characteristic not only during oxide network growth and development; but afterwards, as an integral finish for aluminum and aluminum alloys, the surface of the pore provides active bonding sites for dye molecules, seal reactions and for bonding of adhesives, paints, and other ionic coatings. The ionic nature of the pore surface, as a stand-alone feature, can also be useful as a sensor. Indeed, hydrogen and other selected ionic species can be detected using the anodic oxide. The pores in the highly ordered AAO network are used as templates for electroplating uniformly thick nanowires, the network as a whole is used for nano-patterning of magnetic surfaces for computer applications, and as inert surfaces for bio-applications [25]. See Fig. 6.14.

6.5.2 The Formation of Knitlines

With surface reconstruction, the initial network of oxide flakes impinge on one another, such that the outer edge of each oxide flake becomes the outer diameter of each columnar cell. Electric field effects in the forming AAO structure are at a maximum from the polarized centers at which each cell of the oxide network originates, and taper to the edges of each oxide element (flake) that grows to develop the cell [27]. With continued imposition of the current bias, electrostriction stress develops and each individual oxide flake constantly exerts lateral and downward forces against its nearest neighbor, forming distinct seams—*knitlines*—that circumscribe each columnar cell in the oxide network. The regular spacing of the initial nuclei, and outward uniform growth, produce the naturally occurring geometric pattern with the most thermodynamic stability—the hexagon. Knitlines are a product of

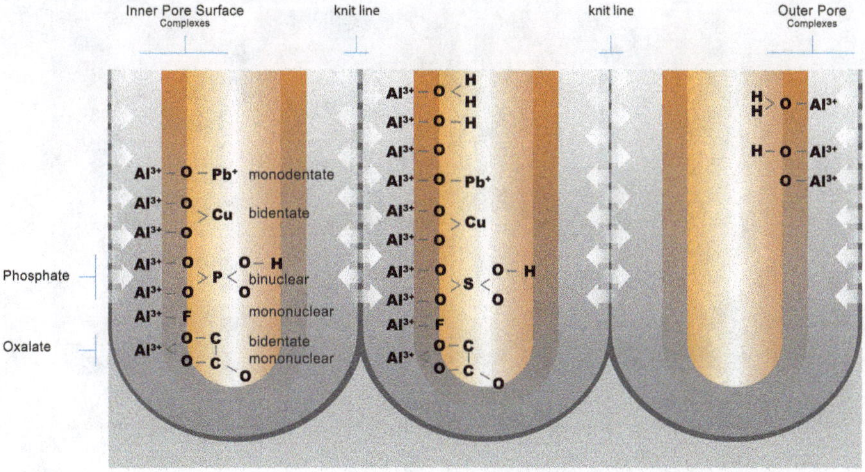

Fig. 6.14 2-D Schematic representing the "network in the network" of the anodic oxide, comprised of individual cells, each with a central pore and circumscribed by a knitline [26]. Inside surface of the pore after anodizing is ionic, offering bonding sites for supplementary treatments such as dyes or adhesives. The ionic nature of the inside surface of the pore supports the sealing reaction by way of reductive dissolution

anodic alumina growth, and continue to form as long as the current is applied and the oxidation reaction proceeds [23].

Prior to surface reconstruction, electrons and hydrogen ions are primarily expelled along the gaps between the oxide flakes, but as the structure knits, ion exchange must occur through the oxide structure in order to maintain the dimensions of the oxide cells, primarily the cell wall thickness (including the base) and pore diameter. As the oxide flakes impinge and the total growth front proceeds into the aluminum, field effects diminish from the polarized centers, to the edges of each discrete flake, where the cells impinge, and there is significantly reduced ion flow across the cell walls and no dynamic flux, as in the pores. Nevertheless, electrostriction produces internal mechanical strain between the cells, which stimulates mass transport across each seam, "knitting" the structure of the network together. Analysis and practice show the cell walls of AAO are fused together, implying that diffusion and/or mass transport occurs across the outer wall of each cell, "knitting" the structure together; hence this structural feature was given the name "knitline." [28] See Fig. 6.15.

6.5.2.1 Function of the Knitline

Knitlines are the only direct avenue from the surface of the oxide to the aluminum substrate, and the avenue of lowest energy from the substrate-oxide interface to the oxide surface. Non-aluminum ions, with their corresponding oxygen vacancies, and intermetallic particles from the substrate migrate outward along the knitlines, during

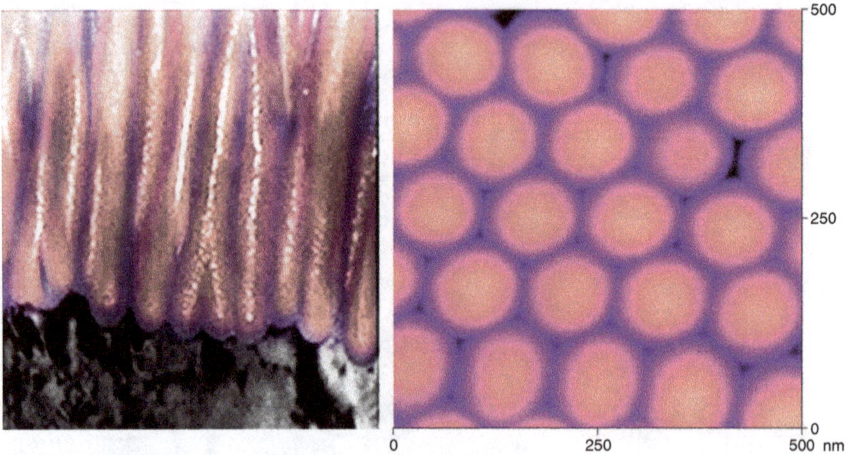

Fig. 6.15 TEM image documenting the presence of knitlines in an industrially produced hard anodic oxide at approximately 100,000×. Knitlines appear darker purple in color (as well as the interface with the aluminum anode) and are digitally enhanced

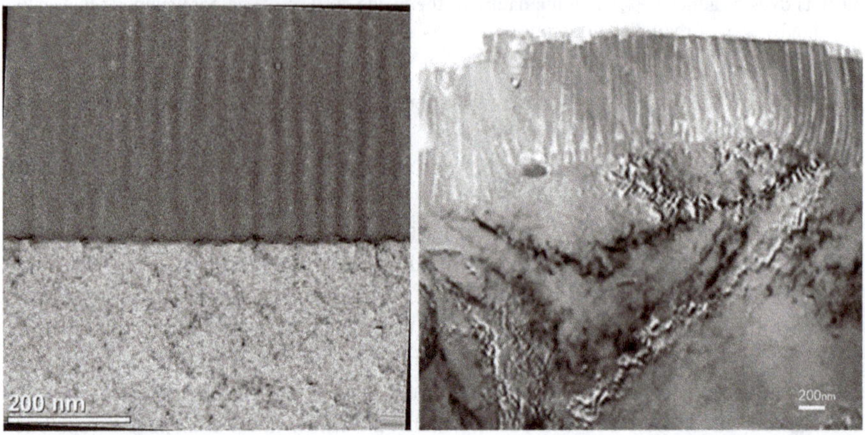

Fig. 6.16 TEM images of anodized substrates showing entrainment and migration of *non-aluminum* alloy constituents' movement through the anodic oxide. The image on the left is a 7XXX series alloy. The dark particles entrained at the *base metal-oxide interface* and moving along the knitlines of the oxide structure are zinc that has become *unmixed* from the alloy due to electrotransport effects. The image on the right is a 6XXX series alloy. The particles disrupting the structural order are Fe-Cr intermetallic particles [28].

oxide growth, toward the electrolyte. Consequently, the knitlines are the location for most alloy impurities and defects within the finished oxide. Structural change of the oxide is often observed due to the egress of alloy constituents as particles and as reactive ions along knitlines, causing bifurcation and voids in the otherwise ordered AAO. Entrapped constituents may cause deleterious changes such as reduced corrosion resistance, to the function of the anodic oxide. As such, the stability and robustness of the total oxide structure depends on the integrity and resident composition of the knitlines [28]. These phenomena will be covered in detail in Chap. 7. See Fig. 6.16.

6.6 The Function of the Electrolyte on Anodic Oxide Growth and Development

The electrolyte functions as the biggest conductor in the anodizing circuit; it is the "conduit" for ionic conduction across the oxide and through itself to the cathode during anodizing. A low level of electron transfer is also facilitated by the electrolyte. The electrolyte serves as the host solvent for the ions that are ejected from the oxide, and with time, the conductivity of the electrolyte is reduced by resistance created by the accumulation of reaction products. This mandates rigorous maintenance of the electrolyte in industrial anodizing lines.

The chemistry of the electrolyte can profoundly influence how the structure of the oxide develops. Depending upon the extent of adsorption, and the type of ion that adsorbs on the outer surface of the oxide structure, the formation of the central pore can be inhibited, due to the development of concentration polarization effects that block ion exchange. The efficiency of ion exchange, measured as a voltage response during constant current anodizing, appears to be related to the electrolyte counterion. This may be useful in explaining the differences in oxide structure that occur with different electrolytes. For example, in chromic acid anodizing, the oxide nucleates and grows from the interface, reconstructing the surface to a point at which AAO structures produced in other electrolytes typically begin to grow upward to form the columnar cells. The more reactive chromate ions oxidize the sites on the inside surface of the developing oxide to the point of completely shutting down ion exchange in the pore, and because the developing column cannot sustain ion exchange, the structure collapses on itself. The wave-like structure of an oxide produced in chromic acid, indicates that the formation proceeds as a sequence of oxide layers with low dielectric strength that rupture from the anode surface shortly after each layer reconstructs surface, breaking apart and reacting with the electrolyte. The effects of the substrate composition and microstructure can be observed in noncellular AAO structures such as barrier-type oxides and oxides grown in chromic acid. See Fig. 6.17.

Characteristically, no foreign anions penetrate the oxide structure, but can adsorb at the more ionic region of the oxide at the interface with the electrolyte. At the oxide-electrolyte interface, the pH is markedly lower, and the interstitial aluminum ions that migrate through the oxide at the interface with electrolyte are ejected into the electrolyte or remain at the interface to react with adsorbed water and electrolyte anions, locally changing the structure to form an oxide "outer layer" that is more ionic. See Fig. 6.18.

It is incorrect to discount dissolution effects of the electrolyte on the oxide structure. While dissolution by the electrolyte is not the source for the central pore, the concept of "like dissolves like" does not escape consideration on the outside surface of the AAO. With extended anodizing times, reduced ion exchange at the outside surface of the AAO enables the structure to approach equilibrium conditions, which

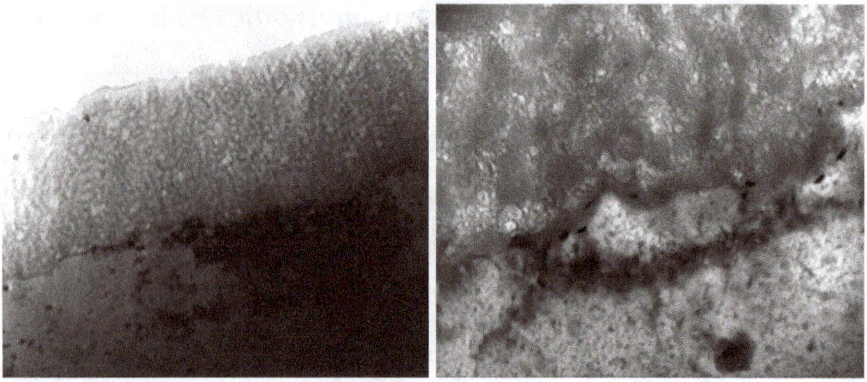

Fig. 6.17 Transmission electron microscope (TEM) photomicrograph of AAO grown in a chromic acid electrolyte (Type I) on a 2024 aluminum alloy substrate

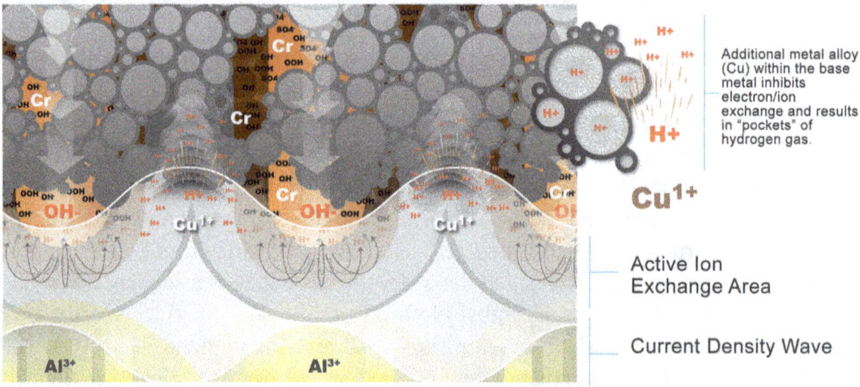

Fig. 6.18 Chromate ions oxidize the sites on the reactive surface of the developing oxide to the point of completely shutting down ion exchange in the pore. Because the developing column cannot sustain ion exchange, the structure collapses on itself. The wave-like oxide structure that consequently forms indicates that the mechanism proceeds as a sequence of oxide layers with low dielectric strength form and rupture from the anode surface shortly after the anode surface is repeatedly reconstructed, breaking apart and reacting with the electrolyte. The effects of the substrate composition and microstructure can be observed in *noncellular* AAO structures such as *barrier-type* oxides and oxides grown in chromic acid

are the conditions determined for aluminum oxide by the Pourbaix diagram. With a pH lower than 2, the electrolyte will dissolve the topmost part of oxide structure, and with time can collapse the structure at the top and close the pore. This takes time, and thicker anodic oxides may exhibit a sealed appearance with dissolution.

6.7 Structure and Properties of the Anodic Oxide

With the establishment of how the AAO network nucleates and grows, it is necessary to continue to discuss the operation of the anodizing circuit with the addition and growth of the anodic oxide, beyond the function of ion exchange across the central pore. Circuit operation through the substrate to the reaction interface with the oxide is possibly the most straightforward to consider, in which all current is carried by electrons to the interface. The reaction at the interface to form the initial oxide and how the structure develops are well discussed in Sect. 6.2. However, the diverse reactions that occur to develop the anodic oxide from surface reconstruction to a fully functional network of individual corrosion cells must be considered from the perspective of how the total oxide functions as a circuit element. In order to do so, the structure and properties of the oxide are explored first by identifying the nature of the AAO structure, and based upon the characteristics inherent to such structures, considering next how the structure will function in an electric field.

6.7.1 Structural Considerations of the AAO and Ionic Conduction

Analysis of the AAO network by way of X-ray diffraction within the TEM shows the anodic oxide exhibits no diffraction contrast and is therefore classified amorphous. Nevertheless, AAO cells have a molecular structure that is important to regard when considering how the anodic oxide grows and develops. This molecular structure comprises the total structure of the oxide network, which must facilitate electron flow and ionic conduction through the AAO in order to complete the circuit during anodizing. Diffusion and mass transport of the non-aluminum constituents of the aluminum alloy substrate and the electrolyte across the dynamic substrate-oxide interface during AAO growth is also facilitated by the AAO structure.

As the insight regarding structural detail of the anodic oxide developed for this section, it was assumed that the cellular network of the anodic oxide is structurally related to the passive layer that naturally forms on aluminum, which is also amorphous, yet contains the η and γ spinel phases of alumina [29]. Kniep, Lamperter, and Steeb [30] determined, by way of Angular Dispersive X-ray Diffraction (ADXD), Energy Dispersive X-ray Diffraction (EDXD), and neutron diffraction at Institut Laue-Langevin in Grenoble, France, that AAO grown in a 200 g/l sulfuric acid electrolyte at 12 °C, with a current density of 1.5 A/dm^2 (15 V), similarly comprises aluminum atoms surrounded by four oxygen atoms, comprising AlO_4^- molecules, forming tetrahedra with short-range order. The structural determination for AAO brings a third dimension to the perspective of how it grows, provides insight to the mechanism by which electrical continuity is established through the oxide

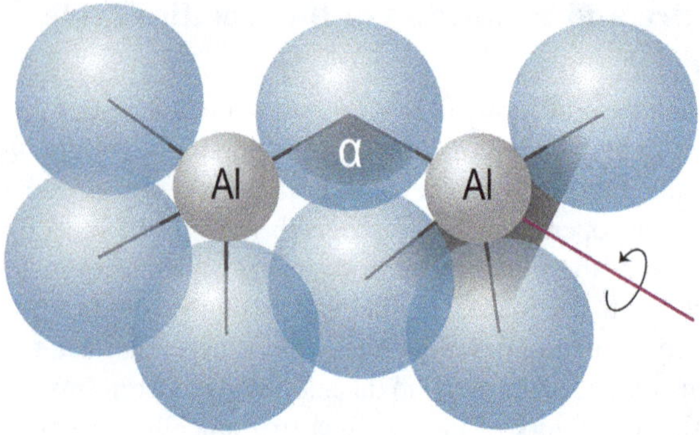

Fig. 6.19 Neutron diffraction done at the *Institut Laue-Langevin* in Grenoble, France by Kniep, Lamperter, and Steeb in 1989 [30] determined each cell of the anodic oxide comprises AlO_4^- tetrahedra with short-range order (periodicity of <5 units). The tetrahedra are the basic building blocks that comprise the internal network of the individual cells, which make up the nanoscale network of the anodic oxide

during the anodizing process, and provides the basis for the reactions that occur on the central pore during operations after anodizing, such as dyeing and sealing. See Figs. 6.19 and 6.20.

By accepting the AAO structure that comprises AlO_4^- tetrahedra, it is also accepted that the oxide network grows as an ionic solid. This has major consequences as to how AAO participates in the anodizing circuit, because, in ionic solids, electricity is conducted primarily through the solid by the diffusion of ions [4]. For an isotropic aluminum substrate, all interstitial ions will be of the same size and charge because they will all be Al^{3+} ions. Therefore, the interstitial spaces on the AAO structure will easily accommodate them and the oxidation reaction on isotropic aluminum will proceed as if there is no resistance to oxide growth at all.

It is critical to remember, when considering how the anodic oxide develops and grows, that ionic diffusion in an ionic solid, behaving as a diode, will only flow one way—in the direction of the applied current.[1] As the electrical circuit continues to operate, a force is exerted by the applied current density, i_a, on the ions in the oxide, and they rearrange themselves so that the current will continue to flow, producing i_o, which may include associated concentration polarization effects that encompass the increase in temperature associated with the heat of oxidation (ohmic resistance) that can build in the oxide over time. Ionic conduction through the oxide is critical to the robustness of

[1] The understanding that the anodic oxide behaves as a diode is as old as one of its first engineering applications, the condensator, invented by Eugene Ducretet in 1875. Recall from Chap. 2 that Ducretet determined that as it was forming, "the oxide (sic) coating allowed for the passage of current in one direction, but resisted the electric current flowing from the other direction." He is credited with assigning the term "valve metal" to metals that exhibit this behavior [31].

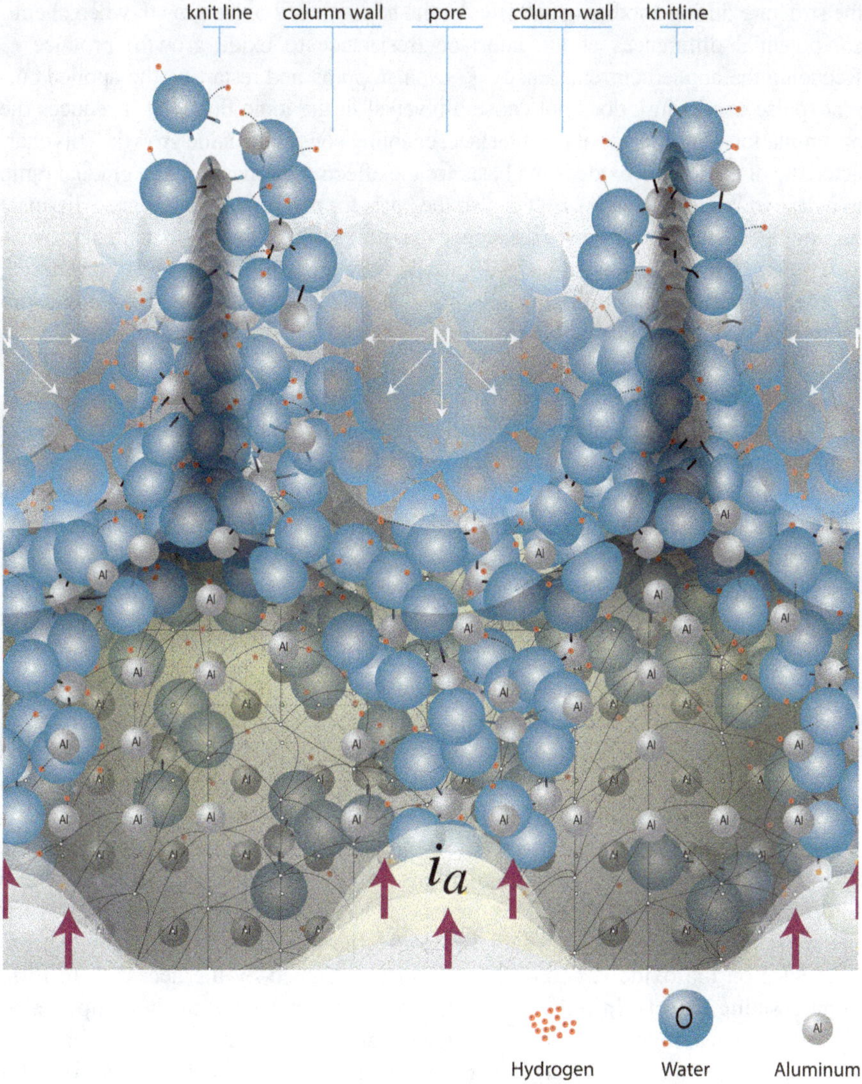

Fig. 6.20 Three-dimensional representation of a developing anodic oxide structure comprising AlO_4^- tetrahedra with short-range order growing from an "ideal" *face-centered cubic (FCC)* aluminum substrate. The orange spheres represent hydrogen atoms while the blue spheres represent oxygen atoms. Growth stress due to oxide growth is represented by the distortion of the oxygen atoms, especially at the knitlines between forming oxide cells. Polarization, N, sustains the pore diameter and the erect character of the oxide cell

the structure during anodizing and is essential to continued oxide growth when chemical potential differences at the interface (resistance to oxide growth) produce i_o. Reducing the applied current density, i_a, even stopping and restarting the applied current (pulse anodizing), does not cause a reversal in the ionic flow, but it reduces the accumulation of resistance at the interface, enabling continued oxide growth. This characteristic of the anodic oxide is the basis for the effective application of a gradual ramp to a desired "oxide growing" i_a. It is also the basis for pulsing of i_a in order to dissipate accumulated Joule's heat due to resistance that may build in the oxide while it grows.

The AAO produced by industrial anodizing is amorphous, and has an ordered, short-range structure comprising tetragonal AlO_4^- molecules that functions as an ionic solid. Polarization enables structural order of the oxide by establishing growth from centers of symmetry. As hydrated aluminum oxide, the bond lengths of the AlO_4^- are longer and weaker. Order is short range (less than 5 unit cells) and electric permittivity is higher (resistance to electric flow is higher). Its ordered structure has larger strain per unit polarization, and therefore can flow under electrostrictive strain. With surface reconstruction, the growth of new oxide continually exerts lateral forces against nearest neighbor oxide flakes and instability of the total oxide increases when polarized oxide elements impinge on one another. A high level of electrostrictive strain arises, increasing the interfacial temperature and the bond lengths on the AlO_4^- structure. This effectively softens the oxide and deformation occurs in the direction of the applied current, opposite to the growth direction of the oxide, which is into the aluminum (alloy) anode. Changes in the structure due to electrostriction results in the development of an exchange current density, i_o, even when anodizing isotropic aluminum. As in all corrosion reactions, i_o governs the kinetics of the anodizing reaction. As oxidation proceeds, and aluminum ions are consumed, most move onto the forming AlO_4^- structure.

Electrostrictive strains act against the bonding in the ionic solid, displacing some of the ions from their equilibrium positions on the oxide structure in areas where interfacial strain is the highest. This creates vacancies on the oxide structure, which are most probably oxygen vacancies because the creation of aluminum (cation) vacancies on the oxide structure is strongly constrained by the necessity to form compensating defects. In order to move, oxygen vacancies are easily compensated by the copious amount of electrons, oxygen and/or hydroxide ions. Vacancies move with the deforming oxide along the column walls, to the surface of the oxide, typically along the knitlines, between each column. The presence of vacancies that remain in the oxide reduces the density of the oxide as it approaches the interface with the electrolyte. The importance of this phenomenon, in addition to electrostriction, is that changes in oxide density from the interface with the metal, along the knitlines, to the interface with the electrolyte, affect AAO structure and properties.

As a result, three basic compositional gradients develop in the anodic oxide as it grows [32]: (1) the oxide formed at the polarized centers of each column exhibits the most stoichiometric oxide. No specific electrolyte ions are present at the interface with the substrate, other than those associated with electrolysis of water. Excess aluminum ions, those not incorporated onto the oxide structure, move to interstitial positions on the oxide structure; and they move with electrons, transported by collisions

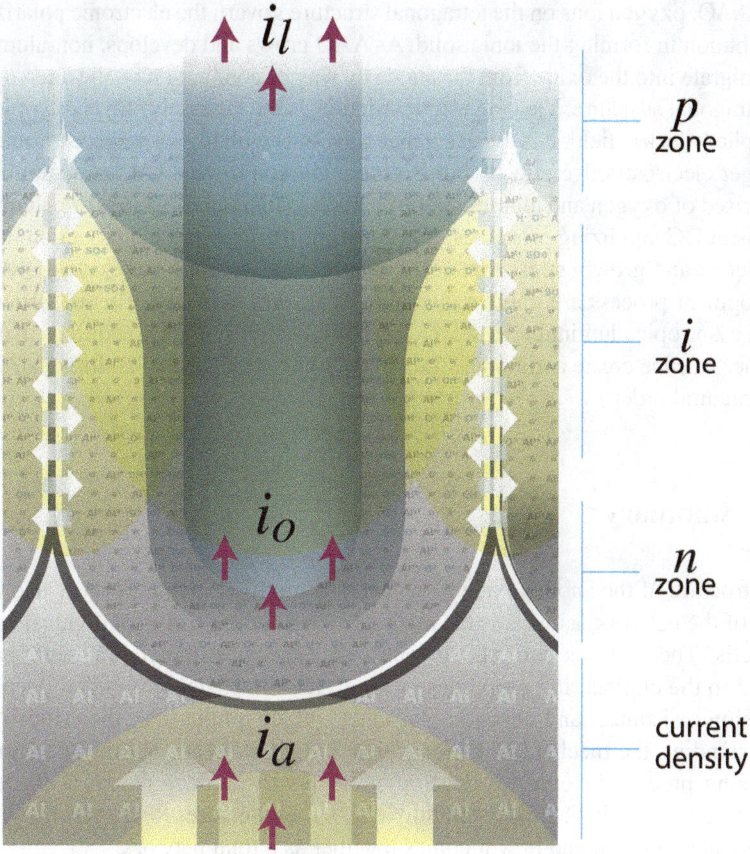

Fig. 6.21 The anodic oxide structure is that of an *aluminum deficient* amorphous ionic solid. It is also an *n-type* semiconductor, forming a p-i-n diode from the surface of the aluminum to the surface of the oxide, due to ion exchange reactions at the *oxide-electrolyte interface* [32, 33]

with the electron flow in the direction of i_a to the outer surface of the oxide, where they dissolve in the electrolyte. The ionic structure in this region of the oxide cell behaves as an n-type semiconductor. (2) The oxide through which the ionic counter-exchange occurs exhibits the most interstitial cations and oxygen vacancies on the oxide structure due to structural distortion caused by dielectrostriction. With its mixed structure, this region of the oxide cell behaves as an i-type semiconductor. (3) The composition of the oxide adjacent to the electrolyte contains foreign anions that have reacted and adsorbed onto the AAO on the inside surface of the central pore above the region where primary oxidation occurs, and exhibits the effects of reactions that occur with defects in the mixed structure from the i-region of the oxide cell. The ionic structure in this region is most like a p-type semiconductor. It is also the oldest oxide formed. For each cell comprising the network, the stability of the oxide varies from the base of the oxide cell, the center from which it grows, to the knitline of each column; and from the base of the central pore to the oxide-electrolyte interface. See Fig. 6.21.

In AAO, oxygen ions on the tetragonal structure govern the electronic polarization contribution in forming the ionic solid. As AAO grows and develops, non-aluminum ions migrate into the oxide from the anode by way of conduction on the interstices of the tetragonal structure. Viewing electrostriction as the interactive response of ions to an applied electric field, it is expected that alloy ions with higher valences would lead to larger electrostrictive effects. For isotropic aluminum, the AAO structure is only comprised of oxygen and aluminum ions and the only interstitial ions are aluminum. In typical DC anodizing processes, i_a is constant, the increase and reduction in electrostrictive and growth stress that occur, respectively, through the oxide growth and development process are uniform, yielding a uniform structure. This is particularly true for isotropic aluminum. However, interstitial cations of alloy additions that have a higher valence create additional strain in the total oxide, and sometimes can distort the structural order.

6.8 Summary

The structure of the anodic oxide alone bears admiration: the uniformity and continuity of the network, as well as the symmetry and function of its individual columnar cells. The features comprising each cell, especially the central pore, can be related to the engineering properties of the AAO: dimensional stability, dyeability, corrosion resistance, and abrasion resistance, which emphasizes the criticality for understanding the mechanism of formation and growth. The ability to control the anodizing process in order to develop strategies to produce oxides with desired properties is based in the understanding of how these features form, starting with the individual oxide cell and how it comes together as a total network.

What separates the growth and development of the anodic oxide from the passive oxide layer that naturally forms on the aluminum surface, or other engineered oxides produced by chemical treatment, is the use of an external source of electrical energy to drive the oxidation reaction. This is where the order begins and engineering can take hold: with the selection of an appropriate current density, i_a, the substrate surface responds with polarization, producing an ordered charge on the surface that produces the basis for the order of the oxidation reactions. Typical anodizing processes yield ordered anodic oxides with predictable, tunable, structural features.

While the anodic oxide is amorphous, it does have a chemical and pre-crystalline structure. The electrochemical and dynamic nature of its growth and development prevent the formation of long-range order; however, AAO has the structure and properties of an ionic solid. The ionic nature of the AAO enables electrical conduction through the oxide and thereby continued oxide growth beyond surface reconstruction. As a finished oxide, the ionic character of the central pore also enables the follow-on processes of dyeing and sealing.

After the applied current density, i_a, polarization, N, is perhaps the most important driving force for the anodizing reaction. Measured in volts, N is a response to

i_a, or N can be set by the potential selected for a constant voltage anodizing process. All of the mechanisms of electric polarization: space charge effects, dipole reorientation, ionic displacements, and electronic polarizability are key mechanisms at play in the growth and development of the anodic aluminum oxide. While not all of the mechanisms occur at once, each one functions to produce the highly ordered, self-assembled nanoscale AAO structure as well as the characteristics and properties of each cell that comprises it.

Initially, polarization functions to set the order for the main nucleation points for the oxide columns that grow and develop to comprise the total anodic oxide network. Together with i_a, polarization produces ordered corrosion of the substrate surface, simultaneously dissolving the aluminum while producing electrolysis in the electrolyte. An ordered charge density wave develops on the surface, producing a space charge on substrate at the interface with the electrolyte. As electrolysis proceeds, areas of localized higher pH develop in the saddles of the charge density wave; periodic concentrations of OH^- ions correspond to ordered concentrations of aluminum ions, all set by polarization of the anode surface. Polarization is therefore the key to the development of stable oxide nodes in an environment that typically would not support aluminum oxide according to the Pourbaix diagram.

Aluminum oxide nodes grow and develop from the point of symmetry set by the polarization of the surface, by way of outward growth stimulated not only by i_a, but by the chemical potential differences between large oxide particles and smaller oxide particles. The chemical potential difference between the oxide and unreacted aluminum also drives the reaction, and thereby, oxide growth consumes the aluminum substrate. As steady-state oxidation proceeds, each node grows outward to meet nearest neighbor oxides that have grown and developed simultaneously. When they impinge on one another, the substrate surface is totally reconstructed with oxide to a uniform depth.

With impingement, a network of equiaxed hexagonal flakes each with uniform volume is produced. On isotropic aluminum, each hexagonal oxide flake comprising the network is of uniform size and thickness. As long as the anode is polarized and current flows, the corrosion reaction continues after surface reconstruction; and growth of the total oxide network continues, into the substrate. Dipole reorientation occurs due to the polarization within the each growing oxide cell, and because anodic alumina is an ionic solid, mechanical strain develops in the structure and increases uniformly in the impinging flakes, along the interface with the substrate and through the oxide thickness. Mechanical forces are counteracted and sustained by the polarization forces on the inside surface of each oxide flake, producing electrostriction within the individual oxide flakes and increasing the stress in the total oxide network.

Electrostriction is especially significant when a reaction product is an ionic solid, and strain is developed with ionic flow along or across an interface or within the structure of the ionic solid. The counteracting polarization forces and mechanical strain produce a virtual die through which the ionic solid flows, which for anodic oxidation is the amorphous AAO structure [16]. Deformation of the ionic solid occurs, producing ionic displacement at the impinging interfaces, producing oxygen vacancies on the AAO structure. Opposing polarization and mechanical forces that span the growing oxide from the oxide-metal interface to the oxide-electrolyte interface produce the central pore of each individual columnar cell in the oxide network; constraint at the interfaces between neighboring cells sustains the structural framework as the oxide growth continues.

Deformation produced by electrostriction in the AAO is permanent and not reversible with a change in the applied current bias. This characteristic is proof that the porous anodic aluminum oxide forms due to solid rather than colloidal flow, although formation of AAO due to colloidal precipitation has been proposed by some researchers [34, 35]. In fact, in the case of the anodic oxide, if the electric field is reversed, rather than the colloids rolling back and the columnar form relaxing, the structural network is destroyed as the oxide dissolves in the electrolyte.

The relationship between polarization, electrostriction, and structure are important to consider when considering engineering properties of the anodic oxide. Hardness, abrasion resistance, corrosion resistance, and clarity can all be "tuned" by manufacturing oxides with different structural features, depending upon the anodizing conditions and process parameters. In general, in response to a constant applied current density, pores are of uniform diameter, linearly proportional to the anodizing potential. Column wall thickness, an extension of the initial oxidation reaction at the polarized centers at the time of surface reconstruction, is typically uniform when measured from the base of the central pore to the interface with the aluminum substrate. The amount of residual stress is directly related to the square of the polarization value for the total oxidation reaction. Exposure time ultimately determines the thickness of the growing oxide.

The consistency in the relationship between polarization, electrostriction stress, and the anodizing process parameters makes anodizing an engineering process, a reliable design option for a number of applications. Whether requirement is for long-lasting clarity, decorative color, or a functional wear surface, the anodic oxide can be the finish of choice for aluminum and aluminum alloys.

References

1. Runge, J. M. (2008). Interfacial phenomena and anodizing: Ramifications and process solutions. *Proceedings of the 17th Annual AAC Conference, San Francisco, California.*
2. Jones, D. A. (1996). *Principles and prevention of corrosion* (2nd ed.). Upper Saddle River, NJ: Prentice Hall.
3. Lupis, C. H. P. (1983). *Chemical thermodynamics of materials.* New York, NY: Elsevier Science Publishing.
4. Shewmon, P. (1989). *Diffusion in solids.* Salem, MA: TMS Publication.
5. Murr, L. (1975). *Interfacial phenomena in metals and alloys.* Reading, MA: Addison-Wesley Publishing.
6. Rana, F. (2007). "Polarization", Lecture 7, ECE 303. Department of Electrical and Computer Engineering, Cornell University.
7. Csokan, P. (1980). Nucleation mechanism in oxide formation during anodic oxidation of aluminum. In M. G. Fontana (Ed.), *Advances in corrosion science and technology.* New York, NY: Plenum Press.

8. Han, C., & Runge, J. (2002). The future of anodizing. *Proceedings of the Annual Technical Conference and Exposition of the Aluminum Anodizers Council, Oakbrook, Illinois.*
9. Çapraz, Ö., Shrotriya, P., Skeldon, P., Thompson, G., & Hebert, K. (2015). Role of oxide stress in the initial growth of self-organized porous aluminum oxide. *Electrochimica Acta, 167*, 404–411.
10. Barkey, D., & McHugh, J. (2010). Pattern formation in anodic aluminum oxide growth by flow instability and dynamic re-stabilization. *Journal of the Electrochemical Society, 157*(11), C388–C391.
11. Sundar, V., & Newnham, R. (1992). *Electrostriction and polarization, Ferroelectrics* (Vol. 135, pp. 431–446). New York, NY: Gordon and Breach Science Publishers, S.A.
12. Vanhumbeeck, J.-F., & Proost, J. (2008). On the contribution of electrostriction to charge-induced stresses in anodic oxide films. *Electrochimica Acta, 53*, 6165–6172.
13. Lee, H.-Y., Peng, Y., & Shkel, Y. (2005). Strain-dielectric response of dielectrics as foundation for electrostriction stresses. *Journal of Applied Physics, 98*, 074104.
14. Feng, X., Huang, Y., & Rosakis, A. (2007, November). On the stoney formula for a thin film/substrate system with nonuniform substrate thickness. *Transactions of the ASME, 74*, 1276–1281.
15. Van Overmeere, Q., Blaffart, F., La Mantia, F., Di Quarto, F., & Proost, J. (2012). Electromechanical coupling in anodic niobium oxide: Electric field-induced strain, internal stress, and dielectric response. *Journal of Applied Physics, 111*, 113529.
16. Goyon, J., Colin, A., Ovarlez, G., Ajdari, A., & Bocquet, L. (2008, July). Spatial cooperativity in soft glassy flows. *Nature Letters, 454*, 3.
17. Runge, J. M. (2015). Enhancing anodic aluminum oxide for bonding applications. *Proceedings of the 24th Annual AAC Conference, San Diego, California.*
18. Garcia-Vergara, S. J., Skeldon, P., Thompson, G., & Habazaki, H. (2006). A flow model of porous anodic film growth on aluminium. *Electrochimica Acta, 52*, 681–687.
19. Baron-Wiecheć, A., Ganem, J.-J., Garcia-Vergara, S. J., Skeldon, P., Thompson, G., & Vickridge, I. (2010). Tracer study of porous film growth on aluminum in phosphoric acid. *Journal of the Electrochemical Society, 157*(11), C399–C407.
20. Çapraz, Ö., Shrotriya, P., Skeldon, P., Thompson, G., & Hebert, K. (2015). Factors controlling stress generation during the initial growth of porous anodic aluminum oxide. *Electrochimica Acta, 159*, 16–22.
21. Baron-Wiecheć, A., Burke, M., Hashimoto, T., Liu, H., Skeldon, P., Thompson, G., Habazaki, H., Ganem, J.-J., & Vickridge, I. (2013). Tracer study of pore initiation in anodic alumina formed in phosphoric acid. *Electrochimica Acta*, 0013–4686.
22. Skeldon, P., Thompson, G. E., Garcia-Vergara, S. J., Iglesias-Rubianes, L., & Blanco-Pinzon, C. E. (2006). A tracer study of porous anodic alumina. *Electrochemical and Solid-State Letters, 9*(11), B47–B51.
23. Runge, J. M., & Pomis, A. J. (2000). Anodic oxide film formation relating mechanism to composition and structure. *Proceedings ASST, Manchester, England.*
24. Paschanka, M., & Schneider, J. (2011). Origin of self-organisation in porous anodic alumina films derived from analogy with Rayleigh-Bénard convection cells. *Journal of Materials Chemistry, 21*, 18761.
25. Toccafondi, C., Stępniowski, W. J., Leoncini, M., & Salerno, M. (2014). Advanced morphological analysis of patterns of thin anodic porous alumina. *Materials Characterization, 94*, 26–36.
26. Runge, J. (2007). Formation of porous anodic oxide finishes: A new approach and theory. *From the Conference Proceedings of Aluminium 2000, Florence.*
27. Parkhutik, V. P., & Shershulsky, V. I. (1992). Theoretical modeling of porous oxide growth on aluminium. *Journal of Physics D: Applied Physics, 25*, 1258–1263.
28. Runge, J. M., & Hossain, T. (2015). Interfacial phenomena in 7000 series alloys and their impact on the anodic oxide. *Proceedings of Aluminium Two Thousand World Congress and International Conference on Extrusion and Benchmark ICEB 2015, Materials Today Proceedings.*

29. Roth, A. (1940). Ein Beitrag zur Kenntnis der Struktur des elektrolyktisch erzeugten Aluminiumoxyds. *Zeitschrift für anorganische und allgemeine Chemie*, *244*, 48–56.
30. Kniep, R., Lamperter, P., & Steeb, S. (1989). Structure of anodic oxide coatings on aluminum. *Advanced Materials, 1*(7), 229–231.
31. Ducretet, M. E. (1875). Note Sur un Rhéotome Liquide a direction Constante, Fondé sur une Propriété Nouvelle de L'Aluminium. *Journal of Theoretical and Applied Physics, 4*(1), 84–85.
32. Lerner, M. (2003). Is the barrier layer responsible for the cellular structure of aluminum oxide film anodized in sulfuric acid? *Proceedings of AESF Surfin Conference, Chicago.*
33. Uhlig, H. H. (1985). *Corrosion and corrosion control, an introduction to corrosion science and engineering* (2nd ed.). New York: Wiley & Sons.
34. Murphy, J. D., & Michelson, C. E. (1961). A theory for the formation of anodic oxide coatings on aluminium. *Proceedings of the Conference on Anodizing Aluminium, Nottingham, UK* (pp. 83–95).
35. Pashchanka, M., & Schneider, J. (2013). Experimental validation of the novel theory explaining self-organization in porous anodic alumina films. *Physical Chemistry Chemical Physics, 15*, 7070.

Chapter 7
Interfacial Phenomena and Anodizing

7.1 Introduction

The initial interface created between the aluminum substrate surface and the electrolyte is the most important combination of variables that govern the fundamental structure and properties of the anodic oxide. The condition of the bulk microstructure of the substrate is most often different from the condition of the microstructure at the surface, where the effects of manufacturing and thermal processing are most concentrated; in fact, in the hierarchy of microstructural defects, the surface alone can be counted as the largest defect.

As the anodic oxide grows, the initial aluminum surface is consumed; therefore, a new surface with a new set of defects is presented continuously as the interface changes from which the oxide grows. With oxide growth, additional interfaces develop: the knitlines, which are the interfaces between each cell comprising the total anodic oxide; and the interface between the oxide and the electrolyte. As additional interfaces develop with the growth of the anodic oxide, aluminum and alloy ions that are not incorporated onto the AAO structure, may be entrained at the metal-oxide interface, conducted through it along knitlines, or, ejected into the electrolyte, making these interfaces as important to continued oxide growth as the anode-oxide interface.

Therefore, in order to understand the total effects of interfacial phenomena on the dynamic nature of the anodizing reaction, it is necessary to consider both sides of each interface as ever-changing solutions. Important questions to consider are: how does the substrate change as the oxide grows? What impact do the movement of electrons and the corresponding diffusion of alloy ions, with different migration rates, and transport of intermetallic particles have at each interface, on the oxide structure, and on the electrolyte? How do different types of interfacial phenomena change the oxide structure? The answers to these questions can be found by using the corrosion model for the anodizing process, the Tafel Equation, which was introduced in Chap. 5, as well as the concepts for oxide growth and development from

J.M. Runge, *The Metallurgy of Anodizing Aluminum*,
https://doi.org/10.1007/978-3-319-72177-4_7

Chap. 6. These principles will be developed further in this chapter in order to address the reactions that occur when anodizing real world substrates.

As the parameters of the anodizing process present, first and foremost, the thermodynamic conditions for the formation of hydrated aluminum oxide, it must be recognized that even the purest aluminum alloys (ideal substrates) contain non-idealities in the form of trace elements, atomic level defects, and other microstructural features and defects that have been introduced by manufacturing, possibly even surface contamination. Of the various microstructural features and defects at the substrate surface, some react with the process parameters to form anodic oxide, and some don't; some even confound the process causing side reactions to occur. Polarization effects, however, introduced by the external power source utilized for anodizing, usually dominate the surface reaction, making anodizing even the most complex alloys possible. Nevertheless, because side reactions that change how the oxide nucleates and grows cannot always be avoided, an understanding of microstructural and external surface defects and how they react at the substrate-oxide interface during anodizing to impact the exchange current density, i_o, as well as their effects as they move through the oxide and into the electrolyte on the limiting current density, i_L, are the keys to being able to address interfacial phenomena and minimize their effects.

As the oxide grows and develops, it becomes an element in the anodizing circuit. The semiconductor characteristics of the oxide, which enable ionic conduction through the oxide structure, together with the effects of the polarization/current density wave at the anode-oxide surface, keep the anodizing circuit operating. Interfacial phenomena at all interfaces including the knitlines between the columns that comprise the "network in the network" of self-assembled, highly ordered nanoscale electrochemical cells are important considerations, especially with anodes of increased alloy complexity, whose intermetallic particles and elements in solid solution locally react to change the order of the anodic aluminum oxide, or completely block its formation, which change its functionality as a finished oxide.

In this chapter, the impact of various interfacial phenomena is explored, at the substrate-anodic oxide boundary, at the oxide-electrolyte boundary, and along knitlines in the oxide structure, as it grows and develops during the anodizing process. Defects are presented according to size, from the smallest to the largest, regarding their effects on the finished anodic oxide. Intrinsic phenomena from the atomic level, such as dislocations and vacancies, to the microscopic level such as microstructural intermetallic phases, particles, and other defects are considered, as well as the effects of extrinsic interfacial phenomena, such as surface texture, component features, and burrs and laps induced by mechanical surface treatments, and the effects of surface contamination. By explaining the reactions that occur due to individual interfacial phenomena, the impact of these phenomena on the kinetics of finish formation can be related to the quality of the total oxide structure over the entire anodized substrates.

7.2 The Role of the Electric Field and Ionic Conduction in the Anodizing Circuit

Recalling the Tafel equation from Chap. 5, presented again as Eq. 7.1, the conditions for the total electrochemical circuit are considered simultaneously for the oxidation reaction: Surface Activation Polarization, N_{Surface}, which includes the characteristics of the anode, including the surface and near-surface microstructure; and Concentration Polarization, N_{Conc}, which includes anode surface texture and any possible surface contamination, as well as the parameters of the anodizing process: electrolyte concentration, distance between the anode and cathode, and the heat of reaction. The total polarization of the anode, N_{Total}, is the necessary value that dictates whether or not anodizing will take place at all, and is always driven by the applied current density, i_a.

$$N_{\text{Total}} = N_{\text{Surface}} + N_{\text{Conc}}$$
$$N_{\text{Total}} = \beta_a \log\left(\frac{i_a}{i_o}\right) + \frac{2.3 * RT}{nF}\log\left(1 - \frac{i_a}{i_L}\right) \tag{7.1}$$

When considering how the practical anodizing process can connect to the science of the corrosion reactions that occur at the various interfaces, which develop during anodic oxidation, it is helpful to consider how the current flows within each element in the anodizing circuit with respect to the structure and properties of each circuit element. Understanding the applied current density, i_a, as the source for the movement of all electrons, ions and particles in all parts of the anodizing circuit facilitates understanding of what, how, why, and where resistance occurs in the circuit, developing exchange current density, i_o, which aids in all aspects of anodizing process development as well as troubleshooting.

Oxide growth can only proceed if the current in the anodizing circuit can flow; therefore, all possible sources for resistance, including the thickness of the developing oxide itself, need to be considered. Continued imposition of the applied current density, i_a, to the anode stimulates, by way of electrotransport, the diffusion of metal ions and intermetallic particles through the anode to the interface with the oxide, through the forming oxide and into the electrolyte. For isotropic aluminum, and for all aluminum and aluminum alloy substrates, the mechanism for electrotransport changes from the face-centered cubic (FCC) metal structure, in which electrons carry the current, to the oxide structure, which comprises tetragonal AlO_{4-} molecules with short-range order, in which ions primarily carry the current. Therefore, the interfaces modeled by the Tafel equation: (1) between the substrate and the oxide, with its corresponding exchange current density, i_o, and corresponding surface activation polarization response, N_{Surface}; and (2) through the oxide thickness, along knitlines and at the oxide surface with the electrolyte, with its corresponding limiting current density, i_L, and corresponding concentration polarization response, N_{Conc}, must be kept in mind when considering how the current will flow through the total anodizing circuit and how it can be maintained to grow an oxide of desired thickness. These are presented again as Eqs. 7.1a and 7.1b.

$$N_{\text{Surface}} = \beta_{\text{a}} \log\left(\frac{i_{\text{a}}}{i_{\text{o}}}\right) \tag{7.1a}$$

and

$$N_{\text{Conc}} = \frac{2.3 * RT}{nF} \log\left(1 - \frac{i_{\text{a}}}{i_{\text{L}}}\right) \tag{7.1b}$$

How the applied current density, i_a, flows from the anode to the interface with the oxide and then through the oxide as it grows and develops is based upon the fundamental change of material comprising the components of the anodizing circuit: the anode is metal and the oxide is an ionic solid that behaves like a semiconductor (see Chap. 6). In other words, i_a is conducted by way of electron flow in the anode and primarily by way of ions in the AAO, with some help from interstitial electrons. The various ions that are carried by electrotransport, and defects and particles that flow and move by way of mass transport from the anode to the interface with the oxide and then into the AAO by way of i_a, pose various types of resistance challenges that develop exchange current density, i_o, and sometimes limiting current density, i_L, at the interface with the anode, creating polarization effects that not only govern the kinetics of oxide growth but cause local disorder in the developing anodic oxide structure.

Resistance develops in the anode as electrons, driven by i_a, collide with interstitial aluminum ions, which impart momentum and bias their flow, creating distinct thermal effects that increase the temperature of the anode as the current flows. Resistance, and therefore the heat of oxidation, increases with the diversity of alloying elements and intermetallic particles that are present in the anode microstructure, when anodizing aluminum alloys. The oxidation reaction becomes more complex with diverse microstructural features such as grains with different sizes and orientation, various atomic level defects, elements in solution with the matrix, and various intermetallic particles of different size, shape, population, and distribution, all presenting different chemical potentials at the surface that develop exchange current density, i_o.

The heat of reaction, caused by resistance in the anodizing circuit, can increase significantly as the oxide grows. This value is represented in the Tafel Equation by T, the heat of oxidation, which has important ramifications on the performance of the anodizing circuit in terms of Concentration Polarization Effects (CPE). By increasing resistance in the anodizing circuit, interfacial phenomena increase the heat of the oxidation reaction, T, by way of increasing the exchange current density, i_o, due to microstructural characteristics of the anode, and by way of increasing the limiting current density, i_L, by hindering the reaction kinetics at the anode-oxide interface, through the oxide or at the oxide-electrolyte interface. Therefore, understanding how to control interfacial phenomena is important to successful anodizing.

As the oxide grows, the reactions that occur during anodizing are impacted not only by the quality of the substrate surface but by the quality of the near-surface microstructure. This concept is important to keep in mind while considering the distances over which the various dynamic reactions occur during anodizing. The amount of substrate consumed by oxide during anodizing is often small in comparison to the bulk of the substrate, but this makes the quality of the near-surface microstructure even more significant, as the source for ions, intermetallic particles and other defects that can interfere with anodic oxide quality. This is particularly important when considering the engineering demands on the anodic oxide that require uniformity and continuity at the oxide surface and cohesion throughout the total oxide thickness.

Egress of i_a, from the AAO, across the interface with the electrolyte and into the electrolyte to the cathodes, is also important to consider because the anodizing circuit must be completed, if the oxide is to grow. Ions also carry i_a in the electrolyte, and the reactions that occur with continued electrolysis and the transport of various metallic ions from the anode through the oxide to the electrolyte provide the means to complete the circuit but also pose interesting side reactions that also create concentration polarization effects by changing i_L.

7.2.1 The Polarization Wave

The polarization wave, which sets the spacing of the initial oxide nuclei that grow to develop the network of cells that comprise the AAO, is sustained throughout the anodizing reaction, tracking at the interface of the substrate with the growing oxide. The polarization wave is the result of electrolysis and the corresponding separation and concentration of charge. The wave comprises electrolyzed H^+ and OH^- ions from the electrolyte, as well as electrons that don't have enough energy to escape the polarized surface and migrate to the cathodes. The wave not only provides the means for a stable anodic oxide growth under conditions that typically dissolve aluminum oxide, it can also provide a "trap" for metallic alloy and aluminum ions ejected from the growing oxide.

The energy necessary to exceed the polarization wave must be at least that of the applied current density, i_a. Therefore, resistance caused by interfacial phenomena, that reduce the exchange current density, may result in the local reduction in energy that prevents the escape of ions from the wave. This reduction in energy can also be the result of CPE, in the form of surface contamination or topographic features on the anode such as sharp corners. Changes in the N_{Surf} and/or in N_{Conc} affect the total polarization, N_{Total}. Any time N_{Conc} increases, the total polarization, N_{Total}, decreases, which can change the dimensions of the growing oxide structure, even locally shutting down the anodic circuit. If N_{Conc} is the result of an intermetallic particle, or extreme surface roughness, the effect is transient, with the duration and impact of the effects depending upon the size and solubility of the defect at the substrate-oxide interface. When the defect exceeds the interface with the oxide, either dissolved by the electrolyte or carried into the oxide, N_{Total} is restored, resistance is

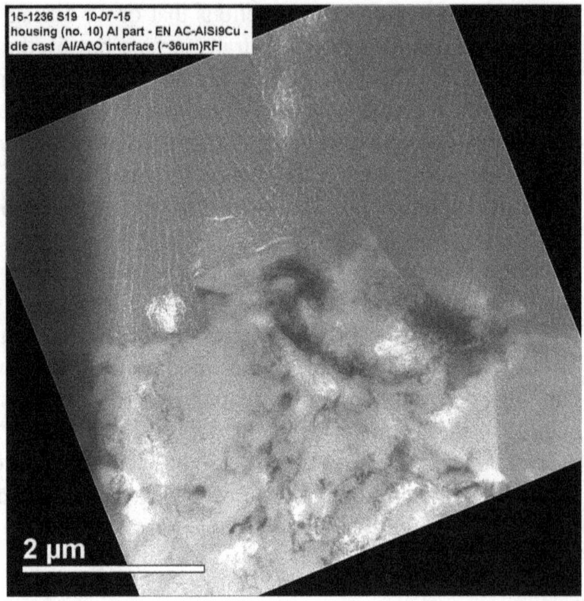

Fig. 7.1 Anodized die-cast alloy Al-Si9-Cu. The effects of a rough surface are manifested in the various oxide growth fronts established at various orientations. The effects of exchange current density and CPE are often significant because they disrupt the structural order of the AAO; columns branch, the oxide surface roughness increases, and other discontinuities develop through the oxide thickness that can impact the function and engineering performance of the total anodized component. (TEM Image courtesy of CeriumLabs)

reduced, and the effects of polarization fully restore the kinetics of the oxidation reaction. The effects of changes in exchange current density and CPE on N_{Total} are often significant because they disrupt the structural order of the AAO; columns branch, the oxide surface roughness increases, and other discontinuities develop through the oxide thickness that can impact the function and engineering performance of the total anodized component. See Fig. 7.1.

Most of the ions in the oxidation circuit that flow from the oxide to the cathodes are hydrogen, by far the smallest positive ions. Hydrogen ions, not trapped in the polarization wave, migrate to the cathodes and complete the anodizing circuit. However, the effects that occur with other ions that pass through the growing oxide structure, and migrate at the same rate or faster than aluminum ions, are distinct and important [1, 2]. Alloy ions are much larger than hydrogen ions, and most are larger than aluminum ions. Those that migrate as fast as, or faster than aluminum orient themselves at the interface with the oxide and "hop" through interstitial spaces in the forming oxide structure, and are ejected into the electrolyte at the base of the pore. The resistance effects created as the ions rearrange themselves to move into the structure are reflected in a decrease in exchange current density, i_o, which impacts surface polarization, N_{Surf}. As they migrate through the oxide toward the cathode, changes in the heat of oxidation create CPE as the oxide thickness increases.

When alloy ions are ejected from the oxide into the electrolyte, they can become trapped in the polarization wave that tracks with the anodizing reaction at the substrate-oxide interface. Ions trapped in the wave can adsorb on the surface of the oxide, and shut down the ion exchange that enables the oxide to grow in thickness through the development of the central pore. The primary alloy ion that develops such oxides is copper; such "high resistance" oxides are grown on 2XXX series

Fig. 7.2 TEM image of the AAO developed on an alloy 2011 anode. The anode surface is heavily cold worked and highly oriented. Copper ions diffuse through the oxide structure and can't exceed the polarization wave that forms at the anode surface, creating concentration polarization effects that prevent the formation of the typical porous, columnar structure. Instead, a net-like structure develops as would-be columns collapse, forming compact cells that limit the oxide thickness. (TEM Image courtesy of CeriumLabs)

wrought alloys and 3XX series cast alloys. The main result created by this phenomenon (adsorption of metal ions on the forming oxide interface with the electrolyte) is the development of profound local CPE with a transient reduction in dielectric withstand at the anode-oxide interface; in fact, with continued imposition of i_a, such defects may cause the oxide to locally rupture from the substrate, by exceeding the dielectric withstand voltage of the forming oxide. As the oxidation reaction continues, the sequence of ion adsorption and voltage breakdown continues, developing a thin, compact oxide without distinct central pores. These oxides rarely grow thicker than 6–8 μ. See Fig. 7.2.

7.3 Surface Activation Polarization Effects

If one considers the expanse of a polarized aluminum surface, it can be concluded that because of its size, it exhibits an average charge density, and that because of its polarized state exhibits a rather uniform charge distribution (space charge). With this situation in mind, it is easy to visualize how oxide nucleation centers are uniformly distributed across the surface and therefore, the resistance term, β_a from the Tafel Equation, can be considered a constant value and a characteristic of the anode (see Chap. 5). However, the various possible chemistries and microstructures for aluminum alloys, presented in Chap. 4, clearly establish that the surface of a real world substrate will behave otherwise; there will inevitably be inherent characteristics and microstructural defects that interrupt the continuity of the base metal because of the polycrystalline and sometimes complex nature of aluminum and its alloys.

If one breaks the surface up into smaller parts, and considers those parts individually, the impact of the microstructural quality is magnified, such that microstructural features can sometimes change from characteristics to defects. The result of these defects is in a cascade of changes: changes in chemical potential at the surface of the aluminum alloy anode produce resistance, which cause changes in the exchange current density, i_o, and increase the heat of reaction, T, which are reflected in changes in oxidation kinetics by way of changes in i_L. Changes in the local polarization values from point to point across the surface, in turn, introduce changes to local rate of oxidation. By considering the ever-changing nature of the interface from which the oxide grows, it is impossible to quantify the impact of each defect, and truly, most are overridden by the polarization effects across the entire surface. Nevertheless, it is important to understand qualitatively how interfacial phenomena can impact the anodizing reaction and change the structure, and therefore the engineering function or performance of the anodic oxide.

7.3.1 The Structural Mismatch

For the case of an isotropic aluminum anode, with an applied current density, i_a, electrons initially carry all of the electric current in the aluminum substrate. All cation interstitials on the FCC lattice are the same size, and carry the same charge, because they are all aluminum. The ratio of electrons to atomic (ionic) currents is high. The flux of electrons, from the applied current, collides with and pulls along all interstitial ions on the lattice. Atomic transport in the anode comprises only aluminum ions that make large diffusive jumps, which are biased by the flux of electrons in the direction of i_a. The continued biased movement of electrons produces momentum that also inhibits jumps from occurring in the opposite direction; consequently, the substrate surface, positively charged as an anode, is flooded with aluminum ions that are focused at polarized centers, producing the first surface for the oxidation reaction.

With anodic oxidation of an isotropic aluminum surface, surface activation polarization effects caused by the exchange current density, i_o, are the result of resistance caused by the *structural mismatch* between the FCC aluminum and the developing tetragonal AAO structure [1]. This is due to the difference between how the current is carried in the anode, by way of electrons, and how it is carried in the oxide, by way of ions. Nevertheless, these effects are uniform across an isotropic aluminum substrate, because physical properties are equal along all crystallographic planes. Therefore, for an isotropic substrate, the anodizing reaction can theoretically occur at the oxide-metal interface continuously, as if the metal surface had not any oxide present at all, until the entire anode is consumed. The uniform and consistent change in how the current is carried across the interface from the anode to the oxide, due to structural changes, as well as the effects of uniform growth stress and electrostriction, which are the result of how the oxide grows and the magnitude of i_a, would produce an ideal AAO, free of any structural deviations in total thickness, pore diameter, and column wall thickness.

With anodic oxidation of industrial aluminum and aluminum alloy substrates, the exchange current density, i_0, changes with surface defects, alloy additions, and contamination. The effects of the structural mismatch can be profound. Alloying elements may have a different crystallographic structure than the parent aluminum, and intermetallic phases and particles therefore exhibit different solubilities in the electrolyte, anodizing at different rates, even confounding oxidation all together. Local differences in oxidation rate produce rough, nonuniform oxides that may yield anodic oxides with limited function. However, the rate of oxidation is not usually impeded on more homogeneous surfaces. Nevertheless, it is observed and recognized that the rate of anodic oxidation on specific grains follows the order of crystallographic orientation (111) > (110) > (100). In other words, the chemical potential is highest on the (111) plane because the atomic packing density is the highest [3]. Other atomic level defects may impede nucleation, such as vacancies, dislocations, steps, or ledges, which develop concentration polarization effects that prevent oxide nucleation because empty space doesn't anodize, or because asperities proud to the anode surface exhibit an increase in charge density that disrupts the polarization wave.

7.3.2 The Compositional Mismatch

The condition of alloying elements in the substrate microstructure is significant as to how their impact on the entire anodizing circuit and therefore how they affect AAO nucleation and growth. An aluminum alloy lattice contains alloy atoms in substitutional solution and a wide range of phases and intermetallic particles that are not coherent with the microstructure. Indeed, the effects that elements in solution with the aluminum have on i_0, and consequently, $N_{Surface}$, are different from the effects of intermetallic particles and precipitates. Since the anodic oxide structure will not comprise alloy atoms or intermetallic particles, in addition to the i_0 effects due to the structural differences between the anode and the oxide, alloying elements and the intermetallic phases, particles and precipitates they form present *a compositional mismatch* between the substrate and the anodic oxide that make i_0 more complex. Intermetallic particles and precipitates also present variations in chemical potential at the anode surface that affect i_0; however, variations in particle size and solubility may also produce concentration polarization effects, CPE, by blocking the flow of hydrogen ions, reducing i_L in the circuit.

For the case of anodes manufactured from any industrial aluminum alloy, from the highest purity to the most complex, the effects of the material characteristics and microstructural features of anisotropic polycrystalline anodes due to manufacturing processes change the local chemical potential across the surface that governs i_0. Grain boundaries, and atomic level defects such as dislocation loops, vacancies, kinks, and ledges, all produce resistance to the flow of the applied current density,

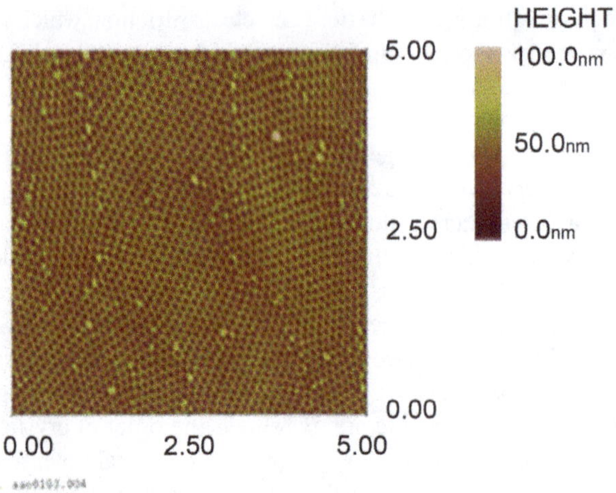

Fig. 7.3 AFM image of a Type III anodic oxide structure on a high purity aluminum substrate shows that the finish "piles up" at the substrate grain boundaries [4]

i_a, producing exchange current density, i_o, across the substrate surface during anodizing. Furthermore, the effects of strain and cold work exacerbate local changes in the surface chemical potential by producing concentrations of atomic level defects. The effects of surface activation polarization, $N_{Surface}$, even on high purity substrates, caused by atomic level and microstructural defects, affect how the oxide structure develops and grows across the surface. See Fig. 7.3.

Alloys bring a new set of electrochemical challenges to the anodizing circuit in the form of various soluble and insoluble constituents that change the chemical potential across the anode surface. Alloying element ions; primary constituent phases that form at the billet level, which have limited solubility in the electrolyte; dispersoids and other intermetallic particles, which vary in coherency and solubility with the substrate; and precipitates, which are incoherent with respect to the substrate microstructure and their interfaces with the substrate matrix, all produce local changes in chemical potential that cause changes in local interfacial resistance, making the exchange current density, i_o, more complex. The Tafel Equation can be used to model these effects.

7.3.3 *Modeling Surface Polarization Effects*

7.3.3.1 Exchange Current Density in Terms of Diffusion and Ionic Conduction

In order to understand the impact of interfacial resistance as producing i_o, as a response to i_a, during the oxidation process, specifically on the transition from the metallic anode to the growing anodic oxide, it is necessary to understand how the applied current will flow across the interface. Because of the structural mismatch, it is necessary to visualize the applied current density, i_a, first as a flow of electrons through the anode to the interface with the forming oxide, where it changes within the oxide, to the diffusion of interstitial ions within the ionic solid in order to continue to build the network structure and complete the anodizing circuit. The value of total current density, representing this change, is the ratio of applied current density, i_a, set by the process, to the exchange current density, i_o, set by the condition of the polarized surface, $N_{Surface}$, in the first half of the Tafel equation, and is presented again as Eq. 7.1a, below. In anodizing, the condition of the surface is transient, because the oxide continuously consumes the aluminum substrate; therefore, interfacial resistance effects are reflected in the ever-changing exchange current density, i_o, which fluctuates with changes in interfacial resistance due to the changing chemical potential at the interface between the aluminum and the growing oxide, governing the surface polarization portion of the Tafel equation [5].

$$N_{Surface} = \beta_a \log\left(\frac{i_a}{i_o}\right) \tag{7.1a}$$

With a fixed i_a, the efficiency of the anode decreases as the alloy surface is reconstructed by oxide (increases in interfacial resistance, which makes i_o increase). As an important part of the electric field forces, polarization of the anode as $N_{surface}$ continues throughout the entire anodizing reaction, and ionic charges, as Al^{3+}, continue to be concentrated at symmetric polarized centers at the changing interface, which are at the base of each pore, of each cell, in the AAO network. If $N_{surface}$ varies across the surface, due to the effects of i_o, polarization may change locally, where microstructural variations are encountered. Oxide nucleation may become locally preferential around significant microstructural defects, producing spots of local disorder in the total oxide structure.

In order to account for the movement of microstructural defects across the metal-oxide interface and into the AAO beyond surface reconstruction, we need to consider the conditions producing the polarization force, $N_{Surface}$. Without an applied current, chemical potential alone could stimulate the flow of ions onto an anodic oxide under equilibrium conditions; however, the reaction would be entirely preferential and oxide would lack order. The applied current density, i_a, is the primary source for the movement of all ions by stimulating ionic movement (diffusion) with the flow of electrons in the polarized substrate, making anodizing a nonequilibrium process.

The force exerted on the ions by the electric field can be accounted for by relating conductivity to the diffusion coefficient, presented by Eq. 7.2. This is done by considering that i_a is initially conducted by electron flow in the metal and then conducted through the AAO by the diffusion of ions [6]. In the Eq. 7.2, the flux, J_i, is equal to the product of the concentration of ions per unit volume, c_i, their mobility, M_i (which is related to the diffusivity, D_i), and the mean force on the particles, acted upon by the chemical potential, μ, presented as a differential because it will change with c_i, and the electric field, ϕ, presented as a differential because it is a response to the ion flow across the changing thickness of the growing oxide, δx. C_i is significant because it represents the concentration of alloying ions that were once in solid solution with the anode, as well as interstitial aluminum ions, that move to intersect the interface with the forming oxide. As c_i increases and decreases with changing substrate homogeneity, it impacts the chemical potential, μ, and the rate of oxide growth, δx.

For anodizing, the polarized centers set by i_a are the points of symmetry where charge, q_i, is most concentrated, and where the force of the electric field, ϕ, is strongest. The subscript i is used, because most of the ion flow will be on the interstitial sites of the substrate and then on the growing AAO.

$$J_i = -M_i c_i \left(\frac{\partial \mu}{\partial x} + q_i \frac{\partial \phi}{\partial x} \right) \tag{7.2}$$

In anodizing, because there is a constant supply of aluminum ions at the anode surface, driven by i_a, the diffusivity of interstitials on the oxide structure, D_i, is a better indicator for diffusion mobility than M_i.

$$M_i RT = D_i \tag{7.3}$$

Therefore, the flux of interstitial ions in terms of diffusion is given by Eq. 7.4:

$$J_i = -\frac{D_i}{RT} c_i \left(\frac{\partial \mu}{\partial x} + q_i \frac{\partial \phi}{\partial x} \right) \tag{7.4}$$

The flux of charge across the interface acting to form oxide or the reacting current per unit area is controlled by the exchange current density, i_o, from the Tafel equation. This can be explained as follows. Most of the aluminum ions at the interface move onto the forming oxide. This means the concentration of interstitial ions, represented by c_i, accounts for the concentration of alloying element ions that were once in solution with the aluminum substrate and have "unmixed" due to i_a, and includes any other aluminum ions, that did not move onto forming AAO structure. The current represented by these relationships is the result of the effects of the applied current density, i_a, on the forming oxide due to the change in structure from the aluminum substrate to the oxide structure, *which is, by definition from section 5.4.1a, the exchange current density,* i_o. The exchange current density can be related to the flux by the equation $i_o = J_i q_i$, therefore,

$$i_o = -\frac{D_i}{RT} q_i c_i \left(\frac{\partial \mu}{\partial x} + q_i \frac{\partial \phi}{\partial x} \right) \tag{7.5}$$

For most industrial aluminum alloys, the concentration of alloying elements with respect to the amount of aluminum is low, therefore, c_i will be low, but increases with alloy complexity. C_i remains low especially when the substrate exhibits microstructural homogeneity, and the distribution of alloying ions across the reaction interface is random and sparse. In cases of an inhomogeneous alloy, with an extremely segregated microstructure, or if a subject defect is large, the developing thickness of the oxide is affected, and δx decreases as c_i increases locally, causing i_o to decrease. Changes in chemical potential, μ, can also cause a decrease in δx, decreasing i_o. The exchange current density, i_o, decreases proportionately with respect to the characteristic μ of the defect, until it has dissolved and/or has passed into the AAO.

These considerations show that as the concentration of interstitial ions, c_i increases and the change in the oxide thickness, δx is reduced, the effect of the chemical potential is greater, making i_o more negative, which makes the surface polarization response, $N_{surface}$, smaller. Changes to i_o typically reduce local oxide growth rate, which reflects an increase in the heat of the oxidation reaction, consequently increasing the surface roughness of the total finish. Changes to the structural features of the oxide also result when $N_{surface}$ is reduced, such as a reduced pore diameter. Therefore, Eq. 7.5 shows that i_o varies as the oxide grows, consuming the alloy substrate, and is affected by the size and solubility of the defect on the surface.

7.3.3.2 Resistance Effects

Although the average $N_{surface}$ governs the total anodizing reaction, and β_a is a characteristic of the metal, variations in chemical potential, μ, induce variations in oxide growth rate, changing $N_{surface}$ locally. The compositional mismatch can be the result of many different types of metallurgical and mechanical features on the anode surface, which can vary in shape, solubility and size. On the microstructural level, the type and concentration of alloy additions, their size and solubility in the electrolyte, the level of alloy constituent segregation, precipitates and other intermetallic particles, all compound the local effects of μ on the substrate by increasing the exchange current density, i_o, and therefore impact the quality of the oxide structure. With diffusion as the result of i_a, atomic level defects also accumulate, quickly making them significant barriers to anodic oxide nucleation and growth. All of these effects, regardless of size, cause changes to how the applied current density, i_a, flows in the anodizing circuit by causing exchange current density, i_o, which impacts $N_{surface}$.

Since i_o is representative of the local and ever-changing reaction kinetics from point to point across the anode surface, it can be written in terms of the local chemical potential, μ_{local}, and the resistance at the corresponding point, R_{Local}.

$$\mu_{local} = i_o R_{Local} \tag{7.6}$$

Substituting Eq. 7.6 into Eq. 7.1a gives:

$$N_{Local} = \beta_a \log\left(\frac{i_a R_{Local}}{\mu_{Local}}\right) \tag{7.7}$$

In which μ_{local} = local chemical potential in an area on the anode, and R_{local} = the local surface resistance due to substrate chemical potential and/or structural change.

If the condition in an area on the anode surface has increased resistance, R_{Local}, caused by insoluble surface contamination or an insoluble nonmetallic inclusion (such as primary silicon inclusions in cast product), the resistance will be infinite and the circuit will shut down local to the contamination causing concentration polarization effects (see Sect. 7.4). Nevertheless, oxidation proceeds around the particle or dirt. However, if the source for R_{Local} is soluble in the electrolyte, the effects of μ_{local} are in how they produce exchange current density, i_o, as well as i_L, which is the measure of the total local reaction kinetics (they are typically slower with more complex alloy anodes) that changes the local polarization, N_{Local}, affecting the spacing of charge across the anode surface. With increased R_{Local} comes a corresponding increase in the heat of the oxidation reaction, T, which means resistance effects are also dependent upon the metallurgical condition of the anode, like i_o, making them both experimental values. From this perspective, the Tafel equation can aid in rating the impact of alloy complexity on the oxidation reaction and can be used as a tool to evaluate comparative surface resistance between alloy lots or even aid in the process of alloy selection.

7.3.3.3 Surface Activation Polarization and Oxide Growth

In addition to the structural and compositional mismatch that can confound surface activation polarization, N_{Surface}, a *thermodynamic* mismatch between the anode and the growing oxide is also present for a variety of reasons: (1) the diffusion rate differences between metals and oxides (diffusion occurs faster in metals); (2) metallic ions have different sizes, and based upon the type of ion as well as its valence, some move faster than others; and (3) particle size and solubility reduces the rate of reaction by developing i_o and a transient change in i_L. Since the substrate is the source for the anodic oxide constituents, and the anodic oxide is the result of the continued reaction of the electrolyte with the substrate surface, the interface between the electrolyte and the anode surface is the key to continued anodic oxide formation and growth.

In nature, energy is required to overcome a mismatch. Energy produces heat, which means the heat of oxidation at the interface will increase. Non-aluminum metallic ions (copper, zinc, and other metal ions in solution with the aluminum alloy substrate) are cathodically protected by the aluminum oxidation reaction and tend to "pile up" at the interface with the forming oxide. As alloy ions, they will not be incorporated onto the AAO structure, and must realign themselves as they move from their positions on the substrate lattice to interstitial spaces within the anodic oxide lattice. Accumulated alloying ions move more slowly than they did in the anode as they migrate by hopping through the interstitial spaces in the anodic oxide network. Copper ions have a higher conductivity than aluminum ions, and therefore migrate faster through the forming structure and are typically ejected into the electrolyte at the base of the oxide cell pore. Zinc ions, that are the largest of the alloy ions and migrate more slowly and typically migrate between columns in the oxide network. Aluminum ion diffusion occurs primarily on the tetragonal molecular structure, which comprises the forming anodic oxide cell. The kinetics of the primary oxidation reaction are therefore retarded during anodizing when alloying elements in solid solution meet the interface at the electrolyte and the substrate surface, producing the resistance that makes i_o [7].

Because they have higher conductivity, copper ions migrate faster than aluminum ions, migrating through spaces in the forming oxide and into the electrolyte through the base of the forming pore. However, their large size, as compared with hydrogen ions, prohibits them from escaping the polarization wave over the anode; copper ions re-adsorb on the surface of the forming oxide, prohibiting further ion exchange, blocking hydrogen ions from completing the circuit. This condition produces concentration polarization effects by making i_L approach zero, and the formation of a columnar structure is inhibited. Would-be column walls collapse on one another, making a more compact oxide, with a more net-like rather than columnar structure that inhibits oxide growth to approximately 6–8 μ. See Figs. 7.2 and 7.4.

7.3.3.4 Industrial Process Note

Actual variations in the initial surface microstructure of the component (anode) are usually controlled during the treatment steps that precede anodizing. Typically, the surface is cleaned and thoroughly rinsed to remove machining and process contamination, yet tenacious residues can be left behind that block the oxidation reaction, producing Concentration Polarization Effects (CPE, see Sect. 7.4). An etch process (caustic or acid) is often performed to remove any native oxide (passive layer) or residual scale from heat treatment that has formed on the surface, followed by a deoxidation step or so-called desmut process to remove any residues of alloy additions and intermetallic particles. This is done to reduce and restore i_o to the fundamental resistance value caused by the structural difference between the aluminum anode and the growing oxide. Chemical polishing, also known as bright dipping, also reduces i_o by reducing the surface roughness of the anode through the chemical attack of surface asperities such as machined edges and burrs. All steps are followed by rinses to remove pretreatment residues, maximizing the amount of free aluminum at the surface and therefore minimizing i_o and maintaining $N_{surface}$.

7.4 Concentration Polarization Effects

As much as the effects of surface activation polarization, $N_{surface}$, are intimately tied to the condition of the anode surface, and corresponding variations in exchange current density, i_o, the effects of concentration polarization, N_{Conc}, are tied to the conditions external to the anode. This includes the actual anodizing process parameters and conditions that occur, which can inhibit the completion of the anodizing circuit by changing the limiting current density, i_L. Concentration polarization effects (CPE) occur when the anodizing circuit is blocked; in other words, hydrogen ions are prohibited from leaving the anode and completing the circuit at the cathode.

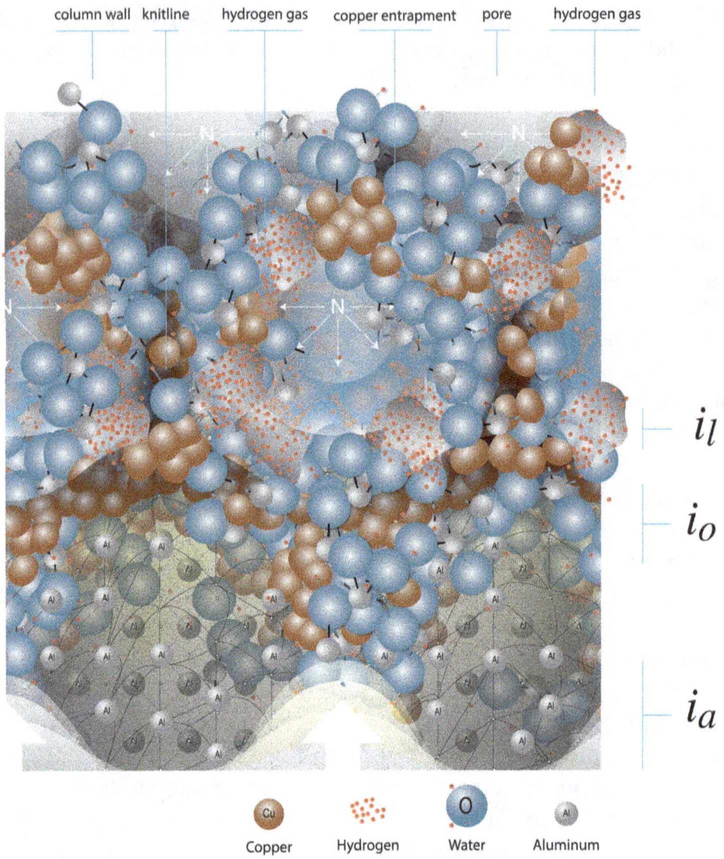

Fig. 7.4 Anodizing aluminum-copper alloys, such as the 2000 series, produces the conditions that develop both exchange current density, i_o, and concentration polarization effects, by reducing the limiting current density, i_L. Alloy unmixing, which occurs in the anode with i_a, produces free copper ions at the interface with the forming oxide that change the local chemical potential at the surface, affecting the exchange current density, i_o. In situations where concentrations of defects intersect the surface, i_o decreases and the effects can manifest as disruptions to the surface polarization, changing how the AAO nucleates, or proceeds to grow, causing deviations in the structural order. In the case of copper-bearing alloys, copper ions are cathodically protected by aluminum during the anodizing reaction. Because they have higher conductivity, copper ions migrate faster than aluminum ions, and diffuse through spaces in the forming oxide and into the electrolyte through the base of the forming pore. Their large size in comparison with hydrogen ions inhibits their escape from the polarization wave over the anode, and some re-adsorb on the surface of the forming oxide, prohibiting the ion exchange necessary to sustain the formation of the central pore and, blocking hydrogen ions from completing the circuit. This condition produces concentration polarization effects by making i_L approach zero. Would-be column walls collapse on one another, making a compact oxide without a columnar structure. For an actual image of this condition, see Fig. 7.2

This can occur by way of three mechanisms: (1) by masking the anode surface, examples of which are a deliberately placed mask, or an intermetallic particle with low solubility encountering the anode-oxide interface, or foreign anions adsorbing on the oxide surface at the oxide-electrolyte interface; (2) an area of charge concentration, such as a sharp corner on the anode, or a burr, or other defect at which charges accumulate; and/or (3) sharp internal corners where like charges repel one another, as with a hole with a high aspect ratio. CPE can be constant throughout the anodizing reaction, as with the placement of a mask, or can be transient, limited to the size of the defect. i_L approaches zero for cases 1 and 3, but i_L exceeds the applied current density for case 2. In all three examples, the circuit is interrupted, and the oxidation reaction stops.

As part of the anodizing reaction, hydrogen is evolved at the interface with the growing oxide; therefore, CPE can be developed over a large surface area across the anode surface, before the oxide forms, or at any point along the growth interface, when an insoluble particle from the near-surface microstructure is encountered. In this sense, limiting current density, i_L, is different from exchange current density i_o, in which chemical potential, μ, has a governing effect. Additionally, and equally important, hydrogen ions must traverse *and exit* the forming oxide, which they do as interstitials, through the ionic structure of the AAO from the central pore and into the electrolyte, where ions must escape the local charge density wave to move to the cathodes to complete the circuit. Therefore, anything that blocks or inhibits the egress of hydrogen ions from the oxide-electrolyte interface, including the thickness of the oxide itself, causes CPE. It is therefore critical to the anodizing process that the effects of concentration polarization are kept to a minimum. See Fig. 7.5.

7.4.1 *Modeling Concentration Polarization Effects*

CPE is modeled by the second half of the Tafel equation and introduces the concept of the limiting current density, i_L, which represents the rate of hydrogen ions proceeding toward the cathodes in order to complete the anodizing circuit. It is presented again as Eq. 7.1b.

$$N_{\text{conc}} = \frac{2.3 * RT}{nF} \log\left(1 - \frac{i_a}{i_L}\right) \tag{7.1b}$$

The anodizing circuit runs best if CPE, represented by N_{conc}, is kept to a value that can be adjusted with the process parameters to accommodate for the inevitable changes in i_o that effect N_{surface}. Typically, CPE can be ignored in equilibrium processes of anodic oxidation, but the high corrosion rate that comes with deliberate anodic dissolution by the applied current density, i_a, demands that i_L be considered as part of the total nonequilibrium oxidation reaction, which is the anodizing circuit, because hydrogen ions are not the only ions that must escape the oxide-electrolyte

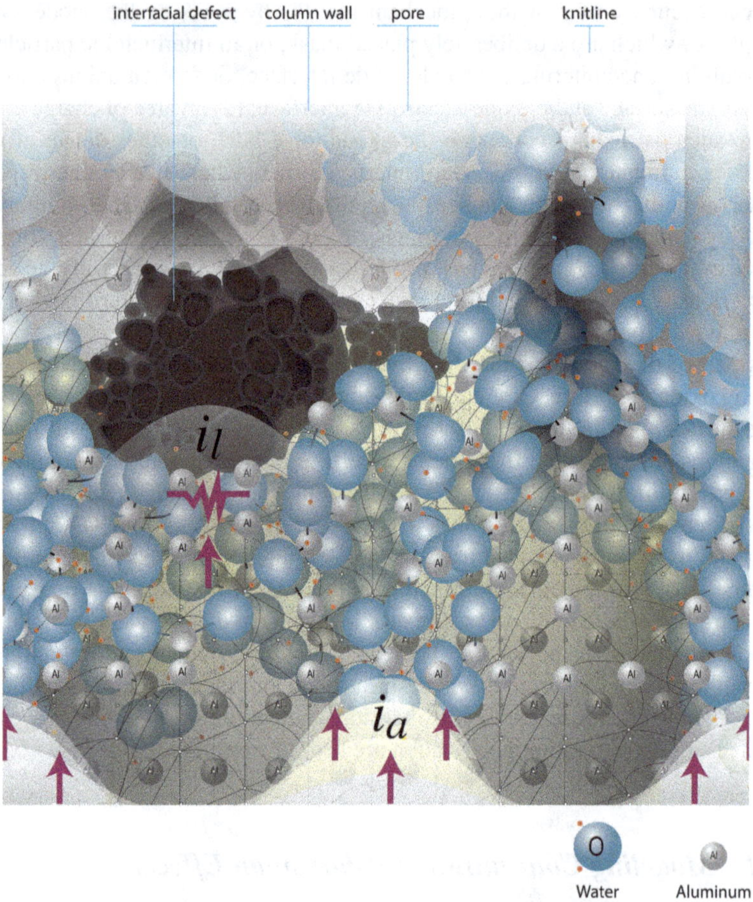

Fig. 7.5 Interfacial and surface defects that block the oxidation reaction, affect the limiting current density, i_L, and cause concentration polarization effects. In situations where concentrations of defects intersect the surface, the effects can manifest as disruptions to the surface polarization, changing how the AAO nucleates or how it continues to grow, causing deviations in the structural order

interface. Alloying element ions must also exceed the current density wave as they move through and out of the oxide toward the cathodes, or they will be trapped at the oxide-electrolyte interface, causing CPE.

7.4.1.1 Limiting Current Density in Terms of Diffusion and Ionic Conduction

If the flow of hydrogen ions through the anodic oxide is considered by way of i_L alone, then the effects of the chemical potential, μ, and the exchange current density are taken care of by $N_{Surface}$, and μ and i_0 can be ignored when strictly considering the effects of concentration polarization. Therefore, c_i from Eq. 7.5 can be viewed as the concentration of hydrogen ions alone, which comprise the limiting current density. The diffusivity of hydrogen through the oxide, D, and R and q_i are constant values, leaving c_i and T to vary with the electric field over the changing total oxide thickness, δx; therefore, the limiting current density is relative to only the concentration of interstitial hydrogen ions, c_i, and the change in electric field effects, $\delta\phi$, that change with the changing oxide thickness, δx, which produce concentration polarization effects.

$$i_L = -\frac{D}{RT} c_i q_i^2 \frac{\partial\phi}{\partial x} \tag{7.8}$$

This relationship shows directly that the limiting current density, i_L, changes with the concentration of hydrogen ions, c_i, and includes the heat of oxidation, T. What is fascinating about the role of the heat of reaction, T, in the total relationship between $N_{Surface}$ and N_{Conc} is that T increases with the resistance to the flow of the applied current density, i_a, at the substrate-oxide interface, producing an increase in exchange current density, i_o, and a corresponding decrease in $N_{Surface}$. Additionally, the same reaction at the surface that caused T to increase effectively reduces i_L, as well as causes concentration polarization effects, CPE, to increase. Therefore, N_{Conc} is linked to $N_{Surface}$ by the heat of oxidation when considering the reactions that occur at the anode-oxide interface. Resistance effects that are carried over from $N_{Surface}$ create a complex relationship that will be considered later in this chapter from the perspective of defective type and source.

7.4.1.2 Process Effects on Concentration Polarization

Jones [8] calculates i_L as the measure of the maximum rate of oxidation that cannot be exceeded due to the limited diffusion rate of hydrogen ions in the electrolyte. Consequently, the anodizing parameters that affect the flow of hydrogen from the anode as the oxide grows: the diffusivity of hydrogen, D_z, the spacing between the anode and the cathode, δ, and the electrolyte concentration, $C_{Electrolyte}$, are included as part of the calculation that expresses the limiting current density, i_L, which determines the rate of hydrogen ions escaping from the interface of the growing oxide, proceeding toward the cathodes in order to complete the anodizing circuit, and is presented again as Eq. 7.9.

$$i_L = \frac{D_z n F C_{Electrolyte}}{\delta} \tag{7.9}$$

where n and F are the number of equivalents exchanged ($n = 1$, because hydrogen is H^{+1}). D_z is the diffusivity of the reacting species, H^+, which is needed to complete the circuit of the anodizing reaction; if anything blocks D_z then $i_L = 0$, and no anodizing will take place. $C_{electrolyte}$ is the concentration of the electrolyte; δ is the distance over which diffusion must occur to complete the circuit; in other words, δ is the virtual distance between the anode and the cathode.

By substituting Eq. 7.9 into Eq. 7.1b, the expression for total concentration polarization, N_{Conc}, can be expressed in terms of the anodizing reaction:

$$N_{\text{conc}} = \frac{2.3 * RT}{nF} \log\left(1 - \frac{i_{\text{a}}\delta}{D_{z} nFC_{\text{Electrolyte}}}\right) \tag{7.10}$$

It is important to note that the heat of the oxidation reaction, T, in this portion of the Tafel equation is what is developed by the part of the anodizing reaction due to CPE. The heat of resistance to oxide growth, developed by surface activation polarization, N_{Surface}, is accounted for by the effects of variations in chemical potential, μ, which develop exchange current density, i_{o}. Therefore, resistance in the anodizing circuit is what connects concentration polarization, N_{Conc}, to N_{Surface}. As an industrial process note: It is generally easier to anticipate the effects of N_{Surface}, because the metallurgical condition of the anode is usually well controlled. Therefore, because resistance effects in the anodizing circuit are manifested thermally, a change in T in the total anodizing circuit can be considered a warning that CPE are happening, indicating problems with the anodizing circuit that produce corresponding changes in anodic oxide growth are occurring, such as burning.

7.4.1.3 Concentration Polarization Effects and Oxide Growth

When considering limiting current density, i_{L}, in terms of the total concentration polarization, N_{Conc}, i_{L} should never be zero, but it should never be equal to or exceed i_{a}, as both conditions shut down the anodizing circuit in areas where they occur across the anode surface. The former condition occurs if the flow of hydrogen is blocked from the anode surface, as if by a mask or an insoluble particle from the microstructure, or from the oxide surface with the electrolyte by way of the adsorption of dissolved cations from the anode alloy. The latter condition, that is, $i_{\text{L}} \geq i_{\text{a}}$, occurs if charge is concentrated at a point on the anode surface, such as at the tip of a burr, a sharp corner, or even at the interface with an intermittent contact, causing the charge density developed by i_{a} to interfere with the flow of hydrogen, i_{o}, from the point of origin. In other words, no current density wave can be established because the concentration of charge is too dense, prohibiting polarization, and the flow of hydrogen is blocked. Both conditions produce finish defects such as a nonuniform oxide thickness, increased oxide surface roughness, variations in surface appearance, and in worst cases, burning.

7.4.1.4 Industrial Process Notes

The Tafel Equation is the only model for anodic oxidation that accounts for the effects of specific process parameters in the anodizing process, specifically by how they impact CPE. CPE are typically manifested as a substantial increase in the heat of the oxidation reaction, T. An understanding of the effects of the process parameters of i_a, δ, and $C_{Electrolyte}$ on the anodic oxide quality enables the development of anodizing strategies that optimize the important process quantities of i_L, and T as the heat of reaction, by keeping N_{Conc} as small as possible and N_{Total} at a potential greater than the oxidation potential for aluminum. Nevertheless, all process changes must be done with attention to side reactions, and demand that engineering testing be performed to assure a process change can be implemented successfully. The summary presented below assumes proper and adequate pretreatment and is based upon a functioning anodizing process that produces quality anodic oxides.

Applied current density, i_a: Adjusting i_a is the single most effective means to change the entire anodizing reaction. When starting from the baseline of a functioning anodizing process, when problems with increasing heat of the anodizing reaction, T, are encountered, especially with complex alloys, manifested as thin, nonuniform, discontinuous oxide growth, all too often the first change to the process is to *increase* i_a or to increase $C_{Electrolyte}$. When one considers that most of the effects encountered to hinder the process are those of CPE, increasing i_a should be counterintuitive, and the best fix is to reduce i_a, to allow the heat of reaction to recover and to ramp slowly to a somewhat lower operating i_a.

Pulsing i_a can also be helpful in relieving CPE. Not only does it aid in discharging resistance heat, T, from the growing oxide, pulsing helps to dissolve loosely bound complexes, such as cations of alloying elements dissolved in the electrolyte that can't escape the current density wave, that have adsorbed on the surface of the AAO at the interface with the electrolyte. By ejecting ions from their adsorbed sites on the AAO, pulse anodizing can aid in developing thicker oxides on copper-containing complex alloys that typically limit oxide growth by developing CPE.

Distance between the anode and the cathode, δ: With an increase in the heat of reaction, T, for a fixed δ, the virtual distance between the anode and the cathode can be reduced by introducing agitation to the electrolyte. Electrolyte agitation is an effective means to help keep the heat of reaction, T, from building at one point at the anode surface and help to keep the temperature at the anode surface uniform. This is good practice for all types of anodizing, even those that typically don't chill the electrolyte. By keeping δ small, i_L will not get too small, reducing CPE.

Concentration of the Electrolyte, $C_{Electrolyte}$: It is commonly viewed that increased solution conductivity (reducing the solution resistance) by increasing $C_{Electrolyte}$ will aid in successful anodizing of complex alloys. Maximum solution conductivity has been measured at 375 g/L for sulfuric acid electrolytes [9]; yet successful anodizing is routinely carried out at concentrations less than half that of maximum conductivity, typically 180 g/L. While increased $C_{Electrolyte}$ is an effective means to draw current away from the oxide surface by reducing the electrolyte resistance, it should not be viewed as a means to accelerate anodic oxide growth. In fact, in the moments when the anode is first placed in the electrolyte, before imposition of i_a, the electrolyte reacts with the anode surface under nonequilibrium conditions, rapidly producing corrosion product barriers to the anodizing reaction. In order to initiate anodizing, a large input of electrical energy may be necessary to break through the corrosion product before a normal surface reaction can go forward. Since surface corrosion product layers are typically discontinuous and nonuniform, the anodic oxides produced in electrolytes with high $C_{\text{Electrolyte}}$ usually have high roughness.

Anodizing with a $C_{\text{Electrolyte}}$ with moderate concentration, 180–220 g/L, coupled with a current density ramp to aid with the diffusion of alloy ions at the anode interface with and through the forming oxide yields the most uniform and continuous oxides on even the most complex alloys [10].

7.5 Anodic Oxide Growth on Alloy Substrates: The Structural Mismatch

The network of equally spaced oxide nuclei that form on the polarized surface of the anode as the oxide develops is amorphous, yet each oxide element that grows to become a columnar cell has a chemical structure. As the surface is reconstructed by oxide during anodizing, this structure provides the path for ionic conduction and continued oxide growth, forming two main interfaces, one at the aluminum-oxide interface and one at the oxide-electrolyte interface. With anodic oxide structural development, continued growth of an ordered network of oxide cells creates additional interfaces between the individual cells that comprise the network. In this section, the reactions that occur at the interfaces in the oxide with various types of defects will be considered as they are important to how anodic oxides can nucleate and grow on complex alloys, whose various alloy ions and intermetallic particles produce resistance to the oxidation reaction. These considerations are also important to troubleshooting, when linking a change in the anode microstructure to the quality of the anodic oxide.

An important concept to keep in mind as interfacial phenomena develop changes in the anodic oxide growth front is that the anodic oxide will never be stoichiometric Al_2O_3. Charge conservation within the ionic solid dictates that the deviation from stoichiometry in anodic alumina, for example from Al_2O_3 to AlO_4^-, is accompanied by the flow of ions and other defects that aid diffusion through the oxide structure. As anodic aluminum oxide grows, charge is carried mainly by interstitial ions across the interface from the anode and through the forming oxide; however, electrons and electron holes left by the deficit in aluminum on the oxide structure also carry current by way of electron holes with positive charge [5]. As AAO transports current by way of holes in the structure, one can visualize how new aluminum ions can be included on the structure of the AAO; as Al^{3+} ions are pumped to the surface by the current bias, new AAO grows at the interface between the oxide and the aluminum substrate, by way of electrotransport. In this way, AAO behaves like a semiconductor as it grows [11].

The transition in oxide structure from the base of each column, from (1) its point of polarized symmetry from which it grows, to (2) the points of impingement during which changes in the structure are introduced by way of electrostriction due to oxide deformation and flow, which produces vacancies at the interfaces

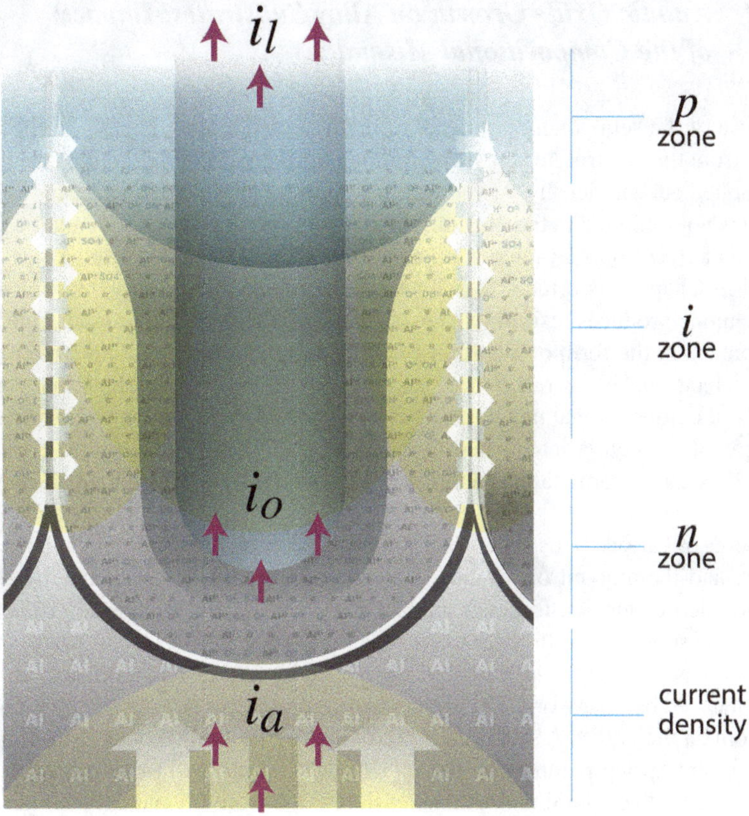

Fig. 7.6 As the oxide grows, the anodic oxide structure is that of an aluminum deficient amorphous ionic solid. It functions as a p-i-n diode from the point of symmetry at its base, where the structure is most stoichiometric (*n*-type semiconductor), to the interfaces between the cells, along the knitlines, where the structure is mixed or *i*-type semiconductor (*n*-type and *p*-type semiconductor) to the surface with the electrolyte, where the structure is most protonated (hydrated) and is most like a p-type semiconductor [12]

between oxide cells as the central pore forms, to (3) the surface with the electrolyte, where additional changes in the structure are produced by way of dissolution and interaction with ions at the oxide-electrolyte interface produces an oxide structure that functions as a p-i-n diode. As the AAO grows and develops its unique structure, sustained by ion exchange, interfacial phenomena conduct along interstitial paths through the oxide, or are dragged along with oxide flow at the knitlines, some ions are ultimately ejected into the electrolyte to complete the anodizing circuit, but others become entrained at the various interfaces that form in the oxide. By understanding the effects of the movement of interfacial phenomena, as they impact i_o and i_L and, ultimately, the total polarization of the anodizing circuit, the anodic oxide can be designed and engineered according to alloy selection. See Fig. 7.6.

7.5.1 Anodic Oxide Growth on Alloy Substrates: Impact of the Compositional Mismatch

Because of the semiconductor nature of the AAO, no ionic tunneling is allowed to occur from the electrolyte across the forming anodic oxide layer against the flow of the applied current density, i_a, yet outward electron and metal ion transfer from the anode is not inhibited in the direction of i_a. This structural concept of anodic aluminum oxide (AAO) as an ionic solid comprising tetragonal AlO_{4-} molecules, introduced in Chap. 5, is critical to addressing the impact of the structural mismatch, which alone produces resistance to AAO growth on isotropic aluminum. Interfacial phenomena in the form of alloy ions, intermetallic particles, and other microstructural defects make the reaction at the interface more complex by producing the effects of compositional mismatch, that is, increased resistance to oxide nucleation and growth, which is manifested in decreasing i_o. To maintain circuit continuity, alloy ions and intermetallic particles must somehow be transported through the forming oxide.

Atoms of various sizes occupy substitutional sites on the aluminum anode crystal lattice, and the amount of resultant distortion introduced to the lattice by way of compression or tension forces is proportional to the size and valence of the atom. This is the source for differences in mechanical properties of the substrate, specifically increased strength and hardness, developed by way of substitutional hardening and enhanced by cold work. The average lattice parameter is governed by the size of the atom on the lattice. A substitutional atom with a different atomic radius changes the adjacent lattice parameters, placing the FCC structure in compression, if the atom is larger than aluminum, and in tension, if the atom is smaller. See Fig. 7.7.

Table 7.1 presents the ionic radii specific to the primary alloying elements of aluminum, as well as the maximum solid solubility (wt. %) at room temperature, relative ionic conductivity, and the oxidation potential in volts. It is interesting to note the sizes of most of the ionic alloy radii are larger when compared with aluminum. The ionic radii specific to the primary alloying elements for aluminum, specifically zinc, copper, and magnesium are 88, 87, and 86 pm, respectively; 22–23% greater than the aluminum ion [14], and manganese is approximately 17% greater. Only silicon, at 54 pm, is smaller. This means that, when in solid solution with the aluminum, most of the ions put the alloy lattice in compression, which accelerates alloying unmixing with the imposition of i_a.

The stability of a solid solution at room temperature (RT) is also important to consider when considering the electrical characteristics of the anode. Magnesium has the greatest solid solubility with aluminum at 2.50% at room temperature; zinc also has significant solid solubility at 2.00% at room temperature. This is in comparison with 0.30% for manganese, 0.10% for silicon, and 0.02% for copper [15]. The significance of reduced solid solubility of the main alloying elements at room temperature is that these elements therefore fall out of solution during solidification to react with aluminum and other trace elements to form phases, particles and precipitates, and all have different chemical potential(s) from the matrix.

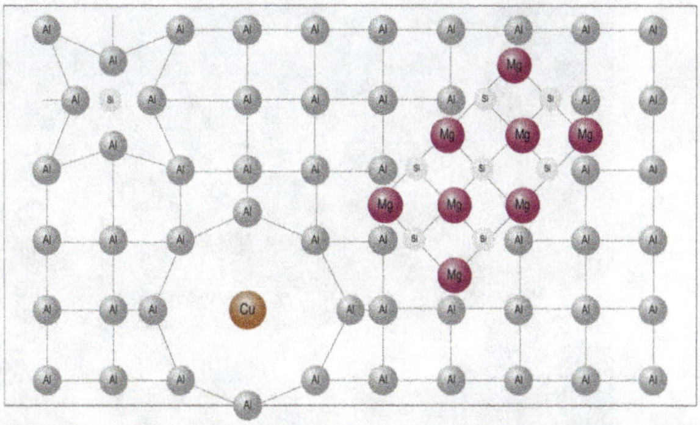

Fig. 7.7 Schematic of an aluminum lattice with substitutional larger (magnesium) and smaller (silicon) atoms. Note how the variation in size causes compressive and tensile stress within the crystal lattice. Also note the corresponding change in the adjacent lattice parameters, both bond length and interstitial space. (Figure courtesy of the AAC [13])

Table 7.1 Characteristics for the ions of the main alloying elements of aluminum

$Ion^{valence}$	Al^{3+}	Cu^{2+}	Mg^{2+}	Mn^{2+}	Si^{4+}	Zn^{2+}
Radius (Å)	0.67	0.87	0.86	0.81	0.54	0.88
Solubility @ RT (wt%)	100	0.02	2.50	0.30	0.10	2.00
Oxidation potential (V)	−1.66	+0.34	−2.38	−1.18	−0.14	−0.76

The oxidation potential for each of the main alloying elements is also important to consider due to the oxidative nature of the anodizing reaction—only magnesium has a more active oxidation potential than aluminum. This means that during aluminum anodizing, magnesium is the only element other than aluminum that will oxidize. The rest will be cathodically protected and remain as metal ions during the anodizing reaction, preserving the exclusive aluminum-oxygen constitutive makeup of the AAO.

The AAO structure produced by industrial anodizing is amorphous, with an ordered structure comprising tetragonal AlO_{4-} molecules that function as an ionic solid as it grows from the FCC aluminum [16]. Therefore, the AAO structure directly adjacent to the substrate would be that of a tetragonal molecule, on which there is one (1) aluminum ion for four (4) oxygen ions. X-ray and neutron diffraction data of nearest neighbor Al-O bonds found them to be 1.80–1.82 Å and the O-O bonds at the base of the pyramid, from the data collected, is expected for the case of two tetrahedra, sharing a common oxygen corner, thereby producing short-range order, to be 2.95 Å; Al-Al spacing was determined to be 3.18 Å. These measurements, reported by Kniep et al., are found to be similar to that for protonated hydrated alumina surfaces modeled by Kubicki and Apitz [17]. If these measurements are strictly considered, the diameter of the largest alloy ion from the substrate that will be presented at the interface, zinc at 0.88 Å, is just able to pass through the AAO short-

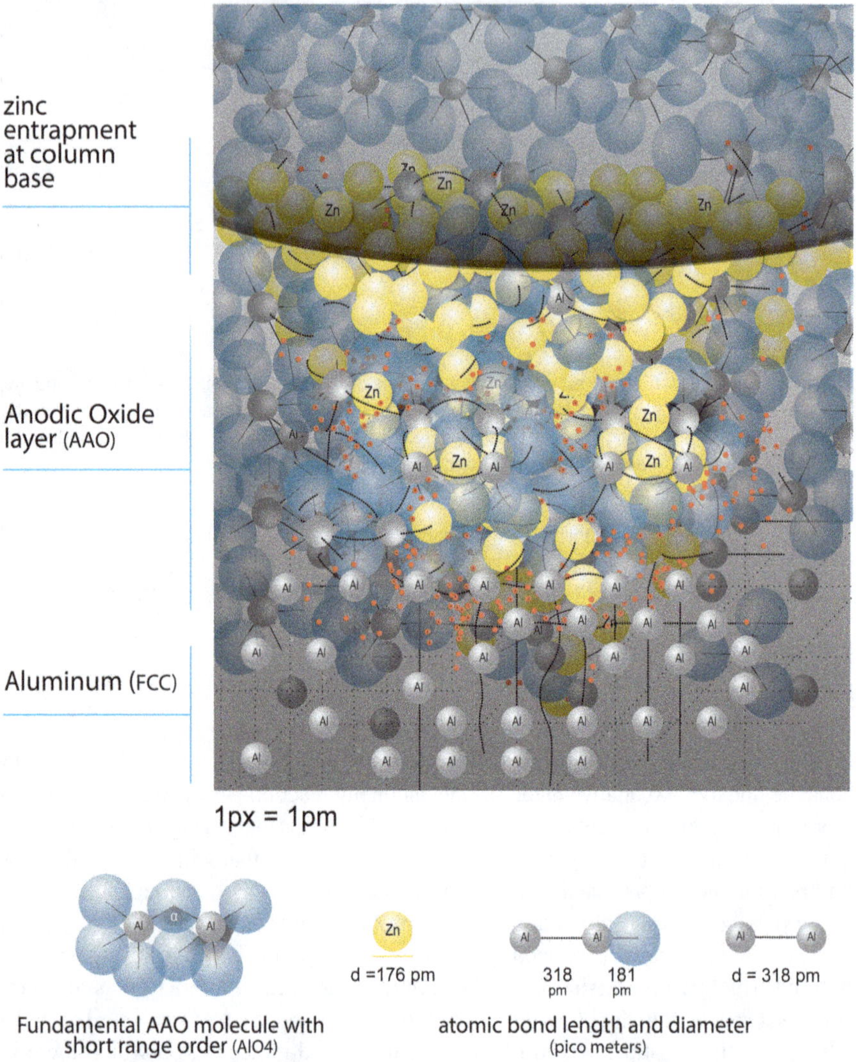

Fig. 7.8 Schematic showing the short-range tetragonal oxide structure of the AAO. Superimposing the oxide structure within an FCC lattice [18] shows how the volumetric spacing between the aluminum substrate (FCC, dark-colored solid lines) and AAO (tetragonal, dotted lines) differ. The bond lengths of the AAO determined by *X*-ray and neutron diffraction support the conclusion that there is space to accommodate the diffusion of alloying elements through the AAO structure. Nevertheless, there is stress produced at the interface between the metal and the AAO as well as between diffusing ions and the AAO structure, when bonds are stretched. The diffusion rate decreases because of the structural mismatch between the AAO and the substrate and because of the close fit between diffusing ions and the AAO caused by the compositional mismatch

range lattice. In other words: zinc, copper, and manganese ions can all fit through the interstitial Al-O spaces in the AAO structure, but the rate of migration will vary with the conductivity and size of the ion. Zinc ions migrate slowest through the developing oxide structure, and copper ions migrate fastest. See Fig. 7.8.

As-grown, anodic aluminum oxide is hydrated (protonated) aluminum oxide; therefore, the bond lengths of the AlO_{4-} are somewhat longer and weaker (1.86–1.89 Å). With electrostriction, a high level of strain develops at the knitlines, increasing the interfacial temperature and the bond lengths on the AlO_{4-} structure, sometimes tearing oxygen ions from the AlO_{4-} structure. Thermal energy creates random enlargements of interstices and the creation of vacancies on the oxide structure, that subsequently facilitate the movement of alloy ions through the AAO [19].

As electrostrictive stress acts against the bonding in the ionic solid, some of the ions are displaced from their equilibrium positions on the oxide structure [20]. Displacement due to electrostriction creates vacancies on the oxide structure, which are most probably oxygen vacancies because the creation of aluminum (cation) vacancies on the oxide structure is strongly constrained by the necessity to form compensating defects. In other words, if a cation vacancy is created, a compensating anion defect must be created to maintain charge neutrality in the solid. The formation of vacancies can be imagined as one of the bonding legs on the tetragonal structure being torn away due to deformation at the interface caused by electrostriction. This effectively softens the oxide and enables flow of the anodic oxide in the direction of the applied current, opposite to the growth direction, which is into the aluminum (alloy) anode. A structural gradient consequently develops in the oxide from the point of symmetry, at the base of the forming column, to the knitline, where the bonding arms are deformed by oxide flow.

In order to move, oxygen vacancies, which have a positive charge, are easily compensated by the polarization of the anode, and they are pushed along the knitline between nearest neighbor oxide cells. Electrons, present in the oxide structure, also compensate for the presence of a vacancy with a positive charge. Vacancies move with the deforming oxide where electrostriction forces are highest, within the column walls along the knitlines, at the interface with the nearest neighbor oxide cells. The importance of this phenomenon is in the changes to the oxide structure due to electrostriction. Longer bonding arms and larger spaces due to the formation of vacancies on the AAO structure provide larger spaces for ionic diffusion, providing a preferential path for the egress of alloy cations that have accumulated at the anode-oxide interface. See Fig. 7.9.

Because of the structural gradient that develops as the AAO grows, diffusion occurs slowly through the oxide interstices, which is characteristic for the semiconductor nature of the AAO. The rate of diffusion is limited not only by the size of the individual ions as they move through the AAO, but also by the inherent differences in diffusivity for different ions at different temperatures, which is compounded by the structural differences between the metal and the oxide. The size of the holes in the AAO structure limits the diffusion of alloy ions through its structure, with the larger ions moving more slowly than the smaller ones. And, even though all metallic

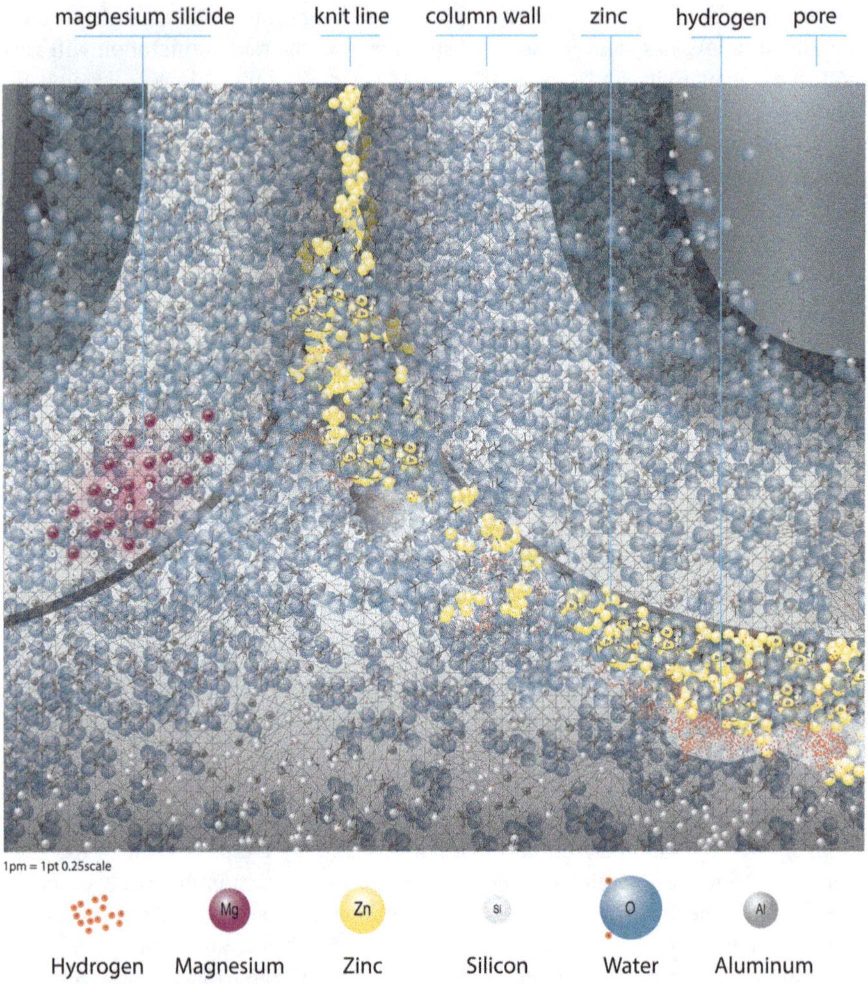

Fig. 7.9 A variety of reactions occur when alloys are anodized. With unmixing of the alloy that occurs with i_a, elements in solution with the aluminum are also part of the electrochemical circuit. All but magnesium will be cathodically protected; therefore, alloying ions remain as metal ions during aluminum anodizing. Magnesium oxide is soluble in most anodizing electrolytes, so it typically doesn't impact the AAO structure. Size-wise all alloying ions can migrate through the forming oxide; however, as the concentration of alloy ions in solution with the aluminum increases, the level of distortion in the oxide developed by electrostriction increases. The size and conductivity of the individual ion governs whether the ion will migrate through the oxide, like copper, or become entrained at the anode-oxide interface, ultimately migrating along the knitline, like zinc

ions can pass through the spaces in the oxide, the rate of diffusion and the competition for space for diffusion through the oxide are different for different ions—some ions will diffuse faster than others, leaving some behind at the interface with the oxide. The ramifications of the diffusion limits at the interface between the substrate and the oxide create the thermodynamic mismatch in anodizing, which is manifested by ion entrainment at the substrate-oxide interface and resistance heat and disruption of the order of the AAO, causing an increase in oxide surface roughness and a decrease in dielectric strength for the total oxide finish.

7.6 Intrinsic Interfacial Phenomena and Oxide Growth

7.6.1 Atomic Level Defects

7.6.1.1 Elements in Alloy Solution as Sources for Surface Activation Polarization

Alloy atoms, because they are mostly larger than aluminum atoms, place compressive stress on the face-centered cubic (FCC) structure as substitutional atoms of the aluminum lattice. Energy is increased so is FCC. With the flux of electrons stimulated by i_a, the diffusivity of the substitutional atoms increases as the diffusivity of the aluminum atoms on the primary lattice increases. Under these conditions, an unmixing of the alloy occurs, and a variety of cations from the various alloy constituents that had been in solution with the aluminum are present as interstitial metallic ions. Therefore, with unmixing of the alloy solid solution due to i_a, the amount of interstitial ions that vary in size and valence (charge) in the polarized anode is markedly increased. At this point in time within the substrate, diffusion is quite high, and is taking place by interstitial mechanisms for all ions, with the smallest, which are the aluminum ions, having the highest rate. The force of the flow of electrons from i_a bombards all ions on the interstices of the FCC lattice, carrying them, in the direction of the applied current density toward the polarized centers at the anode surface [7].

As the aluminum ions are consumed and oxide growth proceeds into the anode, a force is exerted on the remaining cations and charged particles from the anode, in the direction of the applied current density, toward the interface with the oxide. When the metallic ions, carried by i_a, reach the free surface at which oxidation takes place, diffusivity for all atoms decreases. Aluminum ions continue fastest onto the forming tetragonal structure of the AAO. The remaining ions rearrange themselves in order to let the current flow in the total circuit and excess interstitial alloy ions are "injected" at the interface, where their diffusivities immediately decrease [4]. This is when size and population of the ions, which are cathodically protected by the oxidation of aluminum, have the most profound effect on the oxidation reaction, by creating resistance, which produces i_0.

Fig. 7.10 TEM images document actual alloy element entrainment (black-appearing flecks) at the substrate-AAO interface and migrating between AAO columns. STEM analysis of the flecks determined they were zinc

Because the movement of the alloy ions onto the AlO_{4-} tetragonal structure is thermodynamically prohibited, these ions must migrate through interstitial spaces on the oxide structure, or along knitlines. Diffusion rates are therefore controlled by the size of the interstitial space on the oxide structure, and the diameter of the migrating ion. Smaller alloy ions, such as manganese and copper, migrate through the oxide by way of conduction on the interstices of the tetragonal structure, diffusing through holes in the AAO structure, by way of vacancy mechanism, to the electrolyte. Viewing electrostriction as the interactive response of ions to an applied electric field, it is expected that ions of larger size or with higher valences would lead to larger electrostrictive effects that create additional strain at the metal-oxide interface and at the knitlines, as they move between columnar cells in the AAO network, increasing the resistance in the oxide and sometimes distorting the structural order. Therefore, larger ions, such as zinc, create strain in the oxide as they are dragged along knitlines between nearest neighbor oxide columns as the oxide flows due to electrostriction. See Fig. 7.10.

If the ion population is high, the effects of ion conduction can be directly related to defects in the AAO structure. Alloy anodes that contain more than 5% copper present the case for the effects of the migration of smaller ions through the oxide structure to the inside surface of the forming central pore during anodizing. Inasmuch as polarization sustains the pore diameter, ion exchange sustains continued oxide growth. Copper ions, as Cu^{+} or Cu^{2+}, ejected from the oxide, into the pore, are much larger than hydrogen ions, which are also generated by the anodizing reaction, and are necessary to complete the anodizing circuit. Unlike hydrogen ions, copper ions cannot easily escape the current density wave and they react with available electrolyte anions, forming complexes that adsorb on available sites on the inside surface at the base of the pore, shutting down ion exchange. This is an interesting example of how interfacial phenomena can create both i_o and

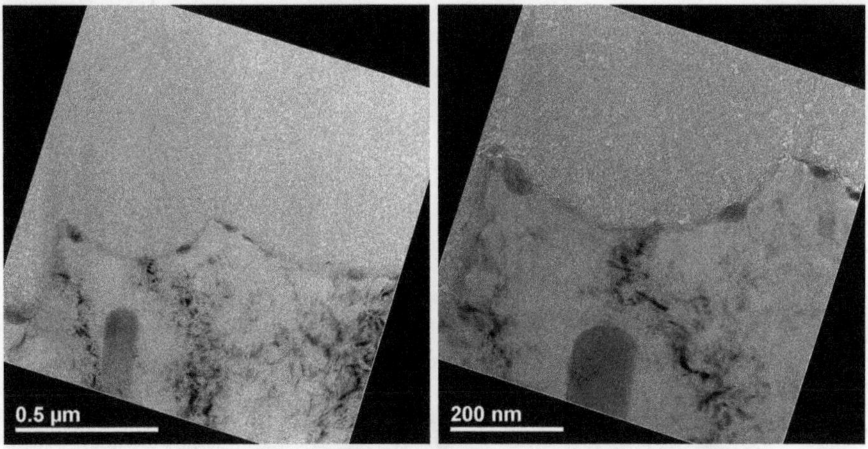

Fig. 7.11 TEM images of the anodic oxide produced on a AlCu4Mg1 (most like AA 2024) threaded socket. The high level of copper in the alloy combined with the unmixing of the major intermetallic particle, $CuAl_2$, within the microstructure limits ion exchange, producing thin, compact anodic oxides. Note the absence of distinct central pores in the columns comprising the AAO. STEM analysis determined the distinct layer entrained at the interface with the oxide comprises copper, which imparts dark-appearing so-called "inherent coloring" to the total oxide finish. (Sample courtesy of Impreglon Surface Engineering, Kaufbeuren, Germany; Image courtesy of Cerium Labs, Austin, Texas)

i_L. The movement of copper ions through the AAO is the source for high resistance in circuits that anodize 2XXX series wrought alloys, or 3XX series castings, which primarily support the growth of thin, compact anodic oxides.

Depending upon the population and size of the ion, as well as the temperature and magnitude of i_a, sometimes ions of one type can become entrained at the interface of the base metal. As these ions accumulate at the interface, they remain metallic, because they are cathodically protected by the oxidation of aluminum, thereby inhibiting the oxidation of underlying aluminum, causing CPE during anodizing, but also cause a variety of effects after anodizing. The most common effect of alloying element entrainment after anodizing is inherent coloring. Anodic aluminum oxide is optically transparent, and once the oxidation reaction is completed, metallic ions react immediately, usually oxidizing themselves. The anodic oxide will take on the color of the oxide of the primary ion entrained at the interfaces with the oxide and at the knitlines (see Chap. 3). See Fig. 7.11.

Long-term corrosion effects due to the entrainment of alloy element ions in the AAO occur when ions trapped in the oxide react with a corroding environment. Transpassive breakdown of the ion-based oxides can occur, forming other corrosion products with an associated change of valence and size [21]. Larger particles that form due to external corrosion reactions with the entrained alloying elements in the oxide increase internal stress between columns, reducing cohesive strength, sometimes causing the finish to rupture. Other corrosion reactions can proceed with alloying elements entrained at the oxide-metal interface, producing a kind of filiform corrosion that may lead to the oxide delaminating from the substrate. See Figs. 7.12 and 7.13.

Fig. 7.12 Schematic of a zinc chloride reaction particle caused by the reaction of a finished anodic oxide with a chlorine-containing atmosphere. Zinc chloride is much larger than the unreacted zinc entrained at the knitlines, developing areas of local swelling near the surface of the oxide, which causes the stress in the total structure of the anodic oxide to increase. Oxides that are sensitive to environmental exposure as a result of base metal alloy chemistry and microstructure exhibit reduced corrosion and wear resistance, and in worst case situations, exhibit spalling and rupture

Another significant effect of ion entrainment that is almost always associated with anodizing zinc-containing alloys from the 7XXX series. As zinc ions accumulate at the interface, they inhibit the oxidation of underlying aluminum, causing CPE, which weakens the interface between the oxide and the substrate. With continued application of the current density, the dielectric withstand voltage for the oxide layer is locally exceeded, which causes the oxide to crack and/or rupture from the substrate, resulting in a localized discontinuity at the interface, which can resemble a blister after anodizing. The size and extent of the accumulation of interstitial alloy atoms at the interface determine if the AAO structure will become discontinuous with the substrate.

Beneath the ruptured oxide, the anodizing reaction will continue so long as the substrate is connected to the anodizing circuit, and more oxide can grow [22]. As the reaction proceeds, the aluminum around the local area of entrained alloying ion concentration will continue to be anodized, supporting the neighboring interface, producing what appears to be a continuous anodic oxide. After anodizing is completed, and the electrochemical driving force for cathodic protection is removed, the accumulation of alloy ions can oxidize or react with the environment, creating localized stress risers across the interface. The oxide appears sound across the entire surface area, but thermal cycling or exposure to high temperatures followed by a

Fig. 7.13 Transpassive breakdown between the columns of the AAO occurs when ions that are trapped along knitlines and at the interface between the metal and the oxide corrode after anodizing. As they lose electrons to the corrosion process, they increase in size, locally swelling the oxide, causing it to rupture and break apart

tape adhesion test easily delaminates the finish, revealing blister-like defects in the oxide. See Figs. 7.14 and 7.15.

7.6.1.2 Atomic Level Defects as Sources for Concentration Polarization

Point Defects: Anodizing Empty Space

The density of point defects such as dislocations and vacancies in metal can be quite high. Such defects are quite mobile and, with the applied current density, i_a, diffuse along with metal ions. With the imposition of i_a, the population of these defects

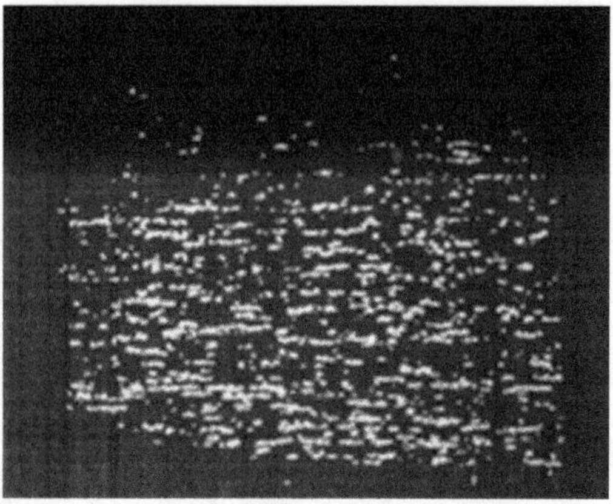

Fig. 7.14 Photo of an anodized and black-dyed 7075 extrusion after tape adhesion testing. The test delaminated the anodic oxide from the anode, exposing the base metal

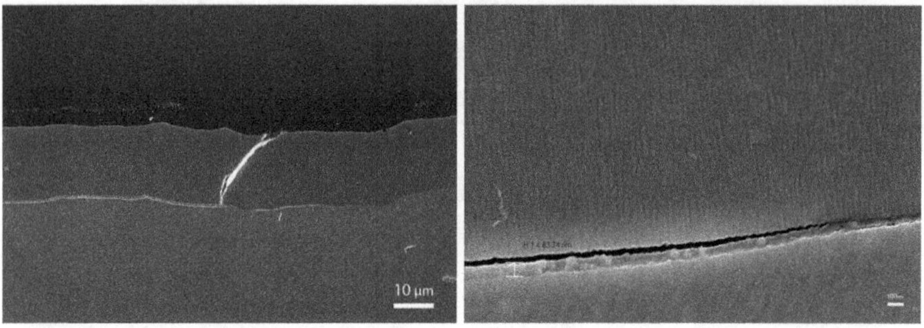

Fig. 7.15 SEM images of AAO, which ruptured due to entrainment of zinc at the substrate-oxide interface. Note, on the right, the thin oxide layer that continued to grow after the oxide ruptured. The layer of entrained zinc reduced the dielectric withstand between the oxide and the anode such that applied current density exceeded the dielectric strength at the interface, causing the oxide to rupture from the anode

increases as the free surface of oxidation is approached, relative to the bulk of the substrate. Grain boundaries are preferential sites for the accumulation of atomic level defects, because they are thermodynamically of lowest energy. Triple points, where three grains come together at a single point, or two grains intersect the surface, serve as pockets for rapid diffusion, vacancy sinks, and sites for rapid transport, which lead to concentrations of atomic level defects at the surface. A heavily deformed substrate exhibits a comparatively higher population of atomic defects as well as areas of defect agglomeration, especially in regions adjacent to and within grain boundaries where the energy is lowest, providing high diffusivity paths for concentrations of point defects toward the surface.

It is important to keep in mind that the accumulation of dislocations and vacancies is an accumulation of empty space. As they move through the aluminum, atomic level defects comprise moving pockets of empty space, which do not support the oxidation reaction that seeks to oxidize aluminum. In the absence of aluminum to anodize, charge accumulates quickly and Concentration Polarization Effects (CPE)

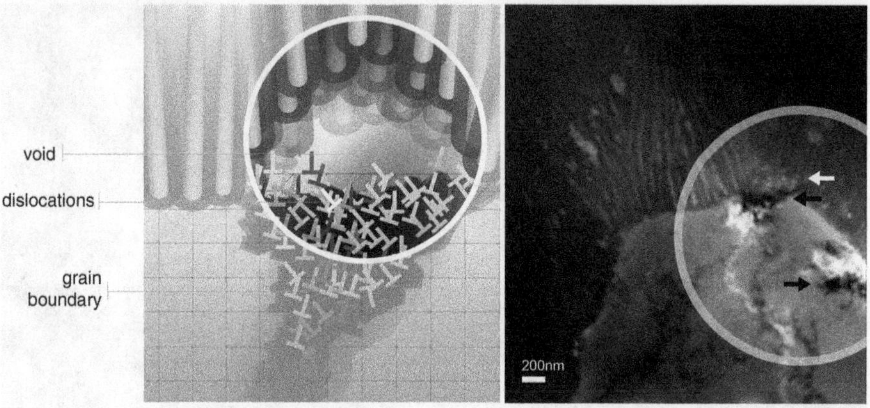

Fig. 7.16 Schematic shows a hole at the interface between the substrate and the forming AAO. TEM photomicrograph of anodized aluminum alloy 6061 T6. A exhibits a "puddle" of dislocations located at a triple point between two grains and the free surface from which the AAO structure is grown. Arrows indicate the defect pocket. Note the discontinuity (hole) at the interface (white arrow); there is no contact between the AAO and the substrate at this point. The black arrows indicate each puddles of dislocations captured in this section. A front of column bases that have detached from the anode in the area of the hole is clearly seen at this point

develop locally, and the oxidation reaction locally stops, but quickly resumes because, typically, the defects are very small. Therefore, defect pockets of dislocations or vacancies exhibit multiple oxide nucleation points that disrupt anodic oxide structural order and generate hydrogen gas, producing discontinuities and voids in the oxide finish [6]. If the density of defects is locally high, the AAO structure above the concentration of atomic level defects will become discontinuous with the base metal. The reaction proceeds through the defect area with some resistance heating, leaving a hole in the AAO at the interface with the substrate, and as the oxidation reaction exceeds the accumulation of defects, the hole is transported into the AAO as a disruption in structural order, such as columnar branching, due to the stopping and restarting of oxide nucleation and growth. See Fig. 7.16.

When pockets of atomic level defects intersect the initial surface from which the oxide grows, depending upon the aspect ratio of the pocket, or pit, charge accumulates and repels adjacent charge over the expanse that the defect occupies. Such defect pockets can be highly geometric, depending upon their source. For example, square-shaped pits are produced with the accumulation of hashtag-shaped Guinier–Preston Zones that house embryonic Mg_2Si precipitates, and triangular-shaped pits may form at a triple point when the reaction encounters a pocket of dislocations. The anodic oxide replicates the shape of the defect, and geometric pits are formed in the surface of the oxide. See Fig. 7.17.

In the case of substrates that have been subject to high levels of deformation at high strain rates, a rather continuous layer of nanoscale defects develops, exhibiting a multiplicity of grain boundaries, vacancies, dislocations, and other defects that result from high levels of cold work [23]. This characteristic layer, which may

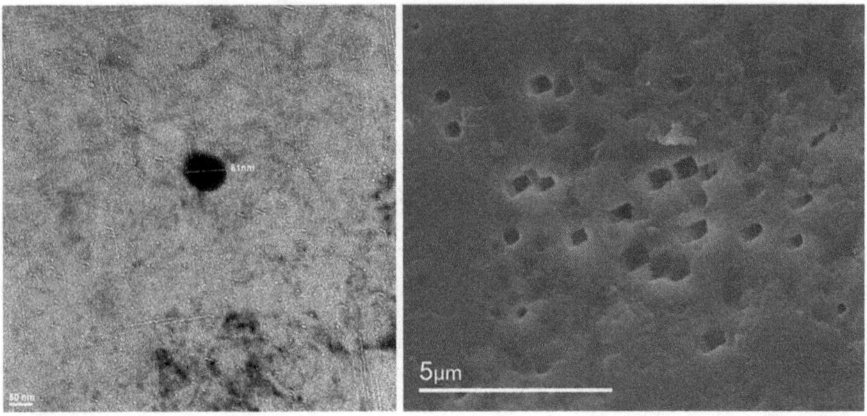

Fig. 7.17 (Left) High density of *Guinier–Preston Zones*, GPZs, (*hashtag-shaped* defects) in the microstructure of a 6063 T5 anode are precursors to the precipitation of Mg_2Si particles. Agglomeration of GPZs at the anode surface produces tiny defect pockets with high aspect ratios. These pockets are regions of charge accumulation during anodizing, which develop CPE, locally preventing the formation of anodic oxide. The result is the formation of *square-shaped* pits in the first oxide to be grown that becomes the oxide at the very top of the oxide finish. These pits typically don't produce gross visual defects but can scatter light, making the surface appear cloudier and less reflective

include nanoscale recrystallization, is known as a Beilby effect or a Beilby layer, named after Sir George Beilby who authored *Aggregation and Flow of Solids* [24], a record of an experimental study of the microstructure and physical properties of solids in various states of aggregation, over the years from 1900 to 1921. Upon consideration of the complexity of a surface comprising so many, individual and aggregated defects, while providing the reason for premature corrosion or mechanical strength failure, the presence of such a defect-rich layer explains some of the complex reactions that occur during anodizing at a highly deformed aluminum surface.

The continuous nucleation and re-nucleation of oxide over a surface rich in nanoscale "empty space" defects, like aggregates of vacancies and/or tangles of dislocations that produce CPE, produces an oxide that is highly disordered at the oxide surface, which is the first oxide to form. The high defect density, indeed, aggregation of any atomic defect can develop actual microstructural defects that impact how the anodic oxide nucleates and grows, either by creating surface activation polarization effects, by decreasing the exchange current density, i_o, at the edges of masked areas, or with anodizing intermetallic particles with varying solubility in the electrolyte; or by creating concentration polarization effects, by decreasing i_L, as with a mask or a particle with low solubility (these cause both i_o and CPE), or by increasing i_L, such as an anode that exhibits the effects of empty space caused by near-surface microstructural conditions such as a Beilby layer. Such oxides can appear visually hazy or cloudy and because of their ultra-low density, and can be fragile, to the point of dissolution and fracture, falling off into the dye or seal baths during subsequent post-treatment, producing a rough surface finish with an irregular appearance. See Figs. 7.18 and 7.19.

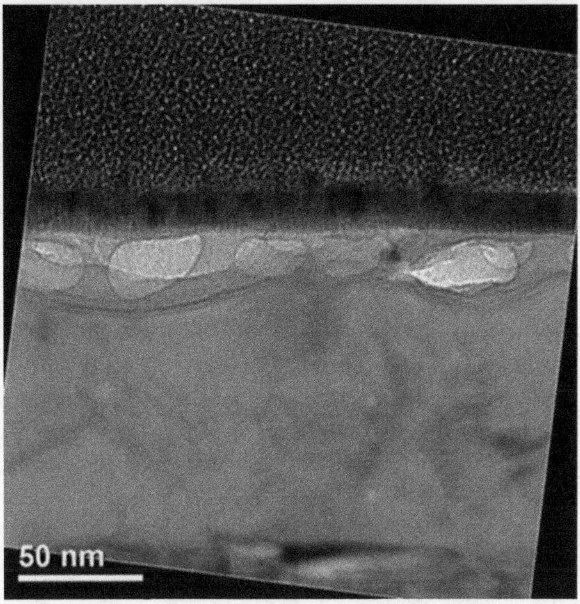

Fig. 7.18 Beilby layer at the surface of a 6063 extrusion. Nanoscale voids are present at the surface due to the strain relief at the surface during deformation

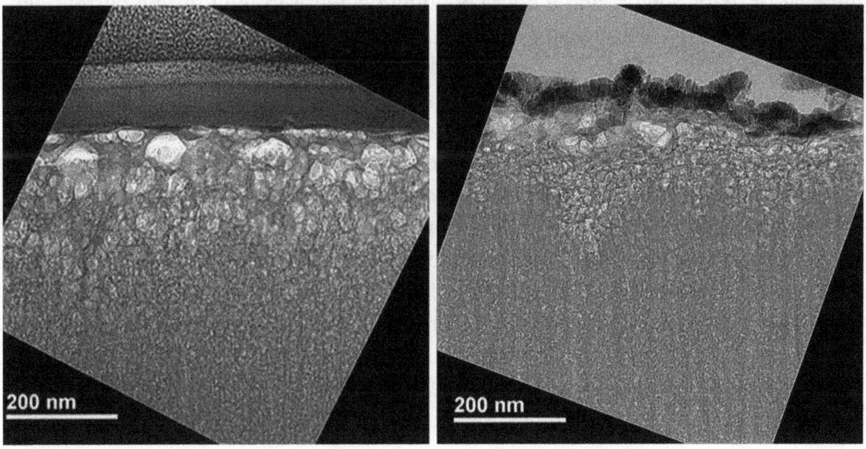

Fig. 7.19 The *Beilby layer* contains concentrations of a variety of atomic level defects at the very surface of highly deformed metallic substrates. With an aluminum substrate, anodizing develops this microstructure, and as it replicates the defects, a fragile oxide is produced. Because the defect layer resides at the very surface of the anode, it becomes the very first oxide formed, and the fragile layer resides at the top of the total oxide structure. At best, this phenomenon produces nothing unusual, perhaps a hazy or cloudy appearance to the total finish. At worst, a Beilby layer can produce a fragile surface structure that breaks off and falls back into subsequent process tanks, increasing surface roughness and exacerbating the cloudy appearance. As a rather ubiquitous surface defect, in the case of thin oxides, the Beilby layer may undermine the overall quality of the finish

7.6.1.3 The Effects of Elements Dissolved in the Electrolyte on Concentration Polarization

A major source for concentration polarization effects: the alloy-related reaction of dissolved cations from the substrate to form loosely bound anion complexes that adsorb and shut down oxide growth within the polarization wave at the metal-oxide interface is well discussed in Sect. 7.2.1 and again in Sect. 7.6.1. Similar effects are produced in electrolytes with a complex or passive counterion, such as chromic acid. Because the anodic aluminum oxide functions as a barrier to external anions, the electrolyte anion never directly reacts with the aluminum, leaving them to adsorb at the oxide-electrolyte interface, sometimes forming an insoluble complex that shuts down ion exchange, producing CPE. In this respect the reaction at the oxide-electrolyte interface that blocks the anodizing reaction is purely related to the limiting current density, i_L, as opposed to the reactions with complex alloys that are subject to the effects of exchange current density, i_o (as a function of chemical potential, μ) and i_L. Therefore, AAO produced in complex electrolytes with low oxide solubility produce so-called "barrier layer oxides" is compact and thin, regardless of the anode alloy, rarely exceed 3 μ. An advantage of anodizing with this type of electrolyte is that it enables the processing of components manufactured from various alloys on the same rack in a single process. See Figs. 7.20 and 7.21.

As the oxide thickness increases, the internal resistance of the oxide increases the heat of oxidation, T, and a corresponding decrease in the limiting current density, i_L, occurs. This effect can be mitigated by reducing i_a temporarily with a "pulse" that quickly drops off and momentarily decreases the rate of the interfacial oxidation reaction. By doing so, the heat of the oxidation reaction, which has accumulated within the oxide, is relieved, possibly by ejecting loosely bound adsorbed compounds from the oxide surface, reestablishing ion exchange at the oxide-electrolyte interface and the reaction is enabled to continue, enabling the production of a thicker anodic oxide. This point in the reaction may never be reached because the desired thickness of the oxide is achieved before it occurs. By way of the same mechanism, complex alloys may benefit from a current density ramp and/or pulses in applied current density during oxide growth, which relieves the interfacial resistance heat, T, and enables non-aluminum ions to escape the oxide structure, the current density wave and proceed into the electrolyte.

7.6.2 Microstructure Level Defects

As the most fundamental microstructural defect, the surface is broken up by grain boundaries. As continuous boundaries that comprise empty space with intermetallic particles, a grain boundary on its own creates CPE. As the number of grains increases and grain size remains fine, as with recrystallization following deformation, grain boundary density increases and changes how the structure of the anodic oxide nucleates and grows. Such oxides appear matte and flat, without reflection, yielding a satin appearance. See Fig. 7.22.

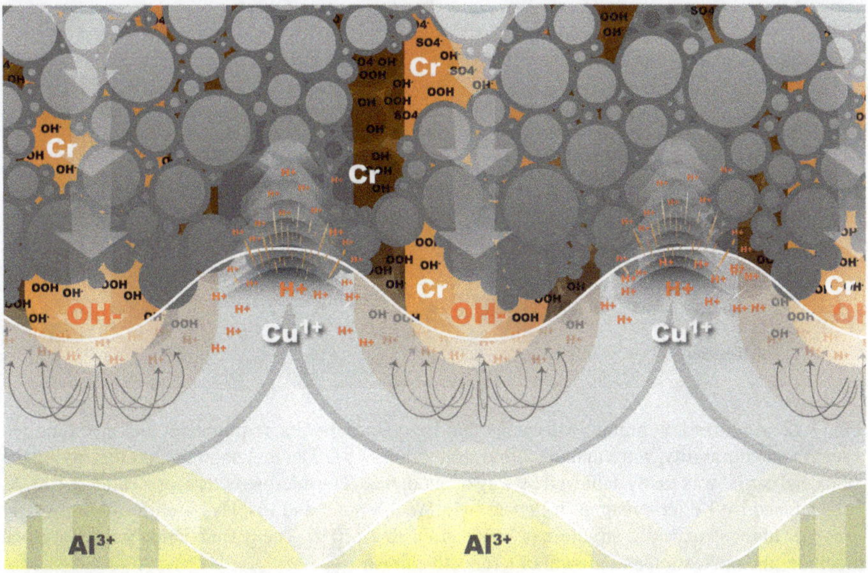

Fig. 7.20 Chromate ions oxidize the sites on the reactive surface of the developing oxide to the point of completely shutting down ion exchange in the pore. Because the developing column cannot sustain ion exchange, the structure collapses on itself. The wave-like oxide structure that consequently forms indicates the mechanism proceeds as a sequence of oxide layers with low dielectric strength that form and rupture from the anode surface shortly after the anode surface is repeatedly reconstructed, breaking apart and reacting with the electrolyte

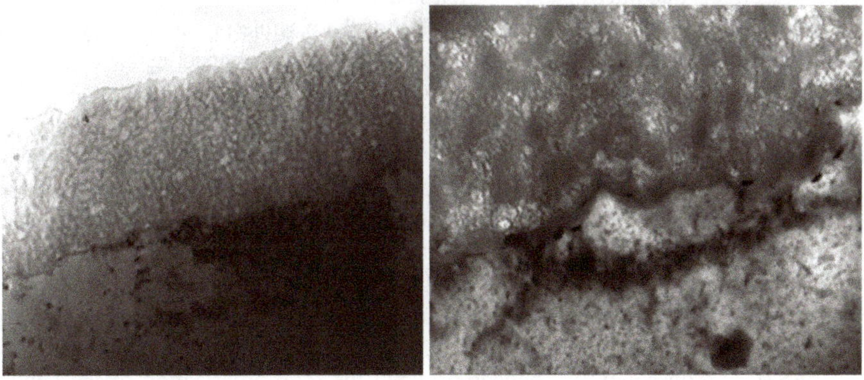

Fig. 7.21 Transmission electron microscope (TEM) photomicrograph of AAO grown in a chromic acid electrolyte (Type I) on a 2024 aluminum alloy substrate

Fig. 7.22 Anodized aluminum extrusions that have a strict color requirement and also required strength and durability, were manufactured from 6061 T651. The image on the left displaying the "flat dead earth" was easily achieved by way of anodizing. The microstructure, presented at 1000X, is developed by Keller's reagent followed by a Weck's stain and displays a uniformly *ultrafine-grained* "fibrous" extruded structure (average grain size <10 µ). The grain boundary density alone precludes the nucleation and growth of a reflective finish

When considering anodic oxide nucleation and growth, it must be kept in mind that intermetallic particles are always larger than the interstitial spaces in the forming oxide. Force exerted by the applied current density, i_a, on intermetallic particles in the anode, cause them to migrate under the influence of the electric field, in the direction of the applied current density toward the anode surface by way of mass transport. The effects that intermetallic particles have on the anodizing circuit are critical, because depending on the size and solubility of the particle in the electrolyte, such particles cause a decrease in exchange current density, i_o, impacting $N_{Surface}$, and a simultaneous decrease in limiting current density, i_L, locally confounding the oxidation reaction, causing CPE.

As intermetallic particles approach the substrate-oxide interface, their size, population, and location in the microstructure depend upon the manufacturing process and thermal history of the anode. Substrate effects such as particle chemistry and solubility in the electrolyte, which affect the exchange current density, i_o, suppress the oxidation kinetics and locally reduce the surface activation polarization, $N_{surface}$. The limiting current density, i_L, may likewise be affected, usually by approaching zero, which is reflected in a corresponding increase in the heat of oxidation, T, as well as N_{conc}. Consequently, changes occur in the total polarization of the anode local to the defect can cause changes in the anodic oxide structure by the altering of the spacing of oxide cells, producing columnar branching and other discontinuities in the structure through the oxide thickness, which increase surface roughness. When the impediment to the oxidation reaction is removed, which typically means that the defect has dissolved or migrated into the AAO by way of mass transport, the

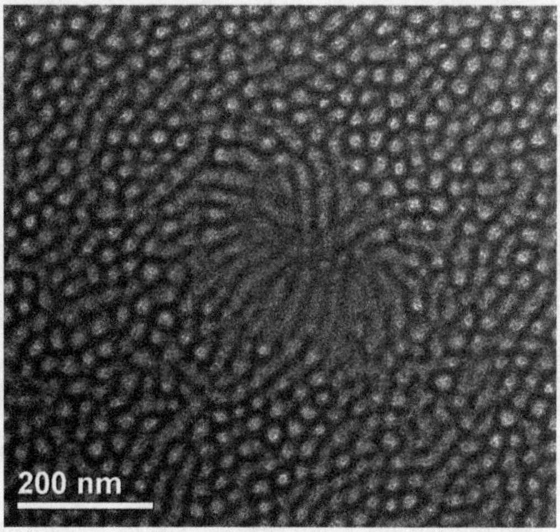

Fig. 7.23 SEM image of transverse section of an AAO layer exhibiting disruption in order due to the inclusion of an *aluminum-containing* intermetallic particle with low solubility in the electrolyte. The particle anodized at a different rate from the surrounding microstructure, producing a distinct change in the AAO structure adjacent to the particle. Such structures will have an increased surface roughness and exhibit differences in dye uptake. The internal structure of the oxide is presented in Fig. 7.25

oxidation process reestablishes itself as the particles exceed the substrate-oxide interface, and oxide growth continues. See Fig. 7.23.

It is important to note that these components of the microstructure—inclusions, intermetallic compounds, precipitates, and other insoluble alloying elements—are not anodized; instead, the primary oxidation reaction proceeds around them, reacting with the aluminum matrix, and insoluble or slightly soluble intermetallic particles are taken up into the structure of the anodic oxide by way of mass transport. The solubility of the particles in the electrolyte varies with the type and is limited by the size (larger particles might not completely dissolve before they are taken up into the oxide). The rate at which intermetallic particles move with i_a varies with size and level of polarizability (whether or not they can carry a charge). Population and orientation of intermetallic particles in the microstructure magnify size effects; because they move slowly, the location with respect to the surface of the anode has a big influence on the impact of intermetallic particles. For example, soluble precipitates such as Mg_2Si are not coherent with the aluminum lattice, and because they do not contain aluminum, they do not anodize under the conditions of a conventional anodizing process, yet because they are soluble in the electrolyte, they don't confound the anodizing reaction to the same extent as other insoluble particles within the microstructure. See Figs. 7.24 and 7.25.

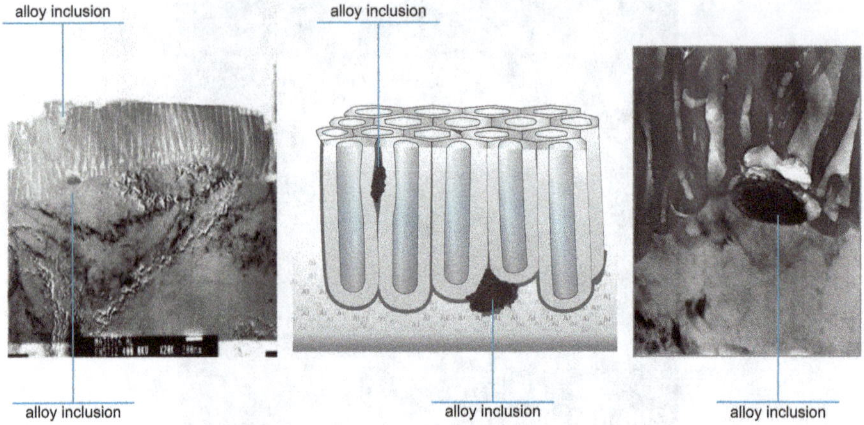

Fig. 7.24 Representation of inclusions transporting through the AAO. TEM images document insoluble alloy intermetallic particles that have moved from the interface into the AAO by way of mass transport. Arrows indicate inclusions have been "anodized around" and taken into the AAO structure. Depending upon their initial size and solubility, particles may be smaller as they dissolve along the knitline as they are transported through the oxide layer. Since they are never metallurgically bound to the AAO, they weaken the cohesive strength of the total finish

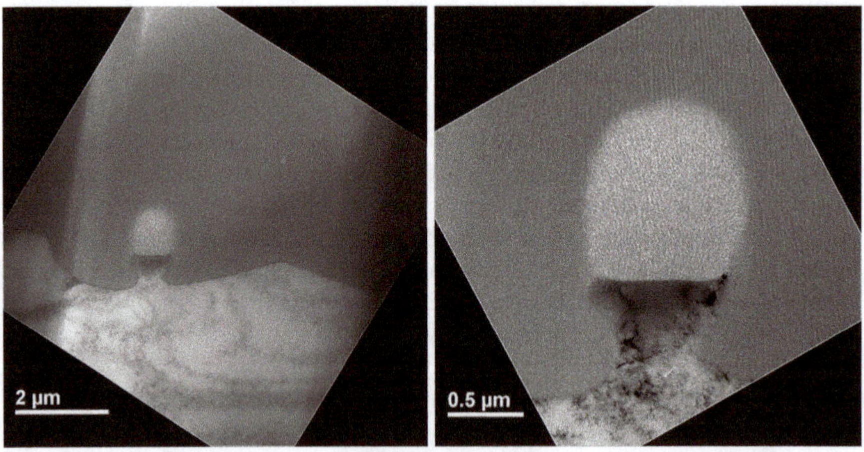

Fig. 7.25 TEM image of a soluble intermetallic particle showing the effects of both surface activation polarization and concentration polarization. The particle is anodizing at a slower rate than the surrounding area; therefore, the particle has masked the oxide thickness over a distance that is about as big as the particle. As the particle dissolves, it is building its own oxide structure, related to the aluminum content of the intermetallic particle

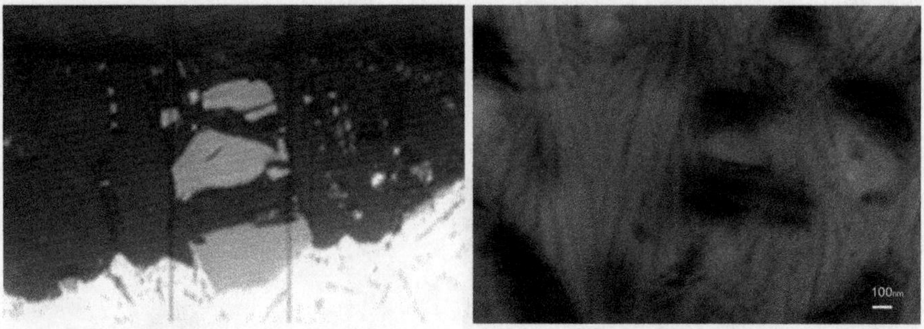

Fig. 7.26 Field Emission SEM image of an anodic oxide on a hypereutectic Al-Si *die-casting*. Note the typically ordered structure is skewed by primary silicon inclusions from the base material. Silicon particles, insoluble in the anodizing electrolyte, blocked the anodizing reaction at the point where they intersected the interface with the anodic oxide, producing transient *concentration polarization effects*; $i_L = 0$ where the silicon inclusion blocks the surface of the cast aluminum anode

The nonequilibrium reaction that continuously provides a supply of aluminum ions to the anode surface assures that CPE, due to intermetallic particles, are most often local and transient. They develop when defects such as primary silicon crystals, other insoluble particles, or intermetallic particles with limited solubility intersect the surface with the developing oxide and temporarily block the anodizing reaction. Therefore, CPE depends upon the size, population, and location of intermetallic particles. In segregated alloy solutions with intermetallic particles that have limited solubility, formation of the anodic oxide proceeds about particles with limited electrolyte solubility and even around them, lifting them into the finish as it grows out from the substrate. The impact of inert or insoluble defects in the aluminum matrix such as lead and hypereutectic silicon on the anodic oxide is in the development of CPE, which causes i_L to approach zero. Therefore, the continuity of the AAO structure is locally disrupted. This disruption in order can produce irregular oxide growth and irregular inter-column spacing. However, as the interface producing CPE is consumed, should a more homogeneous interfacial microstructure be presented for oxidation, oxide growth recovers, and order is restored. Nevertheless, the effects of CPE are significant on oxide growth; such defects are replicated, producing thin, irregular, discontinuous oxides with low cohesive strength. See Figs. 7.26 and 7.27.

Heat treatment, hot isostatic pressing, and other static thermal treatments can change the microstructure of the substrate. Thermal processing can help or hinder anodic oxide formation. It can help by consolidating the microstructure, reducing porosity and forcing soluble precipitates back into solution. It can hinder through the formation of tenacious scales, which, if not removed during pretreatment, both decrease i_0 and cause transient CPE because precipitates are typically limited in size and are soluble in the electrolyte. With time at temperature during tempering, new precipitates form while existing ones coarsen and move by way of mass transport to the grain boundaries, creating a condition referred to as sensitization. Sensitized substrates exhibit preferential corrosion attack at the grain boundaries due to chemical

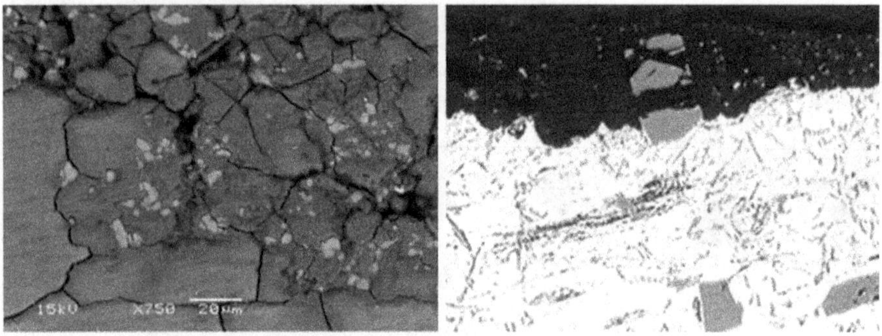

Fig. 7.27 Metallographic *cross section* of anodized die-cast alloy AlSi9Cu3Fe (similar to alloy 383). In the Scanning Electron Microscope (SEM) photomicrograph on the left, the surface of a die-cast component is documented, showing the cracks dotted with *white-appearing* silicon as well as copper deposits appearing as *darker-gray* haze across the surface and concentrated in areas near the cracks. In the optical photomicrograph on the right, large hypereutectic primary silicon platelets from the microstructure as well as smaller eutectic needles are incorporated into the anodic oxide. These insoluble particles reduce the engineering performance of the anodized surface by reducing corrosion and abrasion resistance

potential differences between the bulk of the grain and at the grain boundary. Because the AAO is optically transparent, development of a sensitized microstructure by the anodizing process is often visually objectionable, but the root cause is the deficient surface microstructure of the anode, not the anodizing process. See Fig. 7.28.

7.7 Extrinsic Interfacial Phenomena and Oxide Growth

The anodizing circuit can also be hindered or blocked by the presence of gross surface asperities on the anode surface that are present by design, such as sharp corners, through holes, or an extreme surface texture. They can also be introduced by a faulty machining process, presenting as a lap, seam or a burr, or an extreme surface texture. It is from such defects that concentration polarization was given its name, as areas of charge concentration, accumulate on such asperities or in holes, thereby inhibiting circuit continuity by preventing the formation of uniform polarization and blocking the continuity of the current density wave at that point of the surface. Hydrogen ions, therefore, cannot flow from these points and Concentration Polarization Effects (CPE) result. The same effects occur on the inside surface of sharp corners, where accumulated like charges repel one another. With the accumulation of charge, i_L actually increases and can exceed i_a. A corresponding amount of resistance heat develops, an increase in the local electrolyte temperature and in worst cases, damaging the anode itself by way of burning. See Figs. 7.29 and 7.30.

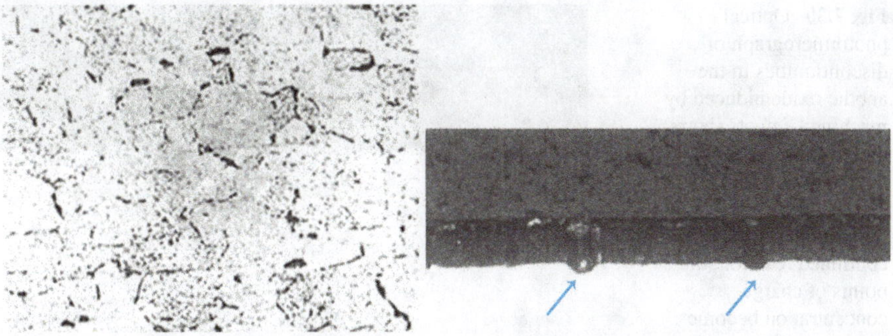

Fig. 7.28 ×400, etchant Tucker's reagent (left) and Keller's reagent (right). Metallurgical micrographs prepared in the long transverse orientation of a 6061 *T-6* extrusion, etched to develop microstructural detail revealed the substrate had been *over-tempered*. The microstructure (left) exhibits coarse intergranular Mg_2Si precipitates as well as high concentrations of dislocation pits across the matrix surface. Anodizing this improperly tempered component (right) following Type II parameters yielded a rough surface that was very difficult to clean. Note the heavily attacked grain boundaries (arrows) that created large discontinuities in the finish

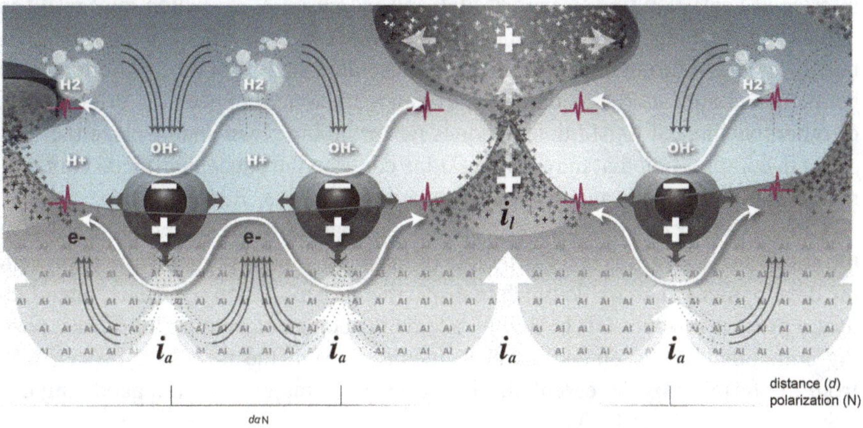

Fig. 7.29 Schematic of the charge density (polarization) wave interrupted by the charge concentration caused by the presence of a burr on the anode surface. The concentration of charge at the defect causes the polarization wave to break down as $i_L \geq i_a$, and $N_{conc} \rightarrow \infty$

Inasmuch as substrate consumption is part of the anodizing reaction, aluminum surface defects introduced by way of mechanical processes are also consumed by oxidation, from as many surfaces as are exposed to the electrolyte. Protrusions exhibit charge concentrations and roots of substrate discontinuities exhibit decreased reactivity due to repulsion forces set up by the proximity of like charges. Oxides formed around aluminum-based defects are prone to chipping and cracking.

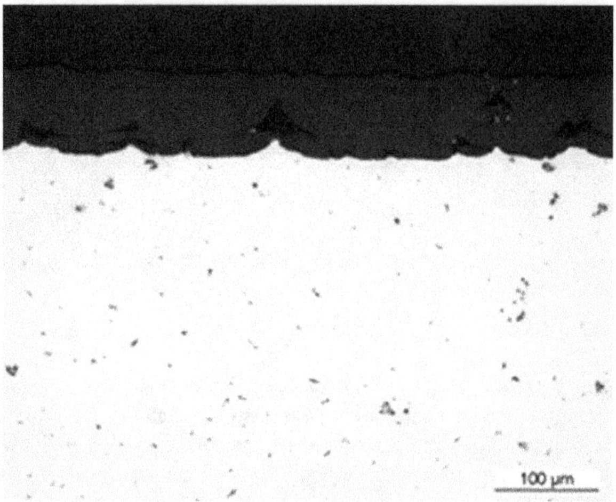

Fig. 7.30 Optical photomicrograph of discontinuities in the anodic oxide induced by machined defects (burrs) on the substrate surface. Note how the voids in the anodic oxide correspond to burrs in the substrate. With continued reaction, the points of charge concentration become blunt and CPE ceases. As oxidation continues, a layer develops with holes that correspond to each burr on the anode surface

The tips of sharp corners and edges created by laps, folds, or cracks introduced by finish machining are areas of charge concentration that result in localized high interfacial heat, resulting in increased T and increased N_{Conc}, which may result in outgassing, burning, and/or the formation of voids in the oxide. For example, if a burr is anodized through its cross section, once detached from the substrate, it will be included in the anodic oxide, much like an inclusion. Shot peening, grit blasting, or other mechanical finishing techniques performed to clean and/or place the substrate in compression that may leave residue can also introduce interfacial contamination that will impede anodic oxide formation. See Figs. 7.31 and 7.32.

CPE always occur when there is a tenacious barrier to the oxidation reaction at the anode surface, which is the principle of masking, placed purposefully to prevent the development of an oxide finish in a specific area by design. However, manufacturing process residues left behind by accident, such as machining lubricants, parting compounds, even embedded blast media, prevent the flow of hydrogen ions from the forming oxide, essentially blocking the completion of the anodizing circuit. See Figs. 7.33 and 7.34.

7.7.1 *Industrial Process Note*

Anodizing leaded alloys such as 6012 and 6062 may be difficult due to residues of smeared lead on the surface from finish machining. These residues produce a mask on the surface that creates CPE. Alkaline cleaners and chemical pretreatments can form corrosion product barriers to further dissolution. For example, sodium hydroxide cleaners can form lead hydroxide, $Pb(OH)_2$, which is stable and will not be removed in the electrolyte. Cleaners that react to form a carbonate, chromate, oxalate, oxide, orthophosphate, sulfate, sulfite, or sulfide as its corrosion product with lead would effectively produce corrosion product barriers on the surface to anodizing. Residual

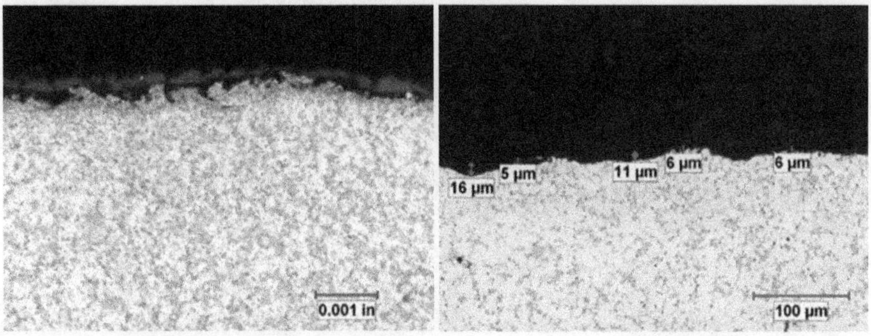

Fig. 7.31 The photomicrographs presented are great examples of CPE in which $i_L \rightarrow i_a$—too much charge concentration between high points and low points of the blasted surface, which prevented adequate anodic oxide growth on a homogeneous alloy that is typically easy to anodize (alloy 518). Compare with machined surface of same alloy, in Fig. 7.32

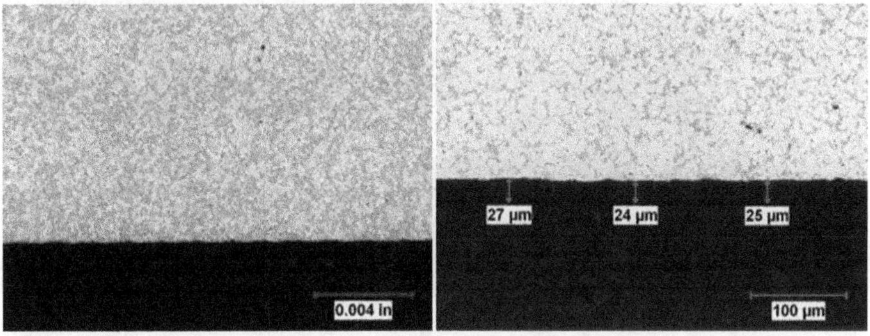

Fig. 7.32 Without asperities, CPE are minimized and a uniform, continuous anodic oxide, two to four times thicker, is developed on the same substrate as that presented in Fig. 7.31

lead will even react with the electrolyte to form a corrosion product barrier with lead that will impede anodizing by increasing the surface resistance. The most effective means of dissolving elemental lead is to have an effective desmut dip and rinse process that utilizes dilute or intermediate strength nitric acid (HNO_3). At concentrations less than 50%, lead nitrate will form; $Pb(NO_3)_2$, which is soluble in water and the removal of the elemental lead removes the anodizing reaction barrier.

7.8 Summary

In order to understand the total effects of interfacial phenomena on the dynamic nature of the anodizing reaction, it is necessary to consider both sides of each interface as ever-changing solutions. The following interfaces: the anode-oxide interface; the knitlines that form in the AAO; and the oxide-electrolyte interface,

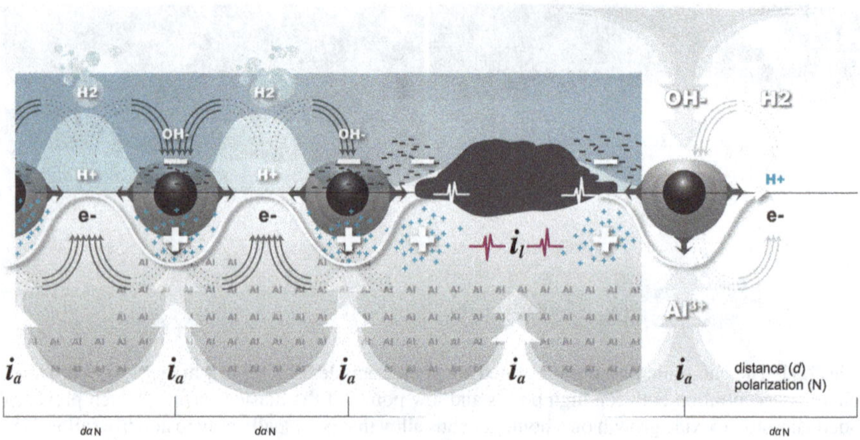

Fig. 7.33 Schematic of the charge density (polarization) wave interrupted by the charge blockage caused by the presence of contamination on the anode surface. The absence of charge at the defect causes the polarization wave to break down as $i_L = 0$, and $N_{conc} \rightarrow \infty$

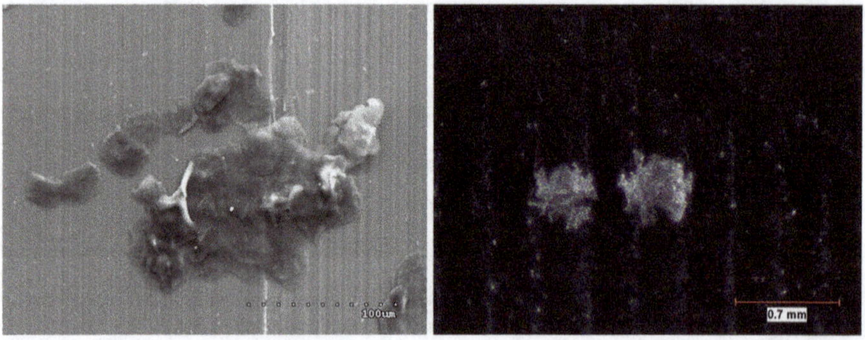

Fig. 7.34 SEM image of residual machining oil on the machined surface of a 7XXX series extrusion that effectively masked the surface during anodizing, producing CPE that blocked the oxidation reaction. The resulting anodic oxide exhibited bright-appearing discontinuities that replicated the size and shape of the contamination. (Photo courtesy of Thomson Cyclery)

must all be considered with respect to the impact of interfacial phenomena on the entire anodizing circuit. The anodic oxide becomes an element in the circuit as it nucleates and grows, and as it develops, because the anode is consumed in the process, the surface of the anode, which forms the anode-oxide interface, continuously changes with oxide growth. Oxide growth results in the formation of distinct additional interfaces: the knitlines between each column in the AAO network (that form multiple oxide-oxide interfaces) and the oxide-electrolyte interface. Interfacial phenomena from the substrate introduce changes to the anodizing circuit that produce resistance or even block the oxidation reaction at all interfaces. Therefore, the quality of the anode surface governs the quality of the oxide.

The kinetics of the reactions that occur during anodizing and the impact that interfacial phenomena have on the polarization at the anode surface as well as the total polarization of the anodizing circuit are well modeled by the Tafel equation, which was introduced in Chap. 5. It is the only model for anodizing that actually takes process parameters into consideration. The basic parameters of applied current density, i_a, the distance between the anode and the cathode, δ, and the concentration of the electrolyte, $C_{Electrolyte}$, are included as key parameters that are fixed externally by the process, in order to produce a polarization value, N_{Total}, at a potential greater than the oxidation potential for aluminum. The response to i_a by resistance elements at the anode surface is addressed by the exchange current density, i_o, which is a characteristic of the condition of the anode surface, and changes with variations in chemical potential, and therefore microstructure. Because the interface with the growing oxide is constantly changing as the oxide consumes the surface, so does i_o. Increases and decreases in the surface activation polarization, $N_{Surface}$, that alone must exceed the standard oxidation potential for aluminum depend upon how interfacial phenomena impact i_o, introducing changes during anodizing from point to point across the anode.

The second part of the Tafel equation brings the rest of the anodizing circuit into consideration. This value, the concentration polarization, N_{Conc}, is linked to the polarization at the surface, and therefore interfacial phenomena, by way of the heat of the oxidation reaction, T. The response to i_a in the rest of the anodizing circuit is accounted for by the limiting current density, i_L, which represents the rate of hydrogen ions proceeding toward the cathodes in order to complete the anodizing circuit. Interfacial phenomena that block i_L are the source for concentration polarization effects that alter the structure of the anodic oxide, sometimes preventing it from developing at all. An understanding of the effects of the process parameters on the anodic oxide quality enables the development of anodizing strategies that optimize the important process quantities of i_L, and T as the heat of reaction, keeping N_{Conc} as small as possible. All of these changes are reflected in the response of the anodizing circuit to the applied current density, i_a, by developing exchange current density, i_o, and limiting current density, i_L; therefore, answers to questions regarding the effects of interfacial phenomena can be found by using the Tafel equation.

In anodizing, polarization is key to the orientation of charge at and directly above the surface, which become the metal-oxide and oxide-electrolyte interfaces. With i_a, anodic oxide formation proceeds as a corrosion process that begins with nucleation at separate and distinct, equally spaced polarized sites across the substrate surface. Therefore, when non-idealities are introduced in the form of alloy ions, intermetallic compounds and other atomic level features and defects such as grain boundaries, dislocations, and vacancies—all non-aluminum interfacial phenomena-retard the anodizing reaction by increasing resistance and changing the surface activation polarization, or confounding it by producing i_L and developing concentration polarization effects, the anodic oxide structure is disrupted and structural order is destroyed.

Differences in alloy chemistry as well as the effects of various manufacturing processes for the aluminum substrate impact how the anodic oxide nucleates and grows. Surface reconstruction is directly related to the initial nucleation of the oxide at polarized centers across the substrate surface set by i_a. How reconstruction and oxide growth proceed is directly related to the concentration and distribution of atomic level defects as well as microstructural defects such as intermetallic particles and grain boundaries. The applied current density stimulates alloy unmixing, which provides alloy ions to the polarized centers, changing the local chemical potential and producing i_o, causing resistance to the oxidation reaction. Intermetallic compounds and precipitates that are soluble in the electrolyte impact chemical potential transiently: resistance effects are limited to the size of the particle, sometimes anodizing at a slower rate than the matrix, until the particle is surrounded by oxide and lifted into the structure. Alloy and intermetallic effects develop and sustain surface activation polarization, or $N_{Surface}$, which should be maintained at a value more anodic than the oxidation potential for aluminum. The anodizing process conditions, especially chemical pretreatment, are developed and set to maximize the amount of free aluminum at the substrate surface, which optimizes $N_{Surface}$.

Concentration polarization effects develop in three ways: (1) as the flow of hydrogen in the anodizing circuit is somehow prevented from reaching the cathodes, which can be caused by interfacial phenomena that mask the surface, such as an insoluble microstructural particle or even an actual mask, or (2) by surface defects on the substrate that are more highly polarized than the rest of the surface, such as corners, edges, holes, the peaks of burrs, or other machined features, or (3) by sharp internal corners and areas of high aspect ratio, such as through holes, where charges repel one another. These defects do not support the nucleation of oxide, and it dissolves as fast as it forms, creating a point of charge concentration on the surface, from which concentration polarization, N_{Conc}, takes its name. Under these conditions, $i_L \geq i_a$ in order for the anodizing reaction to operate efficiently, N_{Conc} must be kept to a minimum.

When considering the knitlines that develop in the anodic oxide, the structural concept of AAO as an ionic solid comprising tetragonal AlO_{4-} molecules, introduced in Chap. 6, is critical to addressing the impact of how interfacial phenomena traverse the oxide layer to make it to the oxide-electrolyte interface. The transition in oxide structure from the base of each column: from its point of symmetry from which it grows; to the points of impingement during which changes are introduced to the structure by electrostriction, forming interfaces between columns as well as the central pore; to the surface with the electrolyte, where additional changes in the structure are produced by way of oxide dissolution and reactions with ions at the oxide-electrolyte interface, produces a unique structure that functions as a p-i-n diode, sustained by ion exchange.

The perspective of the anodic oxide as a semiconductor enables understanding of anodic oxide growth as a part of the complete anodizing circuit: as current flows through the base material, the oxide as it grows becomes a part of the anodizing circuit through which ionic conduction and mass transport occur, continuing through its thickness to the interface with the electrolyte, from which hydrogen ions carry most of the current through the electrolyte to the cathodes, where the entire anodizing circuit is complete. As an ionic solid, electricity is conducted through the oxide by the movement of ions through the interstices of the oxide. Larger ions move slowly through the oxide, primarily along the interfaces or "knitlines" between the columns of the forming oxide due to the transition in structure that occurs due to electrostriction.

Clearly, the interfacial effects of the physical anodizing circuit on the total polarization of the anode, N, impact how the oxide forms, its structural order, and its functionality as an engineering finish for aluminum. The various reactions that occur with interfacial phenomena at the metal-oxide interface to produce exchange current density, i_o, and/or limiting current density, i_L, can change the structural order of the total anodic oxide by changing the surface activation polarization, $N_{Surface}$. Reactions that occur at the oxide-electrolyte interface that block the conduction of ions also reduce develop CPE, which limits anodic oxide growth.

Manufacturing processes affect the level of microstructural homogeneity developed in components and impact the defect population and the variation of chemical potential across the component surface. By understanding the reactions that occur with individual defects, and how they change oxide growth locally, their effects can be understood and addressed across the entire anode surface, through the oxide and into the electrolyte. It is important to understand the nature of interfacial defects such that anodizing process parameters can be modified to effect optimum oxide growth. In Chap. 8, the effects of interfacial phenomena, as they come together to comprise a surface microstructure, are presented that relate component history to anodizing problems and connect manufacturing variations to variations in anodic oxide appearance or function. Such understanding enables best possible alloy selection for manufacturing and metal finishing; appropriate adjustments to the anodizing process, or target areas of manufacturing or metal finishing for corrective action.

References

1. Runge, J. M. (2008). Interfacial phenomena and anodizing: Ramifications and process solutions. In: *Proceedings of the 17th Annual AAC Conference*, San Francisco, CA.
2. Garcia-Vergera, S. J., Habazaki, H., Skeldon, P., & Thompson, G. (2010). Tracer studies relating to alloying element behavior in porous anodic alumina formed in phosphoric acid. *Electrochimica Acta, 55*, 3175–3184.
3. Parsons, R. (1987). Chapter 2: The electric double layer at the solid-solution interface. In W. Stumm (Ed.), *Aquatic surface chemistry*. New York: Wiley.
4. Runge, J., & Han, C. (2002). The future of anodizing. In: *Proceedings of the 10th Annual AAC Conference*, OakBrook, IL.
5. Runge, J. M. (2014). Anodizing for design and function. *Journal of Materials Science and Nanotechnology, 1*, S108.
6. Shewmon, P. (1989). *Diffusion in solids*. Salem, MA: TMS Publication.
7. Runge, J. M. (2013). Base metal microstructural considerations for anodizing wrought aluminum alloys. *The Proceedings of Aluminium 2000*, Milan.
8. Jones, D. A. (1996). *Principles and prevention of corrosion* (2nd ed.). Upper Saddle River, NJ: Prentice Hall.

9. Brace, A. W. (2000). *The technology of anodizing aluminium* (3rd ed.). Modena, Italy: Interall Srl.
10. Chesterfield, L., & Runge, J. (2010). Connecting theory to practice, the science of successfully anodizing aluminum die castings. In: *Proceedings of the 19th Annual Technical Conference and Exposition of the Aluminum Anodizers Council*, Montreal.
11. Uhlig, H. H. (1985). *Corrosion and corrosion control, an introduction to corrosion science and engineering* (2nd ed.). New York: Wiley.
12. Lerner, M. (2003). Is the barrier layer responsible for the cellular structure of aluminum oxide film anodized in sulfuric acid? In: *Proceedings of AESF Surfin Conference*, Chicago.
13. Ellard, B. (2010). Metallurgy basics for aluminum surfaces. *AAC Anodizing Essentials Workshop*.
14. Shannon, R. D. (1976). Revised effective ionic radii and systematic studies of interatomic distances and chalcogenides. *Acta Cryst, A32*, 751–767.
15. Polmear, I. (2006). *Light alloys, from traditional alloys to nanocrystals*. Oxford, UK: Elsevier.
16. Kniep, R., Lamperter, P., & Steeb, S. (1989). Structure of anodic oxide coatings on aluminum. *Advanced Materials, 7*, 229–231.
17. Kubicki, J. D., & Apitz, S. E. (1998). Molecular cluster models of aluminum oxide and aluminum hydroxide surfaces. *American Minerologist, 83*, 1045–1066.
18. Runge, J. M., & Hossain, T. (2015). Interfacial phenomena in 7000 series alloys and their impact on the anodic oxide. In: *Proceedings of Aluminium Two Thousand World Congress and International Conference on Extrusion and Benchmark ICEB 2015, Materials Today Proceedings*.
19. Moulson, A. J., & Herbert, J. M. (2003). *Electroceramics: Materials, properties, applications*. New York: Wiley.
20. Vanhumbeeck, J.-F., & Proost, J. (2008). On the contribution of electrostriction to charge-induced stresses in anodic oxide films. *Electrochimica Acta, 53*, 6165–6172.
21. MacDonald, D. D. (1999). Passivity—The key to our metals-based civilization. *Pure and Applied Chemistry, 71*(6), 951–978.
22. Runge, J. M. (2016, October). Anodizing complex 7000 series alloys. *Light Metal Age*, pp. 52–56.
23. Scamans, G. M., Frolish, M. F., Rainforth, W. M., Zhou, Z., Liu, Y., Zhou, X., & Thompson, G. E. (2010). The ubiquitous Beilby layer on aluminium surfaces. *Surface and Interface Analysis, 42*, 175–179.
24. Beilby, G. (1921). *Aggregation and flow of solids, being the records of an experimental study of the micros-structure and physical properties of solids in various states of aggregation from 1900–1921*. London: MacMillan.

Chapter 8
Base Metal Microstructural Considerations for Anodizing Aluminum Substrates

8.1 Introduction

Engineering designs that utilize aluminum and its alloys are driven by a variety of esthetic, technical, and economic criteria. Design engineers may consider the anodic oxide as the finish of choice for a particular application for a variety of reasons; as an integral finish, the anodic oxide is immune to filiform corrosion and doesn't chip or peel like a paint layer or other finishes that are applied to the surface, even an electrodeposit. Anodizing is rather inexpensive and a relatively easy process to run. The highly ordered, columnar structure of the anodic aluminum oxide enables decoration, corrosion protection, and wear resistance; the anodic oxide has interesting electrical properties that enable it to function as a semiconductor and an insulator. With such unique characteristics, the anodic oxide contributes to the total quality of the product on all levels of design. However, the main ramification of specifying the anodic oxide as a design feature for an engineering application is that the component surface must be considered from alloy formulation through all the various casting, deformation, and mechanical finishing processes required to manufacture it, and, often, the method of surface finishing is not considered until after the alloy and manufacturing process are selected.

Producing the anodic aluminum oxide is a multidisciplinary process that combines basic chemical and electrochemical principles with the physical metallurgy of the substrate. Therefore, the process of anodizing comprises engineering issues and procedures, not only in the anodizing plant, but also prior to anodizing, during substrate manufacturing [1]. Defects that compromise the design requirements of clarity, color, wear and corrosion resistance, bonding, and/or the dielectric withstand voltage of the anodic oxide are often not visible in the as-manufactured condition, but are developed by the anodizing process, or later, when the component fails in service. Consequent investigations often determine the root cause for such defects in the substrate microstructure, which can point to a fundamental design error in the alloy selection for the application, or, in cases when components of the same type,

J.M. Runge, *The Metallurgy of Anodizing Aluminum*,
https://doi.org/10.1007/978-3-319-72177-4_8

suddenly fail to meet finish criteria of appearance and/or thickness, may point to critical variations in the manufacturing process.

Alloy composition and microstructure are always the most decisive factors in determining how a metal will corrode and therefore are the greatest impact on how the aluminum (alloy) substrate will anodize [2]. As presented and discussed in Chap. 6, the various interfacial phenomena that impact how the anodic oxide nucleates and grows are, for the most part, presented by the substrate microstructure. Accepting these statements as facts helps to answer the following questions commonly asked in the metal finishing industry: Why do some alloys anodize with fewer finish defects than other alloys? Why some components are manufactured from the same alloy, through the same anodizing process, on the same rack, sometimes at the same time, exhibit different finish appearances, and sometimes exhibit different finish performance? The answers can frequently be found in variations in the metallurgical quality of the substrate.

In order to show the intimate relationship between base metal microstructure and anodic oxide appearance and quality, this chapter discusses various possible design considerations in terms of the impact of the anodizing process and the appearance of anodic oxides grown substrates with different microstructures produced by different manufacturing processes. The purpose of the small case studies in the first section of the chapter is to show the connection between the design application and the appearance of the anodic oxide. More detailed case studies close the chapter, which show how by understanding the impact of critical microstructural factors, a balanced approach to alloy selection, manufacturing, and surface finishing can be taken to optimize desired performance and appearance. Data from unrelated studies that target components manufactured by different processes, both wrought and cast, from different alloys and tempers are presented to help the anodizer, design engineer, or manufacturer, develop and use a virtual picture of the substrate, based on basic physical metallurgy principles, to imagine how the anodic oxide would grow from the surface of an aluminum component with a certain microstructure and how it would appear, even before it would be anodized. Process details are explored that show how differences in the manufacturing process parameters produce differences in the microstructure that are governed by chemistry, composition, and thermal history, and then in turn are developed by anodizing.

8.2 Anodizing for Design and Function

In order to approach challenges for using anodized aluminum alloys in engineering designs, as well as to effectively troubleshoot problems that might arise with anodized aluminum components, it is necessary to view the challenges from two vantage points: (1) from the characteristics of the anodic oxide and (2) from the characteristics of the base material that is the substrate. These basic vantage points, which comprise anodized aluminum, couldn't be more intimately related: the oxide grows from the substrate. Furthermore, there are two basic reasons that an anodized aluminum component can fail: because the appearance is not what the end user expected; and/or, because it does not meet the functional requirements of the design.

In order to be a good partner on the manufacturing team, the anodizing expert for the total production of a component, it is not enough to know with how much difficulty or ease an aluminum alloy anodizes. Knowing the exact composition or type of intermetallic particle is not as important as knowing the microstructural trends that occur in a component during manufacturing: that regardless of the process or the alloy, phases form when the casting is poured, particles nucleate and grow throughout high temperature exposure; that grains recrystallize and grow during deformation processing; and that strain is relieved over the same high temperature excursions that change the microstructure during deformation. Tempering (aging) develops and affects the size, shape, and distribution of precipitates. It is also important to remark that only precipitates form and coarsen when a component is tempered or aged—no grain growth or strain relief occurs.

An effective anodizing expert must make the effort to know the component history as well as the process limits for growing an anodic oxide of the type necessary to meet the design requirements based upon the substrate microstructure. Knowing the component history enables the anodizing process limits to be set and alternatives to be suggested, based upon sound reasoning that depends upon the metallurgical condition of the component and details of the anodizing process. If this is not done, the rest of the manufacturing team has the right to expect that everything will be OK, especially if the parts are accepted for anodizing. Knowing the component history and understanding the impact it has on the anodizing process enables effective troubleshooting, as well as, in the case of a failure, enables the participation of the anodizer as a discussion partner, not a victim of circumstance. Knowing and understanding the substrate to be anodized demonstrates the capability and knowledge that makes the anodizing expert a good partner in the total manufacturing process.

8.2.1 Relating Process to Anodizing Quality and Performance

8.2.1.1 Appearance, Clarity, and Color

All engineering designs include attributes that are in conflict with one another. With anodizing, the attributes that are most frequently in conflict are surface appearance with corrosion and wear resistance. Because the oxide grows and develops by way of substrate consumption, the alloy composition as well as the function and thickness of the anodic oxide must be a design consideration. Fundamental microstructural details, such as alloy homogeneity, must be considered when the appearance of the anodized finish is driving the design application. The high grain boundary density that comprises complex, high strength alloy substrates, makes producing an esthetically pleasing anodic oxide at any thickness a challenge, let alone thicknesses that provide the highest wear resistance; yet, with appropriate substrate pretreatment on a homogeneous substrate, a thin oxide will yield a bright, reflective finish,

Fig. 8.1 Three bicycle headset bases. The base metal for each is 7075 T6. All are of acceptable appearance and operate to the design requirements. Note the distinctness of image for the reflection in the component is reduced with additional oxide thickness and color. The oxide on the clear component measures ~2 μ, the oxide on the red component measures ~11 μ, and the oxide on the magenta pink component measures ~16 μ. (Photo courtesy of Cane Creek Cyclery)

but for a more complex alloy substrate with the equivalent oxide thickness and pretreatment, the finish will not have the same clarity.

Therefore, design strategies must balance the oxide thickness with the complexity of the microstructure to arrive at the most intelligent compromise, governed first by the most important design property. Trade-offs between surface esthetic considerations such as a reflective or matte appearance and oxide durability in day-to-day wear applications are not uncommon. The thickness of the oxide itself diminishes surface reflectivity; thicker anodic oxides exhibit less clarity and distinctness of image in reflection than thin oxides, even with the most homogeneous substrates. In applications in which color is an important design characteristic, it must be kept in mind that the anodic oxide must be of a sufficient thickness in order to yield a beautiful, rich, uniform appearance after dyeing. See Figs. 8.1 and 8.2.

"Mottled," "streaky," "pitted," and "cloudy" are some of the words used to describe a variety of unfavorable-appearing anodized oxides. This is because on the macroscopic level, the AAO seen with the naked eye is rather transparent and reveals the underlying microstructural detail of the substrate [3, 4]. Inclusions within the AAO interfere with the light path and change the point of reflection. Bands of defects, aligned in the deformation direction, appear as streaks; particles appear as pits; and concentrations of pits and grain boundaries produce a hazy or cloudy appearance in the anodic oxide finish. Changes in substrate surface topography due to etch rate differences between grains of different crystallographic orientation, sensitized grain boundaries and/or surface pits, produce differences in the way light is transmitted and reflected from the substrate surface. Mottled surfaces exhibit an unfavorable, coarse-grained or spangled appearance [5]; sensitized surfaces will exhibit an intergranular-etched appearance [6]. See Figs. 8.3 and 8.4.

While addressing an unfavorable oxide appearance, it is important to keep in mind that the surface microstructure is chemically attacked and modified before the substrate is anodized, when it is subjected to various preparation steps for the purpose of cleaning, native oxide removal, brightening, and etching. Regardless of the

Fig. 8.2 Anodized aluminum extrusion required strength and durability, manufactured from a 6XXX super alloy; the total alloy content exceeded the typical 8% maximum for wrought alloys. The microstructure, presented at 1000X, is developed by Keller's reagent followed by a Weck's stain and displays a uniformly ultrafine-grained "fibrous" extruded structure (average grain size <10 μ). The grain boundary density alone precludes the nucleation and growth of a reflective finish. The design trade-off enabled the successful incorporation of a satin appearance for the anodic oxide and maintained the more important design requirement of impact strength of the 7XXX series alloy

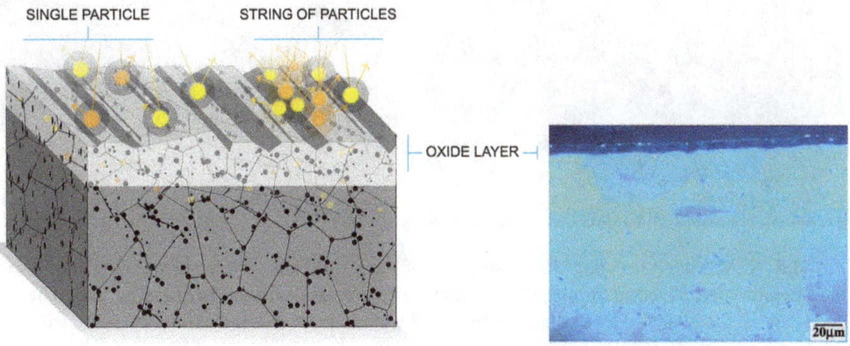

Fig. 8.3 (Left) Schematic of light entering and reflecting through the AAO. Orange and yellow points represent how light is scattered when reflecting through the oxide. Interference caused by a single particle scatters less light than when light is scattered off of a string of particles, causing a "streak." Tabrizian et al. [7]. (Right) Darkfield light optical microscope (LOM) image documenting the clarity of the AAO on a 6061 T6 substrate. The aluminum oxide is transparent and the substrate is the prevailing reflective surface. Zhu et al. [8]

fact that most anodizing pretreatment is processed without an electrical bias, pretreatment steps are electrochemical as well, and are meant to attack the surface in order to clean, brighten, and/or etch. The electrochemical nature of each step of the anodizing process develops the metallurgical condition of the substrate, and areas of heavily segregated material can be accentuated by chemical treatment, yielding an unacceptable appearance. Without polarization, the driving force for these surface

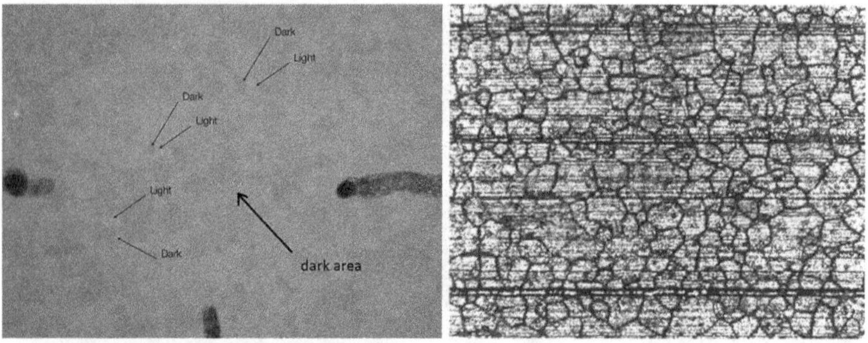

Fig. 8.4 Macrophotographs of samples that exhibit an unacceptable appearance due to microstructural defects developed by anodizing. The image of the sample on the left is from an alloy 6063 extrusion that exhibits a mottled appearance. The image of the sample on the right is from the surface of an alloy 6061 extrusion that exhibits severe intergranular attack from chemical pretreatment before anodizing due to the sensitized condition of the microstructure [5, 6]

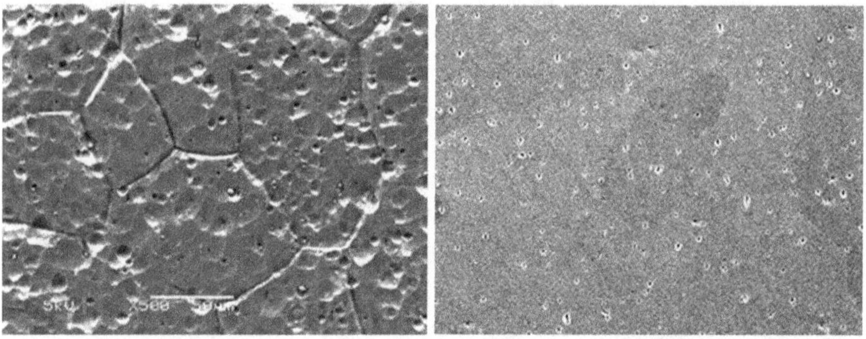

Fig. 8.5 SEM image showing the dissolution of the aluminum matrix surrounding the Fe-rich intermetallic particles due to high electrochemical potential of these intermetallics [8]. The size and shape of the pits, as well as conformance of the oxide to the pit, indicates they preceded anodizing and were most likely developed in pretreatment

reactions is rapid and preferential, beginning at nucleation sites that may have a preference for certain crystallographic surface orientations as well as surface structural features such as grain boundaries, dislocations, steps, or ledges [9]. Metal removal proceeds at a very high rate. On substrates that are compromised by microstructural defects, pretreatment can produce grain boundary grooves, pitting, and other surface defects even before the anodizing reaction takes place. It is critical to note that an uncontrolled pretreatment process can develop surface defects that wouldn't normally be developed during anodizing. Indeed, other defects that are presented as examples in this chapter may have been exacerbated by pretreatment. See Fig. 8.5.

Fig. 8.6 (as-polished) Die-cast alloy AlMg5Si (similar to alloy 518) automotive fuel pump component. The corresponding microstructure, presented as-polished, exhibits a fine, uniform microstructure, yielding a uniform and continuous anodic oxide. At 25 μ thick, the oxide is appropriate for the wear application; the effects of inherent coloring yield a green-black appearance, typical for hard anodized components [10]

8.2.1.2 Corrosion and Wear Resistance

External sources for wear drive the need for anodic oxide durability, from that which occurs on the surface of an automobile piston to what happens to a mobile telephone on the inside of a pocket or a purse. Robustness, in the form of impact and abrasion resistance, is typically a characteristic of thicker oxide finishes that are often grown from high strength substrates with complex alloy chemistries, containing a variety of intermetallic particles. When intermetallic particles from the substrate are included in the AAO, they interfere with the reflection of light through the oxide to and from the substrate surface, to the extent the surface has been developed by the various mechanical and chemical treatment processes prior to anodizing. Thicker finishes will tend to be more corrosion, and wear resistant than thin finishes; however, as more substrate is consumed to develop the anodic oxide, more microstructural features are included in the anodized finish.

Strength and hardness are first developed in the substrate microstructure: cast alloys may exhibit coarse primary silicon crystals or feathery silicon needles; wrought alloys may exhibit fibrous microstructures with highly deformed grains and little-to-no recrystallization. Coarse particles from the microstructure that intersect the oxide surface can produce increases in surface roughness that can make or break engineering performance. Anodic oxides, produced on complex alloys with fine microstructures, tend to be uniform and continuous at all thicknesses and show little effect of the alloy complexity on the anodic oxide except for that of inherent coloring. See Fig. 8.6.

Mass transport of inclusions and insoluble alloy additions such as hypereutectic silicon compete with the oxidation kinetics, resulting in repeated restructuring of the substrate surface, which disrupts the order of the anodic oxide. Depending upon the size and population of insoluble defects, the cohesive strength of the finish can be reduced and paths for environmental ingress can be created, reducing the

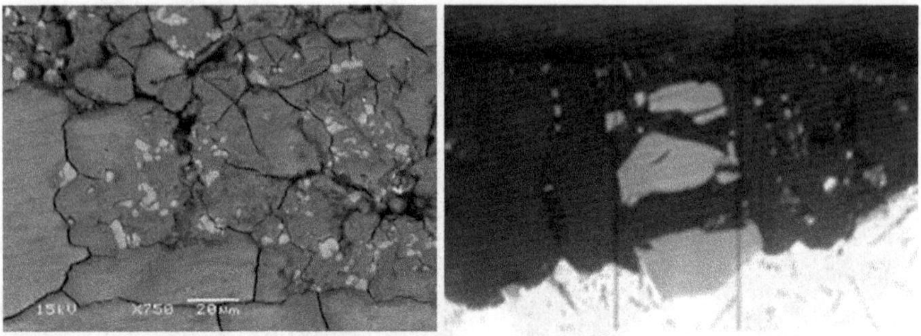

Fig. 8.7 Metallographic cross section of anodized die-cast alloy AlSi9Cu3 (similar to alloy 383). Alloy 383 offers improved strength at elevated temperatures over alloy A380. It also exhibits better flow and is better for intricate components that require improved die filling characteristics. In the scanning electron microscope (SEM) photomicrograph on the left, the surface of an anodized die-cast component is documented, showing the cracks dotted with large, white-appearing silicon crystals that exceed 10 μ in size, as well as copper deposits appearing as darker gray haze across the surface and concentrated in areas near the cracks. In the optical photomicrograph on the right, large hypereutectic primary silicon platelets from the microstructure as well as smaller eutectic needles are incorporated into the anodic oxide. It is apparent that this die-casting cooled slowly, as the presence of coarse silicon platelets is usually associated with alloys that have a composition exceeding the 12.6% silicon eutectic composition

corrosion resistance of the finish; furthermore, the continuity of the porous structure will be interrupted. AAO finishes that include Si particles exhibit lower cohesive and impact strength, the inability to absorb dye, and therefore reduced corrosion protection and wear resistance [5]. See Fig. 8.7.

8.2.1.3 Dielectric Withstand

As an effective insulator, the anodic oxide is often relied on for durability in applications where a plasma or high level of electrical energy must be effectively isolated while other high energy operations take place. The anodic oxide is the finish of choice for sputtering chambers and other critical components for the semiconductor industry. Defects in the substrate microstructure, that disrupt anodic oxide uniformity and continuity across the interface, reduce the ability of the oxide to withstand the energy applied to the base metal that should otherwise be insulated by the oxide. Defects can be microscopic in size, but can also be visible with the naked eye. Surface finishing defects, such as burrs, or other surface texture features developed by blasting or tumbling, develop points of high current density which produce seams, branching, and in the worst case, holes in the AAO structure. See Fig. 8.8.

Fig. 8.8 (as-polished) Metallographic cross section of 6061 T6 extruded material exhibits frequent machining defects, which produce a rough surface texture. At each high point on the surface, the current density was concentrated, producing a great deal of anodic activity—the surface area at the tip of a burr is very small, and the current density very high, preventing the nucleation and growth of an oxide—producing instead, holes in the structure. As the oxidation reaction proceeded, and the burrs became less pointed, oxide grew up to the apex of each burr, producing seams in the oxide layer. Holes, seams, and other discontinuities in the oxide structure reduce the insulating quality of the total oxide, allowing the finish to rupture when, with applied voltage, the dielectric withstand at these points is exceeded

8.2.2 Where Do Microstructural Defects Go During Anodizing?

A new, ever-changing interface is presented to the advancing oxide, and every microstructural detail that might enhance or confound its progress is encountered. Alloying elements in solution with the aluminum substrate, as well as phases that are coherent with the alloy system, and incoherent dispersoids and precipitates, locally change the substrate chemical potential at the interface between the substrate and the electrolyte. Therefore, all non-aluminum interfacial phenomena confound the anodizing reaction, retard it, and disrupt the anodic oxide structure.

Alloying element ions and aluminum-free intermetallic particles don't anodize, while aluminum-containing intermetallic particles have a reduced rate of oxidation compared with the matrix. As the oxidation reaction proceeds, non-aluminum metal ions entrain at the interface and eventually diffuse through the oxide network by hopping through spaces in the network or along intercolumnar knitlines, causing inherent coloring effects and imparting a cloudy appearance. Oxidation occurs around particles at the interface that don't anodize and temporarily mask the surface until they are taken into the AAO structure, disrupting its order. A large population of particles can lead to irregular oxide growth, irregular inter-column spacing, even delamination of the oxide at the interface with the substrate. Most of the time, as the surrounding surface continues to be consumed by oxide growth and a more homogeneous surface is presented for anodizing, beneath the defect, recovery occurs, and

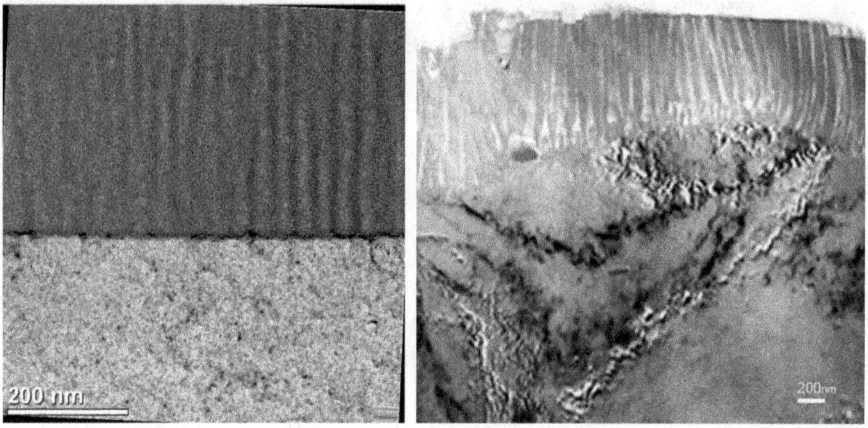

Fig. 8.9 Transmission electron microscope (TEM) photomicrographs anodized wrought substrates displaying the effects of alloying elements and intermetallic particles on the anodic oxide structure. The image on the left is of a 7XXX series alloy. The dark-appearing particles entrained at the interface of the substrate and the AAO are zinc, from the 7XXX series alloy substrate. Note how they progress along the column walls, especially on the right side of the photomicrograph where they are concentrated most. The image on the right is of hard anodized 6061 T6 alloy. Note two intermetallic particles that have been included in the AAO during oxide growth. Note how the reaction is proceeding around the larger particle at the interface. A smaller particle is directly above the large particle, captured in the AAO structure [12]

structural order is restored. However, the oxide replicates the defect, making the surface rougher or the appearance cloudier, sometimes compromising the function of the finish [11].

The increase in the local surface resistance, caused by the difference in chemical potential between a particle and the matrix, alters the level of chemical attack by the electrolyte, as well as the corrosion mechanism that sustains anodic oxide nucleation and growth, by changing the exchange current density, i_o. See Chap. 6. This causes disruption in the order of the anodic aluminum oxide (AAO) by the diffusion of metallic ions due to unmixing of the alloy phases as well as mass transport of non-aluminum intermetallic particles due to electrotransport. Together with the aluminum ions that make the AAO, alloying element ions and intermetallic particles are dragged, by the applied current density, with the aluminum ions to the interface with the growing oxide. Non-aluminum metal ions and microstructural defects are incorporated into the anodic oxide structure, but they are never metallurgically bound to the AAO, consequently disrupting its order. See Fig. 8.9.

8.3 Predicting Microstructure and Its Effect on Anodizing Process

Chapter 3 provides the basic metallurgy necessary to understand how the surface microstructure develops for an aluminum substrate that becomes a working anode. In order to properly select an alloy that will, as anodized, function as intended (meaning both the substrate and the oxide), it is important to know *when* in a manufacturing process the thermodynamic events occur that change the microstructure, such as precipitation, particle coarsening, recrystallization, and grain growth, to get an accurate picture of the microstructure. Microstructural changes induced by manufacturing are usually anticipated and can be controlled by strict manufacturing practices. Therefore, components designed for certain applications will typically exhibit microstructures characteristic for the primary design attribute. This section will summarize these events in the context of the specific manufacturing processes of casting, extrusion, and forging.

Alloy chemistry effects are reflected in the microstructure, by way of the development of the complexity of the microstructure due to alloy loading and by way of how the elements come together compositionally, as intermetallic phases and particles. Compositional effects are reflected in the microstructure in ways that exceed complexity due to alloy loading alone; they are reflected in the population and distribution of intermetallic particles, and by how each particle and phase affects mechanical properties and chemical potential. For example, dispersoids function to pin grain boundaries and inhibit recrystallization, thereby increasing alloy strength; and precipitates increase hardness by their nucleation and growth due to coarsening as the result of extended time at temperature. As shown in Chap. 6, each microstructural particle, indeed every feature, from the atomic level, as clusters of dislocations and vacancies, to the macroscopic level, as surface laps and burrs from mechanical finishing, has a different chemical potential than the matrix aluminum. Variations in local chemical potential change the response to the surface oxidation reaction.

The correlation between time, temperature, and microstructural transformation as they relate to specific manufacturing processes for aluminum components that will be anodized as part of their design application, aid in alloy design, alloy selection for specific engineering designs and troubleshooting when things go wrong. More complex alloys require more complex phase diagrams than what is presented in Chap. 3 for the Al-Si alloys, and alloys made with other primary elements have different thermal characteristics. Thermodynamic software is a necessary tool for alloy design, utilized to predict how a microstructure will develop. The knowledge and tools are rarely available in the anodizing plant, but it is nevertheless important for the anodizer to recognize the impact of the base material on the oxidation response; it is not enough to know which alloys are easy or difficult to anodize.

Table 8.1 lists possible design considerations for anodized aluminum with corresponding industry and common alloy choices. Economics is not included on the list because it is *always* an important design consideration. While reviewing the list, it is necessary to keep in mind that engineering designs often incorporate a mixture

Table 8.1 Possible design considerations for anodized aluminum substrates

Design consideration	Industry	Typical alloy and type
Appearance, including clarity, color, and reflectivity	Consumer electronics, automotive; architecture (inside and outside); sporting goods; transportation; household appliances; jewelry	6XXX, 7XXX extrusions and forgings; 5XXX rolled product; 3XXX rolled product
Corrosion resistance	Automotive and other transport industries; architecture (inside and outside); aerospace; sporting goods; household appliances; defense; semiconductor; machine/automation; most industries that use anodizing	3XX die-castings; 6XXX and 7XXX extrusions and forgings; 5XXX forgings and rolled product; 1XXX and 3XXX rolled product; 2XXX extrusions, forgings, and rolled product
Weathering and wear resistance	Automotive and other transport industries; architecture (inside and outside); aerospace; sporting goods; consumer electronics; defense; semiconductor; machine/automation; most industries that use anodizing	3XX die-castings; 6XXX and 7XXX extrusions and forgings; 5XXX forgings and rolled product; 1XXX and 3XXX rolled product; 2XXX extrusions, forgings, and rolled product
Pressure resistance	Automotive and other transport industries; aerospace; defense; semiconductor	4XX castings (all types); 5XXX and 6XXX rolled, forged, and extruded product
Dielectric withstand	Semiconductor; any engineering design that may use the anodic oxide as a mask for a subsequent anodizing step	6XXX and 5XXX rolled, forged, and extruded product
Surface for bonding, including primers, paints, lacquers, and adhesives	Automotive and other transport industries; machine/automation; aerospace; defense; any industry that may require a label be fastened to an anodized surface	6XXX and 5XXX extrusions; 2XXX and 7XXX rolled, forged, and extruded product

of the listed considerations, requiring intelligent compromises that should be driven by the most important application property for the component and its associated assembly. Finally, higher costs might have to be accepted to arrive at a component that is as beautiful and durable as the design engineer intended for the application. In detail, as Table 8.1 is considered, the various wrought products included may originate as rolled, forged, or extruded product and be subject to follow-on processes such as deep drawing, spinning, or other processes to bring them to their final shape. Almost all products are finish machined and surface treated prior to any finishing process such as anodizing. All of these follow-on processes impact the microstructure, especially at the surface. The process flow for various cast and wrought products is provided as Fig. 8.10.

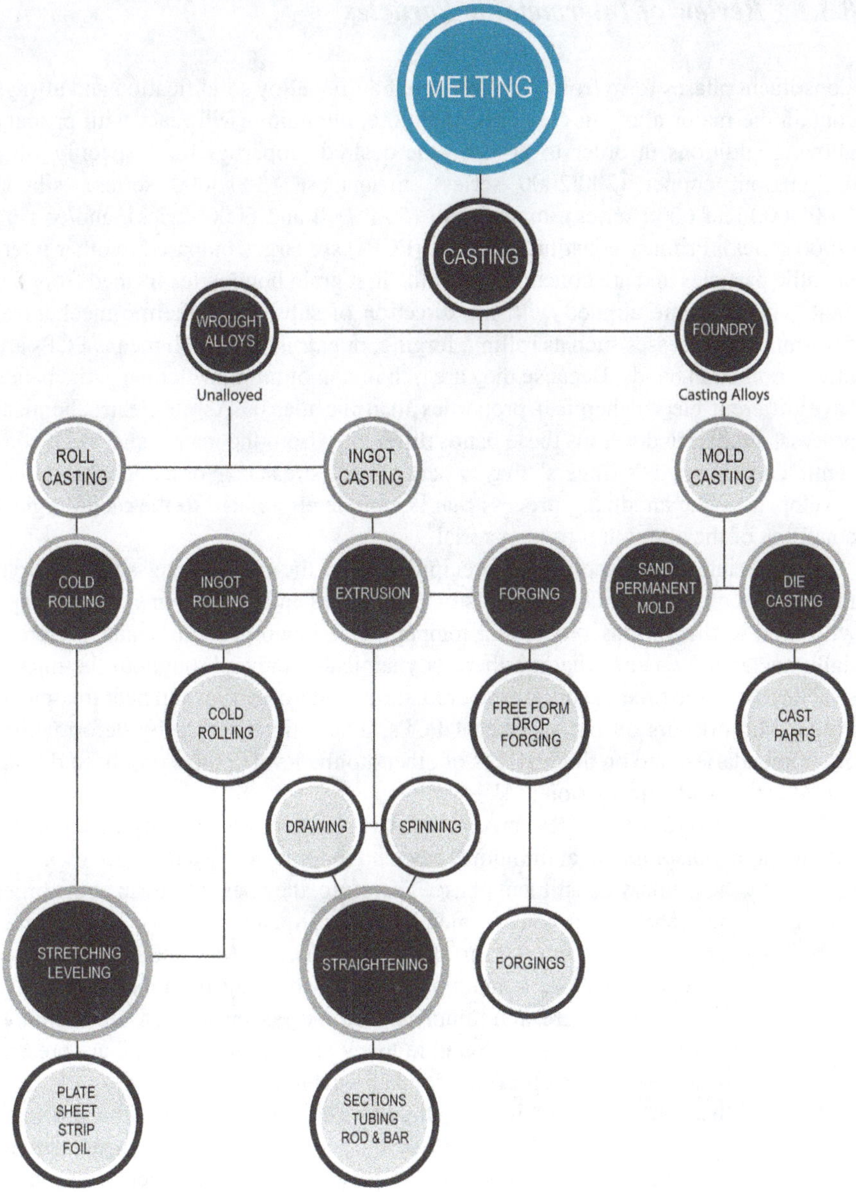

Fig. 8.10 Flow summary of the various manufacturing methods for aluminum and aluminum alloys

8.3.1 *Review of Intermetallic Particles*

Constituent phases form from the parent melt during alloy solidification and always contain the major alloying elements; therefore, aluminum will react with primary alloying additions in order to produce the desired properties for a specific alloy designation: copper (200/2000 series), manganese (300/3000 series), silicon (400/4000 and 6000 series), magnesium (500/5000 and 6000 series), and/or zinc (7000 series). Primary constituent phases (PCPs) are large compared to other intermetallic particles and are concentrated at the first grain boundaries formed from the melt. They become aligned with the direction of subsequent thermo-mechanical deformation processes such as rolling, forging, or extrusion, which means PCPs are often grouped in bands. Because they are rich in non-aluminum elements, the bands have different electrochemical properties than the matrix. As an electrochemical process, anodizing develops these bands differently from the matrix and when concentrated in aligned "stringers" they appear as dark streaks, an objectionable defect developed by the anodizing process that is completely related to the metallurgical condition of the aluminum base material.

Smaller intermetallic particles precipitate from the melt during solidification. Precipitates originate as Guinier–Preston (GP) zones and develop in size and shape with time at the various processing temperatures, heterogeneously and preferentially nucleating on sites that are thermodynamically active throughout the microstructure from the time the billet is poured, through deformation and heat treatment. Precipitation occurs on the surface of PCPs, dislocations set up by deformation, grain boundaries, and on the surfaces of other atomic level defects produced during solidification and deformation [14].

Dispersoids are the smallest precipitate particles comprising alloying elements that are highly *insoluble* in aluminum. They tend to form from excess trace elements rejected by the primary constituent phase. Therefore, they begin to form at the billet level or during preheat, prior to deformation processing, and tend to contain aluminum. They range in size from 0.05 to 0.5 μm. Cr, Ti, Fe, Mn, and Zr are typical dispersoid formers with those containing Fe and Mn being the most chemically active in environments where aluminum is used. Al_3Fe and Al_6Mn are common dispersoid particle stoichiometries found in many aluminum alloys and are present in both non-heat treatable and precipitation hardenable alloys.

In alloys that harden by way of precipitation hardening, precipitates that are soluble in the matrix originate as GP zones and develop as the metal is deformed, but their development is accelerated and controlled by tempering or aging after deformation processing. In many aluminum alloys, primarily, the 6000 series, the composition of the precipitate is Mg_2Si. Precipitation proceeds at a temperature lower than annealing; therefore, no grain growth or stress relief occurs during tempering. Precipitates progressively "ripen" during aging (tempering) from GP zones to spheres to needles, to rods, and to platelets. Corresponding to precipitate ripening is an increase in hardness, with peak hardness reached just before platelets are formed. A decrease in hardness is noted after the formation of the equilibrium phase. With extended tempering

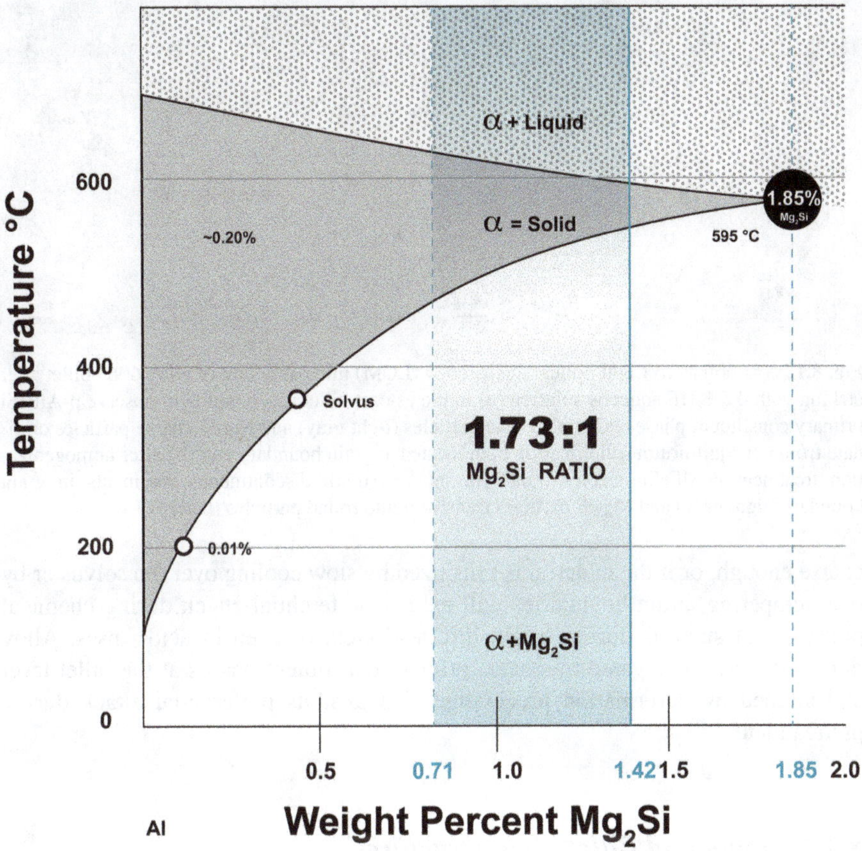

Fig. 8.11 Pseudo-binary Al-Mg2Si phase diagram for an Mg:Si weight ratio of 1.73:1. The displacement of the average composition of the solid under nonequilibrium conditions corresponding to industrial solidification practices is shown [15]

or over-aging, precipitates migrate to the grain boundaries, creating a sensitized condition, predisposing the alloy to intergranular corrosion and stress corrosion cracking, which appears as grain boundary relief through chemical treatment. Figure 8.11 presents the pseudo Al-Mg_2Si phase diagram that is specifically constructed by way of the thermodynamics for the formation of Mg_2Si precipitates, which form in most precipitation hardenable alloys, primarily in the 6000 series [15]. Typical precipitates that form in other precipitation hardenable alloys are: Mg_2Si, $MgZn_2$, Ag_2Al_3, and Al_2Cu.

Dispersoids, precipitates, and other intermetallic particles can increase the pitting potential locally where they intersect the surface, and, depending upon their population and distribution through the microstructure can react rapidly during anodizing pretreatment, developing pits in the substrate surface that cause the anodic oxide to appear cloudy, or change functional requirements of the oxide, such as reducing the expected dielectric withstand voltage. If microstructural defects are

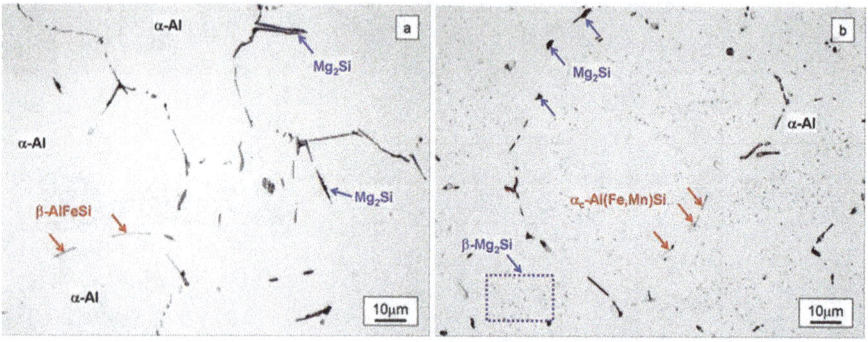

Fig. 8.12 (as-polished) Light optical microscopy (LOM) microstructure of alloy 6063 billet after etching with 0.5% HF aqueous solution (**a**) in the as-cast billet it is possible to observe β-AlFeSi primary constituent phase (PCP) plate-like particles (light gray) and Mg_2Si coarse particles originate from nonequilibrium solidification both located at grain boundary and (**b**) after homogenization treatment β-AlFeSi needles transform into α-AlFeSi discontinuous fragments in grain boundary (light gray) and Mg_2Si particles transform into round particles (black) [15]

coarse enough, or if the material is sensitized by slow cooling over the solvus or by over-tempering, grain boundaries will exhibit preferential attack during chemical pretreatment such as during bright dip, acid etch, or even in acid rinses. Alloy microstructure, segregated by coarse primary constituent phases at the billet level and aligned by deformation processing, also exhibits preferential attack during pretreatment.

8.3.2 *Ingot and Billet Microstructures*

Almost all aluminum products begin as castings and microstructural development begins with solidification from the molten state. The surface microstructure of wrought product is first developed at the billet level. The homogeneity of the billet, developed by (1) the alloy chemistry and (2) the cooling rate of the cast billet will determine the size and shape of the primary constituent phase. Rapid cooling (quenching) always yields a more homogeneous billet as primary phases remain in solution; however, a slack (too slow) quench following homogenization or cycling the billet through multiple preheat operations prior to deformation processes such as extrusion or rolling enables phases to develop and coarsen, giving ample surface area for other intermetallic particles to nucleate, which increases hardness, restricts flow during deformation, and produces a surface with widely varying chemical potential. Coarse intermetallic particles and phases change the flow stress of the metal and increase wear on the extrusion die. See Fig. 8.12.

The solidification process of DC casting results in alloy macrosegregation near the ingot surface. This can affect the outer few millimeters on 6XXX billet to over a centimeter on 2XXX and 7XXX rolling ingot. An X-ray traverse of an ingot cross

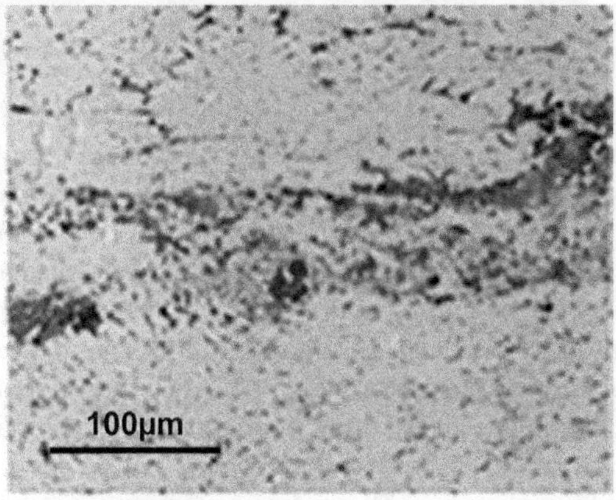

Fig. 8.13 (as-polished) Microstructure of roll cast aluminum alloy 1050 showing segregation of Al-Fe-Si primary constituent phase to the center of the flat-rolled product due to the roll casting process pushing lower melting temperature intermetallic compounds from between higher melting temperature primary aluminum dendrites. With thin gage foil, when intermetallic compounds intersect the surface, they can produce pinholes in the foil as well as inhibit the oxidation reaction during anodizing [16]

section shows the chemical composition changes from alloy rich to alloy depleted and subsequently levels off to the nominal composition through the remaining cross section. Sheet ingots are usually scalped to remove this condition, and some alloys such as 5005 are scalped aggressively to prevent "structural streaking," sometimes manifested after anodizing. Regardless of whether the ingot has been scalped, ingots should be homogenized prior to rolling, extrusion, and forging. Although billet surfaces can be machined off, 6XXX series billets are frequently not, and are extruded directly. Extrusion design and process control can effectively contain this metal in the "dead zone" illustrated in Fig. 8.21 in Sect. 3.5.

Another method for billet production that incorporates rapid solidification, specifically for flat product, is roll casting. The process reduces capital costs and enables shorter production times for slabs, sheet product, and foil. With deformation immediately following casting, smaller particles are developed within the billet and more elements remain in solid solution. However, shearing action of the rolls on formed dendrites can squeeze interdendritic liquid to the center of a rolled shape, enriching the center with low melting temperature intermetallic phases, which are coarse and hard.

In the case of foils that are roll cast, these phases can produce variations in surface chemical potential, as well as other physical defects that disrupt microstructural homogeneity. The occurrence of such defects increases with alloy content. Roll cast defects such as pinholes in the foil surface, alter the appearance of the microstructure through the cross section which consequently alters the appearance of the anodized finish, when aluminum coil product is used for reflectors [16]. See Fig. 8.13.

8.3.3 Characteristic Cast Microstructures

Aluminum castings have been used since aluminum metal has been available and are the foundation for all aluminum products. As a premier example, one needs only to recall from Chap. 1, Napoleon III's aluminum dining service ware, cast in the mid 1850s, on which he served only his most honored guests, leaving the gold for the more "ordinary" people in attendance. Aluminum die-castings have been commercially available since the beginning of the twentieth century. Cast processes precede industrial wrought processes such as forging, rolling, and extrusion because cast billet functions as the basis for all wrought processes. Cast products continue to be important today; aluminum and aluminum alloy castings continue to be used for decorative applications, for sculptures and jewelry, but complex machine components are also cast from aluminum alloys, from automotive pistons to engine blocks.

During the process of alloy development, the ability of the surface to be finished by way of anodizing is often not considered. First and foremost, cast alloys are formulated for mechanical properties of strength, hardness, and resistance to wear and fatigue followed by castability and structural integrity. Alloys are formulated for maximum fluidity, minimum gassing when molten, and the ability to remove the casting from the mold when solid [17]. Mechanical properties are insured first by the integrity of the casting; therefore, the casting alloy and process are designed to minimize porosity and maximize surface integrity. Actual mechanical properties of strength, hardness, and resistance to wear and fatigue are produced metallurgically in aluminum castings two ways: (1) by solid solution hardening; that is: by the substitution of aluminum atoms with alloying atoms in the aluminum crystal structure and (2) by precipitation hardening: the dispersion of second phase constituents or elements in solution and precipitating them out as small intermetallic compounds, incoherent with the microstructure, which inhibit material deformation. The sequence of microstructure development: primary constituent phase formation, dispersoid nucleation, and precipitate nucleation and growth, depends on the alloy composition, cooling rates, and tempering cycles.

Cast microstructure dictates the component quality as well as the quality of the anodic oxide. Because of the higher alloy element content, in addition to high strength and fatigue resistance, castings exhibit more complex surfaces with less free aluminum than wrought alloys and present the widest variability in surface chemical potential, which make castings more difficult to anodize. The composition, casting method, and quality of the casting governs the response to anodizing. Therefore, producing a casting with the most homogeneous microstructure and the finest particle size is of greatest importance when an anodized finish is desired for a particular application [3].

The solubility of the various alloying elements in molten aluminum as well as solid solubility must be considered when producing a casting that will be anodized. Therefore, casting process optimization for metal finishing includes strategies to yield the most homogeneous microstructures. For castings, rapidly cooled structures appear more homogeneous and fine grained than those that are slow cooled,

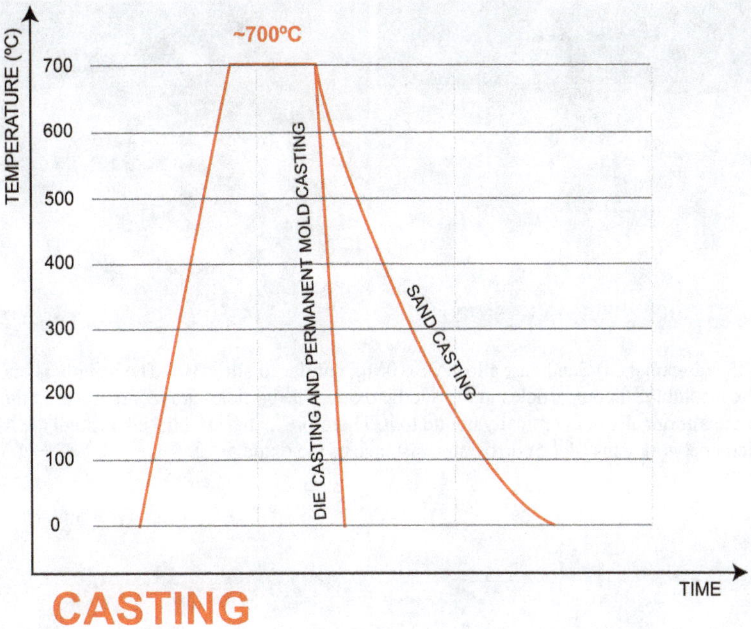

Fig. 8.14 Process-temperature-time diagram. Cooling rate differences produce different cast microstructures

which exhibit coarser intermetallic particles and coarser grained structure. Generally speaking, in order of best to worst response to the anodizing process, permanent mold castings and die-castings tend to exhibit the best response to anodizing, followed by sand casting. But, if the quality of the casting is poor, any of these casting methods may yield a poor response to the anodizing process. A schematic for the time-temperature and process is provided as Fig. 8.14.

Precipitation and growth of intermetallic compounds, phase formation and grain coarsening will occur if the casting is slow cooled. For this reason, sand casting alloys tend to be less complex and probably exhibit the most homogeneous microstructures; however, because of slower cooling, intermetallic particle size tends to be larger. Because of the rapidity of the die-casting process, die-cast microstructures are finer than components cast by other methods. Die-casting dies are designed with vents and cooling lines to facilitate rapid cooling. In addition, components can be ejected into a cooling tank to rapidly quench finished components. Typically, the microstructure of die-castings tends to be fine, with most of the alloying elements remaining in solution. See Figs. 8.15 and 8.16.

In general for casting, coarsening of phases and unintentional tempering of precipitates can occur if cooling media is not turned over as rapidly as components are loaded, developing inadequate cooling and developing coarse microstructures. In die or permanent mold casting, if vents become blocked and cooling is nonuniform, phase segregation can occur, corresponding to the area where cooling was inhibited,

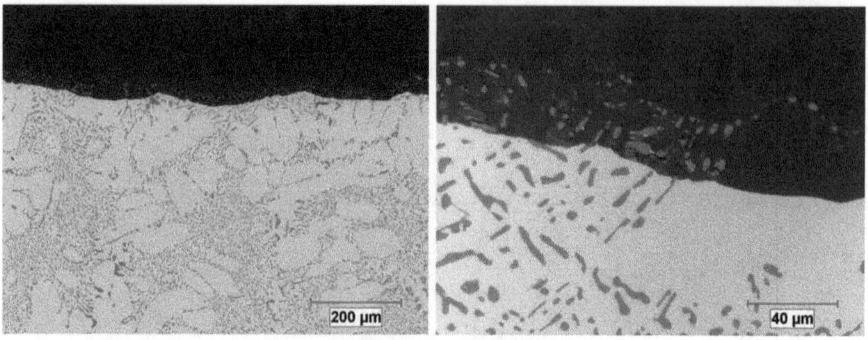

Fig. 8.15 (as-polished) Sand cast alloy AlSi10Mg, similar to alloy 360. The sample is anodized. Note the insoluble silicon particles in the anodic oxide. The particles are incorporated in the AAO, but not metallurgically or chemically bound to it. Therefore, paths established through the AAO by the silicon network can reduce corrosion resistance of the oxide

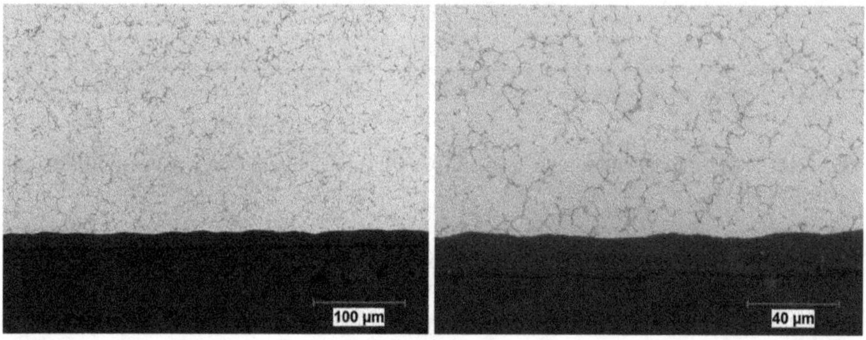

Fig. 8.16 (as-polished) Die-cast microstructure from alloy ADC6, similar to alloy 516. The images are inverted, with the anodic oxide at the bottom. It is very important to note the uniformity, cleanliness, and therefore the continuity of the AAO on this alloy. This is due to the solid solubility of magnesium in the alloy as well as the reduced amount of silicon. Because magnesium is the only element more anodic than aluminum—it oxidizes during the anodizing reaction. Magnesium oxide, as MgO, is soluble in sulfuric acid (the electrolyte), it dissolves from the reaction interface when it is in contact with the electrolyte. This alloy exhibits an excellent anodizing response [11]

changing the surface chemical potential and locally altering the anodizing response. Uncontrolled tempering effects can also occur if die-castings reheat after ejection from the die to a temperature sufficient to temper the microstructure. This can occur with large castings or with batches of smaller castings, in situations when the center of the component or batch of components remains warm after quench and reheats the rest of the component or batch after removal from the quench media. Uncontrolled tempering yields nonuniform microstructure effects, particles of various sizes and shapes develop, depending on the length of reheating and cooling time.

Wear-resistant cast aluminum alloys are based on the presence of the hard, brittle silicon phase. Alloying elements such as iron, manganese, and copper increase the

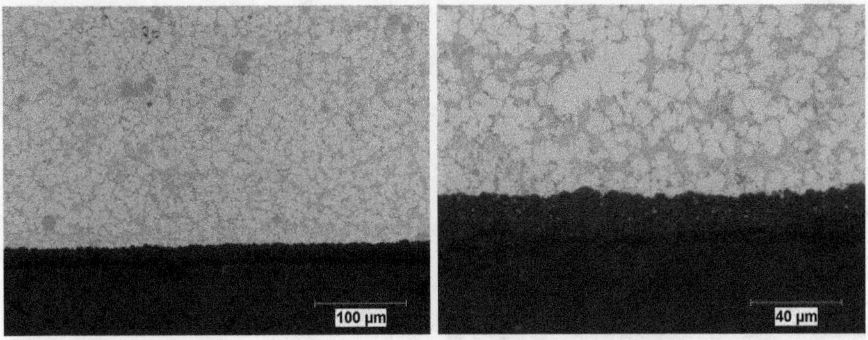

Fig. 8.17 (as-polished) Microstructure of anodized AlSi9Cu3Fe (A383) alloy die-casting; the images are inverted with oxide at the bottom. Iron is added to keep the structure fine by tying up silicon as Al-Fe-Si primary constituent phase. The modified structure is finer and more homogeneous without coarse primary silicon crystals (rounded, gray particles in the image on the left), which also indicates the casting rapidly quenched. Although the fine silicon network still produces discontinuities in the oxide finish, this structure provides good wear resistance in applications that place the surface in abrasion

volume fraction of the intermetallic silicon-bearing phases, contributing to wear resistance. When the silicon content exceeds 12.6%, the eutectic composition for aluminum and silicon, primary silicon crystals are present, and, if fine and well distributed, enhance wear resistance. A rapid cooling process insures that crystals are fine and the microstructure as homogeneous as possible. Alloy modifiers such as phosphorus may be added to the alloy melt to prevent the formation of large primary silicon crystals. The anodic oxides grown on these substrates will tend to be more uniform and continuous, with a higher cohesive strength than oxides grown on castings of the unmodified alloy composition exhibiting the coarser microstructure. Nevertheless, cooling rates dominate the homogeneity of the microstructure and the particle size. See Fig. 8.17.

Complex alloys, such as die-castings, offer mixtures of inclusions, phases, and intermetallic compounds, which confound the anodizing reaction. With as many various defects the die-casting contains, various side reactions occur to impede the surface reaction. The realignment of the non-aluminum substrate elements in solution, such as copper, zinc, and magnesium, at the interface with the forming oxide increases interfacial resistance, resulting in AAO with a rougher surface finish and with nonuniform finish thickness. The kinetics of elemental dissolution and mass transport of inclusions and insoluble hypereutectic silicon compete with the oxidation kinetics, resulting in repeated restructuring of the substrate surface, which in turn disrupts the order of the anodic oxide.

Because die-casting is a closed process, the amount of gas dissolved in melt can "precipitate" out as porosity and migrate toward the surface of the casting as it cools. This is typically avoided by degassing the melt prior to injection and with effective die design. Porosity can also be affected by the presence of residual oxide on the surface of the melt. In the event that porosity is close to or intersects the

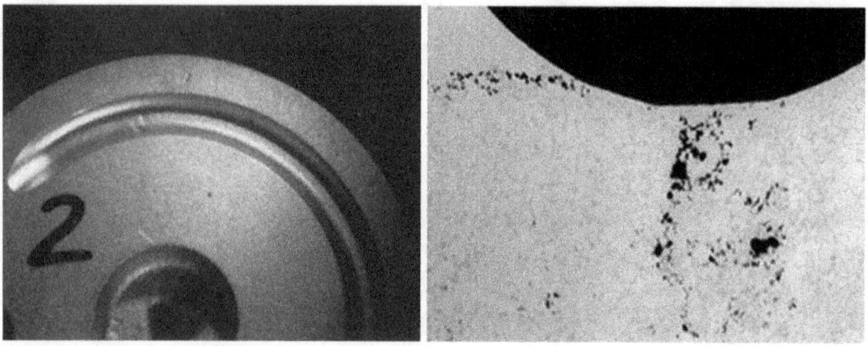

Fig. 8.18 (as-polished) Alloy 518 die-casting with near-surface porosity. The high magnesium content may have exacerbated the surface oxidation of the melt, causing problems with flow, resulting in porosity in the final component. Note the fine-grained cast microstructure [19]

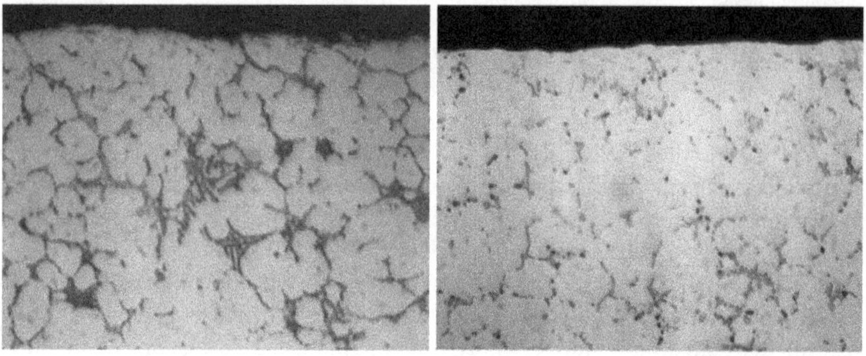

Fig. 8.19 (as-polished) Die-cast alloy ADC 6 (similar to alloy 516) microstructure before hot isostatic pressing (HIP) (left) and after HIP (right). Note how the needle-like script phase, Mg(Fe, $Mn)_3SiAl_2$, is re-dissolved in the aluminum matrix. Mg_2Si precipitates (small black-appearing dots throughout microstructure) are slightly coarser after HIP treatment

surface, the ramifications of chemical processing such as anodizing can develop the pores and create discontinuities in the anodic oxide. In order to insure a uniform cast microstructure, aluminum castings can be subjected to post-heat treatment processes such as solution treatment, quench, and temper, or through hot isostatic pressing (HIP). In the latter, the castings are placed in a container under pressure and heated close to the solution temperature. Phases are re-dissolved in the aluminum matrix, porosity consolidated, and in most cases, quality restored [18]. See Figs. 8.18 and 8.19.

8.3.4 Characteristic Wrought Microstructures

Wrought aluminum alloys must possess a minimum ductility commensurate with the manufacturing process under consideration; otherwise, they crack when deformed [13]. Therefore, because of the increase in hardness produced by making alloy additions, wrought alloys have significantly less alloying elements constitutively and have less complex compositions than cast alloys. For most wrought aluminum alloys, the total alloy element loading is less than or equal to 8% the total alloy content, with each alloy addition contributing attributes to the mechanical and chemical performance of the finished product [15]. Microstructures developed in wrought aluminum alloys can be complex, with the complexity in direct proportion to the alloy loading, with super alloy exceeding the typical 8% maximum alloy content. Regardless of complexity, all aluminum microstructures incorporate a combination of equilibrium and nonequilibrium phases and intermetallic compounds. As with castings, depending upon the alloy composition, intermetallic compounds can be present as precipitates, dispersoids, or phases [20].

All wrought processes such as extrusion and forging change the orientation and spacing of microstructural features, including grain boundaries, and through the introduction of deformation and strain energy, additional atomic level defects in the form of dislocations and vacancies. The anisotropy of the base material, reflected in the microstructure, are replicated in the anodic oxide. However, with increased deformation, it can be concluded that the effects of replication at the surface of the anodic oxide will be compounded due to increased microstructural distortion, changing the surface roughness and the appearance of the finish. Changes in microstructural orientation due to shear in the material as it flows against itself in the die are also replicated in the anodic oxide. The level of deformation as the material flows through the die can break up inclusions and intermetallic compounds, causing appearance differences due to differences in local chemical potential.

Grains originate and grow from the cast microstructure, with the primary constituent phase typically defining the initial grain boundaries. Deformation, regardless of the process used, further develops the grain structure of the aluminum: orienting it in the process direction, causing grain growth at the surface where the workpiece contacts the deformation tool, which leads to the recrystallization of grains, and nucleation of new grains. The amount of grain growth and recrystallization depends upon the speed of the deformation process, and a coarse peripheral phase may result.

Whether the process is hot or cold, all material responds to deformation by flowing in the direction of the applied force, following a flow pattern that results in minimum energy expenditure. The stress developed within the material due to this flow is called the flow stress. For a component to have the most uniform-appearing surface, the flow stress must be maintained and monitored for the temperature, strains, and strain rates for the specific process, including the alloy and the tool. Each aspect will affect the surface microstructure from the size and distribution of phases, precipitates and intermetallic compounds, grain size to surface defects induced by friction and discontinuous yielding [9].

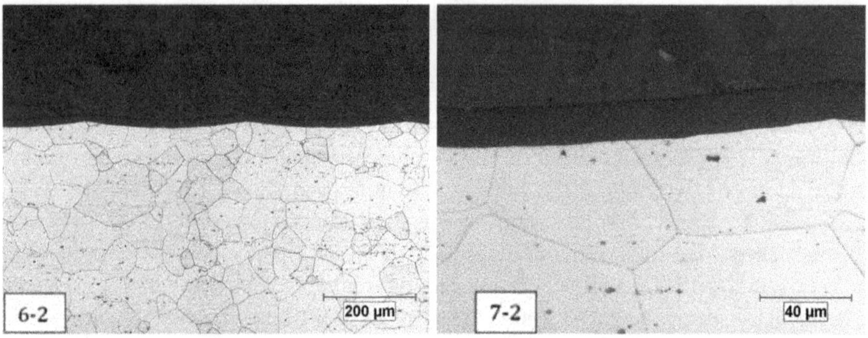

Fig. 8.20 (etchant: Keller's reagent) Metallographic images of alloy 7129 showing an equiaxed microstructure with normal levels of precipitation and residual directionality from deformation processing. The anodic oxide is uniform and continuous, displaying no visual defects

During deformation, primary constituent phases and intermetallic particles will be carried along with the material flow, and, depending upon the amount of strain energy developed through the material, from the surface to the core, these microstructural particles may align in bands forming "stringers." A large extrusion ratio will serve to break up stringers and stretch them out along the length of the workpiece. If the extrusion ratio is small, the stringers will remain in bands. These bands have a different chemical potential than the aluminum matrix, which cause streaks in the anodic oxide. In the case of bands of large PCPs, there will be dramatic differences in the surface roughness on the normally smooth surface of the anodized finish and reduce its cohesive strength.

The speed of the wrought process impacts flow stress and therefore, surface microstructure. A rapid strain rate in both rolling and extrusion will affect recrystallization sooner, producing a finer grain size at the surface and preventing grain growth from occurring on the press. Combining a homogenized billet, rapidly processed through the temperature range over which precipitation occurs by way of a strain rate with an optimized extrusion ratio will produce a microstructure that is as homogeneous as possible, with a fine grain size and little grain boundary precipitation of intermetallic compounds and precipitates, providing possibly the *best* microstructure for anodizing. See Fig. 8.20.

8.3.5 Connecting Extrusion Process and Microstructure to Appearance and Quality of Anodic Oxide

The Aluminum Association defines extrusion as the process of shaping material, such as aluminum, by forcing it to flow through a shaped opening in a die [21]. The extruded material emerges as an elongated piece with the same profile as the die opening. See Fig. 8.21.

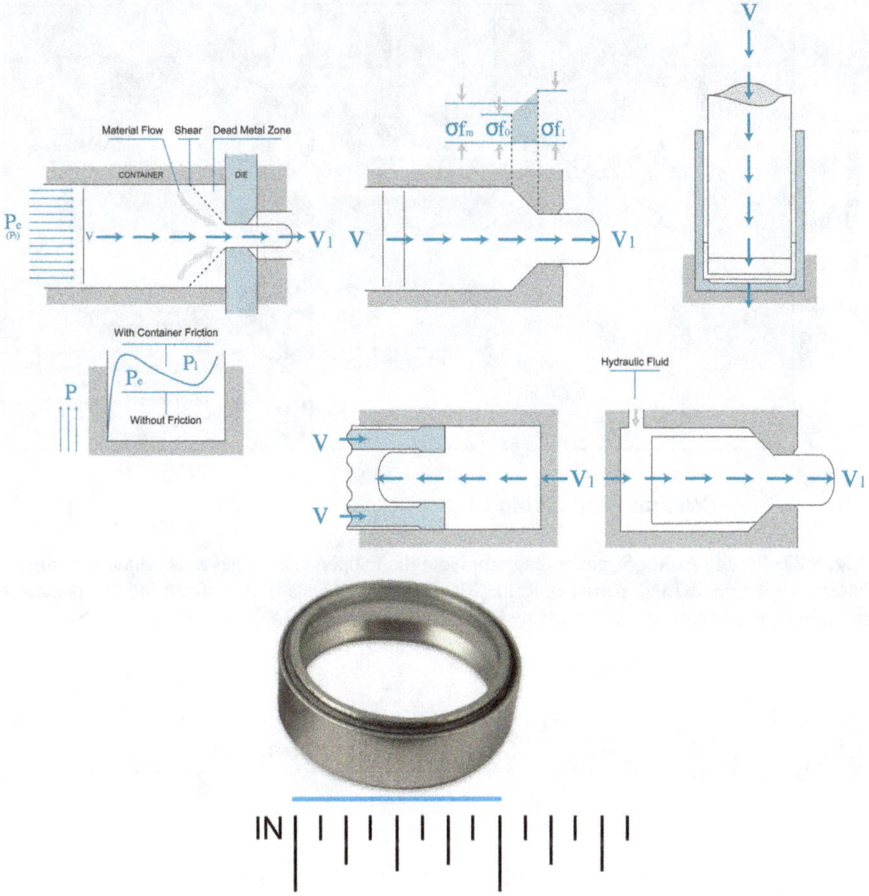

Fig. 8.21 Schematic representation of extrusion according to Schey [17]. Note the directionality in the surface of the anodized ring, provided as an example. This is due to the microstructure conforming to the directionality of the extrusion process

In order to control microstructure, it is critical to understand time-temperature relationships that impact precipitation and grain growth during deformation processing. Although not identical, thermodynamic trends for the temperature ranges over which other intermetallic particles precipitate, including Al-Fe-Si primary constituent phase, $AlFe_3$ and other dispersoids and precipitates presented in the pseudo Al-Mg_2Si phase diagram, Fig. 8.22, are consistent for other precipitation hardenable alloys, too. This means that by tracking the temperature and dwell times for the various steps of a manufacturing process, a rather good picture of the microstructure can be developed, regardless of the wrought alloy under consideration.

The heat treatment effects that occur during deformation develop more metallurgical features and develop existing ones within the microstructure, which further develop variations in the surface chemical potential. A helpful schematic review of

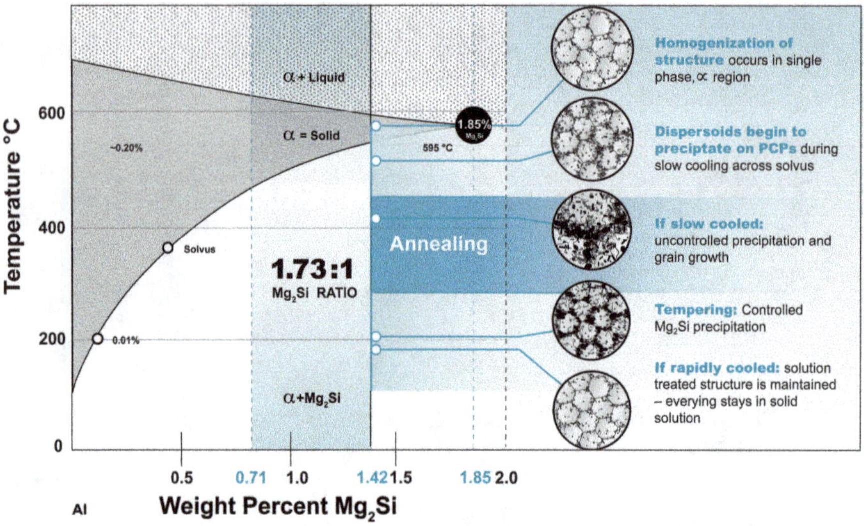

Fig. 8.22 Pseudo Al-Mg_2Si phase diagram. Note the temperature ranges over which the various intermetallic phases and particles form. Most overlap standard manufacturing temperatures achieved during deformation processing [10]

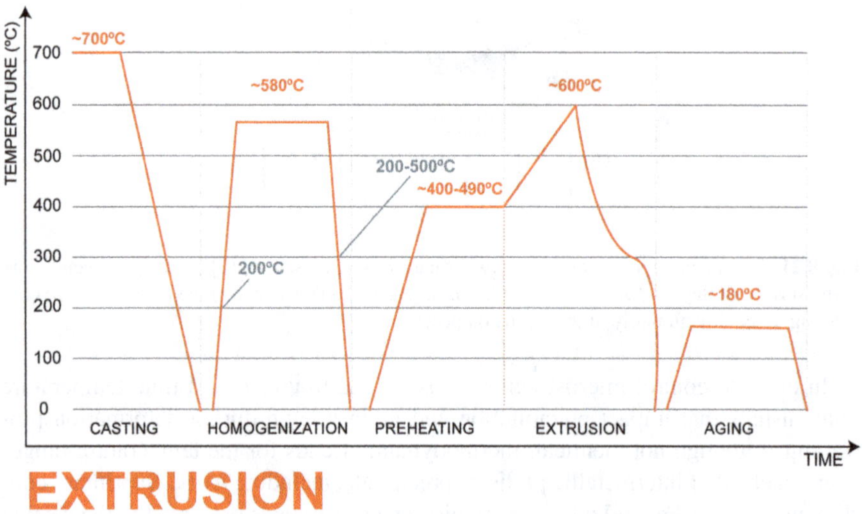

Fig. 8.23 Temperature-Time-Process schematic for extruding 6XXX series aluminum alloys [22]. Approximate dwell times are given by the width of each process segment. For example, homogenization occurs after billet casting, the billet is heated rapidly and held at ~580 °C for a sufficient amount of time to solution treat; this is followed by a rapid cooling of 200–500 °C per hour. Following preheat, which can be done at temperatures sufficient to homogenize the microstructure, the heat of the extrusion process produces a peak temperature of ~600 °C that rapidly drops off with deformation through the temperatures at which precipitation, stress relief, recrystallization, and grain growth occur. When rapidly processed, the condition of the component is called a "press quench," and no follow-on solution treatment is done prior to aging. Controlled precipitation occurs during aging

the sequence of temperature excursions for the various steps of the extrusion process for 6XXX series alloys is presented as Fig. 8.23. For most wrought processes for precipitation hardenable alloys, homogenization of the billet, log, or preform occurs after billet casting, the workpiece is heated rapidly and held at ~580 °C for a sufficient amount of time to solution treat, followed by a rapid cooling of 200–500 °C per hour. Following preheat, during which the billet is held within the temperature range over which precipitation occurs, the heat of the extrusion process produces a peak temperature of ~600 °C that is rapidly cooled such that a homogeneous as possible microstructure is produced. Aging takes place at ~180 °C for the longest amount of time to produce precipitation that provides the specified temper tensile strength/hardness. Annealing effects: recovery, recrystallization, and grain growth, can all happen during deformation. This is interesting to consider when most industrial deformation processes typically last less than 5 min. For low alloy wrought material, the entire extrusion process can last less than 3 min; for aluminum super alloys, the extrusion process can exceed 7 min.

8.3.5.1 Strain Rate Effects: Slow

Mechanical and chemical properties of the workpiece change with the microstructure. The amount of recrystallization and grain growth depends upon the speed of the deformation process. Extrusions that exhibit large grain size and high levels of precipitation throughout the workpiece microstructure may have been processed at a lower homogenization temperature, or at a slow strain rate, or both. Recrystallization and grain growth following deformation occur, and with a slow strain rate, there is enough time for the grains to become very large. Enlarged grains exhibit a larger chemical potential between the grain boundary and the "meat" of the grain, as well as a high level of intermetallic precipitation across the grain. Hardness of the base material is also reduced when grain size is large, producing a corresponding reduction in strength. Regardless of the age hardening that might follow such a process, temper specifications are usually violated. Machinability is altered; intermetallic particles are easily pulled from the surface during buffing or polishing, producing surface pits. See Fig. 8.24.

The precipitation, recrystallization, and grain growth all produced by slower strain rates, due to longer process time through temperature ranges high enough to produce these effects, exacerbate chemical attack significantly, and highlight differences in crystallographic grain orientation by way of preferential etching. Grain orientation effects are exacerbated when grain size is enlarged; with (1,1,1) grains exhibiting a faster etch rate than (1,1,0) or (1,0,1) grains, causing a mottled surface appearance [23]. See Figs. 8.25 and 8.26.

If the preheat temperature prior to wrought processing is at the low end of the indicated range, more precipitation will occur, especially as a component is held in the temperature range just below the solvus on the phase diagram (Fig. 8.20). A discerning microstructural feature develops that indicates slow cooling across the solvus temperature, over the single-phase solid solution region of the phase diagram

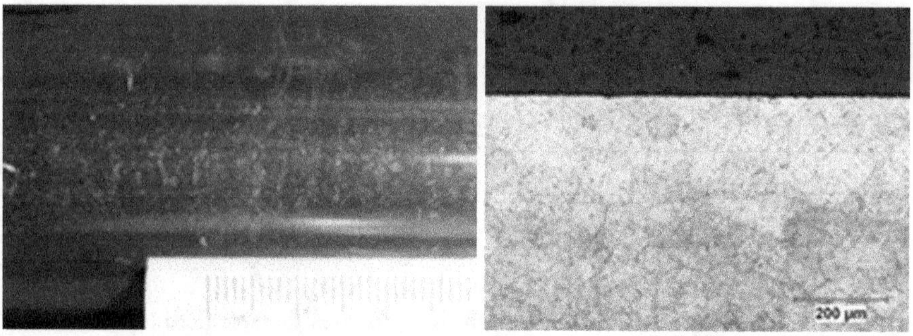

Fig. 8.24 Documentary photograph of the anodized finish on a 6463 extruded and buffed tube with its corresponding surface microstructure. The component, a tube (left), was bright dipped, a chemical brightening process, before anodizing. The surface exhibits "clouds" of small pits and milky-appearing streaks. The tube microstructure (right), developed by Keller's reagent, exhibits a coarse microstructure with many small particles that were pulled out during buffing, etched out during chemical brightening or included within the anodic oxide finish during anodizing. Many large precipitates are noted in the upper, coarse peripheral layer. Because dispersoids nucleate in numbers about the primary constituent phases, concentrations of dispersoids can yield "clouds" of fine pits through chemical treatment that appear milky to the naked eye. Al_3Fe and Al_6Mn intermetallic particles are well documented as causing surface pits during anodizing [8]. As a particle or spherical-shaped defect, a pit can form in the surface where a precipitate once resided whether mechanically, through pullout during polishing or buffing, or through chemical treatment as in cleaning or anodizing. Courtesy of MD Building Products

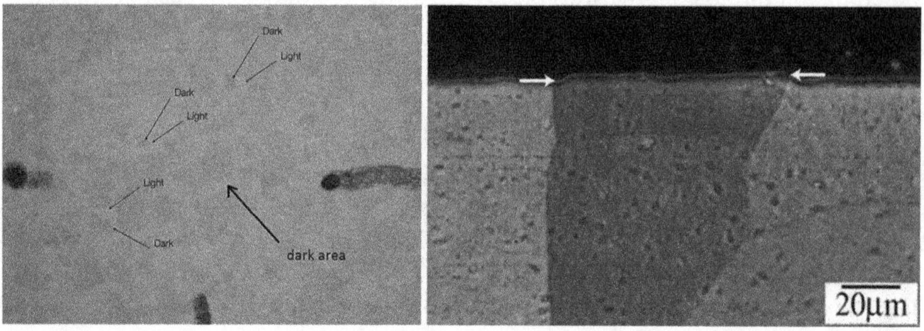

Fig. 8.25 Mottled appearance developed by anodizing. Darkfield light optical microscope (LOM) image documenting an etch step corresponding to the mottled appearance. Different grain orientations exhibit different etch rates with (1,1,1) grains etching fastest. These effects are magnified when grain size is excessively coarse. White arrows bracket anodic oxide. Base material is 6063 T6

into the two-phase region, called the "precipitate-free zone" (PFZ) [24]. PFZs develop at the grain boundaries, as solute is "drained" from the surrounding metal into the grain boundaries during slow cooling, radically changing the local chemical potential at the grain boundaries, causing increased chemical sensitivity at the grain boundaries, and exacerbating a mottled surface appearance, which becomes enhanced with chemical treatment and anodizing. PFZs are a kind of fingerprint,

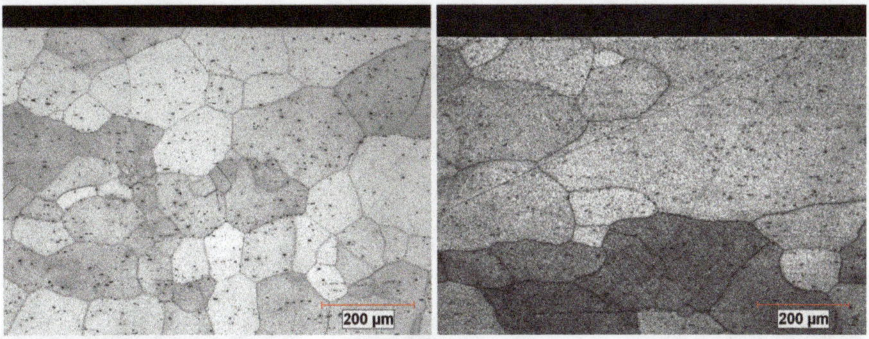

Fig. 8.26 (etchant: Keller's Reagent) Extruded microstructures displaying varying levels of coarse grain size. PCPs are small and well dispersed throughout the alloy 6063 microstructure, but the grains are saturated with precipitates and GPZs. Crystallographic orientation effects are magnified when the grains are large; this is why the surface appears dark and mottled after anodizing. Courtesy of MD Building Products

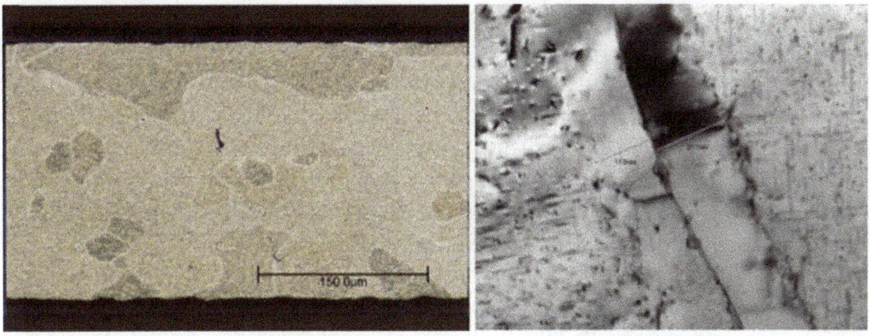

Fig. 8.27 (etchant: Keller's reagent) Microstructure of a 6063 extrusion, which had been over-temperature processed through extrusion, producing enlarged grains with corresponding precipitate-free zones (white outlines around the grains developed by etching the cross section). On the right, a corresponding TEM microstructure highlights a PFZ with a high level of dispersoid precipitation and GP zones (hashtags-shaped particles outside of PFZ). Anodizing develops this microstructure, producing a mottled appearance

which tells that the defect occurred during slow cooling, most likely while the billet or log was held at a temperature below the solvus, during preheat, rather than during extrusion. This is because extrusion typically occurs much too quickly to effect the formation of a PFZ. See Fig. 8.27.

8.3.5.2 Strain Rate Effects: Fast

Rapid processing can cause discontinuous yielding along the length of a component, producing differences in microstructure that are developed by chemical treatment and anodizing. The component presented as Fig. 8.25 was processed by way

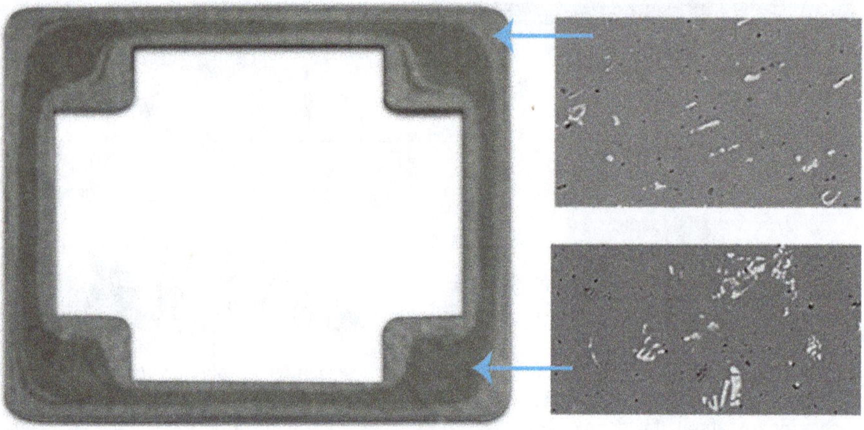

Fig. 8.28 Impact extruded component exhibited surface appearance differences after anodizing. Analysis disclosed orientation differences between light and dark-appearing areas confirmed by differences in orientation and condition of the $(Fe,Mn)_3SiAl_2$ intermetallic phase (the so-called script phase). Courtesy of Impreglon Surface Engineering

of impact extrusion, and perhaps because of a tooling defect, the outside of the component was deformed at a faster rate than the inside. This was manifested in the microstructure in the breaking up and alignment of the Al-Fe-Si primary constituent phase (PCP) in the area of the surface that appeared bright. The inner portion of the component was not deformed to the same extent, and the PCP remained as script phase, creating a different appearance after anodizing. See Fig. 8.28.

The heat of processing at the tool-workpiece interface due to high strain rates can also produce a thin layer of recrystallization at the surface. The complex near-surface microstructure, called the Beilby layer, is only nanometers thick. In addition to exhibiting recrystallization of nano-sized grains, the Beilby layer exhibits a high level of atomic level defects including tangles of dislocations and vacancies due to the high level of deformation at the interface with the tool. Because of its fine grain size, and increased grain boundary density, the Beilby layer exhibits high chemical sensitivity [25]. Anodizing develops this structure. The multiplicity of defects comprising the Beilby layer initially confounds the oxidation reaction, producing a skewed oxide that, because the anodic oxide consumes the Beilby layer at the oxidation process outset, as it grows and resumes its columnar structure, the distorted portion of the oxide is at the top, and the final oxide finish exhibits a cloudy visual appearance. See Fig. 8.29.

Complex tempers that are designated with additional digits such as T651 or T654 (refer to Sect. 4.4.5.2) are also subjected to stretching or compression to a prescribed level in order to produce a stable condition or a "permanent set." Depending upon the amount of retained strain energy from the deformation process, dynamic recrystallization can occur during stretching or compression. Residual strain due to high strain rate processing in extrusions can be swiftly relieved by follow-on straightening post deformation. In microstructures of complex alloys in which strain

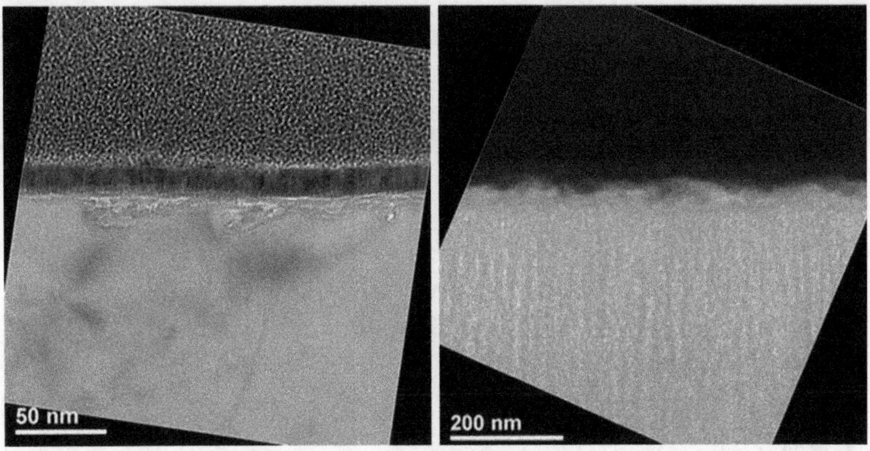

Fig. 8.29 Transmission electron microscope (TEM) images of near-surface microstructure (Beilby layer) of extruded alloy 6063 before (left) and after anodizing (right). The Beilby layer contains pockets of voids and high grain boundary density at a nanoscale. Note the disrupted appearance of the oxide surface due to the haphazard nucleation of the oxide caused by the complexities of the Beilby layer. The anodic oxide appears cloudy to the naked eye

rate was particularly high, strain relief produces bands of small, dynamically recrystallized grains that form bands throughout the structure corresponding to the deformation direction. Anodizing develops this condition, producing a streaky or muddy appearance. Strain relief after deformation by way of stretching, to produce a complex temper, can also produce disproportionately large grains within the microstructure when residual strain in clusters of smaller grains is relieved. This significantly changes the surface appearance after anodizing because the visual interference, provided by the high grain boundary density of several small grains, is eliminated by the presence of a large grain. See Figs. 8.30 and 8.31.

8.3.5.3 Temper Effects

In alloys that develop strength by way of precipitation hardening, strengthening is accelerated and controlled by tempering or aging after deformation processing. Precipitation proceeds at a temperature lower than annealing; therefore, *no grain growth or stress relief occurs during tempering*. Tempering is the final, low temperature heat treatment performed on precipitation hardenable alloys to deliberately develop precipitates in the microstructure in order to produce hardness by way of a controlled process.

In 6XXX alloys, the precipitate formed is Mg_2Si. Precipitation occurs during deformation, and with good process control, a microstructure with a predictable microstructure and hardness can be produced on the tool. This typically results in a T4 or T5 designation. Often, a follow-on temper step is performed at 150–180 °C for

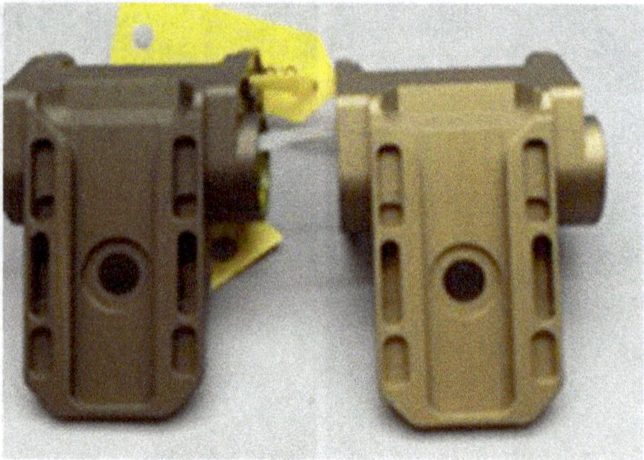

Fig. 8.30 Anodized alloy 6061 T651 extrusions. The sample on the left exhibits the desired appearance, "flat dead earth." The appearance after dyeing is due to the fibrous microstructure, desired to give the necessary strength to the component. A change in color meant a change in microstructure, which was documented metallographically. Differences in strain rate were reflected in the gross development of surface grains due to stress relief during stretching to produce the T651 temper in the part on the right. Dyeing produced a more golden-appearing color due to the reduced grain boundary density. This part was most likely processed at a higher strain rate. Photo courtesy of IHC Corporation, Detroit, Michigan

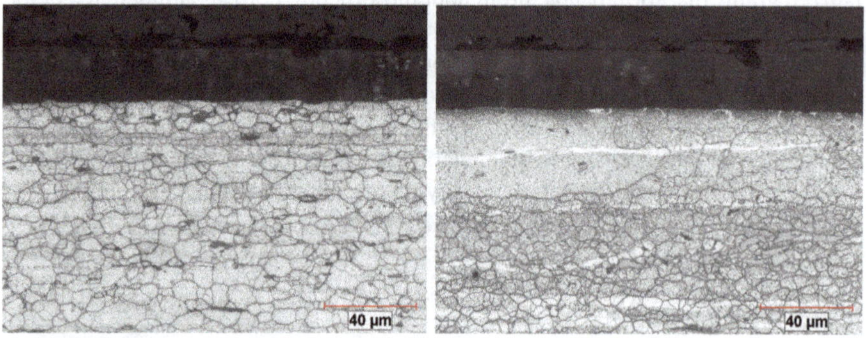

Fig. 8.31 (etchant: Keller's Reagent) Alloy 6061 T651 extrusions that exhibited drastically different appearances after dyeing. Comparative metallography (the microstructure on the left corresponds to the dark-appearing part on the left), determined the extrusions with the more gold appearance exhibited patches of enlarged grains at the surface due to the re-solution of residual strain energy during stretching. The dark muddy-gold appearance observed on the part on the left was the desired color for the components, made possible by the high grain boundary density of the fibrous, high strength microstructure

a fixed amount of time to produce a higher level of hardness, such as T6. A complete list and descriptions of various tempers are provided in section 4.4.5.2 of Chap. 4.

With extended tempering or over-aging, precipitates migrate to the grain boundaries, creating a sensitized condition, predisposing the alloy to intergranular corrosion, which appears as grain boundary relief through chemical treatment. As with

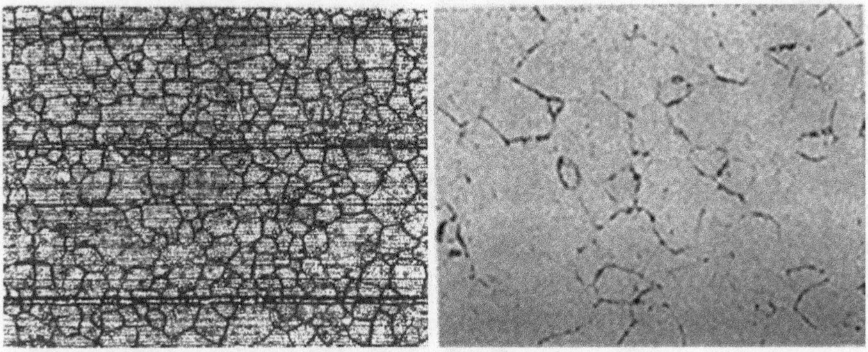

Fig. 32 (left) ×100, as anodized appearance of a 6061 T6 extrusion. Significant grain boundary relief developed during the anodizing process suggests the base material was over-tempered, producing a sensitized microstructure. Deep parallel streaks along the extrusion direction of an otherwise equiaxed structure indicate the extrusion surface was skived by discontinuities on the extrusion die surface. (right) ×400, etchant: Tucker's Reagent. Corresponding microstructure of the alloy 6061 T6 extruded material exhibits severe sensitization [6]

microstructures sensitized by way of slow cooling, that exhibit PFZs, sensitization produced by over-tempering radically changes the local chemical potential at the grain boundaries. Substrates that are sensitized during tempering do not exhibit precipitate-free areas surrounding the grain boundaries, and typically, the precipitates are larger in temper-sensitized microstructures. See Fig. 32.

8.3.5.4 Microstructural Development in Aluminum Super Alloys

As alloy complexity increases for a specific component, so does its microstructure. Aluminum super alloys have been developed, primarily for the aerospace and aircraft industries that are alloyed beyond what is typically achieved when the alloy content is within conventional wrought compositions, that is, at a maximum of 8%. The purpose of the various alloy additions is to produce enhanced material properties, in particular, hardness, strength, and increased fatigue resistance. Consequently, the high alloy content limits the deformation processability of aluminum super alloys, and aluminum super alloys tend to be extremely strain rate sensitive. The microstructures produced appear fibrous and exhibit high levels of precipitation and fine grain size, which, by virtue of the high population of defects that intersect the surface, produce a surface with frequent sites of localized differences in chemical potential that change the anodizing response.

High strength alloys, such as the higher alloyed precipitation hardenable alloys: 6061, 6082, 7055, 7136, and 7475, can be processed through deformation to yield an ultrafine-grained, fibrous structure (grain size <10 μ). Frequent microstructural defects at the substrate surface disrupt the structure of the anodic oxide, producing a nonreflective appearance that often exhibits a level of inherent coloring (see Sect. 2.5.1). The high grain boundary density alone yields a dark-appearing anodic oxide when

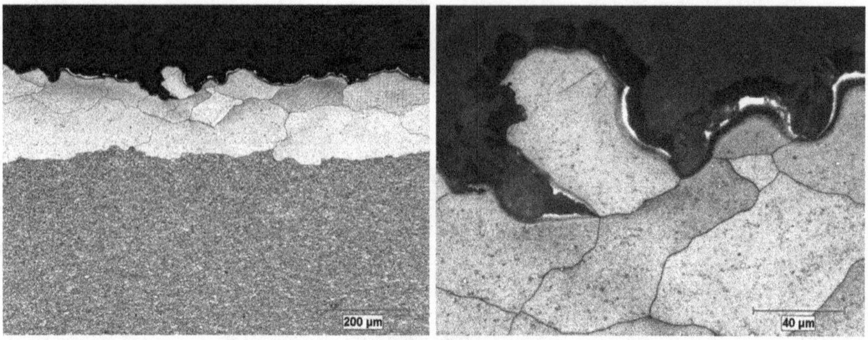

Fig. 8.33 (etchant: Keller's reagent with Weck's stain) Surface of an anodized 7XXX extrusion that exhibits high temperature corrosion at the surface due to intergranular incipient melting and tearing. Anodizing did not contribute to this failure mode. Courtesy of Universal Alloy Association

anodized. This makes it difficult to use a high strength alloy for appearance-driven applications that expect high levels of reflectivity, clarity, and distinctness of image. However, when the design priorities are high strength and durability, and the anodic oxide characteristics desired for the application are corrosion protection and wear resistance, the design compromise is a matte or satin appearance, rather than a clear, reflective finish.

As alloy complexity increases, and strain rate is reduced to avoid cracking, a press quench is insufficient to achieve stress relief and maximum development of mechanical properties. Therefore, some aluminum super alloys, usually the 7XXX series because they are most complex, are subjected to a complete heat treat cycle following deformation processing, including solution treatment prior to aging, in order to produce as homogeneous, recrystallized microstructure as possible. For some products, multiple heat treatment cycles are allowed, when preceded by solution treatment. Some complex alloys, due to the formation of low temperature intermetallic compounds, cannot be solution treated within the single-phase solid solution region of the phase diagram because incipient melting of low melting intermetallic compounds may occur at temperatures in excess of the solvus temperature. Therefore, heat treatment within the annealing temperature range is typically done to achieve recrystallization and stress relief. Microstructures that exhibit incomplete recrystallization are usually produced. See Figs. 8.33 and 8.34.

A condition called "high-temperature oxidation" (HTO) may occur when aluminum alloys are heated to solution heat treatment temperatures in a furnace that has too much moisture in the atmosphere. If moisture is present in the atmosphere, when oxygen present reacts with aluminum or magnesium to form their respective oxides, residual atomic hydrogen diffuses through the aluminum lattice and recombines to form molecular hydrogen at locations of lattice discontinuity near the surface, which is common in wrought product that has much residual cold work. In order to prevent high temperature oxidation during solution treatment, a moisture-free atmosphere can be used [26, 27]. Normal tempering can effectively diffuse residual hydrogen into and out of the microstructure. However, voids that form at

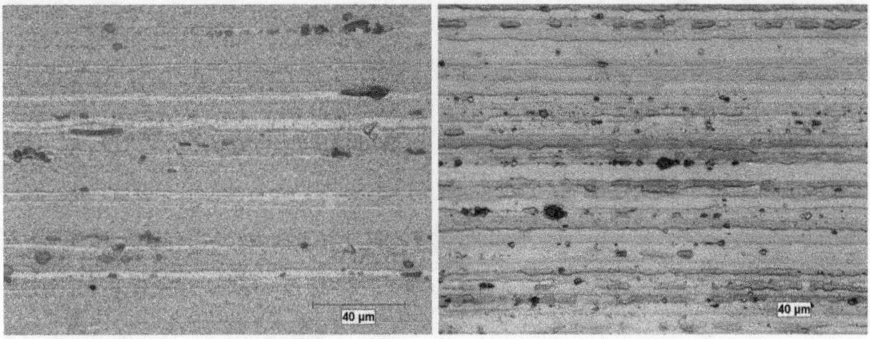

Fig. 8.34 (etchant: Keller's reagent) Comparative photomicrographs of two different extruded aluminum "super alloys." The microstructure on the left is that of a 6082 T5 alloy, which by definition has a press quench. The microstructure of the right is that of a 7XXX T651 super alloy, which has been through three cycles of heat treatment following deformation. The 6082 microstructure is finer with a higher level of precipitation than the 7XXX series alloy, which exhibits some recrystallization and appears coarser. Particle size, in general, is larger for the 7XXX series alloy. Images courtesy of SGS in Melrose Park, Illinois

high temperature may contain an amount of internal oxidation, which persist after heat treatment. By purging the atmosphere of moisture, and using an atmosphere containing sodium fluoroborate or ammonium fluoroborate, a reaction layer of magnesium fluoride forms on the surface of the workpiece, producing what appears to be a barrier to inward diffusion of hydrogen. This reaction layer must be completely removed from the surface of the workpiece prior to anodizing or it will interfere with the anodic oxidation reaction. See Case Study 5.

Diffusion reactions consistently occur during heat treatment. Magnesium, as the most reactive element that is in solution with the alloy, diffuses toward the surface where it will react with the atmosphere. If oxygen is present in the atmosphere, magnesium oxide will form a layer preferentially at the surface. Rapid diffusion of magnesium from the alloy can leave behind Kirkendall voids, lined with oxide, in the near-surface microstructure of the substrate as a result of high temperature oxidation. The formation of voids can produce what looks like blisters at the substrate surface.

8.3.6 *Connecting Forging Process and Microstructure to Appearance and Quality of Anodic Oxide*

Forging is defined as a manufacturing process where metal is pressed, pounded, or squeezed under great pressure to produce high-strength parts [21]. Forged aluminum is ideal for applications where performance and safety are critical but a lightweight metal is needed for speed or energy efficiency [9]. See Figs. 8.35 and 8.36.

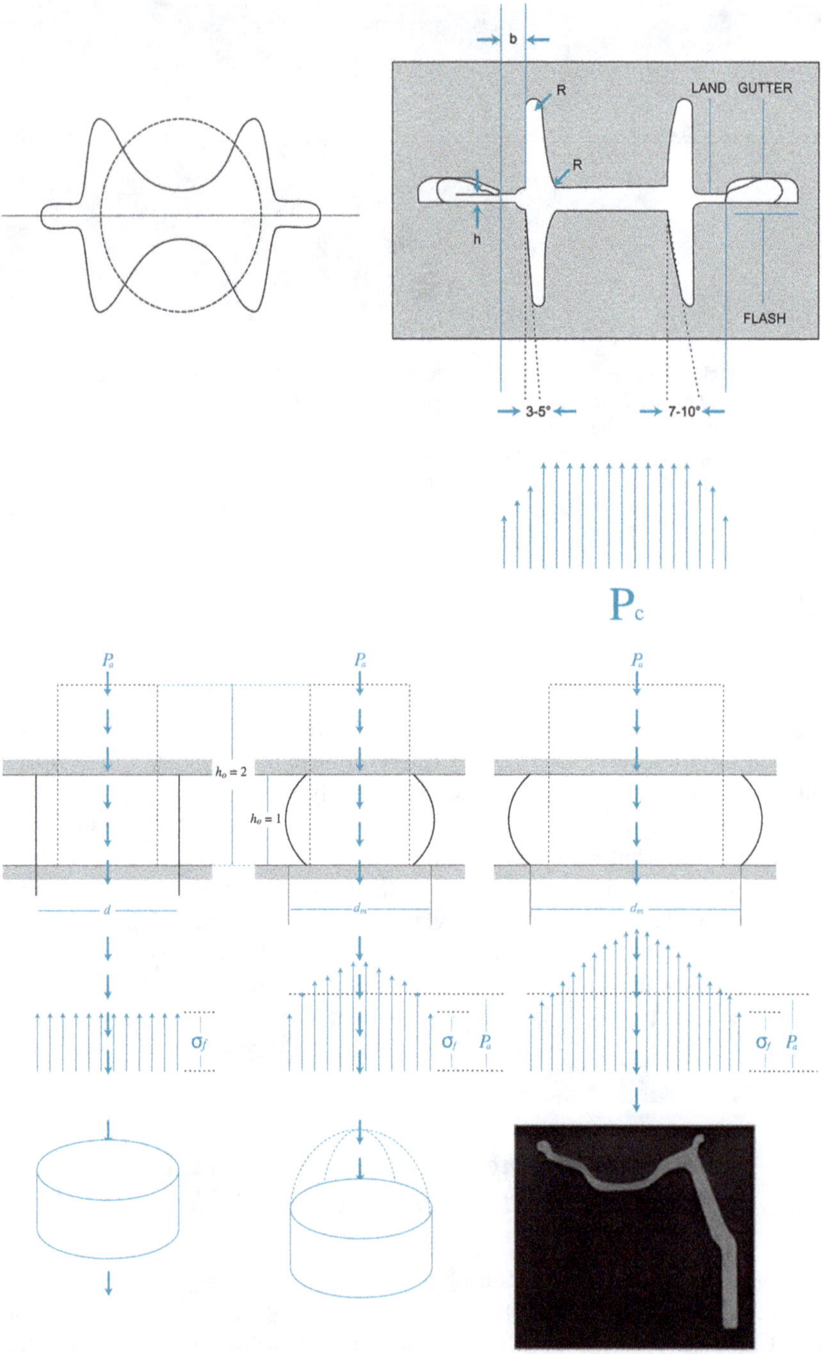

Fig. 8.35 Schematic representation of forging according to Schey [17]. The schematic on the left represents open die forging, and the schematic on the right represents closed die forging. Photo courtesy of Accuride

Fig. 8.36 Photograph of a forged and anodized marine turbine

The forging process puts the metal through significant temperature excursions that overlap microstructural changes that can be correlated to the phase diagram. To help relate how the microstructure of an aluminum component develops over the sequence of the forging process, and how those temperatures and dwell times overlap on the Al-Mg_2Si phase diagram, a similar temperature-process diagram was constructed to that for generic aluminum extrusion processes. See Fig. 8.37.

The biggest difference between the extrusion and forging time-temperature-process schematics can be found in the cooling rate following deformation. In forging, the cooling rate is slower because the workpiece comes in repeated contact with the die during forging. A possible process concern in forging is the possibility of incipient melting of low melting intermetallic compounds between the primary grains. Because cooling rates are much slower in forging, microstructures are finer and more fibrous, exhibiting much more precipitation and high temperature effects such as coring or precipitate-free zones (PFZs), similar to extrusions from complex super alloys. The rate of deformation during extrusion competes less with cooling rate because the cross section of the workpiece is in transient contact with the die; consequently, less precipitation, recrystallization, and grain growth occur in extrusions. See Fig. 8.38.

Forgings, because of the special process situation that results from the workpiece coming in repeated contact with the die, display the effects of precipitation, recrystallization, and grain growth throughout the microstructure. Cooling rates are much slower, yielding microstructures with much more precipitation and high temperature effects such as precipitate-free zones. Dynamic recrystallization occurs during

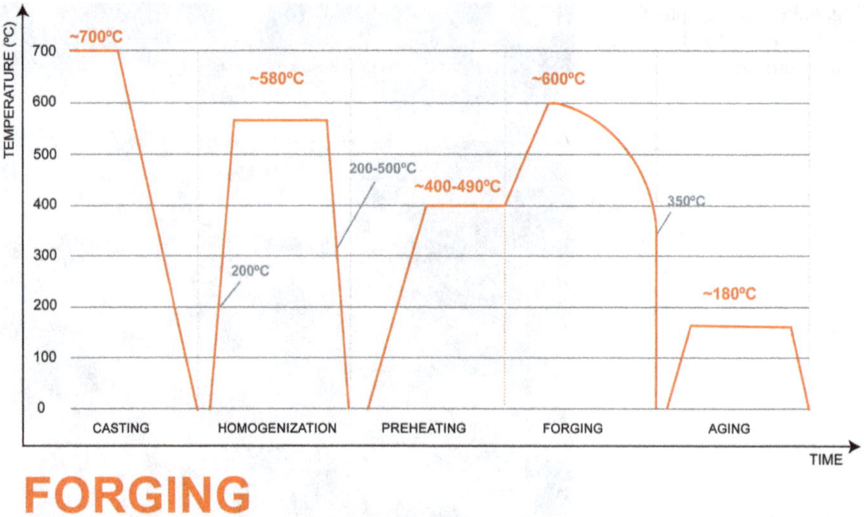

Fig. 8.37 Temperature-Time-Process schematic for forging 6XXX series aluminum alloys. Approximate dwell times are given by the width of each process segment. Note the difference between the schematics for extrusion (Fig. 8.20) and forging is the rate of cooling following the peak temperature at the point of forging. For forging, the cooling rate is much slower during deformation than it is for extrusion. This is because the rate of deformation during extrusion competes less with cooling rate because the cross section of the workpiece is in transient contact with the die; consequently, less precipitation, recrystallization, and grain growth occur in extrusions

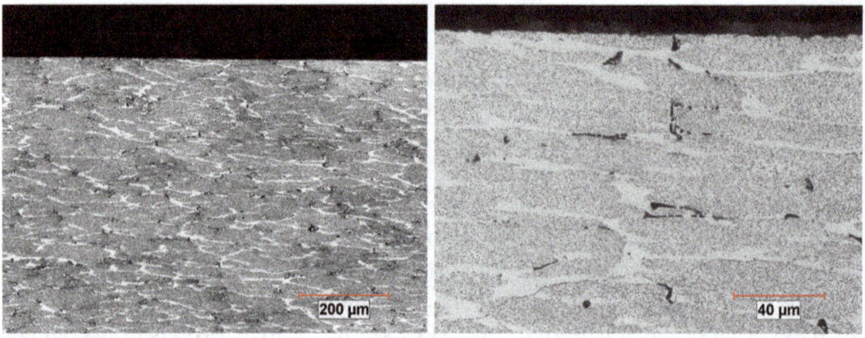

Fig. 8.38 (etchant: Keller's Reagent) Microstructures for the alloy 6061 T6 wheel sample presented as Fig. 8.20. The high level of segregation that is displayed in these microstructures is due to high temperature processing. Slow cooling enabled high levels of dispersoid and Mg_2Si precipitation across the grains, and solute was actually sucked away to form precipitates at the grain boundaries causing precipitate-free zones surrounding the grain boundaries. The etched appearance is due to large solute-rich areas with high levels of precipitation, divided by aluminum-rich areas that contain the primary constituent phase for the alloy. The high level of precipitation produces a cloudy appearance to the anodic oxide. (Images courtesy of Accuride)

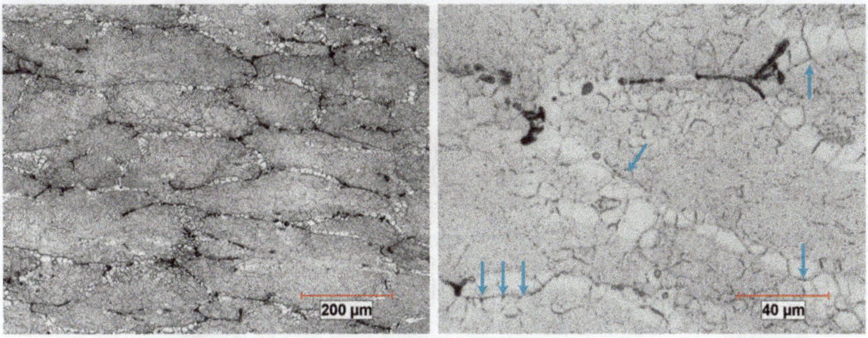

Fig. 8.39 (Etchant: Keller's Reagent) Microstructure of a 6061 T6 forging displaying clusters of small, dynamic recrystallized grains. The primary structure exhibits what appears to be coring: large primary grains with high levels of precipitation are divided by aluminum-rich grain boundaries containing the black-appearing Al-Fe-Si primary constituent phase. The recrystallized grain boundaries in the image on the right exhibit precipitation, indicating the structure in these areas is sensitized, and will exhibit reduced chemical resistance. Blue arrows indicate representative areas of precipitation at dynamically recrystallized grain boundaries. The design intention was to produce a clear and highly reflective finish on this forging. (Images courtesy of Accuride)

deformation, by way of annealing effects that occur as the workpiece cools, but the annealing process is interrupted with the next hit of the forging hammer; therefore, transformation of the grain structure is not continuous, and a nonuniform microstructure is produced. The structures that form are subject to normal temper effects and can become sensitized, changing the chemical potential in areas where microstructural discontinuities intersect the surface. Forgings are often re-solution treated and quenched prior to tempering to assure the most homogeneous microstructures. See Fig. 8.39.

8.4 Relating Anodic Oxide to Function and Appearance

In the following section, the case studies present specific design applications in more detail that show how base metal chemistry, composition, manufacturing process, and thermal history come together to form a microstructure; and, in the context of the function and appearance of the anodic oxide, what microstructural details need to be considered to achieve the design goals.

Fig. 8.40 Three bicycle headset bases. The base metal for each is 7075 T6. All are of acceptable appearance and operate to the design requirements. Note the distinctness of image for the reflection in the component is reduced with additional oxide thickness and color. The oxide on the clear component measures ~2 μ, the oxide on the red component (sample A) measures ~11 μ, and the oxide on the magenta pink (sample B) component measures ~16 μ. (Photo courtesy of Cane Creek Cyclery). The corresponding microstructures are presented in Fig. 8.41

8.4.1 Case Study 1

8.4.1.1 Clarity and Color: Extrusion [9]

In industries such as architecture, sporting goods and consumer electronics, in which the design attributes of clarity and color are as important as the requirement of mechanical strength, trade-offs between appearance and durability must be considered in terms of microstructure. For example, when considering the applications for the part of an extruded bicycle headset and a forged turbine, it is clear that they are designed to function differently, but the performance requirements are similar. Both parts are expected to provide a level of mechanical strength sufficient for the application and function well in high cycle fatigue. However, the bicycle component has an additional requirement that the turbine doesn't have: as part of a decorative headset, the bicycle part must appear bright and reflective and match the other small parts that comprise it. Without compromising clarity, color, or strength, wrought alloy 7075 can be successfully used in this application, provided the extruded microstructures for each part of the set were uniform and the anodic oxide thickness were uniform and consistent with the need for optical transparency or beautiful, reflective color. See Figs. 8.40 and 8.41.

However, what needs to be considered when other components in the handlebar sets exhibit gross differences in the appearance quality? Components extruded from the same alloy and temper, 7075 T6, although not from the same billet or at the same time, exhibited obscured clarity and reflectivity after anodizing as well as dull and muddy color in streaks over the surface after dyeing. See Fig. 8.42.

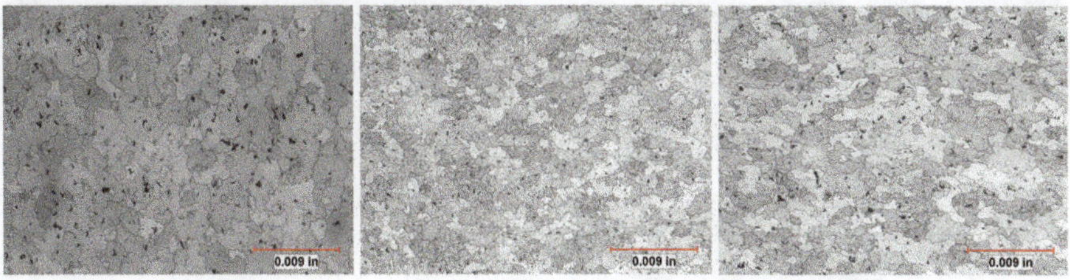

Fig. 8.41 (etchant: Keller's Reagent) Microstructures for each bicycle handlebar base in Fig. 8.3 correspond with each component presented from left to right. Note the microstructures are fully recrystallized and rather equiaxed with globular primary consituent phase (appears as black spots of varying size and population). The chemistries for each component were similar, and apparently, so were the extrusion parameters

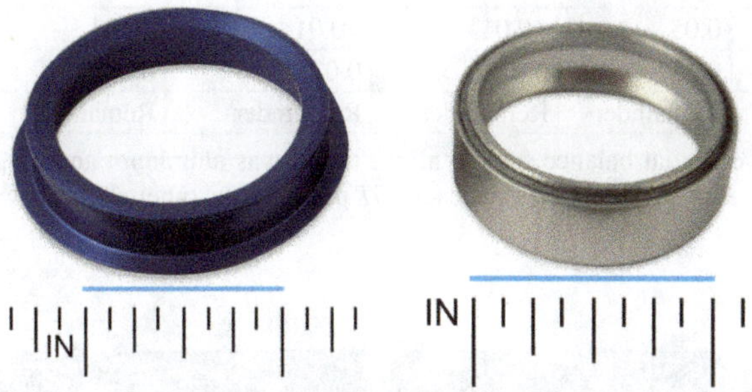

Fig. 8.42 Anodized extruded bicycle handlebar components manufactured from alloy 7075 T6. Note the streaky, dark appearance. A bright, reflective appearance that matched other components in the handlebar set, see Fig. 8.40, was expected. The blue sample was identified as sample C and the clear, streaky sample, as sample D

Chemical analysis and hardness measurements of the base metal determined that all components comply with the chemistry and temper requirements for alloy 7075 T6. See Table 8.2. However, metallographic analysis determined distinct differences in microstructure that elucidated the differences in the appearance of the components that anodized clear and reflective, even when dyed, and those that anodized with a mottled, streaked surface. The unfavorable-appearing samples exhibited large areas of small, deformed grains oriented in the direction of extrusion. Small, dynamically recrystallized grains could be resolved in the microstructures of both samples, which indicates that the temper designation for the sample was more complex than T6, and most likely T651. To contrast, the grain structure of the samples that anodized with a favorable appearance appeared equiaxed, with almost total recrystallization. See Fig. 8.41 and compare with the structures presented in Fig. 8.43.

Table 8.2 Chemical analysis results—Case Study 1

Element	Standard	Sample A—red	Sample B—magenta	Sample C—blue	Sample D—clear
Silicon	0.40	0.11	0.06	0.06	0.29
Iron	0.50	0.23	0.22	0.21	0.13
Copper	1.2–2.0	1.2	1.4	1.5	1.4
Manganese	0.30	0.04	0.01	0.01	0.09
Magnesium	2.1–2.9	2.7	2.5	2.4	2.4
Chromium	0.18–0.28	0.16	0.17	0.16	0.18
Nickel	…	<0.01	<0.01	<0.01	<0.01
Zinc	5.1–6.1	5.5	5.5	5.6	5.5
Gallium	…	…	…	…	…
Vanadium	…	0.0038	0.0053	0.0069	0.0046
Zirconium + Titanium	0.025	0.037	0.021	0.021	0.021
Others each	0.05	<0.01	<0.01	<0.01	<0.01
Others total	0.15	0.038	0.028	0.025	0.026
Aluminum	Remainder	Remainder	Remainder	Remainder	Remainder

In all cases, the essential balance for the alloys tested was aluminum and ranged from 90.0 to 90.1%. *OE* other alloy elements, percent each. *OT* percent total other elements within the alloy

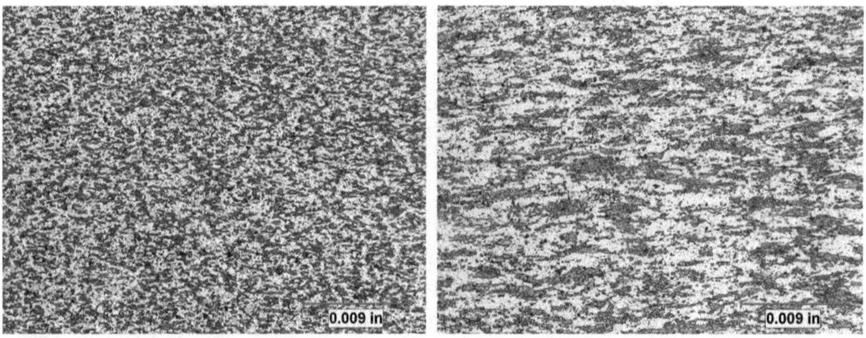

Fig. 8.43 (etchant: Keller's Reagent) Microstructures for the extruded bicycle handlebar components presented in Fig. 8.42, Sample C, on the left and Sample D on the right. Microstructures exhibit tiny, dynamically recrystallized grains from the stretching process following extrusion used to produce a T651 temper. This type of recovery is typical for extrusions produced at high strain rates and strain relieved by straightening on the press at the end of the deformation process. The microstructure of sample D, that which yielded a clearer oxide, albeit streaky, exhibits more recovery than that on the right

Comparative analysis of the various components determined differences in trace element concentrations—those elements that make dispersoid particles that nucleate at grain boundaries and restrict recrystallization—as well as apparent differences in the amount of retained strain energy from the extrusion process developed by follow-on stress relief treatment, by way of stretching, produced drastic differences in microstructure that produced differences in the anodized appearance. In this case, microstructural differences could not only be attributed to apparent temper conditions for the extrusions, but also in the amount of dispersoid-forming trace elements that inhibited recrystallization with straightening. The samples with the less equiaxed structures exhibited higher trace amounts of vanadium, and iron plus manganese, all known dispersoid formers. The microstructures of the failed parts suggest that the components would have higher mechnical strength, but for this application, the appearance and color of the parts was the driving design attribute [20].

Complex tempers that are designated with additional digits such as T651 or T654 (refer to Sect. 4.4.5.2) are also subjected to stretching or compression to a prescribed level in order to produce a stable condition or a "permanent set." Depending upon the amount of retained strain energy from the deformation process, dynamic recrystallization can occur during stretching or compression. Residual strain in extrusions can be swiftly relieved by follow-on straightening post deformation, yielding values that are comparable with a T6 temper without follow-on mechanical stress relief treatment. Bands of small, dynamically recrystallized grains can form throughout the structure, or disproportionately large grains can form. In both cases, the appearance of the surface is radically altered through chemical treatment and anodizing. Therefore, when ordering material per a specification, it is necessary to keep in mind that for material of a specific shape, such as rod, bar, or tube, additional mechanical stress relief treatment may be done as part of tempering to relieve residual strain.

8.4.2 Case Study 2

8.4.2.1 Corrosion Protection: Die-Casting [11]

For all engineering applications, each component within an assembly must function according to the design. Often components from different metals and alloys comprise an assembly. Alloy selection must be driven by the interactions between different materials, such as the galvanic couple between iron (steel) and aluminum (alloys). When this couple is part of an engineering design, the aluminum component must be anodized to inhibit the galvanic action. Using the anodic oxide to isolate the aluminum from the steel also has advantages: AAO imparts corrosion protection and is wear resistant and electrically insulating. The automotive/transportation industry often takes advantage of the speed and efficiency and therefore

Fig. 8.44 Example of an automotive assembly that utilizes a hard anodized aluminum die-casting as the housing. The anodic oxide provides corrosion protection and wear resistance against the steel impellers. (Photo courtesy of Impreglon, Kaufbeuren, Germany)

economy of die-cast aluminum alloys, and many under-the-hood automotive products are anodized die-castings. Microstructure is critical in these applications due to the requirements of impact and wear resistance. In this case study, the assembly is a diesel fuel pump. A special requirement of the pump is that the finish on the housing should not chip or spall, creating particles that can possibly cause the pump to seize. See Fig. 8.44.

High silicon content 300 series alloys, such as alloy 380 and modified versions such as alloy 383, are preferred for die-casting automotive components such as fuel pump parts, brake saddles, pistons, and piston rings. The coefficient of thermal expansion decreases with increasing silicon, which prevents component shrinkage in the die as the component cools and maintains the net shape of the component as-cast. A low expansion coefficient is beneficial for engine applications such as pistons and cylinder blocks, where tolerances are important.

Wear-resistant cast aluminum alloys are based on the presence of the hard, brittle silicon phase. Alloying elements such as iron, manganese, and copper increase the volume faction of the intermetallic silicon-bearing phases, contributing to wear resistance. When the silicon content exceeds 12.6%, the eutectic composition for aluminum and silicon, primary silicon crystals are present, and, if fine and well distributed, enhance wear resistance. A rapid cooling process will insure that the crystals are fine and the microstructure as homogeneous as possible. Alloy modifiers such as phosphorus may be added to the alloy melt to prevent the formation of large primary silicon crystals. The anodic oxides grown on these substrates will tend to be more uniform and continuous, with a higher cohesive strength than oxides grown on castings of the unmodified alloy composition exhibiting the coarser microstructure. Nevertheless, cooling rates dominate the homogeneity of the microstructure and the particle size.

Examination of the anodic oxide on a tested housing in the scanning electron microscope determined a network of cracks intersecting coarse silicon particles. Metallographic analysis showed conclusively the complex microstructure: Coarse primary silicon particles that exceeded 10 μ in diameter were present together with

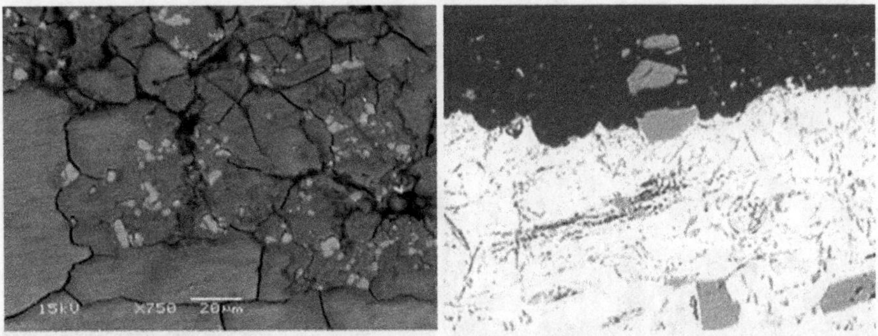

Fig. 8.45 Metallographic cross section of anodized die-cast alloy AlSi9Cu3 (similar to alloy 383). In the scanning electron microscope (SEM) photomicrograph on the left, the surface of an anodized die-cast component is documented, showing the cracks dotted with large, white-appearing silicon crystals that exceed 10 μ in size, as well as copper deposits appearing as darker gray haze across the surface and concentrated in areas near the cracks. In the optical photomicrograph on the right, large hypereutectic primary silicon platelets from the microstructure as well as smaller eutectic needles are incorporated into the anodic oxide, reducing the cohesive strength of the finish, especially in abrasive wear. Images courtesy of SGS in Melrose Park, Illinois

finer eutectic Al-Si particles and silicon platelets. Because silicon does not anodize under the conditions that anodize aluminum, crack equivalents that reduced cohesive strength of the oxide were established within the anodic oxide. These crack equivalents similarly provided the path for capillary action from the surface of the oxide to the interface with the die-casting, which posed a risk for corrosion attack with prolonged service life of the diesel pump. The microstructure of the die-casting also indicated that it was cooled slowly, as the presence of coarse silicon platelets is usually associated with alloys that have a composition exceeding the 12.6% silicon eutectic composition. See Fig. 8.45.

The analysis results indicated the need to reconsider the alloy selection for the housing as well as the use of an anodic oxide. However, anodizing was preferred because it was efficient and inexpensive. Therefore, the supplier of the castings was challenged with how to change the alloy, such that the mechanical requirements of the application were met as well as the finish requirements of abrasion and wear resistance. Because Alloy 383 is one of the best for casting detailed components, and because the cast house also cast alloy A383, it was the first trial as a replacement for the application as well as what proved to be most successful. Other benefits of using A383 are the stability that the alloy shows at high temperatures. The finer, less complex microstructure yields a more coherent anodic oxide. In spite of small silicon particles that are included throughout the finish, wear resistance requirements were met. In fact, the silicon particles appeared to enhance the wear performance within the pump cavity whenever the electroless nickel-coated impeller contacted the housing surface. See Fig. 8.46.

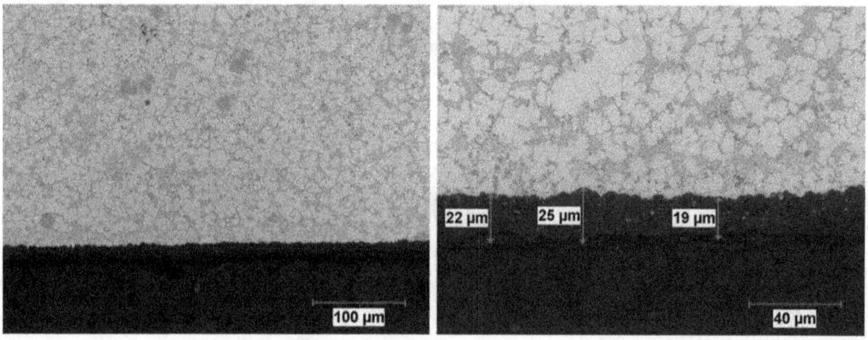

Fig. 8.46 (as-polished) Microstructure of an anodized AlSi9Cu3Fe (A383) alloy die-casting; the images are inverted with oxide at the bottom and the anodic oxide thickness indicated. By reducing iron content, and maintaining the manganese content, the structure remains fine because the iron is consumed as it ties up silicon that may form coarse primary silicon particles or Al-Si eutectic particles as Al-Fe-Si primary constituent phase. The modified structure is finer and more homogeneous without coarse primary silicon crystals (rounded, gray particles in the image on the left), which also indicates the casting was more rapidly quenched than the structure presented in Fig. 8.10. Although the fine silicon network still produces discontinuities in the oxide finish, this structure provides good wear resistance in applications that place the component in abrasion. This became the alloy of choice for the fuel pump housing. Images courtesy of SGS in Melrose Park, Illinois

8.4.3 Case Study 3

8.4.3.1 Dielectric Withstand: Roll Forging

The anodic oxide is the finish of choice for sputtering chambers and other critical components for the semiconductor industry. Typically, semiconductor grade 6XXX series, primarily 6061 T6, extrusions are machined to produce these large chambers. Consequently, in order to save time and money, the use of forgings was considered as a replacement for machined extrusions because the amount of labor and time needed to manufacture a chamber would be reduced. The anodic oxide would continue to be the finish of choice because of its successful track record on 6XXX series extrusions in the same application. The use of a 5XXX series forging is an interesting replacement. Because it would be forged, the required strength for the chamber would be sufficient; and because of the high level of solid solubility of magnesium in aluminum, which is the hallmark of the 5XXX series alloys, the uniformity and continuity of the anodic oxide is enhanced. This would feasibly result in an increase in the dielectric withstand for the total oxide layer, an important characteristic for successful operation of semiconductor manufacturing equipment.

The roll-forged samples provided for comparison were from two different suppliers, and identified as Samples 1 and 2. They were both manufactured from alloy 5052, but exhibited distinctly different appearances and performance after anodizing. Examination of the anodized surfaces disclosed that Sample 1 exhibited far less surface pitting than Sample 2. See Fig. 8.47.

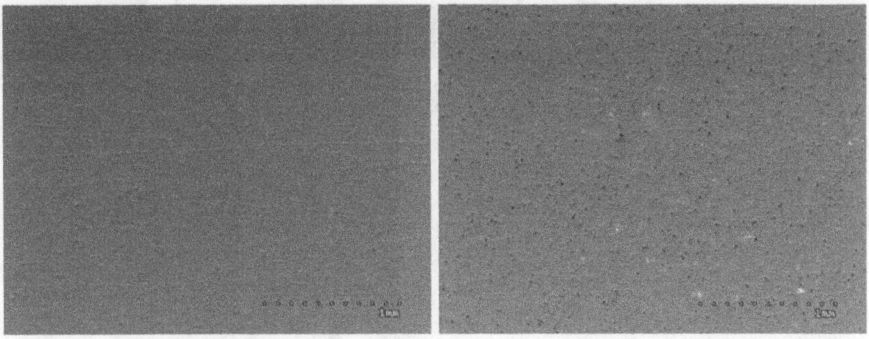

Fig. 8.47 Scanning electron microscope images of the anodized surfaces of Sample 1 (left) and Sample 2 (right). Note the distinct increase the amount of pits on the surface of sample 2. Sample 1 also exhibits frequent circular depressions in the oxide, which most likely conform to the microstructure, but they don't have the depth exhibited by those on Sample 2. Courtesy of Anderson Shumaker

Metallographic analysis of the samples disclosed a big difference in the appearance of the microstructures between samples 1 and 2. Sample 1 appeared more homogeneous, with smaller primary constituent phase formation and much well-dispersed precipitation throughout. Sample 1 exhibited no clear working direction, suggesting the possibility that it was indeed forged, but not roll forged. Sample 2 exhibited directionality, corresponding to the working direction and coarser primary constituent phase (PCP). Large, precipitate-free areas were also observed, but these did not appear to be precipitate-free zones (PFZs) because they were not local to the areas surrounding the grain boundaries. The precipitate-free areas may be the result of coring, which can be related to forging at higher temperatures, close to the liquidus. The increased number of coarse PCPS in Sample 2 indicated differences in the billet quality used to make the samples, which suggested either a difference in alloy chemistry or perhaps a difference in how the billet was cooled. See Fig. 8.48.

Well-defined precipitation at the grain boundaries was observed in the microstructure of Sample 2. The microstructure of the component was oriented with the direction of the material flow orthogonal to the component surface; therefore, end-grain attack during chemical pretreatment prior to anodizing was enabled. Pits at the interface with the anodic oxide corresponded to the grain boundaries, indicating a sensitized condition. The presence of coarse precipitation at the grain boundaries indicates that the workpiece was allowed to cool slowly during and after deformation processing, or that it was deformed at a very high temperature. The appearance of the microstructures also indicate cooling time was proportionately longer for Sample 2, as compared to Sample 1, whose more homogeneous microstructure indicates processing at a lower temperature with much more rapid cooling following deformation.

Comparative chemical analysis by way of optical emission spectrometry was performed to determine if the results of the metallography could be confirmed and reinforced with actual data. The results, presented in Table 8.3, determined both

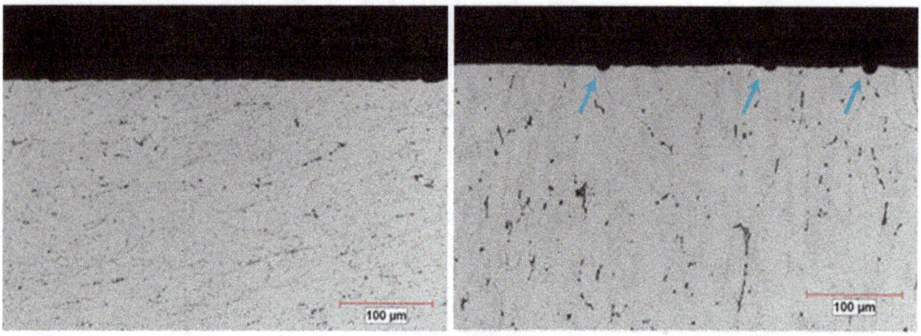

Fig. 8.48 (etchant: Keller's Reagent) SEM image of the anodized surface of a rolled and forged 5052 alloy that was intended for use as a semiconductor processing chamber. Each pit in the surface was a potential point of voltage breakdown. Note the distinct hemispherical pits at the grain boundaries as they intersect surface of the sample on the right (arrows). Alloy 5052 has approximately 2.4% magnesium as the primary alloying element; meaning most will remain in solid solution, even at room temperature. However, in 5000 series alloys that contain more than 1.9% magnesium, precipitates of Al_3Mg_2 can preferentially precipitate at grain boundaries, sensitizing the microstructure, thereby causing pitting to be developed by the anodizing process [26]. Courtesy of Anderson Shumaker

Table 8.3 Chemical analysis results—Case Study 3

Element	Sample 1—good	Sample 2—bad
Silicon	0.06	0.06
Iron	0.13	0.13
Copper	0.006	0.002
Manganese	0.015	0.004
Magnesium	2.52	2.47
Zinc	0.01	0.02
Nickel	<0.002	0.003
Chromium	0.24	0.16
Lead	<0.002	<0.002
Tin	<0.002	<0.002
Titanium	0.017	0.008
Others each	<0.05	<0.05
Others total	<0.15	<0.15
Aluminum	96.9	97.0

All samples tested conform to JIS H4140, Table 2 Alloy No. 5052

samples conform to the JIS specification for alloy 5052. However, sample 1 was more highly alloyed with manganese, chromium, and titanium. As presented in Chap. 4, trace elements can have a marked effect on alloy microstructure.

Review of the role of each trace element in the microstructure gave insight to the overall performance of the different samples in the application; the most significant being manganese and its effect on the formation of primary constituent phases (PCPs). The presence of manganese in the melt reduces the formation of large

AlFeSi PCPs, which on their own can be quite brittle. Manganese binds with Fe to stabilize the $FeAl_6$ phase over the high temperatures the alloy is exposed to during casting and deformation. Manganese modifies the microstructure by keeping the $FeAl_6$ intermetallic phase fine, preventing it from becoming needle-like or plate-like. By keeping brittle phases small, the addition of Mn as a modifier acts to reduce die wear during forging and makes the manganese containing alloys easier to anodize; it also increases ductility and toughness (impact strength). The presence of chromium and titanium inhibits recrystallization, therefore maintaining a fine microstructure; both elements were of higher amount in the good sample.

In concluding this case study, the condition of the microstructure of Sample 2, which exhibits grain boundary sensitization as well as evidence of coring, indicates that slow cooling occurred during deformation. Coring can only occur during the forging process and would have occurred during over-temperature processing. This conclusion is also supported by the sensitized condition of the microstructure because slow cooling, over a greater temperature range, would produce the coarse precipitates that are found in the microstructure. This effect is important to remark because, even as a non-heat treatable alloy, thermal effects still impact the alloy 5052 substrate. The reduced amount of manganese in the alloy exacerbated the condition by enabling the formation of primary constituent phase at the billet level. It is also important to mention that the failure did not preclude the use of forgings for semiconductor and other high pressure, high dielectric withstand applications; the material and process must be closely monitored to make sure the microstructure is optimized for the total manufacturing process, which includes anodizing.

8.4.4 Case Study 4

8.4.4.1 Impact and Wear Resistance: High Alloy Extrusion

Many under-the-hood automotive applications utilize anodized aluminum for a variety of applications that require impact and wear resistance. Alloy selection ranges from complex cast alloys to high strength wrought alloys, which are mostly extruded products. High alloy 6XXX series extrusions are most typically selected with 6061 and 6082 being the most common. Most high strength extrusions can be used in the T5 condition, which means the material is quenched on the press.

The size and distribution of the primary constituent phase, as well as the development of dispersoid particles that function to inhibit recrystallization, are all determined early in the manufacturing process. The size of PCPs is determined at the billet level, where the chemistry effects of trace elements have the biggest impact. As seen in all of the case studies presented so far, the importance of tight control of chemistry cannot be stressed enough, as slight variations in trace elements. High strength alloys, because of the complex chemistries, are processed at slower strain rates to attain the desired level of deformation without cracking. Consequently, the press quench typically maintains a fine-grained, fibrous microstructure.

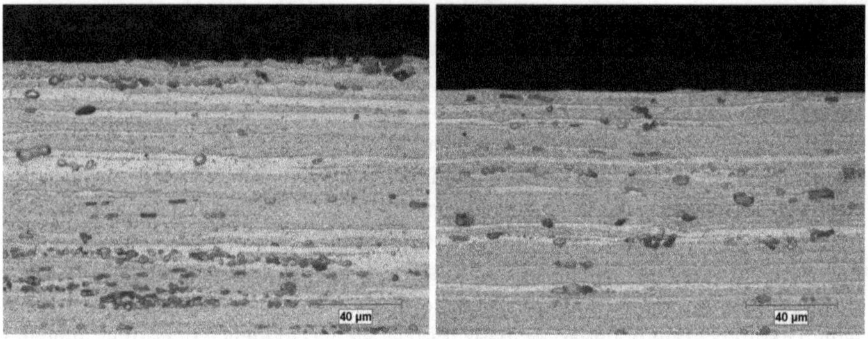

Fig. 8.49 (etchant: Keller's Reagent) Microstructures corresponding to anodized 6082 T6 extruded automotive extrusion. The fibrous microstructures are characteristic for more complex alloys with higher mechanical strength, both are developed using Keller's reagent. The microstructure on the left contains a high population of "stringers," or primary constituent phases (PCPs), broken up and aligned parallel to the direction of deformation. Upon intersecting the surface during anodizing, the stringers were included in the AAO, producing an increased surface roughness, causing a functional design failure. The microstructure of an acceptable part exhibited fewer PCPs and no stringers, consequently did not present the same level of disruption to the surface finish after anodizing. Courtesy of Impreglon Surface Engineering

In this case study, the surface finish of some 6082 T5 extrusions, that were machined and anodized for an under-the-hood application, exhibited an unacceptably high surface roughness after anodizing. Others, processed at the same time, through the same anodizing process, on the same rack, were of acceptable quality. Comparative metallographic analysis determined that the acceptable parts exhibited fewer PCPs, without stringer-type concentrations. The bad parts contained large PCPs in stringers throughout the microstructure. The grain structures of the parts were basically the same, indicating that the failure was not in the extrusion process, but pointed upstream from deformation to the billet quality. Microhardness testing qualified all parts as having a T5 temper. See Fig. 8.49.

Chemical analysis of the base material determined the failed part exhibited the following alloy element differences from the good part: elevated iron content by more than 2×; slightly higher magnesium content, which pushed the magnesium to silicon ratio close to 1:1; and reduced chromium content in the alloy chemistry by more than 3×. It is important, even at trace element amounts, to regard the basic amount of each element, the ratio in comparison to what they form, in order to understand their function in the microstructure. See Table 8.4.

For these samples, keeping in mind that the failure mode was based upon the formation of coarse primary constituent phases, the reduced amount of iron in the good sample effectively reduced the amount of PCP formation as AlFeSi in the billet, as well as reduced the amount of dispersoid precipitation as $AlFe_3$ in the extrusions that didn't fail (microstructure on the left). Additionally, the higher amount of chromium in the "good" component favored the formation of FeCr dispersoids, which also reduced the amount of iron available for PCP formation. By pushing the

Table 8.4 Chemical analysis results—Case Study 4

Element	Sample 1—failed	Sample 2—good
Silicon	0.94	0.96
Iron	0.45	0.20
Copper	0.06	0.06
Manganese	0.79	0.75
Magnesium	0.74	0.90
Chromium	0.04	0.14
Nickel	0.01	<0.01
Zinc	0.08	0.05
Titanium	0.02	0.03
Lead	<0.01	<0.01
Tin	<0.01	<0.01
Others each	<0.05	<0.05
Others total	<0.15	<0.15
Aluminum	Remainder	Remainder

ratio of magnesium to silicon closer to 1:1, most of the silicon would be tied up as Mg_2Si precipitates, which also reduced the amount of PCP formation.

Overall, both samples complied with the chemistry specified for 6082, but differences in trace element amounts had a big impact on the microstructure. Characterizing the base material by way of microstructure and hardness gave the picture of the surface that was anodized and produced the failure, in spite of hardness that complies with a T5 temper; but it was the chemical analysis that actually determined the root cause for failure, which occurred at the billet level. By tightening the chemical requirements for the alloy, the manufacturer of the component took an appropriate step to prevent future failures caused by increased surface roughness.

8.4.5 *Case Study 5*

8.4.5.1 Bonding: Aluminum Super Alloy Extrusion

The unsealed surface of the anodic oxide is often used for bonding. The attachment surface for a primer, paint, lacquer, or adhesive to the high surface area provided by the porous, columnar structure of the total AAO layer is reliable and offers not only an enhanced mechanical attachment, but ionic bonding as well. Combined with the qualities that are characteristics of the anodic oxide, the total finish developed with primer, paint, and the anodic oxide is decorative and more corrosion resistant than primer and paint alone. There is no open avenue at product edges to enable the ingress and procedure of tunneling filaments of corrosion by way of filiform corrosion because the anodic oxide is metallurgically integrated with the aluminum alloy substrate. Therefore, interfacial failures due to corrosion, delamination, and blistering are rare at the interface between anodic aluminum oxide and its parent substrate.

The development of aluminum alloys has grown with history of the aircraft industry, which has long employed anodized aluminum as the substrate for primers, paints, and printed adhesive labels. Today, the aircraft industry is possibly the top user of complex aluminum super alloys, whose composition exceeds the 8% typically considered as a maximum for wrought processing. The increased strength and fatigue resistance afforded by super alloys makes them the newest design choice for applications in other industries such as automotive, consumer electronics, and sporting goods. Super alloys are provided in a strict T6 or T651 temper, meaning that they are taken through solution treatment to provide a controlled, as homogeneous as possible, microstructure for age hardening or tempering.

The combination of time at elevated temperature during solution treatment stimulates diffusion of magnesium to the surface of the workpiece, which is even more reactive when it comes to oxygen in the atmosphere. If moisture is present in the atmosphere, atomic hydrogen can diffuse along grain boundaries into the surface microstructure, which may cause voids to form when hydrogen reacts to form molecular hydrogen, or even water at the grain boundaries, causing high temperature oxidation (HTO). In order to avoid HTO, alloys are usually solution treated in an atmosphere containing either sodium fluoroborate or ammonium fluoroborate. The atmosphere decomposes in the furnace at the workpiece surface to form a compound layer of reaction products, which mitigates the diffusion of hydrogen along the grain boundaries of the alloy [27].

Magnesium, as the most reactive component of most of the more complex alloys, diffuses to the surface and reacts with the atmosphere to form a rather continuous layer of magnesium fluoride, MgF_2, in intimate contact with the workpiece. The reaction products that form with the sodium or ammonium counterions deposit on the surface of the MgF_2 layer. The layers that form are soluble in some of the pretreatment chemistry used in the anodizing process, particularly those that incorporate hydrofluoric acid (HF) or ammonium bifluoride (NH_4HF_2). Such formulations can be used to effectively descale components from the reaction layer even before they go to the anodizing line; however, because solution treatment may be repeated to relieve stress and/or change the grain structure, the reaction layer that forms during heat treatment can become thick and more difficult to remove.

Blistering and delamination failures are complex and require detailed analysis to confirm their nature and cause. Understanding at what interface failure is occurring, in the sequence of layers on top of the aluminum alloy substrate, is important in order to discern which step in the manufacturing process to target to determine the root cause. Interfacial discontinuities, which manifest as delamination and blistering, can be at any interface, even between the anodic oxide finish and the aluminum base material. These types of defects are well documented, particularly with anodized 7XXX series alloys.

This case study deals with extrusions from a 7XXX series super alloy that exhibited blistering and delamination from the surface of the anodic oxide after priming and painting. Analysis of the interfacial phenomena that promulgated this failure

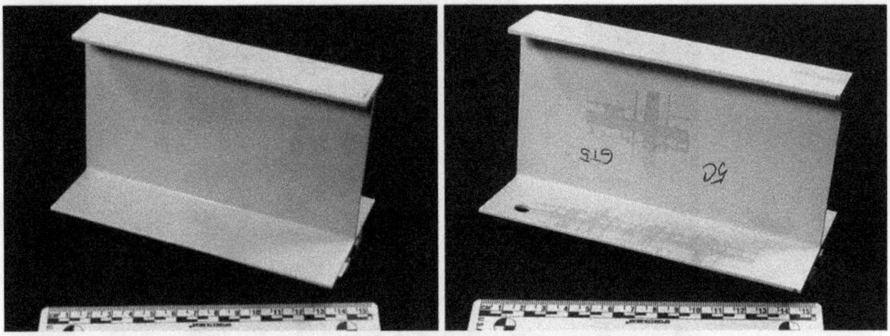

Fig. 8.50 Documentary photograph of the 7XXX super alloy extrusions that exhibit differences in primer/paint adhesion. The sample on the left exhibits acceptable adhesion, GT0, following tape adhesion testing per ASTM D 3359, Standard Test Methods for Measuring Adhesion by Tape Test, while the sample on the right exhibits a GT5 failure at the interface between the primer layer and the anodic oxide. Courtesy of Universal Alloy Corporation

Table 8.5 Comparative XPS results—Case Study 5

Number of heat treat cycles	MgF_2 (%)	MgO (%)	MgF_2/MgO	C (%)
1×	13	9	1.4	11.9
2×	15	7	2.2	31.1
3×	10.50	4	2.6	49.7

exceeded the usual characterization of metallographic and chemical analysis. In this case, careful metallographic preparation of the failed parts was necessary in order to perform: scanning electron microscopy with energy dispersive X-ray analysis (SEM/EDS), X-ray photoelectric spectrophotometry (XPS), and transmission electron microscopy (TEM) were utilized to analyze the failed samples and to compare with good samples. See Fig. 8.50.

For this analysis, the rare opportunity to analyze raw material from the same extrusion lots as those that failed was available. There were two sources for the material, and all samples analyzed exhibited similar results: all exhibited a layer of mixed MgF_2 and MgO, with some samples having thicker reaction layers than others. This was documented by way of SEM/EDS and XPS. The thickness of the reaction layer could be correlated qualitatively to the number of times the workpiece had been through heat treatment. It is noteworthy that while the MgF_2 layer increased, and then dropped off with apparent exposure to the heat treat cycles, the reaction layer formed by the reaction of the counterion by-products at the surface continued to grow thicker. Nevertheless, the thickest total reaction layer was that had been exposed to three heat treat cycles. Table 8.5 presents a summary of the XPS analysis results.

TEM/STEM analysis documented the near-surface microstructure as having a thin layer of recrystallized grains, or a Beilby layer, due to the strain rate sensitivity of the complex alloy and the extrusion parameters inducing localized recrystallization at the surface. The reaction layer from solution treatment was duplex in nature,

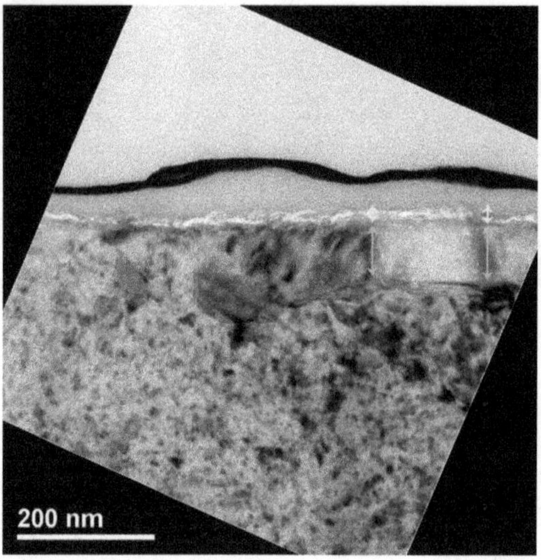

Fig. 8.51 TEM image of cross section prepared at the surface of a complex 7XXX alloy extrusion. The extrusion has been through solution heat treatment and tempering three times. The near-surface microstructure exhibits a layer of recrystallized grains, also known as a Beilby layer, bracketed by arrows. The white-appearing layer at the surface of the Beilby Layer is the MgF_2 reaction layer from the heat treat atmosphere). The amorphous-appearing gray layer between the MgF_2 layer and the layer placed on the surface to preserve the cross section and has the carbon-containing reaction layer that has the sodium counterion from the atmosphere. Courtesy of Universal Alloy Corporation

with the MgF_2 layer residing adjacent to the base metal surface and the decomposition products from the atmosphere on top of the MgF_2 layer. See Fig. 8.51.

Metallographic analysis of the samples documented in Fig. 8.50 disclosed, when imaged at high magnification within the SEM, the presence of the thin MgF_2 layer at the interface between the anodic oxide and the heavily filled primer layer. The anodic oxide on the failed sample was significantly rougher in appearance when compared to the good sample. EDS analysis confirmed the interfacial layer contained a significant amount of Mg and F. The elements that comprise the base material, the anodic oxide (Al, Zn, O, and S) and the primer layer (Ba, Ca, and C) were also detected. See Figs. 8.52 and 8.53.

When considering this failure mode, it must be kept in mind that the anodic oxide grows from and consumes the aluminum surface. Therefore, the presence of MgF_2 scale, which is a slightly soluble in acid solutions, as a reaction layer on the surface of the aluminum, will delay the oxidation reaction as long as it is dissolving in the electrolyte. As the film is dissolved in the electrolyte, and areas across the substrate are exposed, different starting points for AAO nucleation and growth develop across the contaminated surface. This means that the resulting AAO will be as rough as the variations in thickness of the MgF_2 film.

TEM analysis of the interfaces was more difficult to achieve, due to the primer delaminating from the surface of the anodic oxide during preparation on the bad

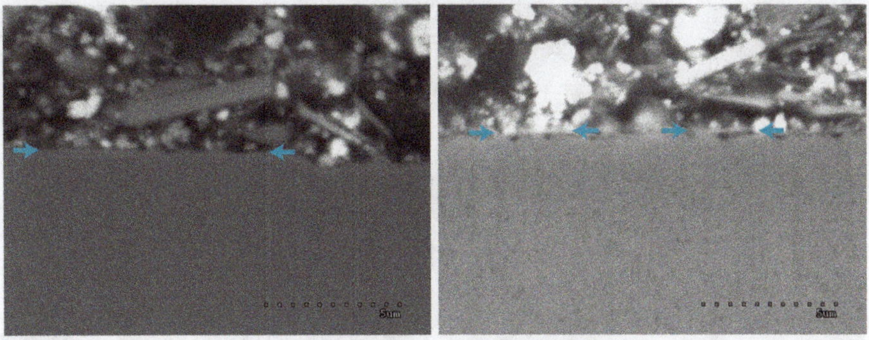

Fig. 8.52 SEM images of the anodized aluminum-primer interface of representative good (left) and bad (right) samples. The anodic oxide emits somewhat lighter than the base aluminum (see arrows). The primer layer is heavily mineral filled, accounting for the particles and needle-like constituents of the top layer on both photomicrographs. Note the smooth, continuous appearance of the interface on the good sample. The bad sample, on the right, exhibits a thin layer above the irregular surface of the anodic oxide. Courtesy of Universal Alloy Corporation

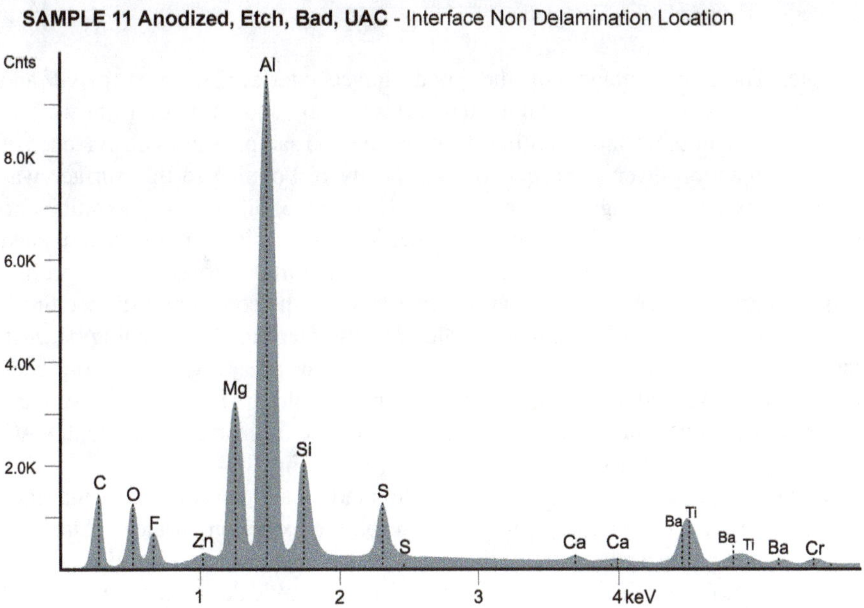

Fig. 8.53 EDS spectrum for the interfacial layer between the anodic oxide and the primer layer from the bad sample documented in this figure. Note the presence of silicon, barium, and calcium from the fillers in the primer layer, which has a carbon-based binder. What is most significant is the substantial amount of magnesium and fluorine, which is from the reaction layer that formed during heat treatment. Aluminum and zinc are from the base metal; sulfur and oxygen are from the anodic oxide

Fig. 8.54 TEM images of the AAO-primer interface of the good extrusion (left) and the surface of the AAO of the bad extrusion (right). Note the filler particle size of the primer in the image on the left, and the surface roughness of the AAO on the right. The roughness is due to variations in the MgF_2 contamination film thickness and competing film dissolution–anodic oxidation reactions. Courtesy of Universal Alloy Corporation

sample. Therefore, images of the good, intact interface were prepared and documented as well as the delaminated surface of the anodic oxide from the bad sample. The smooth, continuous interface on the good sample shows no evidence of an MgF_2 reaction layer, consequently the quality of bonding to this surface was high. The extremely rough surface of the AAO from the failed sample exhibited no residual MgF_2 reaction layer, indicating that it dissolved into and/or delaminated with the primer layer. If one considers the size of the filler particles in the primer, and compares the size of a filler particle to a space in the convoluted surface finish of the bad oxide, the filler particles could feasibly interfere with bonding because they would not be able to fill and conform to the surface. Outgassing from gaps between the AAO and the primer layer could produce blisters in the painted surface. No blistering in the aluminum alloy or at the interface of the base metal with AAO due to high temperature oxidation was observed. See Fig. 8.54.

In this study, the root cause for delamination failure at the AAO-primer interface was the presence of residual heat treat scale on the extrusion surfaces. Although soluble in the pretreat chemistry of the anodizing process and in the anodizing electrolyte, the scale was of a thickness that was not removable during the conventional exposure times set by the anodizing process.

The formation of an MgF_2 reaction layer, or scale, during solution heat treatment in a sodium fluoroborate or ammonium fluoroborate atmosphere is a common phenomenon well documented in the industry. Nevertheless, it may be necessary to characterize the surface of extrusions, post-heat treatment and prior to anodizing, by way of X-ray fluorescence, to document the scale layer and its thickness. In operations where anodizing is captive, it may be easier to utilize the appropriate acid deoxidizing formulation to remove the scale prior to starting the anodizing process typically used for the heat-treated components. Care must be exercised with acid treatment of component with fine detail such as threads, through holes or other dimensionally sensitive component features. It is important to remark that due to the chemical principle of "like dissolves like," alkaline treatment will not effectively remove fluorinated reaction products.

This particular case study is representative of a failure that can happen with all complex aluminum alloys subject to solution treatment in an atmosphere that produces a scale. Addressing the phenomenon by way of chemical removal is only one way to mitigate the problem. As part of the total manufacturing process, heat treat strategies can also be implemented to limit the amount of scale formed, not only with atmosphere control, but also by considering time at temperature.

8.5 Summary and Conclusion

The various examples and case studies presented throughout the chapter demonstrate how, by using basic physical metallurgy principles, together with a basic understanding of manufacturing processes, the multiplicity of microstructural features that form near and at the surface of an aluminum alloy component, all come together to develop the surface of the component. Consequently, when presented with such information, a virtual picture of the surface can be developed, even before the substrate is anodized. Because variations in anodic oxide finish quality can be tied directly to variations in substrate chemistry and microstructure of the anode, understanding at which step during manufacturing that alloy chemistry, composition, deformation processing, and heat treatment effects manifest as microstructural features is therefore a useful tool for developing successful anodizing process strategies and for developing troubleshooting strategies.

The following criteria are basic truths that can be used to relate the functionality of an anodized product to a design application and/or to relate the metallurgy of the base material to the anodized finish, which enables the successful integration of the anodic oxide into component designs and aid in troubleshooting if things go wrong after anodizing:

1. The process of anodizing develops the surface microstructure of the substrate.
2. The anodic oxide is a rather transparent lens through which the substrate surface, including microstructural defects, can be viewed.
3. The anodic oxide, when uniform and continuous, will impart a level of corrosion protection and wear resistance to the substrate.
4. A bright, homogeneous substrate will yield a bright-appearing oxide—and the opposite is also true. A bright-appearing oxide is grown from a bright-homogeneous substrate.

5. Alloy phases, intermetallic compounds, inclusions, and other atomic level defects DO NOT ANODIZE. They confound the anodizing reaction, retard it, and disrupt the anodic oxide structure.
6. The primary oxidation reaction proceeds around intermetallic particles and takes them up into the oxide structure. Particles included within the anodic oxide interfere with incident light and disrupt the distinctness of a reflected image.
7. Manufacturing processes affect the level of microstructural homogeneity developed in components by impacting the defect population and the metallurgical cleanliness of the substrate surface—all the things that are developed by the anodizing process and visible through the anodic oxide.
8. A homogeneous billet is required to yield a homogeneous substrate, but a deficient deformation process can nevertheless produce inhomogeneous or defective components from the homogeneous billet.
9. A deficient or out-of-control anodizing process will produce an anodic oxide of inferior quality.
10. If any one of the steps is out of control that precedes actual anodizing, such as cleaning and pretreatment, or that succeeds actual anodizing, such as dyeing and sealing, an anodic oxide of inferior quality is produced.

To successfully implement anodized aluminum into engineering designs, the characteristics of the substrate microstructure and the anodic oxide are equally important to bear in mind. From this perspective, design strategies must include all of the above criteria in order to optimize the quality of the surface presented for anodic oxide growth, as well as the details of the anodizing process. A metallurgical approach to anodizing is therefore necessary in order to obtain the most robust and the most esthetically pleasing anodized finishes for any alloy manufactured by any process.

Acknowledgements All data and photographic images in this chapter are presented with the expressed written consent of the industrial partners, together with whom this work was accomplished. The author is deeply grateful for their permission.

References

1. Runge, J. M. (2014). Anodizing for design and function. *Journal of Materials Science and Nanotechnology, 1*, S108.
2. Song, G.-L. (2014). The grand challenges in electrochemical corrosion research. *Frontiers in Materials, 1*, 4. https://doi.org/10.3389/fmats.2014.00002.
3. Chesterfield, L., & Runge, J. (2010). Connecting theory to practice, the science of successfully anodizing aluminum die castings. In: *Proceedings of the 19th Annual Technical Conference and Exposition of the Aluminum Anodizers Council*, Montreal.
4. Runge, J. M. Base metal microstructural considerations for anodizing aluminum. In: *Proceedings of the Iran International Aluminum Conference (IIAC2016)*.

5. Runge, J. (2008). Interfacial phenomena and anodizing: ramifications and process solutions. In: *Proceedings of the 17th Annual AAC Conference*, San Francisco, CA.
6. Runge, J. (2016, May). Base metal microstructural considerations for anodizing aluminum extrusions. In: *Proceedings of Eleventh International Aluminum Extrusion Technology Seminar, (ET 2016)*, Chicago, IL. Extrusion Technology for Aluminum Profiles Foundation.
7. Tabrizian, N., Ambat, R., Møller, P., & Hansen, H. N. (2008). Effect of microstructure of AlMg0.5Cu and AlMg0.7Si on the Optical Quality of Anodic Film. In: *Proceedings of the 17th Annual Anodizing Conference and Expo, the Aluminum Anodizers Council*, San Francisco, CA.
8. Zhu, H., Dahle, A. K., Zhang, X., & Couper, M. J. (2008). Effect of extrusion microstructure on formation of streaking defects on the surface of anodized aluminum extrusions. In: *Proceedings of the 17th Annual Anodizing Conference and Expo, the Aluminum Anodizers Council*, San Francisco, CA.
9. Runge, J. (2012). Trace elements and their impact on surface finishing characteristics of aluminum extrusions. In: *Proceedings of ET 2012*, Miami.
10. Runge, J. (2016). Base metal microstructural considerations for anodizing aluminum extrusions. In: *Proceedings of Eleventh International Aluminum Extrusion Technology Seminar, (ET 2016)*, Chicago, IL. Extrusion Technology for Aluminum Profiles Foundation.
11. Runge, J. Understanding the interface: Meeting the challenges of anodizing cast substrates. In: *Proceedings of the 2016 AAC Annual Technical Conference and Exposition*, Montreal, Canada.
12. Runge, J. M., & Hossain, T. (2015). Interfacial phenomena in 7000 series alloys and their impact on the anodic oxide. In: *Proceedings of Aluminium Two Thousand World Congress and International Conference on Extrusion and Benchmark ICEB 2015*; Elsevier Science Direct.
13. Schey, J. A. (1977). *Introduction to manufacturing processes*. New York: McGraw-Hill.
14. Runge, J. M. (2002). The use of hot isostatic pressing as a means to enhance metal finishing. In: *Proceedings of the IHAA Biannual Technical Conference*, Quebec City, QC, Canada.
15. Polmear, I. (2006). *Light alloys, from traditional alloys to nanocrystals*. Oxford, UK: Elsevier.
16. Birbilis, N., & Buchheit, R. (2005). Electrochemical characteristics of intermetallic phases in aluminum alloys and experimental survey and discussion. *Journal of the Electrochemical Society, 152*(4), B140–B151.
17. Asensio-Lozano, J., Suárez-Peña, B., & Vander Voort, G. (2014). Effect of processing steps on the mechanical properties and surface appearance of 6063 aluminium extruded products. *Materials*, 4224–4242.
18. Sanders, R. E., Hollinshead, P. A., & Simielli, E. A. (2004). Industrial development of non-heat treatable aluminum alloys. *Materials Forum, 28*, 53–64.
19. Porter, D. A., & Easterling, K. E. (1992). *Phase transformations in metals and alloys* (2nd ed.). Chapman and Hall.
20. Jones, D. A. (1996). *Principles and prevention of corrosion* (2nd ed.). Upper Saddle River, NJ: Prentice Hall.
21. The Aluminum Association. *"Extrusion" and "Forging"*. Retrieved from www.aluminum.org.
22. Altenpohl, D. G. (1998). *Aluminum: Technology, application and environment: A profile of a modern metal, aluminum from within* (6th ed.). Washington, DC: The Aluminum Association.
23. Strohmeier, B. (1989). Surface characterization of aluminum foil annealed in the presence of ammonium fluoborate. *Applied Surface Science, 40*, 249–263.
24. Parsons, R. (1987). Chapter 2: The electric double layer at the solid-solution interface. In W. Stumm (Ed.), *Aquatic surface chemistry*. New York: Wiley.
25. Scamans, G. M., Frolish, M. F., Rainforth, W. M., Zhou, Z., Liu, Y., Zhou, X., & Thompson, G. E. (2010). The ubiquitous beilby layer on aluminium surfaces. *Surface and Interface Analysis, 42*, 175–179.
26. Searles, J., et al. (2001). Stress corrosion cracking of sensitized AA5083 (Al-4.5Mg-1.0Mn). *Metallurgical and Materials Transactions A, 32*(11), 2259–2867.
27. Hatch, J. (1999). *Aluminum: Properties and physical metallurgy* (9th ed.). Cleveland: Aluminum Association/ASM International.

Index

J.M. Runge, *The Metallurgy of Anodizing Aluminum*,
https://doi.org/10.1007/978-3-319-72177-4

H

Printed in the USA
CPSIA information can be obtained
at www.ICGtesting.com
CBHW052229021124
16840CB00003B/12

9 783319 721750